2009 INTERNATIONAL ENERGY CONSERVATION CODE® AND

ANSI/ASHRAE/IESNA STANDARD 90.1-2007 ENERGY STANDARD FOR BUILDINGS EXCEPT LOW-RISE RESIDENTIAL BUILDINGS

INTERNATIONAL CODE COUNCIL®

2009 International Energy Conservation Code and ANSI/ASHRAE/IESNA Standard 90.1-2007
Energy Standard for Buildings Except Low-Rise Residential Buildings

Publication Date in Dual Format: May 2009

ISBN: 978-1-58001-799-2

2009
by

International Code Council, Inc.
500 New Jersey Avenue, NW, 6th Floor
Washington, D.C. 20001

and

American Society of Heating, Refrigerating and
Air-Conditioning Engineers, Inc.
1791 Tullie Circle NE
Atlanta, GA 30329

PRINTED IN THE U.S.A.

FOREWORD

The publication of the 2009 *International Energy Conservation Code* and *ANSI/ASHRAE/IESNA Standard 90.1-2007, Energy Standard for Buildings Except Low-Rise Residential Buildings* in one book is a positive step forward in the efforts of the International Code Council® (ICC®) and the American Society of Heating, Refrigerating and Air-Conditioning Engineers (ASHRAE) to increase awareness and application of energy-efficient building requirements.

This publication came about as a direct result of the American Recovery and Reinvestment Act of 2009 (ARRA), passed in February 2009. ARRA was designed to stimulate economic recovery by providing stimulus funding to various sectors of the economy and to accomplish a policy goal of creating more energy-efficient buildings.

There is almost universal agreement that increasing the energy efficiency of buildings is a simple and effective way to reduce overall energy use, and, ultimately, reduce carbon emissions. More than 70 percent of all electricity in the United States, and about 40 percent of the total energy worldwide, is consumed by residential and commercial buildings. Consequently, even small increases in building efficiency result in big reductions of energy and carbon emissions.

Both ICC and ASHRAE are proud of the processes they administer to produce the *International Energy Conservation Code* and ASHRAE/IESNA Standard 90.1. They bring together experts, government officials from all levels and industry representatives who manufacture, service and maintain the systems and products that go into energy-efficient buildings. These open and transparent processes produce documents that are respected and usable by all communities.

These two documents are recognized in the ARRA as the benchmarks for the energy efficiency of residential and commercial buildings. They address the same issues and because both may overlap in their coverage of building systems and designs, it makes sense to publish these two documents together—for the benefit of building designers, engineers and building code compliance personnel. In some cases, having both documents in one place will make it easier to choose between different design options. In all cases, this dual edition will make it easier to ensure that newly built and renovated buildings are in compliance with the latest references available, with local requirements and with the goal of ARRA to achieve 90 percent compliance with these target codes in all 50 states by 2017.

Richard P. Weiland
Chief Executive Officer
International Code Council, Inc.

A MEMBER OF THE INTERNATIONAL CODE FAMILY®

ICC
INTERNATIONAL
CODE COUNCIL®

IECC®

INTERNATIONAL ENERGY CONSERVATION CODE®

2009

Receive **FREE** updates, excerpts of code references, technical articles, and more when you register your code book. Go to

www.iccsafe.org/CodesPlus today!

2009 International Energy Conservation Code®

First Printing: January 2009
Second Printing: March 2009

ISBN: 978-1-58001-742-8 (soft-cover edition)

PRINTED IN THE U.S.A.

PREFACE

Introduction

Internationally, code officials recognize the need for a modern, up-to-date energy conservation code addressing the design of energy-efficient building envelopes and installation of energy efficient mechanical, lighting and power systems through requirements emphasizing performance. The *International Energy Conservation Code®*, in this 2009 edition, is designed to meet these needs through model code regulations that will result in the optimal utilization of fossil fuel and nondepletable resources in all communities, large and small.

This comprehensive energy conservation code establishes minimum regulations for energy efficient buildings using prescriptive and performance-related provisions. It is founded on broad-based principles that make possible the use of new materials and new energy efficient designs. This 2009 edition is fully compatible with all the *International Codes®* (I-Codes®) published by the International Code Council (ICC)®, including: the *International Building Code®*, *International Existing Building Code®*, *International Fire Code®*, *International Fuel Gas Code®*, *International Mechanical Code®*, ICC *Performance Code®*, *International Plumbing Code®*, *International Private Sewage Disposal Code®*, *International Property Maintenance Code®*, *International Residential Code®*, *International Wildland-Urban Interface Code™* and *International Zoning Code®*.

The *International Energy Conservation Code* provisions provide many benefits, among which is the model code development process that offers an international forum for energy professionals to discuss performance and prescriptive code requirements. This forum provides an excellent arena to debate proposed revisions. This model code also encourages international consistency in the application of provisions.

Development

The first edition of the *International Energy Conservation Code* (1998) was based on the 1995 edition of the *Model Energy Code* promulgated by the Council of American Building Officials (CABO) and included changes approved through the CABO Code Development Procedures through 1997. CABO assigned all rights and responsibilities to the International Code Council and its three statutory members at that time, including Building Officials and Code Administrators International, Inc. (BOCA), International Conference of Building Officials (ICBO) and Southern Building Code Congress International (SBCCI). This 2009 edition presents the code as originally issued, with changes reflected in the 2000, 2003 and 2006 editions and further changes approved through the ICC Code Development Process through 2008. A new edition such as this is promulgated every three years.

This code is founded on principles intended to establish provisions consistent with the scope of an energy conservation code that adequately conserves energy; provisions that do not unnecessarily increase construction costs; provisions that do not restrict the use of new materials, products or methods of construction; and provisions that do not give preferential treatment to particular types or classes of materials, products or methods of construction.

Adoption

The *International Energy Conservation Code* is available for adoption and use by jurisdictions internationally. Its use within a governmental jurisdiction is intended to be accomplished through adoption by reference in accordance with proceedings establishing the jurisdiction's laws. At the time of adoption, jurisdictions should insert the appropriate information in provisions requiring specific local information, such as the name of the adopting jurisdiction. These locations are shown in bracketed words in small capital letters in the code and in the sample ordinance. The sample adoption ordinance on page vii addresses several key elements of a code adoption ordinance, including the information required for insertion into the code text.

Maintenance

The *International Energy Conservation Code* is kept up to date through the review of proposed changes submitted by code enforcing officials, industry representatives, design professionals and other interested parties. Proposed changes are carefully considered through an open code development process in which all interested and affected parties may participate.

The contents of this work are subject to change both through the Code Development Cycles and the governmental body that enacts the code into law. For more information regarding the code development process, contact the Code and Standard Development Department of the International Code Council.

While the development procedure of the *International Energy Conservation Code* assures the highest degree of care, ICC, its members and those participating in the development of this code do not accept any liability resulting from compliance or noncompliance with the provisions because ICC and its members do not have the power or authority to police or enforce compliance with the contents of this code. Only the governmental body that enacts the code into law has such authority.

Marginal Markings

Solid vertical lines in the margins within the body of the code indicate a technical change from the requirements of the 2006 edition. Deletion indicators in the form of an arrow (➡) are provided in the margin where an entire section, paragraph, exception or table has been deleted or an item in a list of items or a table has been deleted.

Italicized Terms

Selected terms set forth in Chapter 2, Definitions, are italicized where they appear in code text. Such terms are not italicized where the definition set forth in Chapter 2 does not impart the intended meaning in the use of the term. The terms selected have definitions which the user should read carefully to facilitate better understanding of the code.

Effective Use of the International Energy Conservation Code

The *International Energy Conservation Code* (IECC) is a model code that regulates minimum energy conservation requirements for new buildings. The IECC addresses energy conservation requirements for all aspects of energy uses in both commercial and residential construction, including heating and ventilating, lighting, water heating, and power usage for appliances and building systems.

The IECC is a design document. For example, before one constructs a building, the designer must determine the minimum insulation *R*-values and fenestration *U*-factors for the building exterior envelope. Depending on whether the building is for residential use or for commercial use, the IECC sets forth minimum requirements for exterior envelope insulation, window and door *U*-factors and SHGC ratings, duct insulation, lighting and power efficiency, and water distribution insulation.

Arrangement and Format of the 2009 IECC

Before applying the requirements of the IECC it is beneficial to understand its arrangement and format. The IECC, like other codes published by ICC, is arranged and organized to follow sequential steps that generally occur during a plan review or inspection. The IECC is divided into five different parts:

Chapters	Subjects
1–2	Administration and definitions
3	Climate zones and general materials requirements
4	Energy efficiency for residential buildings
5	Energy efficiency for commercial buildings
6	Referenced standards

The following is a chapter-by-chapter synopsis of the scope and intent of the provisions of the *International Energy Conservation Code*:

Chapter 1 Administration. This chapter contains provisions for the application, enforcement and administration of subsequent requirements of the code. In addition to establishing the scope of the code, Chapter 1 identifies which buildings and structures come under its purview. Chapter 1 is largely concerned with maintaining "due process of law" in enforcing the energy conservation criteria contained in the body of the code. Only through careful observation of the administrative provisions can the building official reasonably expect to demonstrate that "equal protection under the law" has been provided.

Chapter 2 Definitions. All terms that are defined in the code are listed alphabetically in Chapter 2. While a defined term may be used in one chapter or another, the meaning provided in Chapter 2 is applicable throughout the code.

Additional definitions regarding climate zones are found in Tables 301.3(1) and (2). These are not listed in Chapter 2.

Where understanding of a term's definition is especially key to or necessary for understanding of a particular code provision, the term is show in *italics* wherever it appears in the code. This is true only for those terms that have a meaning that is unique to the code. In other words, the generally understood meaning of a term or phrase might not be sufficient or consistent with the meaning prescribed by the code; therefore, it is essential that the code-defined meaning be known.

Guidance regarding tense, gender and plurality of defined terms as well as guidance regarding terms not defined in this code is provided.

Chapter 3 Climate Zones. Chapter 3 specifies the climate zones that will serve to establish the exterior design conditions. In addition, Chapter 3 provides interior design conditions that are used as a basis for assumptions in heating and cooling load calculations, and provides basic material requirements for insulation materials and fenestration materials.

Climate has a major impact on the energy use of most buildings. The code establishes many requirements such as wall and roof insulation *R*-values, window and door thermal transmittance requirement (*U*-factors) as well as provisions that affect the mechanical systems based upon the climate where the building is located. This chapter will contain the information that will be used to properly assign the building location into the correct climate zone and will then be used as the basis for establishing requirements or elimination of requirements.

Chapter 4 Residential Energy Efficiency. Chapter 4 contains the energy-efficiency-related requirements for the design and construction of residential buildings regulated under this code. It should be noted that the definition of a *residential building* in this code is unique for this code. In this code, a *residential building* is an R-2, R-3 or R-4 building three stories or less in height. All other buildings, including residential buildings greater than three stories in height, are regulated by the energy conservation requirements of Chapter 5. The applicable portions of a residential building must comply with the provisions within this chapter for energy efficiency. This chapter defines requirements for the portions of the building and building systems that impact energy use in new residential construction and promotes the effective use of energy. The provisions within the chapter promote energy efficiency in the building envelope, the heating and cooling system and the service water heating system of the building.

Chapter 5 Commercial Energy Efficiency. Chapter 5 contains the energy-efficiency-related requirements for the design and construction of most types of commercial buildings and residential buildings greater than three stories in height above grade. Residential buildings, townhouses and garden apartments three stories or less in height are covered in Chapter 4. Like Chapter 4, this chapter defines requirements for the portions of the building and building systems that impact energy use in new commercial construction and new residential construction greater than three stories in height, and promotes the effective use of energy. The provisions within the chapter promote energy efficiency in the building envelope, the heating and cooling system and the service water heating system of the building.

Chapter 6 Referenced Standards. The code contains numerous references to standards that are used to regulate materials and methods of construction. Chapter 6 contains a comprehensive list of all standards that are referenced in the code. The standards are part of the code to the extent of the reference to the standard. Compliance with the referenced standard is necessary for compliance with this code. By providing specifically adopted standards, the construction and installation requirements necessary for compliance with the code can be readily determined. The basis for code compliance is, therefore, established and available on an equal basis to the code official, contractor, designer and owner.

Chapter 6 is organized in a manner that makes it easy to locate specific standards. It lists all of the referenced standards, alphabetically, by acronym of the promulgating agency of the standard. Each agency's standards are then listed in either alphabetical or numeric order based upon the standard identification. The list also contains the title of the standard; the edition (date) of the standard referenced; any addenda included as part of the ICC adoption; and the section or sections of this code that reference the standard.

ORDINANCE

The International Codes are designed and promulgated to be adopted by reference by ordinance. Jurisdictions wishing to adopt the 2009 *International Energy Conservation Code* as an enforceable regulation governing energy efficient building envelopes and installation of energy efficient mechanical, lighting and power systems should ensure that certain factual information is included in the adopting ordinance at the time adoption is being considered by the appropriate governmental body. The following sample adoption ordinance addresses several key elements of a code adoption ordinance, including the information required for insertion into the code text.

SAMPLE ORDINANCE FOR ADOPTION OF THE *INTERNATIONAL ENERGY CONSERVATION CODE* ORDINANCE NO.________

An ordinance of the **[JURISDICTION]** adopting the 2009 edition of the *International Energy Conservation Code*, regulating and governing energy efficient building envelopes and installation of energy efficient mechanical, lighting and power systems in the **[JURISDICTION]**; providing for the issuance of permits and collection of fees therefor; repealing Ordinance No. ______ of the **[JURISDICTION]** and all other ordinances and parts of the ordinances in conflict therewith.

The **[GOVERNING BODY]** of the **[JURISDICTION]** does ordain as follows:

Section 1. That a certain document, three (3) copies of which are on file in the office of the **[TITLE OF JURISDICTION'S KEEPER OF RECORDS]** of **[NAME OF JURISDICTION]**, being marked and designated as the *International Energy Conservation Code*, 2009 edition, as published by the International Code Council, be and is hereby adopted as the Energy Conservation Code of the **[JURISDICTION]**, in the State of **[STATE NAME]** for regulating and governing energy efficient building envelopes and installation of energy efficient mechanical, lighting and power systems as herein provided; providing for the issuance of permits and collection of fees therefor; and each and all of the regulations, provisions, penalties, conditions and terms of said Energy Conservation Code on file in the office of the **[JURISDICTION]** are hereby referred to, adopted, and made a part hereof, as if fully set out in this ordinance, with the additions, insertions, deletions and changes, if any, prescribed in Section 2 of this ordinance.

Section 2. The following sections are hereby revised:

Section 101.1. Insert: **[NAME OF JURISDICTION]**.

Section 108.4. Insert: **[DOLLAR AMOUNT]** in two places.

Section 3. That Ordinance No. ______ of **[JURISDICTION]** entitled **[FILL IN HERE THE COMPLETE TITLE OF THE ORDINANCE OR ORDINANCES IN EFFECT AT THE PRESENT TIME SO THAT THEY WILL BE REPEALED BY DEFINITE MENTION]** and all other ordinances or parts of ordinances in conflict herewith are hereby repealed.

Section 4. That if any section, subsection, sentence, clause or phrase of this ordinance is, for any reason, held to be unconstitutional, such decision shall not affect the validity of the remaining portions of this ordinance. The **[GOVERNING BODY]** hereby declares that it would have passed this ordinance, and each section, subsection, clause or phrase thereof, irrespective of the fact that any one or more sections, subsections, sentences, clauses and phrases be declared unconstitutional.

Section 5. That nothing in this ordinance or in the Energy Conservation Code hereby adopted shall be construed to affect any suit or proceeding impending in any court, or any rights acquired, or liability incurred, or any cause or causes of action acquired or existing, under any act or ordinance hereby repealed as cited in Section 3 of this ordinance; nor shall any just or legal right or remedy of any character be lost, impaired or affected by this ordinance.

Section 6. That the **[JURISDICTION'S KEEPER OF RECORDS]** is hereby ordered and directed to cause this ordinance to be published. (An additional provision may be required to direct the number of times the ordinance is to be published and to specify that it is to be in a newspaper in general circulation. Posting may also be required.)

Section 7. That this ordinance and the rules, regulations, provisions, requirements, orders and matters established and adopted hereby shall take effect and be in full force and effect **[TIME PERIOD]** from and after the date of its final passage and adoption.

TABLE OF CONTENTS

2009 INTERNATIONAL ENERGY CONSERVATION CODE

ANSI/ASHRAE/IESNA STANDARD 90.1-2007, ENERGY STANDARD FOR BUILDINGS EXCEPT LOW-RISE RESIDENTIAL BUILDINGS (I-P EDITION)

ANSI/ASHRAE/IESNA ADDENDA TO ANSI/ASHRAE/IESNA STANDARD 90.1-2007, ENERGY STANDARD FOR BUILDINGS EXCEPT LOW-RISE RESIDENTIAL BUILDINGS

Section

CHAPTER 1

ADMINISTRATION

PART 1—SCOPE AND APPLICATION

SECTION 101
SCOPE AND GENERAL REQUIREMENTS

101.1 Title. This code shall be known as the *International Energy Conservation Code* of [NAME OF JURISDICTION], and shall be cited as such. It is referred to herein as "this code."

101.2 Scope. This code applies to *residential* and *commercial buildings*.

101.3 Intent. This code shall regulate the design and construction of buildings for the effective use of energy. This code is intended to provide flexibility to permit the use of innovative approaches and techniques to achieve the effective use of energy. This code is not intended to abridge safety, health or environmental requirements contained in other applicable codes or ordinances.

101.4 Applicability. Where, in any specific case, different sections of this code specify different materials, methods of construction or other requirements, the most restrictive shall govern. Where there is a conflict between a general requirement and a specific requirement, the specific requirement shall govern.

101.4.1 Existing buildings. Except as specified in this chapter, this code shall not be used to require the removal, *alteration* or abandonment of, nor prevent the continued use and maintenance of, an existing building or building system lawfully in existence at the time of adoption of this code.

101.4.2 Historic buildings. Any building or structure that is listed in the State or National Register of Historic Places; designated as a historic property under local or state designation law or survey; certified as a contributing resource with a National Register listed or locally designated historic district; or with an opinion or certification that the property is eligible to be listed on the National or State Registers of Historic Places either individually or as a contributing building to a historic district by the State Historic Preservation Officer or the Keeper of the National Register of Historic Places, are exempt from this code.

101.4.3 Additions, alterations, renovations or repairs. Additions, alterations, renovations or repairs to an existing building, building system or portion thereof shall conform to the provisions of this code as they relate to new construction without requiring the unaltered portion(s) of the existing building or building system to comply with this code. Additions, alterations, renovations or repairs shall not create an unsafe or hazardous condition or overload existing building systems. An addition shall be deemed to comply with this code if the addition alone complies or if the existing building and addition comply with this code as a single building.

Exception: The following need not comply provided the energy use of the building is not increased:

1. Storm windows installed over existing fenestration.
2. Glass only replacements in an existing sash and frame.
3. Existing ceiling, wall or floor cavities exposed during construction provided that these cavities are filled with insulation.
4. Construction where the existing roof, wall or floor cavity is not exposed.
5. Reroofing for roofs where neither the sheathing nor the insulation is exposed. Roofs without insulation in the cavity and where the sheathing or insulation is exposed during reroofing shall be insulated either above or below the sheathing.
6. Replacement of existing doors that separate *conditioned space* from the exterior shall not require the installation of a vestibule or revolving door, provided, however, that an existing vestibule that separates a *conditioned space* from the exterior shall not be removed,
7. Alterations that replace less than 50 percent of the luminaires in a space, provided that such alterations do not increase the installed interior lighting power.
8. Alterations that replace only the bulb and ballast within the existing luminaires in a space provided that the *alteration* does not increase the installed interior lighting power.

101.4.4 Change in occupancy or use. Spaces undergoing a change in occupancy that would result in an increase in demand for either fossil fuel or electrical energy shall comply with this code. Where the use in a space changes from one use in Table 505.5.2 to another use in Table 505.5.2, the installed lighting wattage shall comply with Section 505.5.

101.4.5 Change in space conditioning. Any nonconditioned space that is altered to become *conditioned space* shall be required to be brought into full compliance with this code.

101.4.6 Mixed occupancy. Where a building includes both *residential* and *commercial* occupancies, each occupancy shall be separately considered and meet the applicable provisions of Chapter 4 for *residential* and Chapter 5 for *commercial*.

101.5 Compliance. *Residential buildings* shall meet the provisions of Chapter 4. *Commercial buildings* shall meet the provisions of Chapter 5.

101.5.1 Compliance materials. The *code official* shall be permitted to approve specific computer software, worksheets, compliance manuals and other similar materials that meet the intent of this code.

101.5.2 Low energy buildings. The following buildings, or portions thereof, separated from the remainder of the building by *building thermal envelope* assemblies complying with this code shall be exempt from the *building thermal envelope* provisions of this code:

1. Those with a peak design rate of energy usage less than 3.4 Btu/h·ft^2 (10.7 W/m^2) or 1.0 watt/ft^2 (10.7 W/m^2) of floor area for space conditioning purposes.
2. Those that do not contain *conditioned space*.

SECTION 102 ALTERNATE MATERIALS—METHOD OF CONSTRUCTION, DESIGN OR INSULATING SYSTEMS

102.1 General. This code is not intended to prevent the use of any material, method of construction, design or insulating system not specifically prescribed herein, provided that such construction, design or insulating system has been *approved* by the *code official* as meeting the intent of this code.

102.1.1 Above code programs. The *code official* or other authority having jurisdiction shall be permitted to deem a national, state or local energy efficiency program to exceed the energy efficiency required by this code. Buildings *approved* in writing by such an energy efficiency program shall be considered in compliance with this code. The requirements identified as "mandatory" in Chapters 4 and 5 of this code, as applicable, shall be met.

PART 2—ADMINISTRATION AND ENFORCEMENT

SECTION 103 CONSTRUCTION DOCUMENTS

103.1 General. Construction documents and other supporting data shall be submitted in one or more sets with each application for a permit. The construction documents shall be prepared by a registered design professional where required by the statutes of the jurisdiction in which the project is to be constructed. Where special conditions exist, the *code official* is authorized to require necessary construction documents to be prepared by a registered design professional.

Exception: The *code official* is authorized to waive the requirements for construction documents or other supporting data if the *code official* determines they are not necessary to confirm compliance with this code.

103.2 Information on construction documents. Construction documents shall be drawn to scale upon suitable material. Electronic media documents are permitted to be submitted when *approved* by the *code official*. Construction documents shall be of sufficient clarity to indicate the location, nature and extent of the work proposed, and show in sufficient detail pertinent data and features of the building, systems and equipment as herein governed. Details shall include, but are not limited to, as applicable, insulation materials and their *R*-values; fenestration *U*-factors and SHGCs; area-weighted *U*-factor and SHGC calculations; mechanical system design criteria; mechanical and service water heating system and equipment types, sizes and efficiencies; economizer description; equipment and systems controls; fan motor horsepower (hp) and controls; duct sealing, duct and pipe insulation and location; lighting fixture schedule with wattage and control narrative; and air sealing details.

103.3 Examination of documents. The *code official* shall examine or cause to be examined the accompanying construction documents and shall ascertain whether the construction indicated and described is in accordance with the requirements of this code and other pertinent laws or ordinances.

103.3.1 Approval of construction documents. When the *code official* issues a permit where construction documents are required, the construction documents shall be endorsed in writing and stamped "Reviewed for Code Compliance." Such *approved* construction documents shall not be changed, modified or altered without authorization from the *code official*. Work shall be done in accordance with the *approved* construction documents.

One set of construction documents so reviewed shall be retained by the *code official*. The other set shall be returned to the applicant, kept at the site of work and shall be open to inspection by the *code official* or a duly authorized representative.

103.3.2 Previous approvals. This code shall not require changes in the construction documents, construction or designated occupancy of a structure for which a lawful permit has been heretofore issued or otherwise lawfully authorized, and the construction of which has been pursued in good faith within 180 days after the effective date of this code and has not been abandoned.

103.3.3 Phased approval. The *code official* shall have the authority to issue a permit for the construction of part of an energy conservation system before the construction documents for the entire system have been submitted or *approved*, provided adequate information and detailed statements have been filed complying with all pertinent requirements of this code. The holders of such permit shall proceed at their own risk without assurance that the permit for the entire energy conservation system will be granted.

103.4 Amended construction documents. Changes made during construction that are not in compliance with the *approved* construction documents shall be resubmitted for approval as an amended set of construction documents.

103.5 Retention of construction documents. One set of *approved* construction documents shall be retained by the *code official* for a period of not less than 180 days from date of completion of the permitted work, or as required by state or local laws.

SECTION 104
INSPECTIONS

104.1 General. Construction or work for which a permit is required shall be subject to inspection by the *code official.*

104.2 Required approvals. Work shall not be done beyond the point indicated in each successive inspection without first obtaining the approval of the *code official.* The *code official,* upon notification, shall make the requested inspections and shall either indicate the portion of the construction that is satisfactory as completed, or notify the permit holder or his or her agent wherein the same fails to comply with this code. Any portions that do not comply shall be corrected and such portion shall not be covered or concealed until authorized by the *code official.*

104.3 Final inspection. The building shall have a final inspection and not be occupied until *approved.*

104.4 Reinspection. A building shall be reinspected when determined necessary by the *code official.*

104.5 Approved inspection agencies. The *code official* is authorized to accept reports of *approved* inspection agencies, provided such agencies satisfy the requirements as to qualifications and reliability.

104.6 Inspection requests. It shall be the duty of the holder of the permit or their duly authorized agent to notify the *code official* when work is ready for inspection. It shall be the duty of the permit holder to provide access to and means for inspections of such work that are required by this code.

104.7 Reinspection and testing. Where any work or installation does not pass an initial test or inspection, the necessary corrections shall be made so as to achieve compliance with this code. The work or installation shall then be resubmitted to the *code official* for inspection and testing.

104.8 Approval. After the prescribed tests and inspections indicate that the work complies in all respects with this code, a notice of approval shall be issued by the *code official.*

104.8.1 Revocation. The *code official* is authorized to, in writing, suspend or revoke a notice of approval issued under the provisions of this code wherever the certificate is issued in error, or on the basis of incorrect information supplied, or where it is determined that the building or structure, premise, or portion thereof is in violation of any ordinance or regulation or any of the provisions of this code.

SECTION 105
VALIDITY

105.1 General. If a portion of this code is held to be illegal or void, such a decision shall not affect the validity of the remainder of this code.

SECTION 106
REFERENCED STANDARDS

106.1 General. The codes and standards referenced in this code shall be those listed in Chapter 6, and such codes and standards shall be considered as part of the requirements of this code to the prescribed extent of each such reference.

106.2 Conflicting requirements. Where the provisions of this code and the referenced standards conflict, the provisions of this code shall take precedence.

106.3 Application of references. References to chapter or section numbers, or to provisions not specifically identified by number, shall be construed to refer to such chapter, section or provision of this code.

106.4 Other laws. The provisions of this code shall not be deemed to nullify any provisions of local, state or federal law.

SECTION 107
FEES

107.1 Fees. A permit shall not be issued until the fees prescribed in Section 107.2 have been paid, nor shall an amendment to a permit be released until the additional fee, if any, has been paid.

107.2 Schedule of permit fees. A fee for each permit shall be paid as required, in accordance with the schedule as established by the applicable governing authority.

107.3 Work commencing before permit issuance. Any person who commences any work before obtaining the necessary permits shall be subject to an additional fee established by the *code official,* which shall be in addition to the required permit fees.

107.4 Related fees. The payment of the fee for the construction, *alteration,* removal or demolition of work done in connection to or concurrently with the work or activity authorized by a permit shall not relieve the applicant or holder of the permit from the payment of other fees that are prescribed by law.

107.5 Refunds. The *code official* is authorized to establish a refund policy.

SECTION 108
STOP WORK ORDER

108.1 Authority. Whenever the *code official* finds any work regulated by this code being performed in a manner either contrary to the provisions of this code or dangerous or unsafe, the *code official* is authorized to issue a stop work order.

108.2 Issuance. The stop work order shall be in writing and shall be given to the owner of the property involved, or to the owner's agent, or to the person doing the work. Upon issuance of a stop work order, the cited work shall immediately cease. The stop work order shall state the reason for the order, and the conditions under which the cited work will be permitted to resume.

108.3 Emergencies. Where an emergency exists, the *code official* shall not be required to give a written notice prior to stopping the work.

108.4 Failure to comply. Any person who shall continue any work after having been served with a stop work order, except such work as that person is directed to perform to remove a violation or unsafe condition, shall be liable to a fine of not less than [AMOUNT] dollars or more than [AMOUNT] dollars.

SECTION 109
BOARD OF APPEALS

109.1 General. In order to hear and decide appeals of orders, decisions or determinations made by the *code official* relative to the application and interpretation of this code, there shall be and is hereby created a board of appeals. The *code official* shall be an ex officio member of said board but shall have no vote on any matter before the board. The board of appeals shall be appointed by the governing body and shall hold office at its pleasure. The board shall adopt rules of procedure for conducting its business, and shall render all decisions and findings in writing to the appellant with a duplicate copy to the *code official*.

109.2 Limitations on authority. An application for appeal shall be based on a claim that the true intent of this code or the rules legally adopted thereunder have been incorrectly interpreted, the provisions of this code do not fully apply or an equally good or better form of construction is proposed. The board shall have no authority to waive requirements of this code.

109.3 Qualifications. The board of appeals shall consist of members who are qualified by experience and training and are not employees of the jurisdiction.

CHAPTER 2
DEFINITIONS

SECTION 201 GENERAL

201.1 Scope. Unless stated otherwise, the following words and terms in this code shall have the meanings indicated in this chapter.

201.2 Interchangeability. Words used in the present tense include the future; words in the masculine gender include the feminine and neuter; the singular number includes the plural and the plural includes the singular.

201.3 Terms defined in other codes. Terms that are not defined in this code but are defined in the *International Building Code*, *International Fire Code*, *International Fuel Gas Code*, *International Mechanical Code*, *International Plumbing Code* or the *International Residential Code* shall have the meanings ascribed to them in those codes.

201.4 Terms not defined. Terms not defined by this chapter shall have ordinarily accepted meanings such as the context implies.

SECTION 202 GENERAL DEFINITIONS

ABOVE-GRADE WALL. A wall more than 50 percent above grade and enclosing *conditioned space*. This includes between-floor spandrels, peripheral edges of floors, roof and basement knee walls, dormer walls, gable end walls, walls enclosing a mansard roof and skylight shafts.

ACCESSIBLE. Admitting close approach as a result of not being guarded by locked doors, elevation or other effective means (see "Readily *accessible*").

ADDITION. An extension or increase in the *conditioned space* floor area or height of a building or structure.

AIR BARRIER. Material(s) assembled and joined together to provide a barrier to air leakage through the building envelope. An air barrier may be a single material or a combination of materials.

ALTERATION. Any construction or renovation to an existing structure other than repair or addition that requires a permit. Also, a change in a mechanical system that involves an extension, addition or change to the arrangement, type or purpose of the original installation that requires a permit.

APPROVED. Approval by the *code official* as a result of investigation and tests conducted by him or her, or by reason of accepted principles or tests by nationally recognized organizations.

AUTOMATIC. Self-acting, operating by its own mechanism when actuated by some impersonal influence, as, for example, a change in current strength, pressure, temperature or mechanical configuration (see "Manual").

BASEMENT WALL. A wall 50 percent or more below grade and enclosing *conditioned space*.

BUILDING. Any structure used or intended for supporting or sheltering any use or occupancy.

BUILDING THERMAL ENVELOPE. The basement walls, exterior walls, floor, roof, and any other building element that enclose *conditioned space*. This boundary also includes the boundary between *conditioned space* and any exempt or unconditioned space.

***C*-FACTOR (THERMAL CONDUCTANCE).** The coefficient of heat transmission (surface to surface) through a building component or assembly, equal to the time rate of heat flow per unit area and the unit temperature difference between the warm side and cold side surfaces (Btu/h ft^2 × °F) [W/(m^2 × K)].

CODE OFFICIAL. The officer or other designated authority charged with the administration and enforcement of this code, or a duly authorized representative.

COMMERCIAL BUILDING. For this code, all buildings that are not included in the definition of "Residential buildings."

CONDITIONED FLOOR AREA. The horizontal projection of the floors associated with the *conditioned space*.

CONDITIONED SPACE. An area or room within a building being heated or cooled, containing uninsulated ducts, or with a fixed opening directly into an adjacent *conditioned space*.

CRAWL SPACE WALL. The opaque portion of a wall that encloses a crawl space and is partially or totally below grade.

CURTAIN WALL. Fenestration products used to create an external nonload-bearing wall that is designed to separate the exterior and interior environments.

DAYLIGHT ZONE.

1. **Under skylights.** The area under skylights whose horizontal dimension, in each direction, is equal to the skylight dimension in that direction plus either the floor-to-ceiling height or the dimension to a ceiling height opaque partition, or one-half the distance to adjacent skylights or vertical fenestration, whichever is least.
2. **Adjacent to vertical fenestration.** The area adjacent to vertical fenestration which receives daylight through the fenestration. For purposes of this definition and unless more detailed analysis is provided, the daylight *zone* depth is assumed to extend into the space a distance of 15 feet (4572 mm) or to the nearest ceiling height opaque partition, whichever is less. The daylight *zone* width is assumed to be the width of the window plus 2 feet (610 mm) on each side, or the window width plus the distance to an opaque partition, or the window width plus one-half the distance to adjacent skylight or vertical fenestration, whichever is least.

DEMAND CONTROL VENTILATION (DCV). A ventilation system capability that provides for the automatic reduction of outdoor air intake below design rates when the actual occupancy of spaces served by the system is less than design occupancy.

DUCT. A tube or conduit utilized for conveying air. The air passages of self-contained systems are not to be construed as air ducts.

DUCT SYSTEM. A continuous passageway for the transmission of air that, in addition to ducts, includes duct fittings, dampers, plenums, fans and accessory air-handling equipment and appliances.

DWELLING UNIT. A single unit providing complete independent living facilities for one or more persons, including permanent provisions for living, sleeping, eating, cooking and sanitation.

ECONOMIZER, AIR. A duct and damper arrangement and automatic control system that allows a cooling system to supply outside air to reduce or eliminate the need for mechanical cooling during mild or cold weather.

ECONOMIZER, WATER. A system where the supply air of a cooling system is cooled indirectly with water that is itself cooled by heat or mass transfer to the environment without the use of mechanical cooling.

ENERGY ANALYSIS. A method for estimating the annual energy use of the *proposed design* and *standard reference design* based on estimates of energy use.

ENERGY COST. The total estimated annual cost for purchased energy for the building functions regulated by this code, including applicable demand charges.

ENERGY RECOVERY VENTILATION SYSTEM. Systems that employ air-to-air heat exchangers to recover energy from exhaust air for the purpose of preheating, precooling, humidifying or dehumidifying outdoor ventilation air prior to supplying the air to a space, either directly or as part of an HVAC system.

ENERGY SIMULATION TOOL. An *approved* software program or calculation-based methodology that projects the annual energy use of a building.

ENTRANCE DOOR. Fenestration products used for ingress, egress and access in nonresidential buildings, including, but not limited to, exterior entrances that utilize latching hardware and automatic closers and contain over 50-percent glass specifically designed to withstand heavy use and possibly abuse.

EXTERIOR WALL. Walls including both above-grade walls and basement walls.

FAN BRAKE HORSEPOWER (BHP). The horsepower delivered to the fan's shaft. Brake horsepower does not include the mechanical drive losses (belts, gears, etc.).

FAN SYSTEM BHP. The sum of the fan brake horsepower of all fans that are required to operate at fan system design conditions to supply air from the heating or cooling source to the *conditioned space(s)* and return it to the source or exhaust it to the outdoors.

FAN SYSTEM DESIGN CONDITIONS. Operating conditions that can be expected to occur during normal system operation that result in the highest supply fan airflow rate to conditioned spaces served by the system.

FAN SYSTEM MOTOR NAMEPLATE HP. The sum of the motor nameplate horsepower of all fans that are required to operate at design conditions to supply air from the heating or cooling source to the *conditioned space(s)* and return it to the source or exhaust it to the outdoors.

FENESTRATION. Skylights, roof windows, vertical windows (fixed or moveable), opaque doors, glazed doors, glazed block and combination opaque/glazed doors. Fenestration includes products with glass and nonglass glazing materials.

***F*-FACTOR.** The perimeter heat loss factor for slab-on-grade floors (Btu/h × ft × °F) [W/(m × K)].

HEAT TRAP. An arrangement of piping and fittings, such as elbows, or a commercially available heat trap that prevents thermosyphoning of hot water during standby periods.

HEATED SLAB. Slab-on-grade construction in which the heating elements, hydronic tubing, or hot air distribution system is in contact with, or placed within or under, the slab.

HIGH-EFFICACY LAMPS. Compact fluorescent lamps, T-8 or smaller diameter linear fluorescent lamps, or lamps with a minimum efficacy of:

1. 60 lumens per watt for lamps over 40 watts,
2. 50 lumens per watt for lamps over 15 watts to 40 watts, and
3. 40 lumens per watt for lamps 15 watts or less.

HUMIDISTAT. A regulatory device, actuated by changes in humidity, used for automatic control of relative humidity.

INFILTRATION. The uncontrolled inward air leakage into a building caused by the pressure effects of wind or the effect of differences in the indoor and outdoor air density or both.

INSULATING SHEATHING. An insulating board with a core material having a minimum *R*-value of R-2.

LABELED. Equipment, materials or products to which have been affixed a label, seal, symbol or other identifying mark of a nationally recognized testing laboratory, inspection agency or other organization concerned with product evaluation that maintains periodic inspection of the production of the above-labeled items and whose labeling indicates either that the equipment, material or product meets identified standards or has been tested and found suitable for a specified purpose.

LISTED. Equipment, materials, products or services included in a list published by an organization acceptable to the *code official* and concerned with evaluation of products or services that maintains periodic inspection of production of *listed* equipment or materials or periodic evaluation of services and whose listing states either that the equipment, material, product or service meets identified standards or has been tested and found suitable for a specified purpose.

LOW-VOLTAGE LIGHTING. Lighting equipment powered through a transformer such as a cable conductor, a rail conductor and track lighting.

MANUAL. Capable of being operated by personal intervention (see "Automatic").

NAMEPLATE HORSEPOWER. The nominal motor horsepower rating stamped on the motor nameplate.

PROPOSED DESIGN. A description of the proposed building used to estimate annual energy use for determining compliance based on total building performance.

READILY ACCESSIBLE. Capable of being reached quickly for operation, renewal or inspection without requiring those to whom ready access is requisite to climb over or remove obstacles or to resort to portable ladders or access equipment (see "*Accessible*").

REPAIR. The reconstruction or renewal of any part of an existing building.

RESIDENTIAL BUILDING. For this code, includes R-3 buildings, as well as R-2 and R-4 buildings three stories or less in height above grade.

ROOF ASSEMBLY. A system designed to provide weather protection and resistance to design loads. The system consists of a roof covering and roof deck or a single component serving as both the roof covering and the roof deck. A roof assembly includes the roof covering, underlayment, roof deck, insulation, vapor retarder and interior finish.

***R*-VALUE (THERMAL RESISTANCE).** The inverse of the time rate of heat flow through a body from one of its bounding surfaces to the other surface for a unit temperature difference between the two surfaces, under steady state conditions, per unit area ($h \cdot ft^2 \cdot °F/Btu$) [$(m^2 \cdot K)/W$].

SCREW LAMP HOLDERS. A lamp base that requires a screw-in-type lamp, such as a compact-fluorescent, incandescent, or tungsten-halogen bulb.

SERVICE WATER HEATING. Supply of hot water for purposes other than comfort heating.

SKYLIGHT. Glass or other transparent or translucent glazing material installed at a slope of 15 degrees (0.26 rad) or more from vertical. Glazing material in skylights, including unit skylights, solariums, sunrooms, roofs and sloped walls is included in this definition.

SLEEPING UNIT. A room or space in which people sleep, which can also include permanent provisions for living, eating, and either sanitation or kitchen facilities but not both. Such rooms and spaces that are also part of a dwelling unit are not *sleeping units.*

SOLAR HEAT GAIN COEFFICIENT (SHGC). The ratio of the solar heat gain entering the space through the fenestration assembly to the incident solar radiation. Solar heat gain includes directly transmitted solar heat and absorbed solar radiation which is then reradiated, conducted or convected into the space.

STANDARD REFERENCE DESIGN. A version of the *proposed design* that meets the minimum requirements of this code and is used to determine the maximum annual energy use requirement for compliance based on total building performance.

STOREFRONT. A nonresidential system of doors and windows mulled as a composite fenestration structure that has been designed to resist heavy use. *Storefront* systems include, but are not limited to, exterior fenestration systems that span from the floor level or above to the ceiling of the same story on commercial buildings.

SUNROOM. A one-story structure attached to a dwelling with a glazing area in excess of 40 percent of the gross area of the structure's exterior walls and roof.

THERMAL ISOLATION. Physical and space conditioning separation from *conditioned space(s)*. The *conditioned space*(s) shall be controlled as separate zones for heating and cooling or conditioned by separate equipment.

THERMOSTAT. An automatic control device used to maintain temperature at a fixed or adjustable set point.

***U*-FACTOR (THERMAL TRANSMITTANCE).** The coefficient of heat transmission (air to air) through a building component or assembly, equal to the time rate of heat flow per unit area and unit temperature difference between the warm side and cold side air films ($Btu/h \cdot ft^2 \cdot °F$) [$W/(m^2 \cdot K)$].

VENTILATION. The natural or mechanical process of supplying conditioned or unconditioned air to, or removing such air from, any space.

VENTILATION AIR. That portion of supply air that comes from outside (outdoors) plus any recirculated air that has been treated to maintain the desired quality of air within a designated space.

ZONE. A space or group of spaces within a building with heating or cooling requirements that are sufficiently similar so that desired conditions can be maintained throughout using a single controlling device.

CHAPTER 3
CLIMATE ZONES

SECTION 301
CLIMATE ZONES

301.1 General. Climate *zones* from Figure 301.1 or Table 301.1 shall be used in determining the applicable requirements from Chapters 4 and 5. Locations not in Table 301.1 (outside the United States) shall be assigned a climate *zone* based on Section 301.3.

301.2 Warm humid counties. Warm humid counties are identified in Table 301.1 by an asterisk.

301.3 International climate zones. The climate *zone* for any location outside the United States shall be determined by applying Table 301.3(1) and then Table 301.3(2).

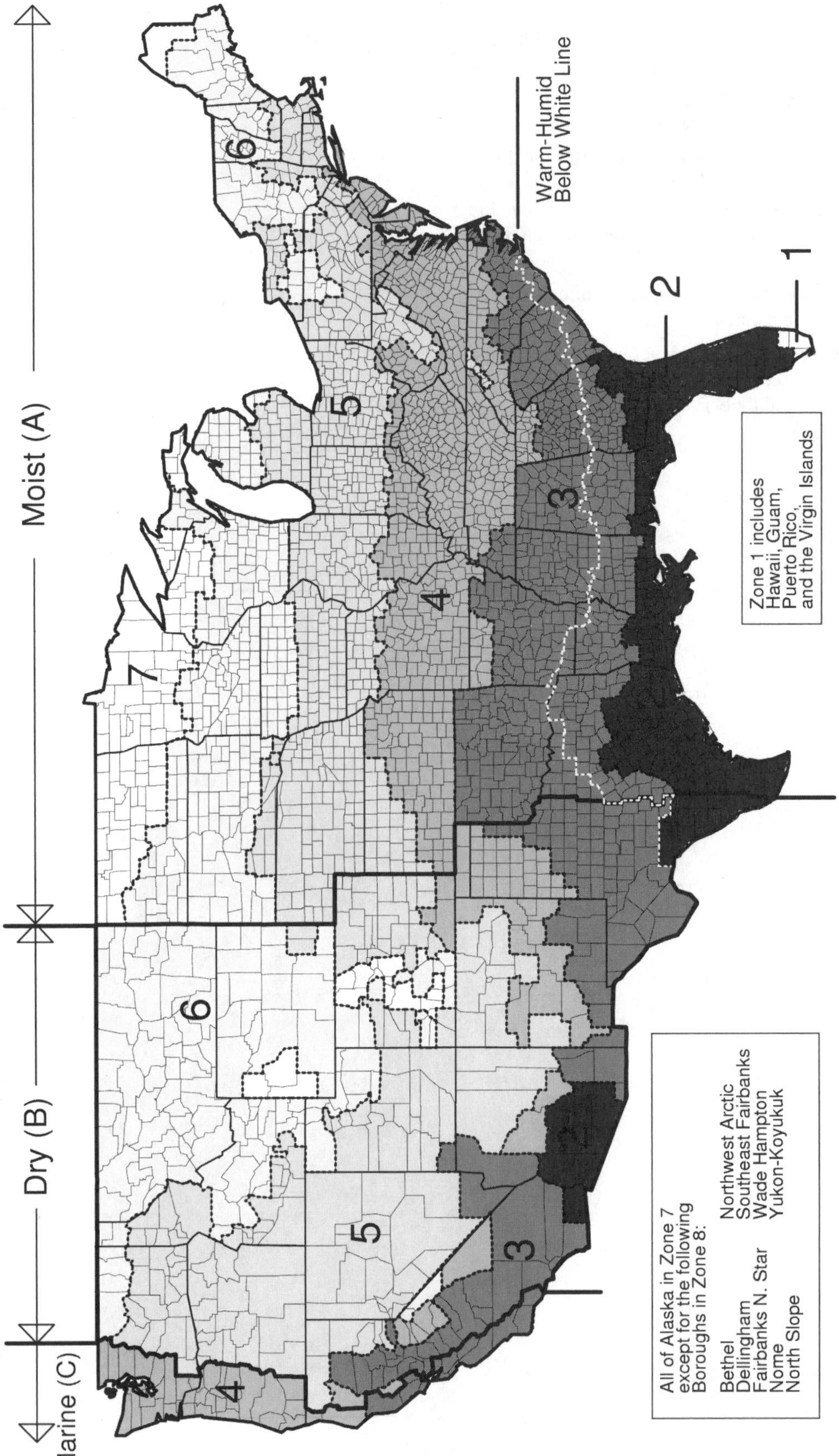
Marine (C)
Dry (B)
Moist (A)
Warm-Humid
Below White Line
All of Alaska in Zone 7
except for the following
Boroughs in Zone 8:
Bethel
Dellingham
Fairbanks N. Star
Nome
North Slope
Northwest Arctic
Southeast Fairbanks
Wade Hampton
Yukon-Koyukuk
Zone 1 includes
Hawaii, Guam,
Puerto Rico,
and the Virgin Islands
1
2
3
4
5
6
7

FIGURE 301.1
CLIMATE ZONES

TABLE 301.1
CLIMATE ZONES, MOISTURE REGIMES, AND WARM-HUMID DESIGNATIONS BY STATE, COUNTY AND TERRITORY

Note: Table 301.1 in the 2006 edition has been replaced in its entirety. Margin lines are omitted for clarity.

Key: A – Moist, B – Dry, C – Marine. Absence of moisture designation indicates moisture regime is irrelevant. Asterisk (*) indicates a warm-humid location.

US STATES

ALABAMA

3A Autauga*
2A Baldwin*
3A Barbour*
3A Bibb
3A Blount
3A Bullock*
3A Butler*
3A Calhoun
3A Chambers
3A Cherokee
3A Chilton
3A Choctaw*
3A Clarke*
3A Clay
3A Cleburne
3A Coffee*
3A Colbert
3A Conecuh*
3A Coosa
3A Covington*
3A Crenshaw*
3A Cullman
3A Dale*
3A Dallas*
3A DeKalb
3A Elmore*
3A Escambia*
3A Etowah
3A Fayette
3A Franklin
3A Geneva*
3A Greene
3A Hale
3A Henry*
3A Houston*
3A Jackson
3A Jefferson
3A Lamar
3A Lauderdale
3A Lawrence
3A Lee
3A Limestone
3A Lowndes*
3A Macon*
3A Madison
3A Marengo*
3A Marion
3A Marshall
2A Mobile*
3A Monroe*
3A Montgomery*
3A Morgan
3A Perry*
3A Pickens
3A Pike*
3A Randolph
3A Russell*
3A Shelby
3A St. Clair
3A Sumter
3A Talladega
3A Tallapoosa
3A Tuscaloosa
3A Walker
3A Washington*
3A Wilcox*
3A Winston

ALASKA

7 Aleutians East
7 Aleutians West
7 Anchorage
8 Bethel
7 Bristol Bay
7 Denali
8 Dillingham
8 Fairbanks North Star
7 Haines
7 Juneau
7 Kenai Peninsula
7 Ketchikan Gateway
7 Kodiak Island
7 Lake and Peninsula
7 Matanuska-Susitna
8 Nome
8 North Slope
8 Northwest Arctic
7 Prince of Wales-Outer Ketchikan
7 Sitka
7 Skagway-Hoonah-Angoon
8 Southeast Fairbanks
7 Valdez-Cordova
8 Wade Hampton
7 Wrangell-Petersburg
7 Yakutat
8 Yukon-Koyukuk

ARIZONA

5B Apache
3B Cochise
5B Coconino
4B Gila
3B Graham
3B Greenlee
2B La Paz
2B Maricopa
3B Mohave
5B Navajo
2B Pima
2B Pinal
3B Santa Cruz
4B Yavapai
2B Yuma

ARKANSAS

3A Arkansas
3A Ashley
4A Baxter
4A Benton
4A Boone
3A Bradley
3A Calhoun
4A Carroll
3A Chicot
3A Clark
3A Clay
3A Cleburne
3A Cleveland
3A Columbia*
3A Conway
3A Craighead
3A Crawford
3A Crittenden
3A Cross
3A Dallas
3A Desha
3A Drew
3A Faulkner
3A Franklin
4A Fulton
3A Garland
3A Grant
3A Greene
3A Hempstead*
3A Hot Spring
3A Howard
3A Independence
4A Izard
3A Jackson
3A Jefferson
3A Johnson
3A Lafayette*
3A Lawrence
3A Lee
3A Lincoln
3A Little River*
3A Logan
3A Lonoke
4A Madison
4A Marion
3A Miller*
3A Mississippi
3A Monroe
3A Montgomery
3A Nevada
4A Newton
3A Ouachita
3A Perry
3A Phillips

(continued)

TABLE 301.1—continued
CLIMATE ZONES, MOISTURE REGIMES, AND WARM-HUMID DESIGNATIONS BY STATE, COUNTY AND TERRITORY

3A Pike
3A Poinsett
3A Polk
3A Pope
3A Prairie
3A Pulaski
3A Randolph
3A Saline
3A Scott
4A Searcy
3A Sebastian
3A Sevier*
3A Sharp
3A St. Francis
4A Stone
3A Union*
3A Van Buren
4A Washington
3A White
3A Woodruff
3A Yell

CALIFORNIA

3C Alameda
6B Alpine
4B Amador
3B Butte
4B Calaveras
3B Colusa
3B Contra Costa
4C Del Norte
4B El Dorado
3B Fresno
3B Glenn
4C Humboldt
2B Imperial
4B Inyo
3B Kern
3B Kings
4B Lake
5B Lassen
3B Los Angeles
3B Madera
3C Marin
4B Mariposa
3C Mendocino
3B Merced
5B Modoc
6B Mono
3C Monterey
3C Napa
5B Nevada
3B Orange
3B Placer
5B Plumas
3B Riverside
3B Sacramento
3C San Benito
3B San Bernardino
3B San Diego
3C San Francisco
3B San Joaquin
3C San Luis Obispo
3C San Mateo
3C Santa Barbara
3C Santa Clara
3C Santa Cruz
3B Shasta
5B Sierra
5B Siskiyou
3B Solano
3C Sonoma
3B Stanislaus
3B Sutter
3B Tehama
4B Trinity
3B Tulare
4B Tuolumne
3C Ventura
3B Yolo
3B Yuba

COLORADO

5B Adams
6B Alamosa
5B Arapahoe
6B Archuleta
4B Baca
5B Bent
5B Boulder
6B Chaffee
5B Cheyenne
7 Clear Creek
6B Conejos
6B Costilla
5B Crowley
6B Custer
5B Delta
5B Denver
6B Dolores
5B Douglas
6B Eagle
5B Elbert
5B El Paso
5B Fremont
5B Garfield
5B Gilpin
7 Grand
7 Gunnison
7 Hinsdale
5B Huerfano
7 Jackson
5B Jefferson
5B Kiowa
5B Kit Carson
7 Lake
5B La Plata
5B Larimer
4B Las Animas
5B Lincoln
5B Logan
5B Mesa
7 Mineral
6B Moffat
5B Montezuma
5B Montrose
5B Morgan
4B Otero
6B Ouray
7 Park
5B Phillips
7 Pitkin
5B Prowers
5B Pueblo
6B Rio Blanco
7 Rio Grande
7 Routt
6B Saguache
7 San Juan
6B San Miguel
5B Sedgwick
7 Summit
5B Teller
5B Washington
5B Weld
5B Yuma

CONNECTICUT

5A (all)

DELAWARE

4A (all)

DISTRICT OF COLUMBIA

4A (all)

FLORIDA

2A Alachua*
2A Baker*
2A Bay*
2A Bradford*
2A Brevard*
1A Broward*
2A Calhoun*
2A Charlotte*
2A Citrus*
2A Clay*
2A Collier*
2A Columbia*
2A DeSoto*
2A Dixie*
2A Duval*
2A Escambia*
2A Flagler*
2A Franklin*
2A Gadsden*
2A Gilchrist*
2A Glades*
2A Gulf*
2A Hamilton*
2A Hardee*
2A Hendry*
2A Hernando*
2A Highlands*
2A Hillsborough*
2A Holmes*
2A Indian River*
2A Jackson*
2A Jefferson*
2A Lafayette*
2A Lake*
2A Lee*
2A Leon*
2A Levy*
2A Liberty*
2A Madison*
2A Manatee*
2A Marion*
2A Martin*
1A Miami-Dade*
1A Monroe*
2A Nassau*
2A Okaloosa*
2A Okeechobee*

(continued)

TABLE 301.1—continued
CLIMATE ZONES, MOISTURE REGIMES, AND WARM-HUMID DESIGNATIONS BY STATE, COUNTY AND TERRITORY

2A Orange*
2A Osceola*
2A Palm Beach*
2A Pasco*
2A Pinellas*
2A Polk*
2A Putnam*
2A Santa Rosa*
2A Sarasota*
2A Seminole*
2A St. Johns*
2A St. Lucie*
2A Sumter*
2A Suwannee*
2A Taylor*
2A Union*
2A Volusia*
2A Wakulla*
2A Walton*
2A Washington*

GEORGIA

2A Appling*
2A Atkinson*
2A Bacon*
2A Baker*
3A Baldwin
4A Banks
3A Barrow
3A Bartow
3A Ben Hill*
2A Berrien*
3A Bibb
3A Bleckley*
2A Brantley*
2A Brooks*
2A Bryan*
3A Bulloch*
3A Burke
3A Butts
3A Calhoun*
2A Camden*
3A Candler*
3A Carroll
4A Catoosa
2A Charlton*
2A Chatham*
3A Chattahoochee*
4A Chattooga
3A Cherokee
3A Clarke
3A Clay*
3A Clayton
2A Clinch*
3A Cobb
3A Coffee*
2A Colquitt*
3A Columbia
2A Cook*
3A Coweta
3A Crawford
3A Crisp*
4A Dade
4A Dawson
2A Decatur*
3A DeKalb
3A Dodge*
3A Dooly*
3A Dougherty*
3A Douglas
3A Early*
2A Echols*
2A Effingham*
3A Elbert
3A Emanuel*
2A Evans*
4A Fannin
3A Fayette
4A Floyd
3A Forsyth
4A Franklin
3A Fulton
4A Gilmer
3A Glascock
2A Glynn*
4A Gordon
2A Grady*
3A Greene
3A Gwinnett
4A Habersham
4A Hall
3A Hancock
3A Haralson
3A Harris
3A Hart
3A Heard
3A Henry
3A Houston*
3A Irwin*
3A Jackson
3A Jasper
2A Jeff Davis*
3A Jefferson
3A Jenkins*
3A Johnson*
3A Jones
3A Lamar
2A Lanier*
3A Laurens*
3A Lee*
2A Liberty*
3A Lincoln
2A Long*
2A Lowndes*
4A Lumpkin
3A Macon*
3A Madison
3A Marion*
3A McDuffie
2A McIntosh*
3A Meriwether
2A Miller*
2A Mitchell*
3A Monroe
3A Montgomery*
3A Morgan
4A Murray
3A Muscogee
3A Newton
3A Oconee
3A Oglethorpe
3A Paulding
3A Peach*
4A Pickens
2A Pierce*
3A Pike
3A Polk
3A Pulaski*
3A Putnam
3A Quitman*
4A Rabun
3A Randolph*
3A Richmond
3A Rockdale
3A Schley*
3A Screven*
2A Seminole*
3A Spalding
4A Stephens
3A Stewart*
3A Sumter*
3A Talbot
3A Taliaferro
2A Tattnall*
3A Taylor*
3A Telfair*
3A Terrell*
2A Thomas*
3A Tift*
2A Toombs*
4A Towns
3A Treutlen*
3A Troup
3A Turner*
3A Twiggs*
4A Union
3A Upson
4A Walker
3A Walton
2A Ware*
3A Warren
3A Washington
2A Wayne*
3A Webster*
3A Wheeler*
4A White
4A Whitfield
3A Wilcox*
3A Wilkes
3A Wilkinson
3A Worth*

HAWAII

1A (all)*

IDAHO

5B Ada
6B Adams
6B Bannock
6B Bear Lake
5B Benewah
6B Bingham
6B Blaine
6B Boise
6B Bonner
6B Bonneville
6B Boundary
6B Butte
6B Camas
5B Canyon
6B Caribou
5B Cassia
6B Clark
5B Clearwater
6B Custer
5B Elmore

(continued)

TABLE 301.1—continued
CLIMATE ZONES, MOISTURE REGIMES, AND WARM-HUMID DESIGNATIONS BY STATE, COUNTY AND TERRITORY

6B Franklin
6B Fremont
5B Gem
5B Gooding
5B Idaho
6B Jefferson
5B Jerome
5B Kootenai
5B Latah
6B Lemhi
5B Lewis
5B Lincoln
6B Madison
5B Minidoka
5B Nez Perce
6B Oneida
5B Owyhee
5B Payette
5B Power
5B Shoshone
6B Teton
5B Twin Falls
6B Valley
5B Washington

ILLINOIS

5A Adams
4A Alexander
4A Bond
5A Boone
5A Brown
5A Bureau
5A Calhoun
5A Carroll
5A Cass
5A Champaign
4A Christian
5A Clark
4A Clay
4A Clinton
5A Coles
5A Cook
4A Crawford
5A Cumberland
5A DeKalb
5A De Witt
5A Douglas
5A DuPage
5A Edgar
4A Edwards
4A Effingham
4A Fayette
5A Ford
4A Franklin
5A Fulton
4A Gallatin
5A Greene
5A Grundy
4A Hamilton
5A Hancock
4A Hardin
5A Henderson
5A Henry
5A Iroquois
4A Jackson
4A Jasper
4A Jefferson
5A Jersey
5A Jo Daviess
4A Johnson
5A Kane
5A Kankakee
5A Kendall
5A Knox
5A Lake
5A La Salle
4A Lawrence
5A Lee
5A Livingston
5A Logan
5A Macon
4A Macoupin
4A Madison
4A Marion
5A Marshall
5A Mason
4A Massac
5A McDonough
5A McHenry
5A McLean
5A Menard
5A Mercer
4A Monroe
4A Montgomery
5A Morgan
5A Moultrie
5A Ogle
5A Peoria
4A Perry
5A Piatt
5A Pike
4A Pope
4A Pulaski
5A Putnam
4A Randolph
4A Richland
5A Rock Island
4A Saline
5A Sangamon
5A Schuyler
5A Scott
4A Shelby
5A Stark
4A St. Clair
5A Stephenson
5A Tazewell
4A Union
5A Vermilion
4A Wabash
5A Warren
4A Washington
4A Wayne
4A White
5A Whiteside
5A Will
4A Williamson
5A Winnebago
5A Woodford

INDIANA

5A Adams
5A Allen
5A Bartholomew
5A Benton
5A Blackford
5A Boone
4A Brown
5A Carroll
5A Cass
4A Clark
5A Clay
5A Clinton
4A Crawford
4A Daviess
4A Dearborn
5A Decatur
5A De Kalb
5A Delaware
4A Dubois
5A Elkhart
5A Fayette
4A Floyd
5A Fountain
5A Franklin
5A Fulton
4A Gibson
5A Grant
4A Greene
5A Hamilton
5A Hancock
4A Harrison
5A Hendricks
5A Henry
5A Howard
5A Huntington
4A Jackson
5A Jasper
5A Jay
4A Jefferson
4A Jennings
5A Johnson
4A Knox
5A Kosciusko
5A Lagrange
5A Lake
5A La Porte
4A Lawrence
5A Madison
5A Marion
5A Marshall
4A Martin
5A Miami
4A Monroe
5A Montgomery
5A Morgan
5A Newton
5A Noble
4A Ohio
4A Orange
5A Owen
5A Parke
4A Perry
4A Pike
5A Porter
4A Posey
5A Pulaski
5A Putnam
5A Randolph
4A Ripley
5A Rush
4A Scott
5A Shelby

(continued)

TABLE 301.1—continued
CLIMATE ZONES, MOISTURE REGIMES, AND WARM-HUMID DESIGNATIONS BY STATE, COUNTY AND TERRITORY

4A Spencer
5A Starke
5A Steuben
5A St. Joseph
4A Sullivan
4A Switzerland
5A Tippecanoe
5A Tipton
5A Union
4A Vanderburgh
5A Vermillion
5A Vigo
5A Wabash
5A Warren
4A Warrick
4A Washington
5A Wayne
5A Wells
5A White
5A Whitley

IOWA

5A Adair
5A Adams
6A Allamakee
5A Appanoose
5A Audubon
5A Benton
6A Black Hawk
5A Boone
6A Bremer
6A Buchanan
6A Buena Vista
6A Butler
6A Calhoun
5A Carroll
5A Cass
5A Cedar
6A Cerro Gordo
6A Cherokee
6A Chickasaw
5A Clarke
6A Clay
6A Clayton
5A Clinton
5A Crawford
5A Dallas
5A Davis
5A Decatur
6A Delaware
5A Des Moines
6A Dickinson
5A Dubuque
6A Emmet
6A Fayette
6A Floyd
6A Franklin
5A Fremont
5A Greene
6A Grundy
5A Guthrie
6A Hamilton
6A Hancock
6A Hardin
5A Harrison
5A Henry
6A Howard
6A Humboldt
6A Ida
5A Iowa
5A Jackson
5A Jasper
5A Jefferson
5A Johnson
5A Jones
5A Keokuk
6A Kossuth
5A Lee
5A Linn
5A Louisa
5A Lucas
6A Lyon
5A Madison
5A Mahaska
5A Marion
5A Marshall
5A Mills
6A Mitchell
5A Monona
5A Monroe
5A Montgomery
5A Muscatine
6A O'Brien
6A Osceola
5A Page
6A Palo Alto
6A Plymouth
6A Pocahontas
5A Polk
5A Pottawattamie
5A Poweshiek
5A Ringgold
6A Sac
5A Scott
5A Shelby
6A Sioux
5A Story
5A Tama
5A Taylor
5A Union
5A Van Buren
5A Wapello
5A Warren
5A Washington
5A Wayne
6A Webster
6A Winnebago
6A Winneshiek
5A Woodbury
6A Worth
6A Wright

KANSAS

4A Allen
4A Anderson
4A Atchison
4A Barber
4A Barton
4A Bourbon
4A Brown
4A Butler
4A Chase
4A Chautauqua
4A Cherokee
5A Cheyenne
4A Clark
4A Clay
5A Cloud
4A Coffey
4A Comanche
4A Cowley
4A Crawford
5A Decatur
4A Dickinson
4A Doniphan
4A Douglas
4A Edwards
4A Elk
5A Ellis
4A Ellsworth
4A Finney
4A Ford
4A Franklin
4A Geary
5A Gove
5A Graham
4A Grant
4A Gray
5A Greeley
4A Greenwood
5A Hamilton
4A Harper
4A Harvey
4A Haskell
4A Hodgeman
4A Jackson
4A Jefferson
5A Jewell
4A Johnson
4A Kearny
4A Kingman
4A Kiowa
4A Labette
5A Lane
4A Leavenworth
4A Lincoln
4A Linn
5A Logan
4A Lyon
4A Marion
4A Marshall
4A McPherson
4A Meade
4A Miami
5A Mitchell
4A Montgomery
4A Morris
4A Morton
4A Nemaha
4A Neosho
5A Ness
5A Norton
4A Osage
5A Osborne
4A Ottawa
4A Pawnee
5A Phillips
4A Pottawatomie
4A Pratt
5A Rawlins
4A Reno
5A Republic

(continued)

TABLE 301.1—continued
CLIMATE ZONES, MOISTURE REGIMES, AND WARM-HUMID DESIGNATIONS BY STATE, COUNTY AND TERRITORY

4A Rice
4A Riley
5A Rooks
4A Rush
4A Russell
4A Saline
5A Scott
4A Sedgwick
4A Seward
4A Shawnee
5A Sheridan
5A Sherman
5A Smith
4A Stafford
4A Stanton
4A Stevens
4A Sumner
5A Thomas
5A Trego
4A Wabaunsee
5A Wallace
4A Washington
5A Wichita
4A Wilson
4A Woodson
4A Wyandotte

KENTUCKY

4A (all)

LOUISIANA

2A Acadia*
2A Allen*
2A Ascension*
2A Assumption*
2A Avoyelles*
2A Beauregard*
3A Bienville*
3A Bossier*
3A Caddo*
2A Calcasieu*
3A Caldwell*
2A Cameron*
3A Catahoula*
3A Claiborne*
3A Concordia*
3A De Soto*
2A East Baton Rouge*
3A East Carroll
2A East Feliciana*
2A Evangeline*
3A Franklin*
3A Grant*
2A Iberia*
2A Iberville*
3A Jackson*
2A Jefferson*
2A Jefferson Davis*
2A Lafayette*
2A Lafourche*
3A La Salle*
3A Lincoln*
2A Livingston*
3A Madison*
3A Morehouse
3A Natchitoches*
2A Orleans*
3A Ouachita*
2A Plaquemines*
2A Pointe Coupee*
2A Rapides*
3A Red River*
3A Richland*
3A Sabine*
2A St. Bernard*
2A St. Charles*
2A St. Helena*
2A St. James*
2A St. John the Baptist*
2A St. Landry*
2A St. Martin*
2A St. Mary*
2A St. Tammany*
2A Tangipahoa*
3A Tensas*
2A Terrebonne*
3A Union*
2A Vermilion*
3A Vernon*
2A Washington*
3A Webster*
2A West Baton Rouge*
3A West Carroll
2A West Feliciana*
3A Winn*

MAINE

6A Androscoggin
7 Aroostook
6A Cumberland
6A Franklin
6A Hancock
6A Kennebec
6A Knox
6A Lincoln
6A Oxford
6A Penobscot
6A Piscataquis
6A Sagadahoc
6A Somerset
6A Waldo
6A Washington
6A York

MARYLAND

4A Allegany
4A Anne Arundel
4A Baltimore
4A Baltimore (city)
4A Calvert
4A Caroline
4A Carroll
4A Cecil
4A Charles
4A Dorchester
4A Frederick
5A Garrett
4A Harford
4A Howard
4A Kent
4A Montgomery
4A Prince George's
4A Queen Anne's
4A Somerset
4A St. Mary's
4A Talbot
4A Washington
4A Wicomico
4A Worcester

MASSACHUSETTS

5A (all)

MICHIGAN

6A Alcona
6A Alger
5A Allegan
6A Alpena
6A Antrim
6A Arenac
7 Baraga
5A Barry
5A Bay
6A Benzie
5A Berrien
5A Branch
5A Calhoun
5A Cass
6A Charlevoix
6A Cheboygan
7 Chippewa
6A Clare
5A Clinton
6A Crawford
6A Delta
6A Dickinson
5A Eaton
6A Emmet
5A Genesee
6A Gladwin
7 Gogebic
6A Grand Traverse
5A Gratiot
5A Hillsdale
7 Houghton
6A Huron
5A Ingham
5A Ionia
6A Iosco
7 Iron
6A Isabella
5A Jackson
5A Kalamazoo
6A Kalkaska
5A Kent
7 Keweenaw
6A Lake
5A Lapeer
6A Leelanau
5A Lenawee
5A Livingston
7 Luce
7 Mackinac
5A Macomb
6A Manistee
6A Marquette
6A Mason
6A Mecosta
6A Menominee
5A Midland
6A Missaukee
5A Monroe
5A Montcalm

(continued)

TABLE 301.1—continued
CLIMATE ZONES, MOISTURE REGIMES, AND WARM-HUMID DESIGNATIONS BY STATE, COUNTY AND TERRITORY

6A Montmorency
5A Muskegon
6A Newaygo
5A Oakland
6A Oceana
6A Ogemaw
7 Ontonagon
6A Osceola
6A Oscoda
6A Otsego
5A Ottawa
6A Presque Isle
6A Roscommon
5A Saginaw
6A Sanilac
7 Schoolcraft
5A Shiawassee
5A St. Clair
5A St. Joseph
5A Tuscola
5A Van Buren
5A Washtenaw
5A Wayne
6A Wexford

MINNESOTA

7 Aitkin
6A Anoka
7 Becker
7 Beltrami
6A Benton
6A Big Stone
6A Blue Earth
6A Brown
7 Carlton
6A Carver
7 Cass
6A Chippewa
6A Chisago
7 Clay
7 Clearwater
7 Cook
6A Cottonwood
7 Crow Wing
6A Dakota
6A Dodge
6A Douglas
6A Faribault
6A Fillmore
6A Freeborn
6A Goodhue
7 Grant
6A Hennepin
6A Houston
7 Hubbard
6A Isanti
7 Itasca
6A Jackson
7 Kanabec
6A Kandiyohi
7 Kittson
7 Koochiching
6A Lac qui Parle
7 Lake
7 Lake of the Woods
6A Le Sueur
6A Lincoln
6A Lyon
7 Mahnomen
7 Marshall
6A Martin
6A McLeod
6A Meeker
7 Mille Lacs
6A Morrison
6A Mower
6A Murray
6A Nicollet
6A Nobles
7 Norman
6A Olmsted
7 Otter Tail
7 Pennington
7 Pine
6A Pipestone
7 Polk
6A Pope
6A Ramsey
7 Red Lake
6A Redwood
6A Renville
6A Rice
6A Rock
7 Roseau
6A Scott
6A Sherburne
6A Sibley
6A Stearns
6A Steele
6A Stevens
7 St. Louis
6A Swift
6A Todd
6A Traverse
6A Wabasha
7 Wadena
6A Waseca
6A Washington
6A Watonwan
7 Wilkin
6A Winona
6A Wright
6A Yellow Medicine

MISSISSIPPI

3A Adams*
3A Alcorn
3A Amite*
3A Attala
3A Benton
3A Bolivar
3A Calhoun
3A Carroll
3A Chickasaw
3A Choctaw
3A Claiborne*
3A Clarke
3A Clay
3A Coahoma
3A Copiah*
3A Covington*
3A DeSoto
3A Forrest*
3A Franklin*
3A George*
3A Greene*
3A Grenada
2A Hancock*
2A Harrison*
3A Hinds*
3A Holmes
3A Humphreys
3A Issaquena
3A Itawamba
2A Jackson*
3A Jasper
3A Jefferson*
3A Jefferson Davis*
3A Jones*
3A Kemper
3A Lafayette
3A Lamar*
3A Lauderdale
3A Lawrence*
3A Leake
3A Lee
3A Leflore
3A Lincoln*
3A Lowndes
3A Madison
3A Marion*
3A Marshall
3A Monroe
3A Montgomery
3A Neshoba
3A Newton
3A Noxubee
3A Oktibbeha
3A Panola
2A Pearl River*
3A Perry*
3A Pike*
3A Pontotoc
3A Prentiss
3A Quitman
3A Rankin*
3A Scott
3A Sharkey
3A Simpson*
3A Smith*
2A Stone*
3A Sunflower
3A Tallahatchie
3A Tate
3A Tippah
3A Tishomingo
3A Tunica
3A Union
3A Walthall*
3A Warren*
3A Washington
3A Wayne*
3A Webster
3A Wilkinson*
3A Winston
3A Yalobusha
3A Yazoo

MISSOURI

5A Adair
5A Andrew
5A Atchison
4A Audrain
4A Barry

(continued)

TABLE 301.1—continued
CLIMATE ZONES, MOISTURE REGIMES, AND WARM-HUMID DESIGNATIONS BY STATE, COUNTY AND TERRITORY

4A Barton
4A Bates
4A Benton
4A Bollinger
4A Boone
5A Buchanan
4A Butler
5A Caldwell
4A Callaway
4A Camden
4A Cape Girardeau
4A Carroll
4A Carter
4A Cass
4A Cedar
5A Chariton
4A Christian
5A Clark
4A Clay
5A Clinton
4A Cole
4A Cooper
4A Crawford
4A Dade
4A Dallas
5A Daviess
5A DeKalb
4A Dent
4A Douglas
4A Dunklin
4A Franklin
4A Gasconade
5A Gentry
4A Greene
5A Grundy
5A Harrison
4A Henry
4A Hickory
5A Holt
4A Howard
4A Howell
4A Iron
4A Jackson
4A Jasper
4A Jefferson
4A Johnson
5A Knox
4A Laclede
4A Lafayette
4A Lawrence
5A Lewis
4A Lincoln
5A Linn
5A Livingston
5A Macon
4A Madison
4A Maries
5A Marion
4A McDonald
5A Mercer
4A Miller
4A Mississippi
4A Moniteau
4A Monroe
4A Montgomery
4A Morgan
4A New Madrid
4A Newton
5A Nodaway
4A Oregon
4A Osage
4A Ozark
4A Pemiscot
4A Perry
4A Pettis
4A Phelps
5A Pike
4A Platte
4A Polk
4A Pulaski
5A Putnam
5A Ralls
4A Randolph
4A Ray
4A Reynolds
4A Ripley
4A Saline
5A Schuyler
5A Scotland
4A Scott
4A Shannon
5A Shelby
4A St. Charles
4A St. Clair
4A Ste. Genevieve
4A St. Francois
4A St. Louis
4A St. Louis (city)
4A Stoddard
4A Stone
5A Sullivan
4A Taney
4A Texas
4A Vernon
4A Warren
4A Washington
4A Wayne
4A Webster
5A Worth
4A Wright

MONTANA

6B (all)

NEBRASKA

5A (all)

NEVADA

5B Carson City (city)
5B Churchill
3B Clark
5B Douglas
5B Elko
5B Esmeralda
5B Eureka
5B Humboldt
5B Lander
5B Lincoln
5B Lyon
5B Mineral
5B Nye
5B Pershing
5B Storey
5B Washoe
5B White Pine

NEW HAMPSHIRE

6A Belknap
6A Carroll
5A Cheshire
6A Coos
6A Grafton
5A Hillsborough
6A Merrimack
5A Rockingham
5A Strafford
6A Sullivan

NEW JERSEY

4A Atlantic
5A Bergen
4A Burlington
4A Camden
4A Cape May
4A Cumberland
4A Essex
4A Gloucester
4A Hudson
5A Hunterdon
5A Mercer
4A Middlesex
4A Monmouth
5A Morris
4A Ocean
5A Passaic
4A Salem
5A Somerset
5A Sussex
4A Union
5A Warren

NEW MEXICO

4B Bernalillo
5B Catron
3B Chaves
4B Cibola
5B Colfax
4B Curry
4B DeBaca
3B Dona Ana
3B Eddy
4B Grant
4B Guadalupe
5B Harding
3B Hidalgo
3B Lea
4B Lincoln
5B Los Alamos
3B Luna
5B McKinley
5B Mora
3B Otero
4B Quay
5B Rio Arriba
4B Roosevelt
5B Sandoval
5B San Juan
5B San Miguel
5B Santa Fe
4B Sierra
4B Socorro
5B Taos
5B Torrance
4B Union
4B Valencia

(continued)

TABLE 301.1—continued
CLIMATE ZONES, MOISTURE REGIMES, AND WARM-HUMID DESIGNATIONS BY STATE, COUNTY AND TERRITORY

NEW YORK

5A Albany
6A Allegany
4A Bronx
6A Broome
6A Cattaraugus
5A Cayuga
5A Chautauqua
5A Chemung
6A Chenango
6A Clinton
5A Columbia
5A Cortland
6A Delaware
5A Dutchess
5A Erie
6A Essex
6A Franklin
6A Fulton
5A Genesee
5A Greene
6A Hamilton
6A Herkimer
6A Jefferson
4A Kings
6A Lewis
5A Livingston
6A Madison
5A Monroe
6A Montgomery
4A Nassau
4A New York
5A Niagara
6A Oneida
5A Onondaga
5A Ontario
5A Orange
5A Orleans
5A Oswego
6A Otsego
5A Putnam
4A Queens
5A Rensselaer
4A Richmond
5A Rockland
5A Saratoga
5A Schenectady
6A Schoharie
6A Schuyler
5A Seneca
6A Steuben
6A St. Lawrence
4A Suffolk
6A Sullivan
5A Tioga
6A Tompkins
6A Ulster
6A Warren
5A Washington
5A Wayne
4A Westchester
6A Wyoming
5A Yates

NORTH CAROLINA

4A Alamance
4A Alexander
5A Alleghany
3A Anson
5A Ashe
5A Avery
3A Beaufort
4A Bertie
3A Bladen
3A Brunswick*
4A Buncombe
4A Burke
3A Cabarrus
4A Caldwell
3A Camden
3A Carteret*
4A Caswell
4A Catawba
4A Chatham
4A Cherokee
3A Chowan
4A Clay
4A Cleveland
3A Columbus*
3A Craven
3A Cumberland
3A Currituck
3A Dare
3A Davidson
4A Davie
3A Duplin
4A Durham
3A Edgecombe
4A Forsyth
4A Franklin
3A Gaston
4A Gates
4A Graham
4A Granville
3A Greene
4A Guilford
4A Halifax
4A Harnett
4A Haywood
4A Henderson
4A Hertford
3A Hoke
3A Hyde
4A Iredell
4A Jackson
3A Johnston
3A Jones
4A Lee
3A Lenoir
4A Lincoln
4A Macon
4A Madison
3A Martin
4A McDowell
3A Mecklenburg
5A Mitchell
3A Montgomery
3A Moore
4A Nash
3A New Hanover*
4A Northampton
3A Onslow*
4A Orange
3A Pamlico
3A Pasquotank
3A Pender*
3A Perquimans
4A Person
3A Pitt
4A Polk
3A Randolph
3A Richmond
3A Robeson
4A Rockingham
3A Rowan
4A Rutherford
3A Sampson
3A Scotland
3A Stanly
4A Stokes
4A Surry
4A Swain
4A Transylvania
3A Tyrrell
3A Union
4A Vance
4A Wake
4A Warren
3A Washington
5A Watauga
3A Wayne
4A Wilkes
3A Wilson
4A Yadkin
5A Yancey

NORTH DAKOTA

6A Adams
7 Barnes
7 Benson
6A Billings
7 Bottineau
6A Bowman
7 Burke
6A Burleigh
7 Cass
7 Cavalier
6A Dickey
7 Divide
6A Dunn
7 Eddy
6A Emmons
7 Foster
6A Golden Valley
7 Grand Forks
6A Grant
7 Griggs
6A Hettinger
7 Kidder
6A LaMoure
6A Logan
7 McHenry
6A McIntosh
6A McKenzie
7 McLean
6A Mercer
6A Morton
7 Mountrail
7 Nelson
6A Oliver
7 Pembina
7 Pierce
7 Ramsey

(continued)

TABLE 301.1—continued
CLIMATE ZONES, MOISTURE REGIMES, AND WARM-HUMID DESIGNATIONS BY STATE, COUNTY AND TERRITORY

6A Ransom
7 Renville
6A Richland
7 Rolette
6A Sargent
7 Sheridan
6A Sioux
6A Slope
6A Stark
7 Steele
7 Stutsman
7 Towner
7 Traill
7 Walsh
7 Ward
7 Wells
7 Williams

OHIO

4A Adams
5A Allen
5A Ashland
5A Ashtabula
5A Athens
5A Auglaize
5A Belmont
4A Brown
5A Butler
5A Carroll
5A Champaign
5A Clark
4A Clermont
5A Clinton
5A Columbiana
5A Coshocton
5A Crawford
5A Cuyahoga
5A Darke
5A Defiance
5A Delaware
5A Erie
5A Fairfield
5A Fayette
5A Franklin
5A Fulton
4A Gallia
5A Geauga
5A Greene
5A Guernsey
4A Hamilton
5A Hancock
5A Hardin
5A Harrison
5A Henry
5A Highland
5A Hocking
5A Holmes
5A Huron
5A Jackson
5A Jefferson
5A Knox
5A Lake
4A Lawrence
5A Licking
5A Logan
5A Lorain
5A Lucas
5A Madison
5A Mahoning
5A Marion
5A Medina
5A Meigs
5A Mercer
5A Miami
5A Monroe
5A Montgomery
5A Morgan
5A Morrow
5A Muskingum
5A Noble
5A Ottawa
5A Paulding
5A Perry
5A Pickaway
4A Pike
5A Portage
5A Preble
5A Putnam
5A Richland
5A Ross
5A Sandusky
4A Scioto
5A Seneca
5A Shelby
5A Stark
5A Summit
5A Trumbull
5A Tuscarawas
5A Union
5A Van Wert
5A Vinton
5A Warren
4A Washington
5A Wayne
5A Williams
5A Wood
5A Wyandot

OKLAHOMA

3A Adair
3A Alfalfa
3A Atoka
4B Beaver
3A Beckham
3A Blaine
3A Bryan
3A Caddo
3A Canadian
3A Carter
3A Cherokee
3A Choctaw
4B Cimarron
3A Cleveland
3A Coal
3A Comanche
3A Cotton
3A Craig
3A Creek
3A Custer
3A Delaware
3A Dewey
3A Ellis
3A Garfield
3A Garvin
3A Grady
3A Grant
3A Greer
3A Harmon
3A Harper
3A Haskell
3A Hughes
3A Jackson
3A Jefferson
3A Johnston
3A Kay
3A Kingfisher
3A Kiowa
3A Latimer
3A Le Flore
3A Lincoln
3A Logan
3A Love
3A Major
3A Marshall
3A Mayes
3A McClain
3A McCurtain
3A McIntosh
3A Murray
3A Muskogee
3A Noble
3A Nowata
3A Okfuskee
3A Oklahoma
3A Okmulgee
3A Osage
3A Ottawa
3A Pawnee
3A Payne
3A Pittsburg
3A Pontotoc
3A Pottawatomie
3A Pushmataha
3A Roger Mills
3A Rogers
3A Seminole
3A Sequoyah
3A Stephens
4B Texas
3A Tillman
3A Tulsa
3A Wagoner
3A Washington
3A Washita
3A Woods
3A Woodward

OREGON

5B Baker
4C Benton
4C Clackamas
4C Clatsop
4C Columbia
4C Coos
5B Crook
4C Curry
5B Deschutes
4C Douglas
5B Gilliam
5B Grant
5B Harney
5B Hood River
4C Jackson
5B Jefferson
4C Josephine

(continued)

**TABLE 301.1—continued
CLIMATE ZONES, MOISTURE REGIMES, AND WARM-HUMID DESIGNATIONS
BY STATE, COUNTY AND TERRITORY**

5B Klamath
5B Lake
4C Lane
4C Lincoln
4C Linn
5B Malheur
4C Marion
5B Morrow
4C Multnomah
4C Polk
5B Sherman
4C Tillamook
5B Umatilla
5B Union
5B Wallowa
5B Wasco
4C Washington
5B Wheeler
4C Yamhill

PENNSYLVANIA

5A Adams
5A Allegheny
5A Armstrong
5A Beaver
5A Bedford
5A Berks
5A Blair
5A Bradford
4A Bucks
5A Butler
5A Cambria
6A Cameron
5A Carbon
5A Centre
4A Chester
5A Clarion
6A Clearfield
5A Clinton
5A Columbia
5A Crawford
5A Cumberland
5A Dauphin
4A Delaware
6A Elk
5A Erie
5A Fayette
5A Forest
5A Franklin
5A Fulton
5A Greene
5A Huntingdon
5A Indiana
5A Jefferson
5A Juniata
5A Lackawanna
5A Lancaster
5A Lawrence
5A Lebanon
5A Lehigh
5A Luzerne
5A Lycoming
6A McKean
5A Mercer
5A Mifflin
5A Monroe
4A Montgomery
5A Montour
5A Northampton
5A Northumberland
5A Perry
4A Philadelphia
5A Pike
6A Potter
5A Schuylkill
5A Snyder
5A Somerset
5A Sullivan
6A Susquehanna
6A Tioga
5A Union
5A Venango
5A Warren
5A Washington
6A Wayne
5A Westmoreland
5A Wyoming
4A York

RHODE ISLAND

5A (all)

SOUTH CAROLINA

3A Abbeville
3A Aiken
3A Allendale*
3A Anderson
3A Bamberg*
3A Barnwell*
3A Beaufort*
3A Berkeley*
3A Calhoun
3A Charleston*
3A Cherokee
3A Chester
3A Chesterfield
3A Clarendon
3A Colleton*
3A Darlington
3A Dillon
3A Dorchester*
3A Edgefield
3A Fairfield
3A Florence
3A Georgetown*
3A Greenville
3A Greenwood
3A Hampton*
3A Horry*
3A Jasper*
3A Kershaw
3A Lancaster
3A Laurens
3A Lee
3A Lexington
3A Marion
3A Marlboro
3A McCormick
3A Newberry
3A Oconee
3A Orangeburg
3A Pickens
3A Richland
3A Saluda
3A Spartanburg
3A Sumter
3A Union
3A Williamsburg
3A York

SOUTH DAKOTA

6A Aurora
6A Beadle
5A Bennett
5A Bon Homme
6A Brookings
6A Brown
6A Brule
6A Buffalo
6A Butte
6A Campbell
5A Charles Mix
6A Clark
5A Clay
6A Codington
6A Corson
6A Custer
6A Davison
6A Day
6A Deuel
6A Dewey
5A Douglas
6A Edmunds
6A Fall River
6A Faulk
6A Grant
5A Gregory
6A Haakon
6A Hamlin
6A Hand
6A Hanson
6A Harding
6A Hughes
5A Hutchinson
6A Hyde
5A Jackson
6A Jerauld
6A Jones
6A Kingsbury
6A Lake
6A Lawrence
6A Lincoln
6A Lyman
6A Marshall
6A McCook
6A McPherson
6A Meade
5A Mellette
6A Miner
6A Minnehaha
6A Moody
6A Pennington
6A Perkins
6A Potter
6A Roberts
6A Sanborn
6A Shannon
6A Spink
6A Stanley
6A Sully
5A Todd
5A Tripp
6A Turner
5A Union
6A Walworth

(continued)

TABLE 301.1—continued
CLIMATE ZONES, MOISTURE REGIMES, AND WARM-HUMID DESIGNATIONS BY STATE, COUNTY AND TERRITORY

5A Yankton
6A Ziebach

TENNESSEE

4A Anderson
4A Bedford
4A Benton
4A Bledsoe
4A Blount
4A Bradley
4A Campbell
4A Cannon
4A Carroll
4A Carter
4A Cheatham
3A Chester
4A Claiborne
4A Clay
4A Cocke
4A Coffee
3A Crockett
4A Cumberland
4A Davidson
4A Decatur
4A DeKalb
4A Dickson
3A Dyer
3A Fayette
4A Fentress
4A Franklin
4A Gibson
4A Giles
4A Grainger
4A Greene
4A Grundy
4A Hamblen
4A Hamilton
4A Hancock
3A Hardeman
3A Hardin
4A Hawkins
3A Haywood
3A Henderson
4A Henry
4A Hickman
4A Houston
4A Humphreys
4A Jackson
4A Jefferson
4A Johnson
4A Knox
3A Lake
3A Lauderdale
4A Lawrence
4A Lewis
4A Lincoln
4A Loudon
4A Macon
3A Madison
4A Marion
4A Marshall
4A Maury
4A McMinn
3A McNairy
4A Meigs
4A Monroe
4A Montgomery
4A Moore
4A Morgan
4A Obion
4A Overton
4A Perry
4A Pickett
4A Polk
4A Putnam
4A Rhea
4A Roane
4A Robertson
4A Rutherford
4A Scott
4A Sequatchie
4A Sevier
3A Shelby
4A Smith
4A Stewart
4A Sullivan
4A Sumner
3A Tipton
4A Trousdale
4A Unicoi
4A Union
4A Van Buren
4A Warren
4A Washington
4A Wayne
4A Weakley
4A White
4A Williamson
4A Wilson

TEXAS

2A Anderson*
3B Andrews
2A Angelina*
2A Aransas*
3A Archer
4B Armstrong
2A Atascosa*
2A Austin*
4B Bailey
2B Bandera*
2A Bastrop*
3B Baylor
2A Bee*
2A Bell*
2A Bexar*
3A Blanco*
3B Borden
2A Bosque*
3A Bowie*
2A Brazoria*
2A Brazos*
3B Brewster
4B Briscoe
2A Brooks*
3A Brown*
2A Burleson*
3A Burnet*
2A Caldwell*
2A Calhoun*
3B Callahan
2A Cameron*
3A Camp*
4B Carson
3A Cass*
4B Castro
2A Chambers*
2A Cherokee*
3B Childress
3A Clay
4B Cochran
3B Coke
3B Coleman
3A Collin*
3B Collingsworth
2A Colorado*
2A Comal*
3A Comanche*
3B Concho
3A Cooke
2A Coryell*
3B Cottle
3B Crane
3B Crockett
3B Crosby
3B Culberson
4B Dallam
3A Dallas*
3B Dawson
4B Deaf Smith
3A Delta
3A Denton*
2A DeWitt*
3B Dickens
2B Dimmit*
4B Donley
2A Duval*
3A Eastland
3B Ector
2B Edwards*
3A Ellis*
3B El Paso
3A Erath*
2A Falls*
3A Fannin
2A Fayette*
3B Fisher
4B Floyd
3B Foard
2A Fort Bend*
3A Franklin*
2A Freestone*
2B Frio*
3B Gaines
2A Galveston*
3B Garza
3A Gillespie*
3B Glasscock
2A Goliad*
2A Gonzales*
4B Gray
3A Grayson
3A Gregg*
2A Grimes*
2A Guadalupe*
4B Hale
3B Hall
3A Hamilton*
4B Hansford
3B Hardeman
2A Hardin*
2A Harris*
3A Harrison*
4B Hartley
3B Haskell

(continued)

TABLE 301.1—continued
CLIMATE ZONES, MOISTURE REGIMES, AND WARM-HUMID DESIGNATIONS BY STATE, COUNTY AND TERRITORY

2A Hays*
3B Hemphill
3A Henderson*
2A Hidalgo*
2A Hill*
4B Hockley
3A Hood*
3A Hopkins*
2A Houston*
3B Howard
3B Hudspeth
3A Hunt*
4B Hutchinson
3B Irion
3A Jack
2A Jackson*
2A Jasper*
3B Jeff Davis
2A Jefferson*
2A Jim Hogg*
2A Jim Wells*
3A Johnson*
3B Jones
2A Karnes*
3A Kaufman*
3A Kendall*
2A Kenedy*
3B Kent
3B Kerr
3B Kimble
3B King
2B Kinney*
2A Kleberg*
3B Knox
3A Lamar*
4B Lamb
3A Lampasas*
2B La Salle*
2A Lavaca*
2A Lee*
2A Leon*
2A Liberty*
2A Limestone*
4B Lipscomb
2A Live Oak*
3A Llano*
3B Loving
3B Lubbock
3B Lynn
2A Madison*
3A Marion*
3B Martin
3B Mason
2A Matagorda*
2B Maverick*
3B McCulloch
2A McLennan*
2A McMullen*
2B Medina*
3B Menard
3B Midland
2A Milam*
3A Mills*
3B Mitchell
3A Montague
2A Montgomery*
4B Moore
3A Morris*
3B Motley
3A Nacogdoches*
3A Navarro*
2A Newton*
3B Nolan
2A Nueces*
4B Ochiltree
4B Oldham
2A Orange*
3A Palo Pinto*
3A Panola*
3A Parker*
4B Parmer
3B Pecos
2A Polk*
4B Potter
3B Presidio
3A Rains*
4B Randall
3B Reagan
2B Real*
3A Red River*
3B Reeves
2A Refugio*
4B Roberts
2A Robertson*
3A Rockwall*
3B Runnels
3A Rusk*
3A Sabine*
3A San Augustine*
2A San Jacinto*
2A San Patricio*
3A San Saba*
3B Schleicher
3B Scurry
3B Shackelford
3A Shelby*
4B Sherman
3A Smith*
3A Somervell*
2A Starr*
3A Stephens
3B Sterling
3B Stonewall
3B Sutton
4B Swisher
3A Tarrant*
3B Taylor
3B Terrell
3B Terry
3B Throckmorton
3A Titus*
3B Tom Green
2A Travis*
2A Trinity*
2A Tyler*
3A Upshur*
3B Upton
2B Uvalde*
2B Val Verde*
3A Van Zandt*
2A Victoria*
2A Walker*
2A Waller*
3B Ward
2A Washington*
2B Webb*
2A Wharton*
3B Wheeler
3A Wichita
3B Wilbarger
2A Willacy*
2A Williamson*
2A Wilson*
3B Winkler
3A Wise
3A Wood*
4B Yoakum
3A Young
2B Zapata*
2B Zavala*

UTAH

5B Beaver
6B Box Elder
6B Cache
6B Carbon
6B Daggett
5B Davis
6B Duchesne
5B Emery
5B Garfield
5B Grand
5B Iron
5B Juab
5B Kane
5B Millard
6B Morgan
5B Piute
6B Rich
5B Salt Lake
5B San Juan
5B Sanpete
5B Sevier
6B Summit
5B Tooele
6B Uintah
5B Utah
6B Wasatch
3B Washington
5B Wayne
5B Weber

VERMONT

6A (all)

VIRGINIA

4A (all)

WASHINGTON

5B Adams
5B Asotin
5B Benton
5B Chelan
4C Clallam
4C Clark
5B Columbia
4C Cowlitz
5B Douglas
6B Ferry
5B Franklin
5B Garfield
5B Grant
4C Grays Harbor
4C Island
4C Jefferson
4C King
4C Kitsap

(continued)

TABLE 301.1—continued
CLIMATE ZONES, MOISTURE REGIMES, AND WARM-HUMID DESIGNATIONS BY STATE, COUNTY AND TERRITORY

5B Kittitas
5B Klickitat
4C Lewis
5B Lincoln
4C Mason
6B Okanogan
4C Pacific
6B Pend Oreille
4C Pierce
4C San Juan
4C Skagit
5B Skamania
4C Snohomish
5B Spokane
6B Stevens
4C Thurston
4C Wahkiakum
5B Walla Walla
4C Whatcom
5B Whitman
5B Yakima

WEST VIRGINIA

5A Barbour
4A Berkeley
4A Boone
4A Braxton
5A Brooke
4A Cabell
4A Calhoun
4A Clay
5A Doddridge
5A Fayette
4A Gilmer
5A Grant
5A Greenbrier
5A Hampshire
5A Hancock
5A Hardy
5A Harrison
4A Jackson
4A Jefferson
4A Kanawha
5A Lewis
4A Lincoln
4A Logan
5A Marion
5A Marshall
4A Mason
4A McDowell
4A Mercer
5A Mineral
4A Mingo
5A Monongalia
4A Monroe
4A Morgan
5A Nicholas
5A Ohio
5A Pendleton
4A Pleasants
5A Pocahontas
5A Preston
4A Putnam
5A Raleigh
5A Randolph
4A Ritchie
4A Roane
5A Summers
5A Taylor
5A Tucker
4A Tyler
5A Upshur
4A Wayne
5A Webster
5A Wetzel
4A Wirt
4A Wood
4A Wyoming

WISCONSIN

6A Adams
7 Ashland
6A Barron
7 Bayfield
6A Brown
6A Buffalo
7 Burnett
6A Calumet
6A Chippewa
6A Clark
6A Columbia
6A Crawford
6A Dane
6A Dodge
6A Door
7 Douglas
6A Dunn
6A Eau Claire
7 Florence
6A Fond du Lac
7 Forest
6A Grant
6A Green
6A Green Lake
6A Iowa
7 Iron
6A Jackson
6A Jefferson
6A Juneau
6A Kenosha
6A Kewaunee
6A La Crosse
6A Lafayette
7 Langlade
7 Lincoln
6A Manitowoc
6A Marathon
6A Marinette
6A Marquette
6A Menominee
6A Milwaukee
6A Monroe
6A Oconto
7 Oneida
6A Outagamie
6A Ozaukee
6A Pepin
6A Pierce
6A Polk
6A Portage
7 Price
6A Racine
6A Richland
6A Rock
6A Rusk
6A Sauk
7 Sawyer
6A Shawano
6A Sheboygan
6A St. Croix
7 Taylor
6A Trempealeau
6A Vernon
7 Vilas
6A Walworth
7 Washburn
6A Washington
6A Waukesha
6A Waupaca
6A Waushara
6A Winnebago
6A Wood

WYOMING

6B Albany
6B Big Horn
6B Campbell
6B Carbon
6B Converse
6B Crook
6B Fremont
5B Goshen
6B Hot Springs
6B Johnson
6B Laramie
7 Lincoln
6B Natrona
6B Niobrara
6B Park
5B Platte
6B Sheridan
7 Sublette
6B Sweetwater
7 Teton
6B Uinta
6B Washakie
6B Weston

US TERRITORIES

AMERICAN SAMOA

1A (all)*

GUAM

1A (all)*

NORTHERN MARIANA ISLANDS

1A (all)*

PUERTO RICO

1A (all)*

VIRGIN ISLANDS

1A (all)*

TABLE 301.3(1)
INTERNATIONAL CLIMATE ZONE DEFINITIONS

MAJOR CLIMATE TYPE DEFINITIONS
Warm-humid Definition—Moist (A) locations where either of the following wet-bulb temperature conditions shall occur during the warmest six consecutive months of the year: 1. 67°F (19.4°C) or higher for 3,000 or more hours; or 2. 73°F (22.8°C) or higher for 1,500 or more hours
Dry (B) Definition—Locations meeting the following criteria: Not marine and $P_{in} < 0.44 \times (TF - 19.5)$ [$P_{cm} < 2.0 \times (TC + 7)$ in SI units] where: P_{in} = Annual precipitation in inches (cm) T = Annual mean temperature in °F (°C)
Moist (A) Definition—Locations that are not marine and not dry.

For SI: °C = [(°F)-32]/1.8; 1 inch = 2.54 cm.

TABLE 301.3(2)
INTERNATIONAL CLIMATE ZONE DEFINITIONS

ZONE NUMBER	THERMAL CRITERIA	
	IP Units	SI Units
1	9000 < CDD50°F	5000 < CDD10°C
2	6300 < CDD50°F ≤ 9000	3500 < CDD10°C ≤ 5000
3A and 3B	4500 < CDD50°F ≤ 6300 AND HDD65°F ≤ 5400	2500 < CDD10°C ≤ 3500 AND HDD18°C ≤ 3000
4A and 4B	CDD50°F ≤ 4500 AND HDD65°F ≤ 5400	CDD10°C ≤ 2500 AND HDD18°C ≤ 3000
3C	HDD65°F ≤ 3600	HDD18°C ≤ 2000
4C	3600 < HDD65°F ≤ 5400	2000 < HDD18°C ≤ 3000
5	5400 < HDD65°F ≤ 7200	3000 < HDD18°C ≤ 4000
6	7200 < HDD65°F ≤ 9000	4000 < HDD18°C ≤ 5000
7	9000 < HDD65°F ≤ 12600	5000 < HDD18°C ≤ 7000
8	12600 < HDD65°F	7000 < HDD18°C

For SI: °C = [(°F)-32]/1.8

SECTION 302
DESIGN CONDITIONS

302.1 Interior design conditions. The interior design temperatures used for heating and cooling load calculations shall be a maximum of 72°F (22°C) for heating and minimum of 75°F (24°C) for cooling.

SECTION 303
MATERIALS, SYSTEMS AND EQUIPMENT

303.1 Identification. Materials, systems and equipment shall be identified in a manner that will allow a determination of compliance with the applicable provisions of this code.

303.1.1 Building thermal envelope insulation. An *R*-value identification mark shall be applied by the manufacturer to each piece of *building thermal envelope* insulation 12 inches (305 mm) or greater in width. Alternately, the insulation installers shall provide a certification listing the type, manufacturer and *R*-value of insulation installed in each element of the *building thermal envelope*. For blown or sprayed insulation (fiberglass and cellulose), the initial installed thickness, settled thickness, settled *R*-value, installed density, coverage area and number of bags installed shall be *listed* on the certification. For sprayed polyurethane foam (SPF) insulation, the installed thickness of the areas covered and *R*-value of installed thickness shall be *listed* on the certification. The insulation installer shall sign, date and post the certification in a conspicuous location on the job site.

303.1.1.1 Blown or sprayed roof/ceiling insulation. The thickness of blown-in or sprayed roof/ceiling insulation (fiberglass or cellulose) shall be written in inches (mm) on markers that are installed at least one for every 300 square feet (28 m^2) throughout the attic space. The markers shall be affixed to the trusses or joists and marked with the minimum initial installed thickness with numbers a minimum of 1 inch (25 mm) in height. Each marker shall face the attic access opening. Spray polyurethane foam thickness and installed *R*-value shall be *listed* on certification provided by the insulation installer.

303.1.2 Insulation mark installation. Insulating materials shall be installed such that the manufacturer's *R*-value mark is readily observable upon inspection.

303.1.3 Fenestration product rating. *U*-factors of fenestration products (windows, doors and skylights) shall be determined in accordance with NFRC 100 by an accredited, independent laboratory, and labeled and certified by the manufacturer. Products lacking such a labeled *U*-factor shall be assigned a default *U*-factor from Table 303.1.3(1) or 303.1.3(2). The solar heat gain coefficient (SHGC) of glazed fenestration products (windows, glazed doors and skylights) shall be determined in accordance with NFRC 200 by an accredited, independent laboratory, and labeled and certified by the manufacturer. Products lacking such a labeled SHGC shall be assigned a default SHGC from Table 303.1.3(3).

TABLE 303.1.3(1)
DEFAULT GLAZED FENESTRATION *U*-FACTOR

FRAME TYPE	SINGLE PANE	DOUBLE PANE	SKYLIGHT	
			Single	Double
Metal	1.20	0.80	2.00	1.30
Metal with Thermal Break	1.10	0.65	1.90	1.10
Nonmetal or Metal Clad	0.95	0.55	1.75	1.05
Glazed Block	0.60			

Insulated means double glazed.

TABLE 303.1.3(2)
DEFAULT DOOR *U*-FACTORS

DOOR TYPE	*U*-FACTOR
Uninsulated Metal	1.20
Insulated Metal	0.60
Wood	0.50
Insulated, nonmetal edge, max 45% glazing, any glazing double pane	0.35

TABLE 303.1.3(3)
DEFAULT GLAZED FENESTRATION SHGC

SINGLE GLAZED		DOUBLE GLAZED		GLAZED BLOCK
Clear	Tinted	Clear	Tinted	
0.8	0.7	0.7	0.6	0.6

303.1.4 Insulation product rating. The thermal resistance (*R*-value) of insulation shall be determined in accordance with the U.S. Federal Trade Commission *R*-value rule (CFR Title 16, Part 460, May 31, 2005) in units of h × ft^2 × °F/Btu at a mean temperature of 75°F (24°C).

303.2 Installation. All materials, systems and equipment shall be installed in accordance with the manufacturer's installation instructions and the *International Building Code*.

303.2.1 Protection of exposed foundation insulation. Insulation applied to the exterior of basement walls, crawlspace walls and the perimeter of slab-on-grade floors shall have a rigid, opaque and weather-resistant protective covering to prevent the degradation of the insulation's thermal performance. The protective covering shall cover the exposed exterior insulation and extend a minimum of 6 inches (153 mm) below grade.

303.3 Maintenance information. Maintenance instructions shall be furnished for equipment and systems that require preventive maintenance. Required regular maintenance actions shall be clearly stated and incorporated on a readily accessible label. The label shall include the title or publication number for the operation and maintenance manual for that particular model and type of product.

CHAPTER 4
RESIDENTIAL ENERGY EFFICIENCY

SECTION 401
GENERAL

401.1 Scope. This chapter applies to residential buildings.

401.2 Compliance. Projects shall comply with Sections 401, 402.4, 402.5, and 403.1, 403.2.2, 403.2.3, and 403.3 through 403.9 (referred to as the mandatory provisions) and either:

1. Sections 402.1 through 402.3, 403.2.1 and 404.1 (prescriptive); or
2. Section 405 (performance).

401.3 Certificate. A permanent certificate shall be posted on or in the electrical distribution panel. The certificate shall not cover or obstruct the visibility of the circuit directory label, service disconnect label or other required labels. The certificate shall be completed by the builder or registered design professional. The certificate shall list the predominant *R*-values of insulation installed in or on ceiling/roof, walls, foundation (slab, *basement wall*, crawlspace wall and/or floor) and ducts outside conditioned spaces; *U*-factors for fenestration and the solar heat gain coefficient (SHGC) of fenestration. Where there is more than one value for each component, the certificate shall list the value covering the largest area. The certificate shall list the types and efficiencies of heating, cooling and service water heating equipment. Where a gas-fired unvented room heater, electric furnace, or baseboard electric heater is installed in the residence, the certificate shall list "gas-fired unvented room heater," "electric furnace" or "baseboard electric heater," as appropriate. An efficiency shall not be *listed* for gas-fired unvented room heaters, electric furnaces or electric baseboard heaters.

SECTION 402
BUILDING THERMAL ENVELOPE

402.1 General (Prescriptive).

402.1.1 Insulation and fenestration criteria. The *building thermal envelope* shall meet the requirements of Table 402.1.1 based on the climate *zone* specified in Chapter 3.

402.1.2 *R*-value computation. Insulation material used in layers, such as framing cavity insulation and insulating sheathing, shall be summed to compute the component *R*-value. The manufacturer's settled *R*-value shall be used for blown insulation. Computed *R*-values shall not include an *R*-value for other building materials or air films.

TABLE 402.1.1
INSULATION AND FENESTRATION REQUIREMENTS BY COMPONENT[a]

CLIMATE ZONE	FENESTRATION *U*-FACTOR[b]	SKYLIGHT[b] *U*-FACTOR	GLAZED FENESTRATION SHGC[b, e]	CEILING *R*-VALUE	WOOD FRAME WALL *R*-VALUE	MASS WALL *R*-VALUE[i]	FLOOR *R*-VALUE	BASEMENT[c] WALL *R*-VALUE	SLAB[d] *R*-VALUE & DEPTH	CRAWL SPACE[c] WALL *R*-VALUE
1	1.2	0.75	0.30	30	13	3/4	13	0	0	0
2	0.65[j]	0.75	0.30	30	13	4/6	13	0	0	0
3	0.50[j]	0.65	0.30	30	13	5/8	19	5/13[f]	0	5/13
4 except Marine	0.35	0.60	NR	38	13	5/10	19	10/13	10, 2 ft	10/13
5 and Marine 4	0.35	0.60	NR	38	20 or 13+5[h]	13/17	30[g]	10/13	10, 2 ft	10/13
6	0.35	0.60	NR	49	20 or 13+5[h]	15/19	30[g]	15/19	10, 4 ft	10/13
7 and 8	0.35	0.60	NR	49	21	19/21	38[g]	15/19	10, 4 ft	10/13

For SI: 1 foot = 304.8 mm.

a. *R*-values are minimums. *U*-factors and SHGC are maximums. R-19 batts compressed into a nominal 2 × 6 framing cavity such that the *R*-value is reduced by R-1 or more shall be marked with the compressed batt *R*-value in addition to the full thickness *R*-value.

b. The fenestration *U*-factor column excludes skylights. The SHGC column applies to all glazed fenestration.

c. "15/19" means R-15 continuous insulated sheathing on the interior or exterior of the home or R-19 cavity insulation at the interior of the basement wall. "15/19" shall be permitted to be met with R-13 cavity insulation on the interior of the basement wall plus R-5 continuous insulated sheathing on the interior or exterior of the home. "10/13" means R-10 continuous insulated sheathing on the interior or exterior of the home or R-13 cavity insulation at the interior of the basement wall.

d. R-5 shall be added to the required slab edge *R*-values for heated slabs. Insulation depth shall be the depth of the footing or 2 feet, whichever is less in Zones 1 through 3 for heated slabs.

e. There are no SHGC requirements in the Marine Zone.

f. Basement wall insulation is not required in warm-humid locations as defined by Figure 301.1 and Table 301.1.

g. Or insulation sufficient to fill the framing cavity, R-19 minimum.

h. "13+5" means R-13 cavity insulation plus R-5 insulated sheathing. If structural sheathing covers 25 percent or less of the exterior, insulating sheathing is not required where structural sheathing is used. If structural sheathing covers more than 25 percent of exterior, structural sheathing shall be supplemented with insulated sheathing of at least R-2.

i. The second *R*-value applies when more than half the insulation is on the interior of the mass wall.

j. For impact rated fenestration complying with Section R301.2.1.2 of the *International Residential Code* or Section 1608.1.2 of the *International Building Code*, the maximum *U*-factor shall be 0.75 in Zone 2 and 0.65 in Zone 3.

TABLE 402.1.3
EQUIVALENT *U*-FACTORS[a]

CLIMATE ZONE	FENESTRATION *U*-FACTOR	SKYLIGHT *U*-FACTOR	CEILING *U*-FACTOR	FRAME WALL *U*-FACTOR	MASS WALL *U*-FACTOR[b]	FLOOR *U*-FACTOR	BASEMENT WALL *U*-FACTOR	CRAWL SPACE WALL *U*-FACTOR[c]
1	1.20	0.75	0.035	0.082	0.197	0.064	0.360	0.477
2	0.65	0.75	0.035	0.082	0.165	0.064	0.360	0.477
3	0.50	0.65	0.035	0.082	0.141	0.047	0.091[c]	0.136
4 except Marine	0.35	0.60	0.030	0.082	0.141	0.047	0.059	0.065
5 and Marine 4	0.35	0.60	0.030	0.057	0.082	0.033	0.059	0.065
6	0.35	0.60	0.026	0.057	0.060	0.033	0.050	0.065
7 and 8	0.35	0.60	0.026	0.057	0.057	0.028	0.050	0.065

a. Nonfenestration *U*-factors shall be obtained from measurement, calculation or an approved source.

b. When more than half the insulation is on the interior, the mass wall *U*-factors shall be a maximum of 0.17 in Zone 1, 0.14 in Zone 2, 0.12 in Zone 3, 0.10 in Zone 4 except Marine, and the same as the frame wall *U*-factor in Marine Zone 4 and Zones 5 through 8.

c. Basement wall *U*-factor of 0.360 in warm-humid locations as defined by Figure 301.1 and Table 301.2.

402.1.3 *U*-factor alternative. An assembly with a *U*-factor equal to or less than that specified in Table 402.1.3 shall be permitted as an alternative to the *R*-value in Table 402.1.1.

402.1.4 Total UA alternative. If the total *building thermal envelope* UA (sum of *U*-factor times assembly area) is less than or equal to the total UA resulting from using the *U*-factors in Table 402.1.3 (multiplied by the same assembly area as in the proposed building), the building shall be considered in compliance with Table 402.1.1. The UA calculation shall be done using a method consistent with the ASHRAE *Handbook of Fundamentals* and shall include the thermal bridging effects of framing materials. The SHGC requirements shall be met in addition to UA compliance.

402.2 Specific insulation requirements (Prescriptive).

402.2.1 Ceilings with attic spaces. When Section 402.1.1 would require R-38 in the ceiling, R-30 shall be deemed to satisfy the requirement for R-38 wherever the full height of uncompressed R-30 insulation extends over the wall top plate at the eaves. Similarly, R-38 shall be deemed to satisfy the requirement for R-49 wherever the full height of uncompressed R-38 insulation extends over the wall top plate at the eaves. This reduction shall not apply to the *U*-factor alternative approach in Section 402.1.3 and the total UA alternative in Section 402.1.4.

402.2.2 Ceilings without attic spaces. Where Section 402.1.1 would require insulation levels above R-30 and the design of the roof/ceiling assembly does not allow sufficient space for the required insulation, the minimum required insulation for such roof/ceiling assemblies shall be R-30. This reduction of insulation from the requirements of Section 402.1.1 shall be limited to 500 square feet (46 m^2) or 20 percent of the total insulated ceiling area, whichever is less. This reduction shall not apply to the *U*-factor alternative approach in Section 402.1.3 and the total UA alternative in Section 402.1.4.

402.2.3 Access hatches and doors. Access doors from conditioned spaces to unconditioned spaces (e.g., attics and crawl spaces) shall be weatherstripped and insulated to a level equivalent to the insulation on the surrounding surfaces. Access shall be provided to all equipment that prevents damaging or compressing the insulation. A wood framed or equivalent baffle or retainer is required to be provided when loose fill insulation is installed, the purpose of which is to prevent the loose fill insulation from spilling into the living space when the attic access is opened, and to provide a permanent means of maintaining the installed *R*-value of the loose fill insulation.

402.2.4 Mass walls. Mass walls for the purposes of this chapter shall be considered above-grade walls of concrete block, concrete, insulated concrete form (ICF), masonry cavity, brick (other than brick veneer), earth (adobe, compressed earth block, rammed earth) and solid timber/logs.

402.2.5 Steel-frame ceilings, walls, and floors. Steel-frame ceilings, walls and floors shall meet the insulation requirements of Table 402.2.5 or shall meet the *U*-factor requirements in Table 402.1.3. The calculation of the *U*-factor for a steel-frame envelope assembly shall use a series-parallel path calculation method.

Exception: In Climate Zones 1 and 2, the continuous insulation requirements in Table 402.2.4 shall be permitted to be reduced to R-3 for steel frame wall assemblies with studs spaced at 24 inches (610 mm) on center.

TABLE 402.2.5
STEEL-FRAME CEILING, WALL AND FLOOR INSULATION (*R*-VALUE)

WOOD FRAME *R*-VALUE REQUIREMENT	COLD-FORMED STEEL EQUIVALENT *R*-VALUE[a]
Steel Truss Ceilings[b]	
R-30	R-38 or R-30 + 3 or R-26 + 5
R-38	R-49 or R-38 + 3
R-49	R-38 + 5
Steel Joist Ceilings[b]	
R-30	R-38 in 2 × 4 or 2 × 6 or 2 × 8 R-49 in any framing
R-38	R-49 in 2 × 4 or 2 × 6 or 2 × 8 or 2 × 10
Steel-Framed Wall	
R-13	R-13 + 5 or R-15 + 4 or R-21 + 3 or R-0 + 10
R-19	R-13 + 9 or R-19 + 8 or R-25 + 7
R-21	R-13 + 10 or R-19 + 9 or R-25 + 8
Steel Joist Floor	
R-13	R-19 in 2 × 6 R-19 + 6 in 2 × 8 or 2 × 10
R-19	R-19 + 6 in 2 × 6 R-19 + 12 in 2 × 8 or 2 × 10

a. Cavity insulation *R*-value is listed first, followed by continuous insulation *R*-value.

b. Insulation exceeding the height of the framing shall cover the framing.

402.2.6 Floors. Floor insulation shall be installed to maintain permanent contact with the underside of the subfloor decking.

402.2.7 Basement walls. Walls associated with conditioned basements shall be insulated from the top of the *basement wall* down to 10 feet (3048 mm) below grade or to the basement floor, whichever is less. Walls associated with unconditioned basements shall meet this requirement unless the floor overhead is insulated in accordance with Sections 402.1.1 and 402.2.6.

402.2.8 Slab-on-grade floors. Slab-on-grade floors with a floor surface less than 12 inches (305 mm) below grade shall be insulated in accordance with Table 402.1.1. The insulation shall extend downward from the top of the slab on the outside or inside of the foundation wall. Insulation located below grade shall be extended the distance provided in Table 402.1.1 by any combination of vertical insulation, insulation extending under the slab or insulation extending out from the building. Insulation extending away from the building shall be protected by pavement or by a minimum of 10 inches (254 mm) of soil. The top edge of the insulation installed between the *exterior wall* and the edge of the interior slab shall be permitted to be cut at a 45-degree (0.79 rad) angle away from the *exterior wall*. Slab-edge insulation is not required in jurisdictions designated by the *code official* as having a very heavy termite infestation.

402.2.9 Crawl space walls. As an alternative to insulating floors over crawl spaces, crawl space walls shall be permitted to be insulated when the crawl space is not vented to the outside. Crawl space wall insulation shall be permanently fastened to the wall and extend downward from the floor to the finished grade level and then vertically and/or horizontally for at least an additional 24 inches (610 mm). Exposed earth in unvented crawl space foundations shall be covered with a continuous Class I vapor retarder in accordance with the *International Building Code*. All joints of the vapor retarder shall overlap by 6 inches (153 mm) and be sealed or taped. The edges of the vapor retarder shall extend at least 6 inches (153 mm) up the stem wall and shall be attached to the stem wall.

402.2.10 Masonry veneer. Insulation shall not be required on the horizontal portion of the foundation that supports a masonry veneer.

402.2.11 Thermally isolated sunroom insulation. The minimum ceiling insulation *R*-values shall be R-19 in Zones 1 through 4 and R-24 in Zones 5 through 8. The minimum wall *R*-value shall be R-13 in all zones. New wall(s) separating a sunroom from *conditioned space* shall meet the *building thermal envelope* requirements.

402.3 Fenestration. (Prescriptive).

402.3.1 *U*-factor. An area-weighted average of fenestration products shall be permitted to satisfy the *U*-factor requirements.

402.3.2 Glazed fenestration SHGC. An area-weighted average of fenestration products more than 50 percent glazed shall be permitted to satisfy the SHGC requirements.

402.3.3 Glazed fenestration exemption. Up to 15 square feet (1.4 m^2) of glazed fenestration per dwelling unit shall be permitted to be exempt from *U*-factor and SHGC requirements in Section 402.1.1. This exemption shall not apply to the *U*-factor alternative approach in Section 402.1.3 and the Total UA alternative in Section 402.1.4.

402.3.4 Opaque door exemption. One side-hinged opaque door assembly up to 24 square feet (2.22 m^2) in area is exempted from the *U*-factor requirement in Section 402.1.1. This exemption shall not apply to the *U*-factor alternative approach in Section 402.1.3 and the total UA alternative in Section 402.1.4.

402.3.5 Thermally isolated sunroom *U*-factor. For Zones 4 through 8, the maximum fenestration *U*-factor shall be 0.50 and the maximum skylight *U*-factor shall be 0.75. New windows and doors separating the sunroom from *conditioned space* shall meet the *building thermal envelope* requirements.

402.3.6 Replacement fenestration. Where some or all of an existing fenestration unit is replaced with a new fenestration product, including sash and glazing, the replacement fenestration unit shall meet the applicable requirements for *U*-factor and SHGC in Table 402.1.1.

402.4 Air leakage (Mandatory).

402.4.1 Building thermal envelope. The *building thermal envelope* shall be durably sealed to limit infiltration. The sealing methods between dissimilar materials shall allow for differential expansion and contraction. The following shall be caulked, gasketed, weatherstripped or otherwise sealed with an air barrier material, suitable film or solid material:

1. All joints, seams and penetrations.

2. Site-built windows, doors and skylights.
3. Openings between window and door assemblies and their respective jambs and framing.
4. Utility penetrations.
5. Dropped ceilings or chases adjacent to the thermal envelope.
6. Knee walls.
7. Walls and ceilings separating a garage from conditioned spaces.
8. Behind tubs and showers on exterior walls.
9. Common walls between dwelling units.
10. Attic access openings.
11. Rim joist junction.
12. Other sources of infiltration.

402.4.2 Air sealing and insulation. Building envelope air tightness and insulation installation shall be demonstrated to comply with one of the following options given by Section 402.4.2.1 or 402.4.2.2:

402.4.2.1 Testing option. Building envelope tightness and insulation installation shall be considered acceptable when tested air leakage is less than seven air changes per hour (ACH) when tested with a blower door at a pressure of 33.5 psf (50 Pa). Testing shall occur after rough in and after installation of penetrations of the building envelope, including penetrations for utilities, plumbing, electrical, ventilation and combustion appliances.

During testing:

1. Exterior windows and doors, fireplace and stove doors shall be closed, but not sealed;
2. Dampers shall be closed, but not sealed, including exhaust, intake, makeup air, backdraft and flue dampers;
3. Interior doors shall be open;
4. Exterior openings for continuous ventilation systems and heat recovery ventilators shall be closed and sealed;
5. Heating and cooling system(s) shall be turned off;
6. HVAC ducts shall not be sealed; and
7. Supply and return registers shall not be sealed.

402.4.2.2 Visual inspection option. Building envelope tightness and insulation installation shall be considered acceptable when the items listed in Table 402.4.2, applicable to the method of construction, are field verified. Where required by the *code official*, an *approved* party independent from the installer of the insulation shall inspect the air barrier and insulation.

402.4.3 Fireplaces. New wood-burning fireplaces shall have gasketed doors and outdoor combustion air.

402.4.4 Fenestration air leakage. Windows, skylights and sliding glass doors shall have an air infiltration rate of no more than 0.3 cfm per square foot (1.5 L/s/m^2), and swinging doors no more than 0.5 cfm per square foot (2.6 L/s/m^2), when tested according to NFRC 400 or AAMA/WDMA/CSA 101/I.S.2/A440 by an accredited, independent laboratory and *listed* and *labeled* by the manufacturer.

Exceptions: Site-built windows, skylights and doors.

402.4.5 Recessed lighting. Recessed luminaires installed in the *building thermal envelope* shall be sealed to limit air leakage between conditioned and unconditioned spaces. All recessed luminaires shall be IC-rated and *labeled* as meeting ASTM E 283 when tested at 1.57 psf (75 Pa) pressure differential with no more than 2.0 cfm (0.944 L/s) of air movement from the *conditioned space* to the ceiling cavity. All recessed luminaires shall be sealed with a gasket or caulk between the housing and the interior wall or ceiling covering.

402.5 Maximum fenestration *U*-factor and SHGC (Mandatory). The area-weighted average maximum fenestration *U*-factor permitted using trade-offs from Section 402.1.4 or 404 shall be 0.48 in Zones 4 and 5 and 0.40 in Zones 6 through 8 for vertical fenestration, and 0.75 in Zones 4 through 8 for skylights. The area-weighted average maximum fenestration SHGC permitted using trade-offs from Section 405 in Zones 1 through 3 shall be 0.50.

SECTION 403
SYSTEMS

403.1 Controls (Mandatory). At least one thermostat shall be provided for each separate heating and cooling system.

403.1.1 Programmable thermostat. Where the primary heating system is a forced-air furnace, at least one thermostat per dwelling unit shall be capable of controlling the heating and cooling system on a daily schedule to maintain different temperature set points at different times of the day. This thermostat shall include the capability to set back or temporarily operate the system to maintain zone temperatures down to 55°F (13°C) or up to 85°F (29°C). The thermostat shall initially be programmed with a heating temperature set point no higher than 70°F (21°C) and a cooling temperature set point no lower than 78°F (26°C).

403.1.2 Heat pump supplementary heat (Mandatory). Heat pumps having supplementary electric-resistance heat shall have controls that, except during defrost, prevent supplemental heat operation when the heat pump compressor can meet the heating load.

403.2 Ducts.

403.2.1 Insulation (Prescriptive). Supply ducts in attics shall be insulated to a minimum of R-8. All other ducts shall be insulated to a minimum of R-6.

Exception: Ducts or portions thereof located completely inside the *building thermal envelope*.

403.2.2 Sealing (Mandatory). All ducts, air handlers, filter boxes and building cavities used as ducts shall be sealed.

Joints and seams shall comply with Section M1601.4.1 of the *International Residential Code*.

Duct tightness shall be verified by either of the following:

1. Postconstruction test: Leakage to outdoors shall be less than or equal to 8 cfm (226.5 L/min) per 100 ft^2 (9.29 m^2) of *conditioned floor area* or a total leakage less than or equal to 12 cfm (12 L/min) per 100 ft^2 (9.29 m^2) of *conditioned floor area* when tested at a pressure differential of 0.1 inches w.g. (25 Pa) across the entire system, including the manufacturer's air handler enclosure. All register boots shall be taped or otherwise sealed during the test.
2. Rough-in test: Total leakage shall be less than or equal to 6 cfm (169.9 L/min) per 100 ft^2 (9.29 m^2) of *conditioned floor area* when tested at a pressure differential of 0.1 inches w.g. (25 Pa) across the roughed in system, including the manufacturer's air handler enclosure. All register boots shall be taped or otherwise sealed during the test. If the air handler is not installed at the time of the test, total leakage shall be less than or equal to 4 cfm (113.3 L/min) per 100 ft^2 (9.29 m^2) of *conditioned floor area*.

Exceptions: Duct tightness test is not required if the air handler and all ducts are located within *conditioned space*.

TABLE 402.4.2
AIR BARRIER AND INSULATION INSPECTION COMPONENT CRITERIA

COMPONENT	CRITERIA
Air barrier and thermal barrier	Exterior thermal envelope insulation for framed walls is installed in substantial contact and continuous alignment with building envelope air barrier. Breaks or joints in the air barrier are filled or repaired. Air-permeable insulation is not used as a sealing material. Air-permeable insulation is inside of an air barrier.
Ceiling/attic	Air barrier in any dropped ceiling/soffit is substantially aligned with insulation and any gaps are sealed. Attic access (except unvented attic), knee wall door, or drop down stair is sealed.
Walls	Corners and headers are insulated. Junction of foundation and sill plate is sealed.
Windows and doors	Space between window/door jambs and framing is sealed.
Rim joists	Rim joists are insulated and include an air barrier.
Floors (including above-garage and cantilevered floors)	Insulation is installed to maintain permanent contact with underside of subfloor decking. Air barrier is installed at any exposed edge of insulation.
Crawl space walls	Insulation is permanently attached to walls. Exposed earth in unvented crawl spaces is covered with Class I vapor retarder with overlapping joints taped.
Shafts, penetrations	Duct shafts, utility penetrations, knee walls and flue shafts opening to exterior or unconditioned space are sealed.
Narrow cavities	Batts in narrow cavities are cut to fit, or narrow cavities are filled by sprayed/blown insulation.
Garage separation	Air sealing is provided between the garage and conditioned spaces.
Recessed lighting	Recessed light fixtures are air tight, IC rated, and sealed to drywall. Exception—fixtures in conditioned space.
Plumbing and wiring	Insulation is placed between outside and pipes. Batt insulation is cut to fit around wiring and plumbing, or sprayed/blown insulation extends behind piping and wiring.
Shower/tub on exterior wall	Showers and tubs on exterior walls have insulation and an air barrier separating them from the exterior wall.
Electrical/phone box on exterior walls	Air barrier extends behind boxes or air sealed-type boxes are installed.
Common wall	Air barrier is installed in common wall between dwelling units.
HVAC register boots	HVAC register boots that penetrate building envelope are sealed to subfloor or drywall.
Fireplace	Fireplace walls include an air barrier.

403.2.3 Building cavities (Mandatory). Building framing cavities shall not be used as supply ducts.

403.3 Mechanical system piping insulation (Mandatory). Mechanical system piping capable of carrying fluids above 105°F (41°C) or below 55°F (13°C) shall be insulated to a minimum of R-3.

403.4 Circulating hot water systems (Mandatory). All circulating service hot water piping shall be insulated to at least R-2. Circulating hot water systems shall include an automatic or readily *accessible* manual switch that can turn off the hot-water circulating pump when the system is not in use.

403.5 Mechanical ventilation (Mandatory). Outdoor air intakes and exhausts shall have automatic or gravity dampers that close when the ventilation system is not operating.

403.6 Equipment sizing (Mandatory). Heating and cooling equipment shall be sized in accordance with Section M1401.3 of the *International Residential Code*.

403.7 Systems serving multiple dwelling units (Mandatory). Systems serving multiple dwelling units shall comply with Sections 503 and 504 in lieu of Section 403.

403.8 Snow melt system controls (Mandatory). Snow- and ice-melting systems, supplied through energy service to the building, shall include automatic controls capable of shutting off the system when the pavement temperature is above 50°F, and no precipitation is falling and an automatic or manual control that will allow shutoff when the outdoor temperature is above 40°F.

403.9 Pools (Mandatory). Pools shall be provided with energy-conserving measures in accordance with Sections 403.9.1 through 403.9.3.

403.9.1 Pool heaters. All pool heaters shall be equipped with a readily *accessible* on-off switch to allow shutting off the heater without adjusting the thermostat setting. Pool heaters fired by natural gas shall not have continuously burning pilot lights.

403.9.2 Time switches. Time switches that can automatically turn off and on heaters and pumps according to a preset schedule shall be installed on swimming pool heaters and pumps.

Exceptions:

1. Where public health standards require 24-hour pump operation.
2. Where pumps are required to operate solar- and waste-heat-recovery pool heating systems.

403.9.3 Pool covers. Heated pools shall be equipped with a vapor-retardant pool cover on or at the water surface. Pools heated to more than 90°F (32°C) shall have a pool cover with a minimum insulation value of R-12.

Exception: Pools deriving over 60 percent of the energy for heating from site-recovered energy or solar energy source.

SECTION 404
ELECTRICAL POWER AND LIGHTING SYSTEMS

404.1 Lighting equipment (Prescriptive). A minimum of 50 percent of the lamps in permanently installed lighting fixtures shall be high-efficacy lamps.

SECTION 405
SIMULATED PERFORMANCE ALTERNATIVE
(Performance)

405.1 Scope. This section establishes criteria for compliance using simulated energy performance analysis. Such analysis shall include heating, cooling, and service water heating energy only.

405.2 Mandatory requirements. Compliance with this section requires that the mandatory provisions identified in Section 401.2 be met. All supply and return ducts not completely inside the *building thermal envelope* shall be insulated to a minimum of R-6.

405.3 Performance-based compliance. Compliance based on simulated energy performance requires that a proposed residence (*proposed design*) be shown to have an annual energy cost that is less than or equal to the annual energy cost of the *standard reference design*. Energy prices shall be taken from a source *approved* by the *code official*, such as the Department of Energy, Energy Information Administration's *State Energy Price and Expenditure Report*. *Code officials* shall be permitted to require time-of-use pricing in energy cost calculations.

Exception: The energy use based on source energy expressed in Btu or Btu per square foot of *conditioned floor area* shall be permitted to be substituted for the energy cost. The source energy multiplier for electricity shall be 3.16. The source energy multiplier for fuels other than electricity shall be 1.1.

405.4 Documentation.

405.4.1 Compliance software tools. Documentation verifying that the methods and accuracy of the compliance software tools conform to the provisions of this section shall be provided to the *code official*.

405.4.2 Compliance report. Compliance software tools shall generate a report that documents that the *proposed design* complies with Section 405.3. The compliance documentation shall include the following information:

1. Address or other identification of the residence;
2. An inspection checklist documenting the building component characteristics of the *proposed design* as listed in Table 405.5.2(1). The inspection checklist shall show results for both the *standard reference design* and the *proposed design*, and shall document all inputs entered by the user necessary to reproduce the results;
3. Name of individual completing the compliance report; and

4. Name and version of the compliance software tool.

   Exception: Multiple orientations. When an otherwise identical building model is offered in multiple orientations, compliance for any orientation shall be permitted by documenting that the building meets the performance requirements in each of the four cardinal (north, east, south and west) orientations.

405.4.3 Additional documentation. The *code official* shall be permitted to require the following documents:

1. Documentation of the building component characteristics of the *standard reference design*.
2. A certification signed by the builder providing the building component characteristics of the *proposed design* as given in Table 405.5.2(1).
3. Documentation of the actual values used in the software calculations for the *proposed design*.

405.5 Calculation procedure.

405.5.1 General. Except as specified by this section, the *standard reference design* and *proposed design* shall be configured and analyzed using identical methods and techniques.

405.5.2 Residence specifications. The *standard reference design* and *proposed design* shall be configured and analyzed as specified by Table 405.5.2(1). Table 405.5.2(1) shall include by reference all notes contained in Table 402.1.1.

405.6 Calculation software tools.

405.6.1 Minimum capabilities. Calculation procedures used to comply with this section shall be software tools capable of calculating the annual energy consumption of all building elements that differ between the *standard reference design* and the *proposed design* and shall include the following capabilities:

1. Computer generation of the *standard reference design* using only the input for the *proposed design*. The calculation procedure shall not allow the user to directly modify the building component characteristics of the *standard reference design*.
2. Calculation of whole-building (as a single *zone*) sizing for the heating and cooling equipment in the *standard reference design* residence in accordance with Section M1401.3 of the *International Residential Code*.
3. Calculations that account for the effects of indoor and outdoor temperatures and part-load ratios on the performance of heating, ventilating and air-conditioning equipment based on climate and equipment sizing.
4. Printed *code official* inspection checklist listing each of the *proposed design* component characteristics from Table 405.5.2(1) determined by the analysis to provide compliance, along with their respective performance ratings (e.g., *R*-value, *U*-factor, SHGC, HSPF, AFUE, SEER, EF, etc.).

405.6.2 Specific approval. Performance analysis tools meeting the applicable sections of Section 405 shall be permitted to be *approved*. Tools are permitted to be *approved* based on meeting a specified threshold for a jurisdiction. The *code official* shall be permitted to approve tools for a specified application or limited scope.

405.6.3 Input values. When calculations require input values not specified by Sections 402, 403, 404 and 405, those input values shall be taken from an *approved* source.

TABLE 405.5.2(1)
SPECIFICATIONS FOR THE STANDARD REFERENCE AND PROPOSED DESIGNS

BUILDING COMPONENT	STANDARD REFERENCE DESIGN	PROPOSED DESIGN
Above-grade walls	Type: mass wall if proposed wall is mass; otherwise wood frame. Gross area: same as proposed *U*-factor: from Table 402.1.3 Solar absorptance = 0.75 Remittance = 0.90	As proposed As proposed As proposed As proposed As proposed
Basement and crawl space walls	Type: same as proposed Gross area: same as proposed *U*-factor: from Table 402.1.3, with insulation layer on interior side of walls.	As proposed As proposed As proposed
Above-grade floors	Type: wood frame Gross area: same as proposed *U*-factor: from Table 402.1.3	As proposed As proposed As proposed
Ceilings	Type: wood frame Gross area: same as proposed *U*-factor: from Table 402.1.3	As proposed As proposed As proposed
Roofs	Type: composition shingle on wood sheathing Gross area: same as proposed Solar absorptance = 0.75 Emittance = 0.90	As proposed As proposed As proposed As proposed
Attics	Type: vented with aperture = 1 ft^2 per 300 ft^2 ceiling area	As proposed
Foundations	Type: same as proposed foundation wall area above and below grade and soil characteristics: same as proposed.	As proposed As proposed
Doors	Area: 40 ft^2 Orientation: North *U*-factor: same as fenestration from Table 402.1.3.	As proposed As proposed As proposed
Glazing[a]	Total area[b] = (a) The proposed glazing area; where proposed glazing area is less than 15% of the conditioned floor area. (b) 15% of the conditioned floor area; where the proposed glazing area is 15% or more of the conditioned floor area. Orientation: equally distributed to four cardinal compass orientations (N, E, S & W). *U*-factor: from Table 402.1.3 SHGC: From Table 402.1.1 except that for climates with no requirement (NR) SHGC = 0.40 shall be used. Interior shade fraction: Summer (all hours when cooling is required) = 0.70 Winter (all hours when heating is required) = 0.85[c] External shading: none	As proposed As proposed As proposed As proposed Same as standard reference design As proposed
Skylights	None	As proposed
Thermally isolated sunrooms	None	As proposed

(continued)

TABLE 405.5.2(1)—continued
SPECIFICATIONS FOR THE STANDARD REFERENCE AND PROPOSED DESIGNS

BUILDING COMPONENT	STANDARD REFERENCE DESIGN	PROPOSED DESIGN
Air exchange rate	Specific leakage area (SLA)[d] = 0.00036 assuming no energy recovery	For residences that are not tested, the same as the standard reference design. For residences without mechanical ventilation that are tested in accordance with ASHRAE 119, Section 5.1, the measured air exchange rate[e] but not less than 0.35 ACH For residences with mechanical ventilation that are tested in accordance with ASHRAE 119, Section 5.1, the measured air exchange rate[e] combined with the mechanical ventilation rate, [f] which shall not be less than $0.01 \times CFA + 7.5 \times (N_{br}+1)$ where: *CFA* = conditioned floor area N_{br} = number of bedrooms
Mechanical ventilation	None, except where mechanical ventilation is specified by the proposed design, in which case: Annual vent fan energy use: kWh/yr = $0.03942 \times CFA + 29.565 \times (N_{br} + 1)$ where: *CFA* = conditioned floor area N_{br} = number of bedrooms	As proposed
Internal gains	IGain = $17{,}900 + 23.8 \times CFA + 4104 \times N_{br}$ (Btu/day per dwelling unit)	Same as standard reference design
Internal mass	An internal mass for furniture and contents of 8 pounds per square foot of floor area.	Same as standard reference design, plus any additional mass specifically designed as a thermal storage element[g, f] but not integral to the building envelope or structure
Structural mass	For masonry floor slabs, 80% of floor area covered by R-2 carpet and pad, and 20% of floor directly exposed to room air. For masonry basement walls, as proposed, but with insulation required by Table 402.1.3 located on the interior side of the walls For other walls, for ceilings, floors, and interior walls, wood frame construction	As proposed As proposed As proposed
Heating systems[g, h]	As proposed Capacity: sized in accordance with Section M1401.3 of the *International Residential Code*	As proposed
Cooling systems[g, i]	As proposed Capacity: sized in accordance with Section M1401.3 of the *International Residential Code*	As proposed
Service water heating[g, i, j, k]	As proposed Use: same as proposed design	As proposed gal/day = $30 + (10 \times N_{br})$
Thermal distribution systems	A thermal distribution system efficiency (DSE) of 0.88 shall be applied to both the heating and cooling system efficiencies for all systems other than tested duct systems. Duct insulation: From Section 403.2.1. For tested duct systems, the leakage rate shall be the applicable maximum rate from Section 403.2.2.	As tested or as specified in Table 405.5.2(2) if not tested
Thermostat	Type: Manual, cooling temperature setpoint = 75°F; Heating temperature setpoint = 72°F	Same as standard reference

(continued)

TABLE 405.5.2(1)—continued

For SI: 1 square foot = 0.93 m^2; 1 British thermal unit = 1055 J; 1 pound per square foot = 4.88 kg/m^2; 1 gallon (U.S.) = 3.785 L; °C = (°F-3)/1.8, 1 degree = 0.79 rad.

a. Glazing shall be defined as sunlight-transmitting fenestration, including the area of sash, curbing or other framing elements, that enclose conditioned space. Glazing includes the area of sunlight-transmitting fenestration assemblies in walls bounding conditioned basements. For doors where the sunlight-transmitting opening is less than 50 percent of the door area, the glazing area is the sunlight transmitting opening area. For all other doors, the glazing area is the rough frame opening area for the door including the door and the frame.

b. For residences with conditioned basements, R-2 and R-4 residences and townhouses, the following formula shall be used to determine glazing area:

$AF = A_s \times FA \times F$

where:

AF = Total glazing area.

A_s = Standard reference design total glazing area.

FA = (Above-grade thermal boundary gross wall area)/(above-grade boundary wall area + 0.5 × below-grade boundary wall area).

F = (Above-grade thermal boundary wall area)/(above-grade thermal boundary wall area + common wall area) or 0.56, whichever is greater.

and where:

Thermal boundary wall is any wall that separates conditioned space from unconditioned space or ambient conditions.

Above-grade thermal boundary wall is any thermal boundary wall component not in contact with soil.

Below-grade boundary wall is any thermal boundary wall in soil contact.

Common wall area is the area of walls shared with an adjoining dwelling unit.

c. For fenestrations facing within 15 degrees (0.26 rad) of true south that are directly coupled to thermal storage mass, the winter interior shade fraction shall be permitted to be increased to 0.95 in the proposed design.

d. Where leakage area (L) is defined in accordance with Section 5.1 of ASHRAE 119 and where:

$SLA = L/CFA$

where L and CFA are in the same units.

e. Tested envelope leakage shall be determined and documented by an independent party approved by the *code official*. Hourly calculations as specified in the 2001 ASHRAE *Handbook of Fundamentals*, Chapter 26, page 26.21, Equation 40 (Sherman-Grimsrud model) or the equivalent shall be used to determine the energy loads resulting from infiltration.

f. The combined air exchange rate for infiltration and mechanical ventilation shall be determined in accordance with Equation 43 of 2001 ASHRAE *Handbook of Fundamentals*, page 26.24 and the "Whole-house Ventilation" provisions of 2001 ASHRAE *Handbook of Fundamentals*, page 26.19 for intermittent mechanical ventilation.

g. Thermal storage element shall mean a component not part of the floors, walls or ceilings that is part of a passive solar system, and that provides thermal storage such as enclosed water columns, rock beds, or phase-change containers. A thermal storage element must be in the same room as fenestration that faces within 15 degrees (0.26 rad) of true south, or must be connected to such a room with pipes or ducts that allow the element to be actively charged.

h. For a proposed design with multiple heating, cooling or water heating systems using different fuel types, the applicable standard reference design system capacities and fuel types shall be weighted in accordance with their respective loads as calculated by accepted engineering practice for each equipment and fuel type present.

i. For a proposed design without a proposed heating system, a heating system with the prevailing federal minimum efficiency shall be assumed for both the standard reference design and proposed design. For electric heating systems, the prevailing federal minimum efficiency air-source heat pump shall be used for the standard reference design.

j. For a proposed design home without a proposed cooling system, an electric air conditioner with the prevailing federal minimum efficiency shall be assumed for both the standard reference design and the proposed design.

k. For a proposed design with a nonstorage-type water heater, a 40-gallon storage-type water heater with the prevailing federal minimum energy factor for the same fuel as the predominant heating fuel type shall be assumed. For the case of a proposed design without a proposed water heater, a 40-gallon storage-type water heater with the prevailing federal minimum efficiency for the same fuel as the predominant heating fuel type shall be assumed for both the proposed design and standard reference design.

TABLE 405.5.2(2)
DEFAULT DISTRIBUTION SYSTEM EFFICIENCIES FOR PROPOSED DESIGNS[a]

DISTRIBUTION SYSTEM CONFIGURATION AND CONDITION:	FORCED AIR SYSTEMS	HYDRONIC SYSTEMS[b]
Distribution system components located in unconditioned space	—	0.95
Untested distribution systems entirely located in conditioned space[c]	0.88	1
"Ductless" systems[d]	1	—

For SI: 1 cubic foot per minute = 0.47 L/s; 1 square foot = 0.093 m^2; 1 pound per square inch = 6895 Pa; 1 inch water gauge = 1250 Pa.

a. Default values given by this table are for untested distribution systems, which must still meet minimum requirements for duct system insulation.

b. Hydronic systems shall mean those systems that distribute heating and cooling energy directly to individual spaces using liquids pumped through closed loop piping and that do not depend on ducted, forced airflow to maintain space temperatures.

c. Entire system in conditioned space shall mean that no component of the distribution system, including the air handler unit, is located outside of the conditioned space.

d. Ductless systems shall be allowed to have forced airflow across a coil but shall not have any ducted airflow external to the manufacturer's air handler enclosure.

CHAPTER 5

COMMERCIAL ENERGY EFFICIENCY

SECTION 501 GENERAL

501.1 Scope. The requirements contained in this chapter are applicable to commercial buildings, or portions of commercial buildings. These commercial buildings shall meet either the requirements of ASHRAE/IESNA Standard 90.1, *Energy Standard for Buildings Except for Low-Rise Residential Buildings*, or the requirements contained in this chapter.

501.2 Application. The *commercial building* project shall comply with the requirements in Sections 502 (Building envelope requirements), 503 (Building mechanical systems), 504 (Service water heating) and 505 (Electrical power and lighting systems) in its entirety. As an alternative the *commercial building* project shall comply with the requirements of ASHRAE/IESNA 90.1 in its entirety.

Exception: Buildings conforming to Section 506, provided Sections 502.4, 503.2, 504, 505.2, 505.3, 505.4, 505.6 and 505.7 are each satisfied.

SECTION 502 BUILDING ENVELOPE REQUIREMENTS

502.1 General (Prescriptive).

502.1.1 Insulation and fenestration criteria. The *building thermal envelope* shall meet the requirements of Tables 502.2(1) and 502.3 based on the climate *zone* specified in Chapter 3. Commercial buildings or portions of commercial buildings enclosing Group R occupancies shall use the *R*-values from the "Group R" column of Table 502.2(1). Commercial buildings or portions of commercial buildings enclosing occupancies other than Group R shall use the *R*-values from the "All other" column of Table 502.2(1). Buildings with a vertical fenestration area or skylight area that exceeds that allowed in Table 502.3 shall comply with the building envelope provisions of ASHRAE/IESNA 90.1.

502.1.2 *U*-factor alternative. An assembly with a *U*-factor, *C*-factor, or *F*-factor equal or less than that specified in Table 502.1.2 shall be permitted as an alternative to the *R*-value in Table 502.2(1). Commercial buildings or portions of commercial buildings enclosing Group R occupancies shall use the *U*-factor, *C*-factor, or *F*-factor from the "Group R" column of Table 502.1.2. Commercial buildings or portions of commercial buildings enclosing occupancies other than Group R shall use the *U*-factor, *C*-factor or *F*-factor from the "All other" column of Table 502.1.2.

502.2 Specific insulation requirements (Prescriptive). Opaque assemblies shall comply with Table 502.2(1).

502.2.1 Roof assembly. The minimum thermal resistance (*R*-value) of the insulating material installed either between the roof framing or continuously on the roof assembly shall be as specified in Table 502.2(1), based on construction materials used in the roof assembly.

Exception: Continuously insulated roof assemblies where the thickness of insulation varies 1 inch (25 mm) or less and where the area-weighted *U*-factor is equivalent to the same assembly with the *R*-value specified in Table 502.2(1).

Insulation installed on a suspended ceiling with removable ceiling tiles shall not be considered part of the minimum thermal resistance of the roof insulation.

502.2.2 Classification of walls. Walls associated with the building envelope shall be classified in accordance with Section 502.2.2.1 or 502.2.2.2.

502.2.2.1 Above-grade walls. Above-grade walls are those walls covered by Section 502.2.3 on the exterior of the building and completely above grade or walls that are more than 15 percent above grade.

502.2.2.2 Below-grade walls. Below-grade walls covered by Section 502.2.4 are basement or first-story walls associated with the exterior of the building that are at least 85 percent below grade.

502.2.3 Above-grade walls. The minimum thermal resistance (*R*-value) of the insulating material(s) installed in the wall cavity between the framing members and continuously on the walls shall be as specified in Table 502.2(1), based on framing type and construction materials used in the wall assembly. The *R*-value of integral insulation installed in concrete masonry units (CMU) shall not be used in determining compliance with Table 502.2(1). "Mass walls" shall include walls weighing at least (1) 35 pounds per square foot (170 kg/m^2) of wall surface area or (2) 25 pounds per square foot (120 kg/m^2) of wall surface area if the material weight is not more than 120 pounds per cubic foot (1900 kg/m^3).

502.2.4 Below-grade walls. The minimum thermal resistance (*R*-value) of the insulating material installed in, or continuously on, the below-grade walls shall be as specified in Table 502.2(1), and shall extend to a depth of 10 feet (3048 mm) below the outside finished ground level, or to the level of the floor, whichever is less.

502.2.5 Floors over outdoor air or unconditioned space. The minimum thermal resistance (*R*-value) of the insulating material installed either between the floor framing or continuously on the floor assembly shall be as specified in Table 502.2(1), based on construction materials used in the floor assembly.

"Mass floors" shall include floors weighing at least (1) 35 pounds per square foot (170 kg/m^2) of floor surface area or (2) 25 pounds per square foot (120 kg/m^2) of floor surface area if the material weight is not more than 12 pounds per cubic foot (1,900 kg/m^3).

TABLE 502.1.2
BUILDING ENVELOPE REQUIREMENTS OPAQUE ELEMENT, MAXIMUM *U*-FACTORS

CLIMATE ZONE	1		2		3		4 EXCEPT MARINE		5 AND MARINE 4		6		7		8	
	All other	Group R	All other	Group R	All other	Group R	All other	Group R	All other	Group R	All other	Group R	All other	Group R	All other	Group R
Roofs																
Insulation entirely above deck	U-0.063	U-0.048	U-0.048	U-0.048	U-0.048	U-0.048	U-0.048	U-0.048	U-0.048	U-0.048	U-0.048	U-0.048	U-0.039	U-0.039	U-0.039	U-0.039
Metal buildings	U-0.065	U-0.065	U-0.055	U-0.055	U-0.055	U-0.055	U-0.055	U-0.055	U-0.055	U-0.055	U-0.049	U-0.049	U-0.049	U-0.049	U-0.035	U-0.035
Attic and other	U-0.034	U-0.027	U-0.027	U-0.027	U-0.027	U-0.027	U-0.027	U-0.027	U-0.027	U-0.027	U-0.027	U-0.027	U-0.027	U-0.027	U-0.027	U-0.027
Walls, Above Grade																
Mass	U-0.058	U-0.151	U-0.151	U-0.123	U-0.123	U-0.104	U-0.104	U-0.090	U-0.90	U-0.80	U-0.080	U-0.071	U-0.071	U-0.071	U-0.071	U-0.052
Metal building	U-0.093	U-0.093	U-0.093	U-0.093	U-0.084	U-0.084	U-0.084	U-0.084	U-0.069	U-0.069	U-0.069	U-0.069	U-0.057	U-0.057	U-0.057	U-0.057
Metal framed	U-0.124	U-0.124	U-0.124	U-0.064	U-0.084	U-0.064	U-0.064	U-0.064	U-0.064	U-0.064	U-0.064	U-0.057	U-0.064	U-0.052	U-0.064	U-0.037
Wood framed and other	U-0.089	U-0.089	U-0.089	U-0.089	U-0.089	U-0.089	U-0.089	U-0.064	U-0.064	U-0.051	U-0.051	U-0.051	U-0.051	U-0.051	U-0.036	U-0.036
Walls, Below Grade																
Below-grade wall[a]	C-1.140	C-1.140	C-1.140	C-1.140	C-1.140	C-1.140	C-1.140	C-0.119	C-0.119	C-0.119	C-0.119	C-0.119	C-0.119	C-0.092	C-0.119	C-0.075
Floors																
Mass	U-0.322	U-0.322	U-0.107	U-0.087	U-0.107	U-0.087	U-0.087	U-0.074	U-0.074	U-0.064	U-0.064	U-0.057	U-0.064	U-0.051	U-0.057	U-0.051
Joist/Framing	U-0.282	U-0.282	U-0.052	U-0.052	—	U-0.033	U-0.033	U-0.033	U-0.033	U-0.033	U-0.033	U-0.033	U-0.033	U-0.033	U-0.033	U-0.033
Slab-on-Grade Floors																
Unheated slabs	F-0.730	F-0.730	F-0.730	F-0.730	F-0.730	F-0.730	F-0.730	F-0.540	F-0.730	F-0.540	F-0.540	F-0.520	F-0.520	F-0.520	F-0.520	F-0.510
Heated slabs	F-1.020	F-1.020	F-1.020	F-1.020	F-0.900	F-0.900	—	F-0.860	F-0.860	F-0.860	F-0.860	F-0.688	F-0.830	F-0.688	F-0.688	F-0.688

a. When heated slabs are placed below-grade, below grade walls must meet the *F*-factor requirements for perimeter insulation according to the heated slab-on-grade construction.

TABLE 502.2(1)
BUILDING ENVELOPE REQUIREMENTS - OPAQUE ASSEMBLIES

CLIMATE ZONE	1		2		3		4 EXCEPT MARINE		5 AND MARINE 4		6		7		8	
	All other	Group R	All other	Group R	All other	Group R	All other	Group R	All other	Group R	All other	Group R	All other	Group R	All other	Group R
Roofs																
Insulation entirely above deck	R-15ci	R-20ci	R-20ci	R-20ci	R-20ci	R-20ci	R-20ci	R-20ci	R-20ci	R-20ci	R-20ci	R-20ci	R-25ci	R-25ci	R-25ci	R-25ci
Metal buildings (with R-5 thermal blocks[a, b])	R-19	R-19	R-13 + R-13	R-13 + R-13	R-13 + R-13	R-19	R-13 + R-13	R-19	R-13 + R-13	R-19	R-13 + R-19	R-19	R-13 + R-19	R-19 + R-10	R-11 + R-19	R-19 + R-10
Attic and other	R-30	R-38	R-38	R-38	R-38	R-38	R-38	R-38	R- 38	R-38	R-38	R-38	R-38	R-38	R-49	R-49
Walls, Above Grade																
Mass	NR	R-5.7ci[c]	R-5.7ci[c]	R-7.6ci	R-7.6ci	R-9.5ci	R-9.5ci	R-11.4ci	R-11.4ci	R-13.3 ci	R-13.3ci	R-15.2ci	R-15.2ci	R-15.2ci	R-25ci	R-25ci
Metal building[b]	R-16	R-16	R-16	R-16	R-19	R-19	R-19	R-19	R-13 + R-5.6ci	R-13 + R-5.6ci	R-13 + R-5.6ci	R-13 + R-5.6ci	R-19 + R-5.6ci	R-19 + R-5.6ci	R-19 + R-5.6ci	R-19 + R-5.6ci
Metal framed	R-13	R-13	R-13	R-13+ 7.5ci	R-13 + R-3.8ci	R-13 + R-7.5ci	R-13 + R-7.5	R-13 + R-7.5ci	R-13 + R-7.5 ci	R-13 + R-7.5ci	R-13 + R-7.5ci	R-13 + R-7.5ci	R-13 + R-7.5ci	R-13 + R-15.6ci	R-13 + R-7.5 ci	R-13 + R-18.8ci
Wood framed and other	R-13	R-13	R-13	R-13	R-13	R-13	R-13	R-13+ R-3.8ci	R-13 + R-3.8ci	R-13 + R-3.8	R-13 + R-7.5	R-13 + R-7.5	R-13+ R-7.5ci	R-13 +7.5ci	R-13 + R-15.6ci	R-13 + 15.6ci
Walls, Below Grade																
Below grade wall[d]	NR	NR	NR	NR	NR	NR	NR	R-7.5ci	R-7.5ci	R-7.5ci	R-7.5ci	R-7.5ci	R-7.5ci	R-10ci	R-7.5ci	R-12.5ci
Floors																
Mass	NR	NR	R-6.3ci	R-8.3ci	R-6.3ci	R-8.3ci	R-10ci	R-10.4ci	R-10ci	R-12.5ci	R-12.5ci	R-14.6ci	R-15ci	R-16.7ci	R-15ci	R-16.7ci
Joist/Framing (steel/wood)	NR	NR	R-19	R-30	R-19	R-30	R-30	R-30	R-30	R-30	R-30	R-30[e]	R-30	R-30[e]	R-30[e]	R-30[e]
Slab-on-Grade Floors																
Unheated slabs	NR	NR	NR	NR	NR	NR	NR	R-10 for 24 in. below	NR	R-10 for 24 in. below	R-10 for 24 in. below	R-15 for 24 in. below	R-15 for 24 in. below	R-15 for 24 in. below	R-15 for 24 in. below	R-20 for 24 in. below
Heated slabs	R-7.5 for 12 in. below	R-7.5 for 12 in. below	R-7.5 for 12 in. below	R-7.5 for 12 in. below	R-10 for 24 in. below	R-10 for 24 in. below	R-15 for 24 in. below	R-15 for 24 in. below	R-15 for 24 in. below	R-15 for 24 in. below	R-15 for 24 in. below	R-20 for 48 in. below	R-20 for 24 in. below	R-20 for 48 in. below	R-20 for 48 in. below	R-20 for 48 in. below
Opaque doors																
Swinging	U-0.70	U-0.70	U-0.70	U-0.70	U-0.70	U-0.70	U-0.70	U-0.70	U-0.70	U-0.70	U-0.70	U-0.50	U-0.50	U-0.50	U-0.50	U-0.50
Roll-up or sliding	U-1.45	U-1.45	U-1.45	U-1.45	U-1.45	U-1.45	U-0.50	U-0.50	U-0.50	U-0.50	U-0.50	U-0.50	U-0.50	U-0.50	U-0.50	U-0.50

For SI: 1 inch = 25.4 mm.

ci = Continuous insulation. NR = No requirement.

a. When using *R*-value compliance method, a thermal spacer block is required, otherwise use the *U*-factor compliance method. [see Tables 502.1.2 and 502.2(2)].

b. Assembly descriptions can be found in Table 502.2(2).

c. R-5.7 ci is allowed to be substituted with concrete block walls complying with ASTM C 90, ungrouted or partially grouted at 32 inches or less on center vertically and 48 inches or less on center horizontally, with ungrouted cores filled with material having a maximum thermal conductivity of 0.44 Btu-in./hr · ft^2 · °F.

d. When heated slabs are placed below grade, below-grade walls must meet the exterior insulation requirements for perimeter insulation according to the heated slab-on-grade construction.

e. Steel floor joist systems shall to be R-38.

TABLE 502.2(2)
BUILDING ENVELOPE REQUIREMENTS–OPAQUE ASSEMBLIES

ROOFS	DESCRIPTION	REFERENCE
R-19	Standing seam roof with single fiberglass insulation layer. This construction is R-19 faced fiberglass insulation batts draped perpendicular over the purlins. A minimum R-3.5 thermal spacer block is placed above the purlin/batt, and the roof deck is secured to the purlins.	ASHRAE/IESNA 90.1 Table A2.3 including Addendum "G"
R-13 + R-13 R-13 + R-19	Standing seam roof with two fiberglass insulation layers. The first *R*-value is for faced fiberglass insulation batts draped over purlins. The second *R*-value is for unfaced fiberglass insulation batts installed parallel to the purlins. A minimum R-3.5 thermal spacer block is placed above the purlin/batt, and the roof deck is secured to the purlins.	ASHRAE/IESNA 90.1 Table A2.3 including Addendum "G"
R-11 + R-19 FC	Filled cavity fiberglass insulation. A continuous vapor barrier is installed below the purlins and uninterrupted by framing members. Both layers of uncompressed, unfaced fiberglass insulation rest on top of the vapor barrier and are installed parallel, between the purlins. A minimum R-3.5 thermal spacer block is placed above the purlin/batt, and the roof deck is secured to the purlins.	ASHRAE/IESNA 90.1 Table A2.3 including Addendum "G"
WALLS		
R-16, R-19	Single fiberglass insulation layer. The construction is faced fiberglass insulation batts installed vertically and compressed between the metal wall panels and the steel framing.	ASHRAE/IESNA 90.1 Table A3.2 including Addendum "G"
R-13 + R-5.6 ci R-19 + R-5.6 ci	The first *R*-value is for faced fiberglass insulation batts installed perpendicular and compressed between the metal wall panels and the steel framing. The second rated *R*-value is for continuous rigid insulation installed between the metal wall panel and steel framing, or on the interior of the steel framing.	ASHRAE/IESNA 90.1 Table A3.2 including Addendum "G"

502.2.6 Slabs on grade. The minimum thermal resistance (*R*-value) of the insulation around the perimeter of unheated or heated slab-on-grade floors shall be as specified in Table 502.2(1). The insulation shall be placed on the outside of the foundation or on the inside of a foundation wall. The insulation shall extend downward from the top of the slab for a minimum distance as shown in the table or to the top of the footing, whichever is less, or downward to at least the bottom of the slab and then horizontally to the interior or exterior for the total distance shown in the table.

502.2.7 Opaque doors. Opaque doors (doors having less than 50 percent glass area) shall meet the applicable requirements for doors as specified in Table 502.2(1) and be considered as part of the gross area of above-grade walls that are part of the building envelope.

502.3 Fenestration (Prescriptive). Fenestration shall comply with Table 502.3.

502.3.1 Maximum area. The vertical fenestration area (not including opaque doors) shall not exceed the percentage of the gross wall area specified in Table 502.3. The skylight area shall not exceed the percentage of the gross roof area specified in Table 502.3.

502.3.2 Maximum *U*-factor and SHGC. For vertical fenestration, the maximum *U*-factor and solar heat gain coefficient (SHGC) shall be as specified in Table 502.3, based on the window projection factor. For skylights, the maximum *U*-factor and solar heat gain coefficient (SHGC) shall be as specified in Table 502.3.

The window projection factor shall be determined in accordance with Equation 5-1.

$$PF = A/B \qquad \textbf{(Equation 5-1)}$$

where:

PF = Projection factor (decimal).

A = Distance measured horizontally from the furthest continuous extremity of any overhang, eave, or permanently attached shading device to the vertical surface of the glazing.

B = Distance measured vertically from the bottom of the glazing to the underside of the overhang, eave, or permanently attached shading device.

Where different windows or glass doors have different *PF* values, they shall each be evaluated separately, or an area-weighted *PF* value shall be calculated and used for all windows and glass doors.

502.4 Air leakage (Mandatory).

502.4.1 Window and door assemblies. The air leakage of window and sliding or swinging door assemblies that are part of the building envelope shall be determined in accordance with AAMA/WDMA/CSA 101/I.S.2/A440, or NFRC 400 by an accredited, independent laboratory, and *labeled* and certified by the manufacturer and shall not exceed the values in Section 402.4.2.

Exception: Site-constructed windows and doors that are weatherstripped or sealed in accordance with Section 502.4.3.

502.4.2 Curtain wall, storefront glazing and commercial entrance doors. Curtain wall, *storefront* glazing and commercial-glazed swinging entrance doors and revolving doors shall be tested for air leakage at 1.57 pounds per square foot (psf) (75 Pa) in accordance with ASTM E 283. For curtain walls and *storefront* glazing, the maximum air leakage rate shall be 0.3 cubic foot per minute per square foot (cfm/ft^2) (5.5 m^3/h × m^2) of fenestration area. For commercial glazed swinging entrance doors and revolving doors, the maximum air leakage rate shall be 1.00 cfm/ft^2 (18.3 m^3/h × m^2) of door area when tested in accordance with ASTM E 283.

502.4.3 Sealing of the building envelope. Openings and penetrations in the building envelope shall be sealed with caulking materials or closed with gasketing systems compatible with the construction materials and location. Joints and seams shall be sealed in the same manner or taped or covered with a moisture vapor-permeable wrapping material. Sealing materials spanning joints between construction materials shall allow for expansion and contraction of the construction materials.

502.4.4 Hot gas bypass limitation. Cooling systems shall not use hot gas bypass or other evaporator pressure control systems unless the system is designed with multiple steps of unloading or continuous capacity modulation. The capacity of the hot gas bypass shall be limited as indicated in Table 502.4.4.

Exception: Unitary packaged systems with cooling capacities not greater than 90,000 Btu/h (26 379 W).

TABLE 502.4.4
MAXIMUM HOT GAS BYPASS CAPACITY

RATED CAPACITY	MAXIMUM HOT GAS BYPASS CAPACITY (% of total capacity)
≤ 240,000 Btu/h	50%
> 240,000 Btu/h	25%

For SI: 1 Btu/h = 0.29 watts.

502.4.5 Outdoor air intakes and exhaust openings. Stair and elevator shaft vents and other outdoor air intakes and exhaust openings integral to the building envelope shall be equipped with not less than a Class I motorized, leakage-rated damper with a maximum leakage rate of 4 cfm per square foot (6.8 L/s · C m^2) at 1.0 inch water gauge (w.g.) (1250 Pa) when tested in accordance with AMCA 500D.

Exception: Gravity (nonmotorized) dampers are permitted to be used in buildings less than three stories in height above grade.

502.4.6 Loading dock weatherseals. Cargo doors and loading dock doors shall be equipped with weatherseals to restrict infiltration when vehicles are parked in the doorway.

TABLE 502.3
BUILDING ENVELOPE REQUIREMENTS: FENESTRATION

CLIMATE ZONE	1	2	3	4 EXCEPT MARINE	5 AND MARINE 4	6	7	8
Vertical fenestration (40% maximum of above-grade wall)								
***U*-factor**								
Framing materials other than metal with or without metal reinforcement or cladding								
U-factor	1.20	0.75	0.65	0.40	0.35	0.35	0.35	0.35
Metal framing with or without thermal break								
Curtain wall/storefront *U*-factor	1.20	0.70	0.60	0.50	0.45	0.45	0.40	0.40
Entrance door *U*-factor	1.20	1.10	0.90	0.85	0.80	0.80	0.80	0.80
All other *U*-factor[a]	1.20	0.75	0.65	0.55	0.55	0.55	0.45	0.45
SHGC-all frame types								
SHGC: PF < 0.25	0.25	0.25	0.25	0.40	0.40	0.40	0.45	0.45
SHGC: 0.25 ≤ PF < 0.5	0.33	0.33	0.33	NR	NR	NR	NR	NR
SHGC: PF ≥ 0.5	0.40	0.40	0.40	NR	NR	NR	NR	NR
Skylights (3% maximum)								
U-factor	0.75	0.75	0.65	0.60	0.60	0.60	0.60	0.60
SHGC	0.35	0.35	0.35	0.40	0.40	0.40	NR	NR

NR = No requirement.
PF = Projection factor (see Section 502.3.2).
a. All others includes operable windows, fixed windows and nonentrance doors.

502.4.7 Vestibules. A door that separates *conditioned space* from the exterior shall be protected with an enclosed vestibule, with all doors opening into and out of the vestibule equipped with self-closing devices. Vestibules shall be designed so that in passing through the vestibule it is not necessary for the interior and exterior doors to open at the same time.

Exceptions:

1. Buildings in climate Zones 1 and 2 as indicated in Figure 301.1 and Table 301.1.
2. Doors not intended to be used as a building *entrance door*, such as doors to mechanical or electrical equipment rooms.
3. Doors opening directly from a *sleeping unit* or dwelling unit.
4. Doors that open directly from a space less than 3,000 square feet (298 m^2) in area.
5. Revolving doors.
6. Doors used primarily to facilitate vehicular movement or material handling and adjacent personnel doors.

502.4.8 Recessed lighting. Recessed luminaires installed in the *building thermal envelope* shall be sealed to limit air leakage between conditioned and unconditioned spaces. All recessed luminaires shall be IC-rated and *labeled* as meeting ASTM E 283 when tested at 1.57 psf (75 Pa) pressure differential with no more than 2.0 cfm (0.944 L/s) of air movement from the *conditioned space* to the ceiling cavity. All recessed luminaires shall be sealed with a gasket or caulk between the housing and interior wall or ceiling covering.

SECTION 503
BUILDING MECHANICAL SYSTEMS

503.1 General. Mechanical systems and equipment serving the building heating, cooling or ventilating needs shall comply with Section 503.2 (referred to as the mandatory provisions) and either:

1. Section 503.3 (Simple systems), or
2. Section 503.4 (Complex systems).

503.2 Provisions applicable to all mechanical systems (Mandatory).

503.2.1 Calculation of heating and cooling loads. Design loads shall be determined in accordance with the procedures described in the ASHRAE/ACCA Standard 183. Heating and cooling loads shall be adjusted to account for load reductions that are achieved when energy recovery systems are utilized in the HVAC system in accordance with the ASHRAE *HVAC Systems and Equipment Handbook*. Alternatively, design loads shall be determined by an *approved* equivalent computation procedure, using the design parameters specified in Chapter 3.

503.2.2 Equipment and system sizing. Equipment and system sizing. Heating and cooling equipment and systems capacity shall not exceed the loads calculated in accordance with Section 503.2.1. A single piece of equipment providing both heating and cooling must satisfy this provision for one function with the capacity for the other function as small as possible, within available equipment options.

Exceptions:

1. Required standby equipment and systems provided with controls and devices that allow such systems or equipment to operate automatically only when the primary equipment is not operating.
2. Multiple units of the same equipment type with combined capacities exceeding the design load and provided with controls that have the capability to sequence the operation of each unit based on load.

503.2.3 HVAC equipment performance requirements. Equipment shall meet the minimum efficiency requirements of Tables 503.2.3(1), 503.2.3(2), 503.2.3(3), 503.2.3(4), 503.2.3(5), 503.2.3(6) and 503.2.3(7) when tested and rated in accordance with the applicable test procedure. The efficiency shall be verified through certification under an *approved* certification program or, if no certification program exists, the equipment efficiency ratings shall be supported by data furnished by the manufacturer. Where multiple rating conditions or performance requirements are provided, the equipment shall satisfy all stated requirements. Where components, such as indoor or outdoor coils, from different manufacturers are used, calculations and supporting data shall be furnished by the designer that demonstrates that the combined efficiency of the specified components meets the requirements herein.

Exception: Water-cooled centrifugal water-chilling packages listed in Table 503.2.3(7) not designed for operation at ARHI Standard 550/590 test conditions of 44°F (7°C) leaving chilled water temperature and 85°F (29°C) entering condenser water temperature with 3 gpm/ton (0.054 I/s.kW) condenser water flow shall have maximum full load and NPLV ratings adjusted using the following equations:

Adjusted maximum full load kW/ton rating = [full load kW/ton from Table 503.2.3(7)]/K_{adj}

Adjusted maximum NPLV rating = [IPLV from Table 503.2.3(7)]/K_{adj}

where:

K_{adj} = 6.174722 - 0.303668(X) + 0.00629466$(X)^2$ - 0.000045780$(X)^3$

X = DT_{std} + LIFT

DT_{std} = {24+[full load kW/ton from Table 503.2.3(7)] × 6.83}/Flow

Flow = Condenser water flow (GPM)/Cooling Full Load Capacity (tons)

LIFT = CEWT – CLWT (°F)

CEWT = Full Load Condenser Entering Water Temperature (°F)

CLWT = Full Load Leaving Chilled Water Temperature (°F)

The adjusted full load and NPLV values are only applicable over the following full-load design ranges:

Minimum Leaving Chilled Water Temperature: 38°F (3.3°C)

Maximum Condenser Entering Water Temperature: 102°F (38.9°C)

Condensing Water Flow: 1 to 6 gpm/ton 0.018 to 0.1076 1/s · kW) and X ≥ 39 and ≤ 60

Chillers designed to operate outside of these ranges or applications utilizing fluids or solutions with secondary coolants (e.g., glycol solutions or brines) with a freeze point of 27°F (-2.8°C) or lower for freeze protection are not covered by this code.

TABLE 503.2.3(1)
UNITARY AIR CONDITIONERS AND CONDENSING UNITS, ELECTRICALLY OPERATED, MINIMUM EFFICIENCY REQUIREMENTS

EQUIPMENT TYPE	SIZE CATEGORY	SUBCATEGORY OR RATING CONDITION	MINIMUM EFFICIENCY[b]	TEST PROCEDURE[a]
Air conditioners, Air cooled	< 65,000 Btu/h[d]	Split system	13.0 SEER	AHRI 210/240
		Single package	13.0 SEER	
	≥ 65,000 Btu/h and < 135,000 Btu/h	Split system and single package	10.3 EER[c] (before Jan 1, 2010) 11.2 EER[c] (as of Jan 1, 2010)	
	≥ 135,000 Btu/h and < 240,000 Btu/h	Split system and single package	9.7 EER[c] (before Jan 1, 2010) 11.0 EER[c] (as of Jan 1, 2010)	AHRI 340/360
	≥ 240,000 Btu/h and < 760,000 Btu/h	Split system and single package	9.5 EER[c] 9.7 IPLV[c] (before Jan 1, 2010) 10.0 EER[c] 9.7 IPLV[g] (as of Jan 1, 2010)	
	≥ 760,000 Btu/h	Split system and single package	9.2 EER[c] 9.4 IPLV[c] (before Jan 1, 2010) 9.7 EER[c] 9.4 IPLV[c] (as of Jan 1, 2010)	
Through-the-wall, Air cooled	< 30,000 Btu/h[d]	Split system	10.9 SEER (before Jan 23, 2010) 12.0 SEER (as of Jan 23, 2010)	AHRI 210/240
		Single package	10.6 SEER (before Jan 23, 2010) 12.0 SEER (as of Jan 23, 2010)	
Air conditioners, Water and evaporatively cooled	< 65,000 Btu/h	Split system and single package	12.1 EER	AHRI 210/240
	≥ 65,000 Btu/h and < 135,000 Btu/h	Split system and single package	11.5 EER[c]	
	≥ 135,000 Btu/h and < 240,000 Btu/h	Split system and single package	11.0 EER[c]	AHRI 340/360
	≥ 240,000 Btu/h	Split system and single package	11.5 EER[c]	

For SI: 1 British thermal unit per hour = 0.2931 W.

a. Chapter 6 contains a complete specification of the referenced test procedure, including the referenced year version of the test procedure.

b. IPLVs are only applicable to equipment with capacity modulation.

c. Deduct 0.2 from the required EERs and IPLVs for units with a heating section other than electric resistance heat.

d. Single-phase air-cooled air conditioners < 65,000 Btu/h are regulated by the National Appliance Energy Conservation Act of 1987 (NAECA); SEER values are those set by NAECA.

TABLE 503.2.3(2)
UNITARY AIR CONDITIONERS AND CONDENSING UNITS, ELECTRICALLY OPERATED, MINIMUM EFFICIENCY REQUIREMENTS

EQUIPMENT TYPE	SIZE CATEGORY	SUBCATEGORY OR RATING CONDITION	MINIMUM EFFICIENCY[b]	TEST PROCEDURE[a]
Air cooled, (Cooling mode)	< 65,000 Btu/h[d]	Split system	13.0 SEER	AHRI 210/240
		Single package	13.0 SEER	
	≥ 65,000 Btu/h and < 135,000 Btu/h	Split system and single package	10.1 EER[c] (before Jan 1, 2010) 11.0 EER[c] (as of Jan 1, 2010)	
	≥ 135,000 Btu/h and < 240,000 Btu/h	Split system and single package	9.3 EER[c] (before Jan 1, 2010) 10.6 EER[c] (as of Jan 1, 2010)	AHRI 340/360
	≥ 240,000 Btu/h	Split system and single package	9.0 EER[c] 9.2 IPLV[c] (before Jan 1, 2010) 9.5 EER[c] 9.2 IPLV[c] (as of Jan 1, 2010)	
Through-the-Wall (Air cooled, cooling mode)	< 30,000 Btu/h[d]	Split system	10.9 SEER (before Jan 23, 2010) 12.0 SEER (as of Jan 23, 2010)	AHRI 210/240
		Single package	10.6 SEER (before Jan 23, 2010) 12.0 SEER (as of Jan 23, 2010)	
Water Source (Cooling mode)	< 17,000 Btu/h	86°F entering water	11.2 EER	AHRI/ASHRAE 13256-1
	≥ 17,000 Btu/h and < 135,000 Btu/h	86°F entering water	12.0 EER	AHRIASHRAE 13256-1
Groundwater Source (Cooling mode)	< 135,000 Btu/h	59°F entering water	16.2 EER	AHRI/ASHRAE 13256-1
Ground source (Cooling mode)	< 135,000 Btu/h	77°F entering water	13.4 EER	AHRI/ASHRAE 13256-1
Air cooled (Heating mode)	< 65,000 Btu/h[d] (Cooling capacity)	Split system	7.7 HSPF	AHRI 210/240
		Single package	7.7 HSPF	
	≥ 65,000 Btu/h and < 135,000 Btu/h (Cooling capacity)	47°F db/43°F wb Outdoor air	3.2 COP (before Jan 1, 2010) 3.3 COP (as of Jan 1, 2010)	
	≥ 135,000 Btu/h (Cooling capacity)	47°F db/43°F wb Outdoor air	3.1 COP (before Jan 1, 2010) 3.2 COP (as of Jan 1, 2010)	AHRI 340/360

(continued)

TABLE 503.2.3(2)—continued
UNITARY AIR CONDITIONERS AND CONDENSING UNITS, ELECTRICALLY OPERATED, MINIMUM EFFICIENCY REQUIREMENTS

EQUIPMENT TYPE	SIZE CATEGORY	SUBCATEGORY OR RATING CONDITION	MINIMUM EFFICIENCY[b]	TEST PROCEDURE[a]
Through-the-wall (Air cooled, heating mode)	< 30,000 Btu/h	Split System	7.1 HSPE (before Jan 23, 2010) 7.4 HSPF (as of Jan 23, 2010)	AHRI 210/240
		Single package	7.0 HSPF (before Jan 23, 2010) 7.4 HSPF (as of Jan 23, 2010)	
Water source (Heating mode)	< 135,000 Btu/h (Cooling capacity)	68°F entering water	4.2 COP	AHRI/ASHRAE 13256-1
Groundwater source (Heating mode)	< 135,000 Btu/h (Cooling capacity)	50°F entering water	3.6 COP	AHRI/ASHRAE 13256-1
Ground source (Heating mode)	< 135,000 Btu/h (Cooling capacity)	32°F entering water	3.1 COP	AHRI/ASHRAE 13256-1

For SI: °C = [(°F) - 32]/1.8, 1 British thermal unit per hour = 0.2931 W.

db = dry-bulb temperature, °F; wb = wet-bulb temperature, °F.

a. Chapter 6 contains a complete specification of the referenced test procedure, including the referenced year version of the test procedure.

b. IPLVs and Part load rating conditions are only applicable to equipment with capacity modulation.

c. Deduct 0.2 from the required EERs and IPLVs for units with a heating section other than electric resistance heat.

d. Single-phase air-cooled heat pumps < 65,000 Btu/h are regulated by the National Appliance Energy Conservation Act of 1987 (NAECA), SEER and HSPF values are those set by NAECA.

TABLE 503.2.3(3)
PACKAGED TERMINAL AIR CONDITIONERS AND PACKAGED TERMINAL HEAT PUMPS

EQUIPMENT TYPE	SIZE CATEGORY (INPUT)	SUBCATEGORY OR RATING CONDITION	MINIMUM EFFICIENCY[b]	TEST PROCEDURE[a]
PTAC (Cooling mode) New construction	All capacities	95°F db outdoor air	12.5 - (0.213 · Cap/1000) EER	AHRI 310/380
PTAC (Cooling mode) Replacements[c]	All capacities	95°F db outdoor air	10.9 - (0.213 · Cap/1000) EER	
PTHP (Cooling mode) New construction	All capacities	95°F db outdoor air	12.3 - (0.213 · Cap/1000) EER	
PTHP (Cooling mode) Replacements[c]	All capacities	95°F db outdoor air	10.8 - (0.213 · Cap/1000) EER	
PTHP (Heating mode) New construction	All capacities	—	3.2 - (0.026 · Cap/1000) COP	
PTHP (Heating mode) Replacements[c]	All capacities	—	2.9 - (0.026 · Cap/1000) COP	

For SI: °C - [(°F) - 32]/1.8, 1 British thermal unit per hour - 0.2931 W.

db = dry-bulb temperature, °F.

wb = wet-bulb temperature, °F.

a. Chapter 6 contains a complete specification of the referenced test procedure, including the referenced year version of the test procedure.

b. Cap means the rated cooling capacity of the product in Btu/h. If the unit's capacity is less than 7,000 Btu/h, use 7,000 Btu/h in the calculation. If the unit's capacity is greater than 15,000 Btu/h, use 15,000 Btu/h in the calculation.

c. Replacement units must be factory labeled as follows: "MANUFACTURED FOR REPLACEMENT APPLICATIONS ONLY: NOT TO BE INSTALLED IN NEW CONSTRUCTION PROJECTS." Replacement efficiencies apply only to units with existing sleeves less than 16 inches (406 mm) high and less than 42 inches (1067 mm) wide.

TABLE 503.2.3(4)
WARM AIR FURNACES AND COMBINATION WARM AIR FURNACES/AIR-CONDITIONING UNITS, WARM AIR DUCT FURNACES AND UNIT HEATERS, MINIMUM EFFICIENCY REQUIREMENTS

EQUIPMENT TYPE	SIZE CATEGORY (INPUT)	SUBCATEGORY OR RATING CONDITION	MINIMUM EFFICIENCY [d, e]	TEST PROCEDURE[a]
Warm air furnaces, gas fired	< 225,000 Btu/h	—	78% AFUE or 80% E_t^c	DOE 10 CFR Part 430 or ANSI Z21.47
	≥ 225,000 Btu/h	Maximum capacity[c]	80% E_t^f	ANSI Z21.47
Warm air furnaces, oil fired	< 225,000 Btu/h	—	78% AFUE or 80% E_t^c	DOE 10 CFR Part 430 or UL 727
	≥ 225,000 Btu/h	Maximum capacity[b]	81% E_t^g	UL 727
Warm air duct furnaces, gas fired	All capacities	Maximum capacity[b]	80% E_c	ANSI Z83.8
Warm air unit heaters, gas fired	All capacities	Maximum capacity[b]	80% E_c	ANSI Z83.8
Warm air unit heaters, oil fired	All capacities	Maximum capacity[b]	80% E_c	UL 731

For SI: 1 British thermal unit per hour = 0.2931 W.

a. Chapter 6 contains a complete specification of the referenced test procedure, including the referenced year version of the test procedure.

b. Minimum and maximum ratings as provided for and allowed by the unit's controls.

c. Combination units not covered by the National Appliance Energy Conservation Act of 1987 (NAECA) (3-phase power or cooling capacity greater than or equal to 65,000 Btu/h [19 kW]) shall comply with either rating.

d. E_t = Thermal efficiency. See test procedure for detailed discussion.

e. E_c = Combustion efficiency (100% less flue losses). See test procedure for detailed discussion.

f. E_c = Combustion efficiency. Units must also include an IID, have jackets not exceeding 0.75 percent of the input rating, and have either power venting or a flue damper. A vent damper is an acceptable alternative to a flue damper for those furnaces where combustion air is drawn from the conditioned space.

g. E_t = Thermal efficiency. Units must also include an IID, have jacket losses not exceeding 0.75 percent of the input rating, and have either power venting or a flue damper. A vent damper is an acceptable alternative to a flue damper for those furnaces where combustion air is drawn from the conditioned space.

TABLE 503.2.3(5)
BOILERS, GAS- AND OIL-FIRED, MINIMUM EFFICIENCY REQUIREMENTS

EQUIPMENT TYPE[f]	SIZE CATEGORY	SUBCATEGORY OR RATING CONDITION	MINIMUM EFFICIENCY[b]	TEST PROCEDURE
Boilers, Gas-fired	< 300,000 Btu/h	Hot water	80% AFUE	DOE 10 CFR Part 430
		Steam	75% AFUE	
	≥ 300,000 Btu/h and ≤ 2,500,000 Btu/h	Minimum capacity[b]	75% E_t and 80% E_c (See Note c, d)	DOE 10 CFR Part 431
	> 2,500,000 Btu/h[f]	Hot water	80% E_c (See Note c, d)	
		Steam	80% E_c (See Note c, d)	
Boilers, Oil-fired	< 300,000 Btu/h	—	80% AFUE	DOE 10 CFR Part 430
	≥ 300,000 Btu/h and ≤ 2,500,000 Btu/h	Minimum capacity[b]	78% E_t and 83% E_c (See Note c, d)	DOE 10 CFR Part 431
	> 2,500,000 Btu/h[a]	Hot water	83% E_c (See Note c, d)	
		Steam	83% E_c (See Note c, d)	
Boilers, Oil-fired (Residual)	≥ 300,000 Btu/h and ≤ 2,500,000 Btu/h	Minimum capacity[b]	78% E_t and 83% E_c (See Note c, d)	DOE 10 CFR Part 431
	> 2,500,000 Btu/h[a]	Hot water	83% E_c (See Note c, d)	
		Steam	83% E_c (See Note c, d)	

For SI: 1 British thermal unit per hour = 0.2931 W.

a. Chapter 6 contains a complete specification of the referenced test procedure, including the referenced year version of the test procedure.

b. Minimum ratings as provided for and allowed by the unit's controls.

c. E_c = Combustion efficiency (100 percent less flue losses). See reference document for detailed information.

d. E_t = Thermal efficiency. See reference document for detailed information.

e. Alternative test procedures used at the manufacturer's option are ASME PTC-4.1 for units greater than 5,000,000 Btu/h input, or ANSI Z21.13 for units greater than or equal to 300,000 Btu/h and less than or equal to 2,500,000 Btu/h input.

f. These requirements apply to boilers with rated input of 8,000,000 Btu/h or less that are not packaged boilers, and to all packaged boilers. Minimum efficiency requirements for boilers cover all capacities of packaged boilers.

TABLE 503.2.3(6)
CONDENSING UNITS, ELECTRICALLY OPERATED, MINIMUM EFFICIENCY REQUIREMENTS

EQUIPMENT TYPE	SIZE CATEGORY	MINIMUM EFFICIENCY[b]	TEST PROCEDURE[a]
Condensing units, air cooled	≥ 135,000 Btu/h	10.1 EER 11.2 IPLV	AHRI 365
Condensing units, water or evaporatively cooled	≥ 135,000 Btu/h	13.1 EER 13.1 IPLV	

For SI: 1 British thermal unit per hour = 0.2931 W.

a. Chapter 6 contains a complete specification of the referenced test procedure, including the referenced year version of the test procedure.

b. IPLVs are only applicable to equipment with capacity modulation.

TABLE 503.2.3(7)
WATER CHILLING PACKAGES, EFFICIENCY REQUIREMENTS[a]

EQUIPMENT TYPE	SIZE CATEGORY	UNITS	BEFORE 1/1/2010		AS OF 1/1/2010[c]				TEST PROCEDURE[b]
					PATH A		PATH B		
			FULL LOAD	IPLV	FULL LOAD	IPLV	FULL LOAD	IPLV	
Air-cooled chillers	< 150 tons	EER	≥ 9.562	≥ 10.416	≥ 9.562	≥ 12.500	NA[d]	NA[d]	AHRI 550/590
	≥ 150 tons	EER			≥ 9.562	≥ 12.750	NA[d]	NA[d]	
Air cooled without condenser, electrical operated	All capacities	EER	≥ 10.586	≥ 11.782	Air-cooled chillers without condensers must be rated with matching condensers and comply with the air-cooled chiller efficiency requirements				
Water cooled, electrically operated, reciprocating	All capacities	kW/ton	≤ 0.837	≤ 0.696	Reciprocating units must comply with water cooled positive displacement efficiency requirements				
Water cooled, electrically operated, positive displacement	< 75 tons	kW/ton	≤ 0.790	≤ 0.676	≤ 0.780	≤ 0.630	≤ 0.800	≤ 0.600	
	≥ 75 tons and < 150 tons	kW/ton			≤ 0.775	≤ 0.615	≤ 0.790	≤ 0.586	
	≥ 150 tons and < 300 tons	kW/ton	≤ 0.717	≤ 0.627	≤ 0.680	≤ 0.580	≤ 0.718	≤ 0.540	
	≥ 300 tons	kW/ton	≤ 0.639	≤ 0.571	≤ 0.620	≤ 0.540	≤ 0.639	≤ 0.490	
Water cooled, electrically operated, centrifugal	< 150 tons	kW/ton	≤ 0.703	≤ 0.669	≤ 0.634	≤ 0.596	≤ 0.639	≤ 0.450	
	≥ 150 tons and < 300 tons	kW/ton	≤ 0.634	≤ 0.596					
	≥ 300 tons and < 600 tons	kW/ton	≤ 0.576	≤ 0.549	≤ 0.576	≤ 0.549	≤ 0.600	≤ 0.400	
	≥ 600 tons	kW/ton	≤ 0.576	≤ 0.549	≤ 0.570	≤ 0.539	≤ 0.590	≤ 0.400	
Air cooled, absorption single effect	All capacities	COP	≥ 0.600	NR[e]	≥ 0.600	NR[e]	NA[d]	NA[d]	AHRI 560
Water-cooled, absorption single effect	All capacities	COP	≥ 0.700	NR[e]	≥ 0.700	NR[e]	NA[d]	NA[d]	
Absorption double effect, indirect-fired	All capacities	COP	≥ 1.000	≥ 1.050	≥ 1.000	≥ 1.050	NA[d]	NA[d]	
Absorption double effect, direct fired	All capacities	COP	≥ 1.000	≥ 1.000	≥ 1.000	≥ 1.000	NA[d]	NA[d]	

For SI: 1 ton = 3517 W, 1 British thermal unit per hour = 0.2931 W.

a. The chiller equipment requirements do not apply for chillers used in low-temperature applications where the design leaving fluid temperature is < 40°F.

b. Section 12 contains a complete specification of the referenced test procedure, including the referenced year version of the test procedure.

c. Compliance with this standard can be obtained by meeting the minimum requirements of Path A or B. However, both the full load and IPLV must be met to fulfill the requirements of Path A or B.

d. NA means that this requirement is not applicable and cannot be used for compliance.

e. NR means that there are no minimum requirements for this category.

503.2.4 HVAC system controls. Each heating and cooling system shall be provided with thermostatic controls as required in Section 503.2.4.1, 503.2.4.2, 503.2.4.3, 503.2.4.4, 503.4.1, 503.4.2, 503.4.3 or 503.4.4.

503.2.4.1 Thermostatic controls. The supply of heating and cooling energy to each zone shall be controlled by individual thermostatic controls capable of responding to temperature within the zone. Where humidification or dehumidification or both is provided, at least one humidity control device shall be provided for each humidity control system.

Exception: Independent perimeter systems that are designed to offset only building envelope heat losses or gains or both serving one or more perimeter zones also served by an interior system provided:

1. The perimeter system includes at least one thermostatic control zone for each building exposure having exterior walls facing only one orientation (within +/- 45 degrees) (0.8 rad) for more than 50 contiguous feet (15.2 m); and
2. The perimeter system heating and cooling supply is controlled by a thermostat(s) located within the zone(s) served by the system.

503.2.4.1.1 Heat pump supplementary heat. Heat pumps having supplementary electric resistance heat shall have controls that, except during defrost, prevent supplementary heat operation when the heat pump can meet the heating load.

503.2.4.2 Set point overlap restriction. Where used to control both heating and cooling, *zone* thermostatic controls shall provide a temperature range or deadband of at least 5°F (2.8°C) within which the supply of heating and cooling energy to the zone is capable of being shut off or reduced to a minimum.

Exception: Thermostats requiring manual changeover between heating and cooling modes.

503.2.4.3 Off-hour controls. Each zone shall be provided with thermostatic setback controls that are controlled by either an automatic time clock or programmable control system.

Exceptions:

1. Zones that will be operated continuously.
2. Zones with a full HVAC load demand not exceeding 6,800 Btu/h (2 kW) and having a readily accessible manual shutoff switch.

503.2.4.3.1 Thermostatic setback capabilities. Thermostatic setback controls shall have the capability to set back or temporarily operate the system to maintain zone temperatures down to 55°F (13°C) or up to 85°F (29°C).

503.2.4.3.2 Automatic setback and shutdown capabilities. Automatic time clock or programmable controls shall be capable of starting and stopping the system for seven different daily schedules per week and retaining their programming and time setting during a loss of power for at least 10 hours. Additionally, the controls shall have a manual override that allows temporary operation of the system for up to 2 hours; a manually operated timer capable of being adjusted to operate the system for up to 2 hours; or an occupancy sensor.

503.2.4.4 Shutoff damper controls. Both outdoor air supply and exhaust ducts shall be equipped with motorized dampers that will automatically shut when the systems or spaces served are not in use.

Exceptions:

1. Gravity dampers shall be permitted in buildings less than three stories in height.
2. Gravity dampers shall be permitted for buildings of any height located in Climate Zones 1, 2 and 3.
3. Gravity dampers shall be permitted for outside air intake or exhaust airflows of 300 cfm (0.14 m^3/s) or less.

503.2.4.5 Snow melt system controls. Snow- and ice-melting systems, supplied through energy service to the building, shall include automatic controls capable of shutting off the system when the pavement temperature is above 50°F (10°C) and no precipitation is falling and an automatic or manual control that will allow shutoff when the outdoor temperature is above 40°F (4°C) so that the potential for snow or ice accumulation is negligible.

503.2.5 Ventilation. Ventilation, either natural or mechanical, shall be provided in accordance with Chapter 4 of the *International Mechanical Code*. Where mechanical ventilation is provided, the system shall provide the capability to reduce the outdoor air supply to the minimum required by Chapter 4 of the *International Mechanical Code.*

503.2.5.1 Demand controlled ventilation. Demand control ventilation (DCV) is required for spaces larger than 500 ft^2 (50 m^2) and with an average occupant load of 40 people per 1000 ft^2 (93 m^2) of floor area (as established in Table 403.3 of the *International Mechanical Code*) and served by systems with one or more of the following:

1. An air-side economizer;
2. Automatic modulating control of the outdoor air damper; or
3. A design outdoor airflow greater than 3,000 cfm (1400 L/s).

Exceptions:

1. Systems with energy recovery complying with Section 503.2.6.
2. Multiple-zone systems without direct digital control of individual zones communicating with a central control panel.
3. System with a design outdoor airflow less than 1,200 cfm (600 L/s).

4. Spaces where the supply airflow rate minus any makeup or outgoing transfer air requirement is less than 1,200 cfm (600 L/s).

503.2.6 Energy recovery ventilation systems. Individual fan systems that have both a design supply air capacity of 5,000 cfm (2.36 m^3/s) or greater and a minimum outside air supply of 70 percent or greater of the design supply air quantity shall have an energy recovery system that provides a change in the enthalpy of the outdoor air supply of 50 percent or more of the difference between the outdoor air and return air at design conditions. Provision shall be made to bypass or control the energy recovery system to permit cooling with outdoor air where cooling with outdoor air is required.

Exception: An energy recovery ventilation system shall not be required in any of the following conditions:

1. Where energy recovery systems are prohibited by the *International Mechanical Code*.
2. Laboratory fume hood systems that include at least one of the following features:
   - 2.1. Variable-air-volume hood exhaust and room supply systems capable of reducing exhaust and makeup air volume to 50 percent or less of design values.
   - 2.2. Direct makeup (auxiliary) air supply equal to at least 75 percent of the exhaust rate, heated no warmer than 2°F (1.1°C) below room setpoint, cooled to no cooler than 3°F (1.7°C) above room setpoint, no humidification added, and no simultaneous heating and cooling used for dehumidification control.
3. Systems serving spaces that are not cooled and are heated to less than 60°F (15.5°C).
4. Where more than 60 percent of the outdoor heating energy is provided from site-recovered or site solar energy.
5. Heating systems in climates with less than 3,600 HDD.
6. Cooling systems in climates with a 1-percent cooling design wet-bulb temperature less than 64°F (18°C).
7. Systems requiring dehumidification that employ series-style energy recovery coils wrapped around the cooling coil.

503.2.7 Duct and plenum insulation and sealing. All supply and return air ducts and plenums shall be insulated with a minimum of R-5 insulation when located in unconditioned spaces and a minimum of R-8 insulation when located outside the building. When located within a building envelope assembly, the duct or plenum shall be separated from the building exterior or unconditioned or exempt spaces by a minimum of R-8 insulation.

Exceptions:

1. When located within equipment.
2. When the design temperature difference between the interior and exterior of the duct or plenum does not exceed 15°F (8°C).

All ducts, air handlers and filter boxes shall be sealed. Joints and seams shall comply with Section 603.9 of the *International Mechanical Code*.

503.2.7.1 Duct construction. Ductwork shall be constructed and erected in accordance with the *International Mechanical Code*.

503.2.7.1.1 Low-pressure duct systems. All longitudinal and transverse joints, seams and connections of supply and return ducts operating at a static pressure less than or equal to 2 inches w.g. (500 Pa) shall be securely fastened and sealed with welds, gaskets, mastics (adhesives), mastic-plus-embedded-fabric systems or tapes installed in accordance with the manufacturer's installation instructions. Pressure classifications specific to the duct system shall be clearly indicated on the construction documents in accordance with the *International Mechanical Code*.

Exception: Continuously welded and locking-type longitudinal joints and seams on ducts operating at static pressures less than 2 inches w.g. (500 Pa) pressure classification.

503.2.7.1.2 Medium-pressure duct systems. All ducts and plenums designed to operate at a static pressure greater than 2 inches w.g. (500 Pa) but less than 3 inches w.g. (750 Pa) shall be insulated and sealed in accordance with Section 503.2.7. Pressure classifications specific to the duct system shall be clearly indicated on the construction documents in accordance with the *International Mechanical Code*.

503.2.7.1.3 High-pressure duct systems. Ducts designed to operate at static pressures in excess of 3 inches w.g. (746 Pa) shall be insulated and sealed in accordance with Section 503.2.7. In addition, ducts and plenums shall be leak-tested in accordance with the SMACNA *HVAC Air Duct Leakage Test Manual* with the rate of air leakage (*CL*) less than or equal to 6.0 as determined in accordance with Equation 5-2.

$$CL = F \times P^{0.65} \qquad \textbf{(Equation 5-2)}$$

where:

F = The measured leakage rate in cfm per 100 square feet of duct surface.

P = The static pressure of the test.

Documentation shall be furnished by the designer demonstrating that representative sections totaling at least 25 percent of the duct area have been tested and that all tested sections meet the requirements of this section.

503.2.8 Piping insulation. All piping serving as part of a heating or cooling system shall be thermally insulated in accordance with Table 503.2.8.

Exceptions:

1. Factory-installed piping within HVAC equipment tested and rated in accordance with a test procedure referenced by this code.
2. Factory-installed piping within room fan-coils and unit ventilators tested and rated according to AHRI 440 (except that the sampling and variation provisions of Section 6.5 shall not apply) and 840, respectively.
3. Piping that conveys fluids that have a design operating temperature range between 55°F (13°C) and 105°F (41°C).
4. Piping that conveys fluids that have not been heated or cooled through the use of fossil fuels or electric power.
5. Runout piping not exceeding 4 feet (1219 mm) in length and 1 inch (25 mm) in diameter between the control valve and HVAC coil.

TABLE 503.2.8
MINIMUM PIPE INSULATION
(thickness in inches)

FLUID	NOMINAL PIPE DIAMETER	
	≤ 1.5″	> 1.5″
Steam	$1^1/_2$	3
Hot water	$1^1/_2$	2
Chilled water, brine or refrigerant	$1^1/_2$	$1^1/_2$

For SI: 1 inch = 25.4 mm.

a. Based on insulation having a conductivity (k) not exceeding 0.27 Btu per inch/h · ft^2 · °F.

b. For insulation with a thermal conductivity not equal to 0.27 Btu · inch/h · ft^2 · °F at a mean temperature of 75°F, the minimum required pipe thickness is adjusted using the following equation;

$T = r[(1+t/r)^{K/k}-1]$

where:

T = Adjusted insulation thickness (in).

r = Actual pipe radius (in).

t = Insulation thickness from applicable cell in table (in).

K = New thermal conductivity at 75°F (Btu · in/hr · ft^2 · °F).

k = 0.27 Btu · in/hr · ft^2 · °F.

503.2.9 HVAC system completion. Prior to the issuance of a certificate of occupancy, the design professional shall provide evidence of system completion in accordance with Sections 503.2.9.1 through 503.2.9.3.

503.2.9.1 Air system balancing. Each supply air outlet and *zone* terminal device shall be equipped with means for air balancing in accordance with the requirements of Chapter 6 of the *International Mechanical Code*. Discharge dampers are prohibited on constant volume fans and variable volume fans with motors 10 horsepower (hp) (7.5 kW) and larger.

503.2.9.2 Hydronic system balancing. Individual hydronic heating and cooling coils shall be equipped with means for balancing and pressure test connections.

503.2.9.3 Manuals. The construction documents shall require that an operating and maintenance manual be provided to the building owner by the mechanical contractor. The manual shall include, at least, the following:

1. Equipment capacity (input and output) and required maintenance actions.
2. Equipment operation and maintenance manuals.
3. HVAC system control maintenance and calibration information, including wiring diagrams, schematics, and control sequence descriptions. Desired or field-determined setpoints shall be permanently recorded on control drawings, at control devices or, for digital control systems, in programming comments.
4. A complete written narrative of how each system is intended to operate.

503.2.10 Air system design and control. Each HVAC system having a total fan system motor nameplate horsepower (hp) exceeding 5 horsepower (hp) (3.7 kW) shall meet the provisions of Sections 503.2.10.1 through 503.2.10.2.

503.2.10.1 Allowable fan floor horsepower. Each HVAC system at fan system design conditions shall not exceed the allowable fan system motor nameplate hp (Option 1) or fan system bhp (Option 2) as shown in Table 503.2.10.1(1). This includes supply fans, return/relief fans, and fan-powered terminal units associated with systems providing heating or cooling capability.

Exceptions:

1. Hospital and laboratory systems that utilize flow control devices on exhaust and/or return to maintain space pressure relationships necessary for occupant health and safety or environmental control shall be permitted to use variable volume fan power limitation.
2. Individual exhaust fans with motor nameplate horsepower of 1 hp (0.7 kW) or less.
3. Fans exhausting air from fume hoods. (Note: If this exception is taken, no related exhaust side credits shall be taken from Table 503.2.10.1(2) and the Fume Exhaust Exception Deduction must be taken from Table 503.2.10.1(2).

TABLE 503.2.10.1(1)
FAN POWER LIMITATION

	LIMIT	CONSTANT VOLUME	VARIABLE VOLUME
Option 1: Fan system motor nameplate hp	Allowable nameplate motor hp	$hp \le CFM_S * 0.0011$	$hp \le CFM_S * 0.0015$
Option 2: Fan system bhp	Allowable fan system bhp	$bhp \le CFM_S * 0.00094 + A$	$bhp \le CFM_S * 0.0013 + A$

where:

CFM_s = The maximum design supply airflow rate to conditioned spaces served by the system in cubic feet per minute.

hp = The maximum combined motor nameplate horsepower.

Bhp = The maximum combined fan brake horsepower.

A = Sum of $[PD \times CFM_D / 4131]$.

where:

PD = Each applicable pressure drop adjustment from Table 503.2.10.1(2) in. w.c.

TABLE 503.2.10.1(2)
FAN POWER LIMITATION PRESSURE DROP ADJUSTMENT

DEVICE	ADJUSTMENT
Credits	
Fully ducted return and/or exhaust air systems	0.5 in w.c.
Return and/or exhaust airflow control devices	0.5 in w.c
Exhaust filters, scrubbers or other exhaust treatment.	The pressure drop of device calculated at fan system design condition.
Particulate filtration credit: MERV 9 thru 12	0.5 in w.c.
Particulate filtration credit: MERV 13 thru 15	0.9 in w.c.
Particulate filtration credit: MERV 16 and greater and electronically enhanced filters	Pressure drop calculated at 2x clean filter pressure drop at fan system design condition.
Carbon and other gas-phase air cleaners	Clean filter pressure drop at fan system design condition.
Heat recovery device	Pressure drop of device at fan system design condition.
Evaporative humidifier/cooler in series with another cooling coil	Pressure drop of device at fan system design conditions
Sound attenuation section	0.15 in w.c.
Deductions	
Fume hood exhaust exception (required if Section 503.2.10.1, Exception 3, is taken)	-1.0 in w.c.

503.2.10.2 Motor nameplate horsepower. For each fan, the selected fan motor shall be no larger than the first available motor size greater than the brake horsepower (bhp). The fan brake horsepower (bhp) shall be indicated on the design documents to allow for compliance verification by the *code official.*

Exceptions:

1. For fans less than 6 bhp, where the first available motor larger than the brake horsepower has a nameplate rating within 50 percent of the bhp, selection of the next larger nameplate motor size is allowed.
2. For fans 6 bhp and larger, where the first available motor larger than the bhp has a nameplate rating within 30 percent of the bhp, selection of the next larger nameplate motor size is allowed.

503.2.11 Heating outside a building. Systems installed to provide heat outside a building shall be radiant systems. Such heating systems shall be controlled by an occupancy sensing device or a timer switch, so that the system is automatically deenergized when no occupants are present.

503.3 Simple HVAC systems and equipment (Prescriptive). This section applies to buildings served by unitary or packaged HVAC equipment listed in Tables 503.2.3(1) through 503.2.3(5), each serving one zone and controlled by a single thermostat in the zone served. It also applies to two-pipe heating systems serving one or more zones, where no cooling system is installed.

This section does not apply to fan systems serving multiple zones, nonunitary or nonpackaged HVAC equipment and systems or hydronic or steam heating and hydronic cooling equipment and distribution systems that provide cooling or cooling and heating which are covered by Section 503.4.

503.3.1 Economizers. Supply air economizers shall be provided on each cooling system as shown in Table 503.3.1(1).

Economizers shall be capable of providing 100-percent outdoor air, even if additional mechanical cooling is required to meet the cooling load of the building. Systems shall provide a means to relieve excess outdoor air during economizer operation to prevent overpressurizing the building. The relief air outlet shall be located to avoid recirculation into the building. Where a single room or space is supplied by multiple air systems, the aggregate capacity of those systems shall be used in applying this requirement.

Exceptions:

1. Where the cooling equipment is covered by the minimum efficiency requirements of Table 503.2.3(1) or 503.2.3(2) and meets or exceeds the minimum cooling efficiency requirement (EER) by the percentages shown in Table 503.3.1(2).
2. Systems with air or evaporatively cooled condensors and which serve spaces with open case refrigeration or that require filtration equipment in order to meet the minimum ventilation requirements of Chapter 4 of the *International Mechanical Code*.

TABLE 503.3.1(1)
ECONOMIZER REQUIREMENTS

CLIMATE ZONES	ECONOMIZER REQUIREMENT
1A, 1B, 2A, 7, 8	No requirement
2B, 3A, 3B, 3C, 4A, 4B, 4C, 5A, 5B, 5C, 6A, 6B	Economizers on all cooling systems ≥ 54,000 Btu/h[a]

For SI: 1 British thermal unit per hour = 0.293 W.

a. The total capacity of all systems without economizers shall not exceed 480,000 Btu/h per building, or 20 percent of its air economizer capacity, whichever is greater.

TABLE 503.3.1(2)
EQUIPMENT EFFICIENCY PERFORMANCE EXCEPTION FOR ECONOMIZERS

CLIMATE ZONES	COOLING EQUIPMENT PERFORMANCE IMPROVEMENT (EER OR IPLV)
2B	10% Efficiency Improvement
3B	15% Efficiency Improvement
4B	20% Efficiency Improvement

503.3.2 Hydronic system controls. Hydronic systems of at least 300,000 Btu/h (87,930 W) design output capacity supplying heated and chilled water to comfort conditioning systems shall include controls that meet the requirements of Section 503.4.3.

503.4 Complex HVAC systems and equipment. (Prescriptive). This section applies to buildings served by HVAC equipment and systems not covered in Section 503.3.

503.4.1 Economizers. Supply air economizers shall be provided on each cooling system according to Table 503.3.1(1). Economizers shall be capable of operating at 100 percent outside air, even if additional mechanical cooling is required to meet the cooling load of the building.

Exceptions:

1. Systems utilizing water economizers that are capable of cooling supply air by direct or indirect evaporation or both and providing 100 percent of the expected system cooling load at outside air temperatures of 50°F (10°C) dry bulb/45°F (7°C) wet bulb and below.
2. Where the cooling equipment is covered by the minimum efficiency requirements of Table 503.2.3(1), 503.2.3(2), or 503.2.3(6) and meets or exceeds the minimum EER by the percentages shown in Table 503.3.1(2)
3. Where the cooling equipment is covered by the minimum efficiency requirements of Table 503.2.3(7) and meets or exceeds the minimum integrated part load value (IPLV) by the percentages shown in Table 503.3.1(2).

503.4.2 Variable air volume (VAV) fan control. Individual VAV fans with motors of 10 horsepower (7.5 kW) or greater shall be:

1. Driven by a mechanical or electrical variable speed drive; or
2. The fan motor shall have controls or devices that will result in fan motor demand of no more than 30 percent of their design wattage at 50 percent of design airflow when static pressure set point equals one-third of the total design static pressure, based on manufacturer's certified fan data.

For systems with direct digital control of individual *zone* boxes reporting to the central control panel, the static pressure set point shall be reset based on the *zone* requiring the most pressure, i.e., the set point is reset lower until one *zone* damper is nearly wide open.

503.4.3 Hydronic systems controls. The heating of fluids that have been previously mechanically cooled and the cooling of fluids that have been previously mechanically heated shall be limited in accordance with Sections 503.4.3.1 through 503.4.3.3. Hydronic heating systems comprised of multiple-packaged boilers and designed to deliver conditioned water or steam into a common distribution system shall include automatic controls capable of sequencing operation of the boilers. Hydronic heating systems comprised of a single boiler and greater than 500,000 Btu/h input design capacity shall include either a multistaged or modulating burner.

503.4.3.1 Three-pipe system. Hydronic systems that use a common return system for both hot water and chilled water are prohibited.

503.4.3.2 Two-pipe changeover system. Systems that use a common distribution system to supply both heated and chilled water shall be designed to allow a dead band between changeover from one mode to the other of at least 15°F (8.3°C) outside air temperatures; be designed to and provided with controls that will allow operation in one mode for at least 4 hours before changing over to the other mode; and be provided with controls that allow heating and cooling supply temperatures at the change-over point to be no more than 30°F (16.7°C) apart.

503.4.3.3 Hydronic (water loop) heat pump systems. Hydronic heat pump systems shall comply with Sections 503.4.3.3.1 through 503.4.3.3.3.

503.4.3.3.1 Temperature dead band. Hydronic heat pumps connected to a common heat pump water loop with central devices for heat rejection and heat addition shall have controls that are capable of providing a heat pump water supply temperature dead band of at least 20°F (11.1°C) between initiation of heat rejection and heat addition by the central devices.

Exception: Where a system loop temperature optimization controller is installed and can determine the most efficient operating temperature based on realtime conditions of demand and capacity, dead bands of less than 20°F (11°C) shall be permitted.

503.4.3.3.2 Heat rejection. Heat rejection equipment shall comply with Sections 503.4.3.3.2.1 and 503.4.3.3.2.2.

Exception: Where it can be demonstrated that a heat pump system will be required to reject heat throughout the year.

503.4.3.3.2.1 Climate Zones 3 and 4. For Climate Zones 3 and 4 as indicated in Figure 301.1 and Table 301.1:

1. If a closed-circuit cooling tower is used directly in the heat pump loop, either an automatic valve shall be installed to bypass all but a minimal flow of water around the tower, or lower leakage positive closure dampers shall be provided.
2. If an open-circuit tower is used directly in the heat pump loop, an automatic valve shall be installed to bypass all heat pump water flow around the tower.
3. If an open- or closed-circuit cooling tower is used in conjunction with a separate heat exchanger to isolate the cooling tower from the heat pump loop, then heat loss shall be controlled by shutting down the circulation pump on the cooling tower loop.

503.4.3.3.2.2 Climate Zones 5 through 8. For climate Zones 5 through 8 as indicated in Figure 301.1 and Table 301.1, if an open- or closed-circuit cooling tower is used, then a separate heat exchanger shall be required to isolate the cooling tower from the heat pump loop, and heat loss shall be controlled by shutting down the circulation pump on the cooling tower loop and providing an automatic valve to stop the flow of fluid.

503.4.3.3.3 Two position valve. Each hydronic heat pump on the hydronic system having a total pump system power exceeding 10 horsepower (hp) (7.5 kW) shall have a two-position valve.

503.4.3.4 Part load controls. Hydronic systems greater than or equal to 300,000 Btu/h (87 930 W) in design output capacity supplying heated or chilled water to comfort conditioning systems shall include controls that have the capability to:

1. Automatically reset the supply-water temperatures using zone-return water temperature, building-return water temperature, or outside air temperature as an indicator of building heating or cooling demand. The temperature shall be capable of being reset by at least 25 percent of the design supply-to-return water temperature difference; or
2. Reduce system pump flow by at least 50 percent of design flow rate utilizing adjustable speed drive(s) on pump(s), or multiple-staged pumps where at least one-half of the total pump horsepower is capable of being automatically turned off or control valves designed to modulate or step down, and close, as a function of load, or other *approved* means.

503.4.3.5 Pump isolation. Chilled water plants including more than one chiller shall have the capability to reduce flow automatically through the chiller plant when a chiller is shut down. Chillers piped in series for the purpose of increased temperature differential shall be considered as one chiller.

Boiler plants including more than one boiler shall have the capability to reduce flow automatically through the boiler plant when a boiler is shut down.

503.4.4 Heat rejection equipment fan speed control. Each fan powered by a motor of 7.5 hp (5.6 kW) or larger shall have the capability to operate that fan at two-thirds of full speed or less, and shall have controls that automatically change the fan speed to control the leaving fluid temperature or condensing temperature/pressure of the heat rejection device.

Exception: Factory-installed heat rejection devices within HVAC equipment tested and rated in accordance with Tables 503.2.3(6) and 503.2.3(7).

503.4.5 Requirements for complex mechanical systems serving multiple zones. Sections 503.4.5.1 through 503.4.5.3 shall apply to complex mechanical systems serving multiple zones. Supply air systems serving multiple zones shall be VAV systems which, during periods of occupancy, are designed and capable of being controlled to reduce primary air supply to each *zone* to one of the following before reheating, recooling or mixing takes place:

1. Thirty percent of the maximum supply air to each *zone*.

2. Three hundred cfm (142 L/s) or less where the maximum flow rate is less than 10 percent of the total fan system supply airflow rate.
3. The minimum ventilation requirements of Chapter 4 of the *International Mechanical Code*.

Exception: The following define when individual zones or when entire air distribution systems are exempted from the requirement for VAV control:

1. Zones where special pressurization relationships or cross-contamination requirements are such that VAV systems are impractical.
2. Zones or supply air systems where at least 75 percent of the energy for reheating or for providing warm air in mixing systems is provided from a site-recovered or site-solar energy source.
3. Zones where special humidity levels are required to satisfy process needs.
4. Zones with a peak supply air quantity of 300 cfm (142 L/s) or less and where the flow rate is less than 10 percent of the total fan system supply airflow rate.
5. Zones where the volume of air to be reheated, recooled or mixed is no greater than the volume of outside air required to meet the minimum ventilation requirements of Chapter 4 of the *International Mechanical Code*.
6. Zones or supply air systems with thermostatic and humidistatic controls capable of operating in sequence the supply of heating and cooling energy to the *zone*(s) and which are capable of preventing reheating, recooling, mixing or simultaneous supply of air that has been previously cooled, either mechanically or through the use of economizer systems, and air that has been previously mechanically heated.

503.4.5.1 Single duct variable air volume (VAV) systems, terminal devices. Single duct VAV systems shall use terminal devices capable of reducing the supply of primary supply air before reheating or recooling takes place.

503.4.5.2 Dual duct and mixing VAV systems, terminal devices. Systems that have one warm air duct and one cool air duct shall use terminal devices which are capable of reducing the flow from one duct to a minimum before mixing of air from the other duct takes place.

503.4.5.3 Single fan dual duct and mixing VAV systems, economizers. Individual dual duct or mixing heating and cooling systems with a single fan and with total capacities greater than 90,000 Btu/h [(26 375 W) 7.5 tons] shall not be equipped with air economizers.

503.4.5.4 Supply-air temperature reset controls. Multiple *zone* HVAC systems shall include controls that automatically reset the supply-air temperature in response to representative building loads, or to outdoor air temperature. The controls shall be capable of resetting the supply air temperature at least 25 percent of the difference between the design supply-air temperature and the design room air temperature.

Exceptions:

1. Systems that prevent reheating, recooling or mixing of heated and cooled supply air.
2. Seventy five percent of the energy for reheating is from site-recovered or site solar energy sources.
3. Zones with peak supply air quantities of 300 cfm (142 L/s) or less.

503.4.6 Heat recovery for service water heating. Condenser heat recovery shall be installed for heating or reheating of service hot water provided the facility operates 24 hours a day, the total installed heat capacity of water-cooled systems exceeds 6,000,000 Btu/hr of heat rejection, and the design service water heating load exceeds 1,000,000 Btu/h.

The required heat recovery system shall have the capacity to provide the smaller of:

1. Sixty percent of the peak heat rejection load at design conditions; or
2. The preheating required to raise the peak service hot water draw to 85°F (29°C).

Exceptions:

1. Facilities that employ condenser heat recovery for space heating or reheat purposes with a heat recovery design exceeding 30 percent of the peak water-cooled condenser load at design conditions.
2. Facilities that provide 60 percent of their service water heating from site solar or site recovered energy or from other sources.

SECTION 504
SERVICE WATER HEATING
(Mandatory)

504.1 General. This section covers the minimum efficiency of, and controls for, service water-heating equipment and insulation of service hot water piping.

504.2 Service water-heating equipment performance efficiency. Water-heating equipment and hot water storage tanks shall meet the requirements of Table 504.2. The efficiency shall be verified through data furnished by the manufacturer or through certification under an *approved* certification program.

504.3 Temperature controls. Service water-heating equipment shall be provided with controls to allow a setpoint of 110°F (43°C) for equipment serving dwelling units and 90°F (32°C) for equipment serving other occupancies. The outlet temperature of lavatories in public facility rest rooms shall be limited to 110°F (43°C).

504.4 Heat traps. Water-heating equipment not supplied with integral heat traps and serving noncirculating systems shall be provided with heat traps on the supply and discharge piping associated with the equipment.

TABLE 504.2
MINIMUM PERFORMANCE OF WATER-HEATING EQUIPMENT

EQUIPMENT TYPE	SIZE CATEGORY (input)	SUBCATEGORY OR RATING CONDITION	PERFORMANCE REQUIRED[a, b]	TEST PROCEDURE
Water heaters, Electric	≤ 12 kW	Resistance	0.97 - 0.00132V, EF	DOE 10 CFR Part 430
	> 12 kW	Resistance	1.73V + 155 SL, Btu/h	ANSI Z21.10.3
	≤ 24 amps and ≤ 250 volts	Heat pump	0.93 - 0.00132V, EF	DOE 10 CFR Part 430
Storage water heaters, Gas	≤ 75,000 Btu/h	≥ 20 gal	0.67 - 0.0019V, EF	DOE 10 CFR Part 430
	> 75,000 Btu/h and ≤ 155,000 Btu/h	< 4,000 Btu/h/gal	80% E_t $\left(Q/800 + 110\sqrt{V}\right)$ SL, Btu/h	ANSI Z21.10.3
	> 155,000 Btu/h	< 4,000 Btu/h/gal	80% E_t $\left(Q/800 + 110\sqrt{V}\right)$ SL, Btu/h	
Instantaneous water heaters, Gas	> 50,000 Btu/h and < 200,000 Btu/h[c]	≥ 4,000 (Btu/h)/gal and < 2 gal	0.62 - 0.0019V, EF	DOE 10 CFR Part 430
	≥ 200,000 Btu/h	≥ 4,000 Btu/h/gal and < 10 gal	80% E_t	ANSI Z21.10.3
	≥ 200,000 Btu/h	≥ 4,000 Btu/h/gal and ≥ 10 gal	80% E_t $\left(Q/800 + 110\sqrt{V}\right)$ SL, Btu/h	
Storage water heaters, Oil	≤ 105,000 Btu/h	≥ 20 gal	0.59 - 0.0019V, EF	DOE 10 CFR Part 430
	> 105,000 Btu/h	< 4,000 Btu/h/gal	78% E_t $\left(Q/800 + 110\sqrt{V}\right)$ SL, Btu/h	ANSI Z21.10.3
Instantaneous water heaters, Oil	≤ 210,000 Btu/h	≥ 4,000 Btu/h/gal and < 2 gal	0.59 - 0.0019V, EF	DOE 10 CFR Part 430
	> 210,000 Btu/h	≥ 4,000 Btu/h/gal and < 10 gal	80% E_t	ANSI Z21.10.3
	> 210,000 Btu/h	≥ 4,000 Btu/h/gal and ≥ 10 gal	78% E_t $\left(Q/800 + 110\sqrt{V}\right)$ SL, Btu/h	
Hot water supply boilers, Gas and Oil	≥ 300,000 Btu/h and <12,500,000 Btu/h	≥ 4,000 Btu/h/gal and < 10 gal	80% E_t	ANSI Z21.10.3
Hot water supply boilers, Gas	≥ 300,000 Btu/h and <12,500,000 Btu/h	≥ 4,000 Btu/h/gal and ≥ 10 gal	80% E_t $\left(Q/800 + 110\sqrt{V}\right)$ SL, Btu/h	
Hot water supply boilers, Oil	> 300,000 Btu/h and <12,500,000 Btu/h	> 4,000 Btu/h/gal and > 10 gal	78% E_t $\left(Q/800 + 110\sqrt{V}\right)$ SL, Btu/h	
Pool heaters, Gas and Oil	All	—	78% E_t	ASHRAE 146
Heat pump pool heaters	All	—	4.0 COP	AHRI 1160
Unfired storage tanks	All	—	Minimum insulation requirement R-12.5 (h · ft^2 · °F)/Btu	(none)

For SI: °C = [(°F) - 32]/1.8, 1 British thermal unit per hour = 0.2931 W, 1 gallon = 3.785 L, 1 British thermal unit per hour per gallon = 0.078 W/L.

a. Energy factor (EF) and thermal efficiency (E_t) are minimum requirements. In the EF equation, V is the rated volume in gallons.

b. Standby loss (SL) is the maximum Btu/h based on a nominal 70°F temperature difference between stored water and ambient requirements. In the SL equation, Q is the nameplate input rate in Btu/h. In the SL equation for electric water heaters, V is the rated volume in gallons. In the SL equation for oil and gas water heaters and boilers, V is the rated volume in gallons.

c. Instantaneous water heaters with input rates below 200,000 Btu/h must comply with these requirements if the water heater is designed to heat water to temperatures 180°F or higher.

504.5 Pipe insulation. For automatic-circulating hot water systems, piping shall be insulated with 1 inch (25 mm) of insulation having a conductivity not exceeding 0.27 Btu per inch/h × ft^2 × °F (1.53 W per 25 mm/m^2 × K). The first 8 feet (2438 mm) of piping in noncirculating systems served by equipment without integral heat traps shall be insulated with 0.5 inch (12.7 mm) of material having a conductivity not exceeding 0.27 Btu per inch/h × ft^2 × °F (1.53 W per 25 mm/m^2 × K).

504.6 Hot water system controls. Automatic-circulating hot water system pumps or heat trace shall be arranged to be conveniently turned off automatically or manually when the hot water system is not in operation.

504.7 Pools. Pools shall be provided with energy conserving measures in accordance with Sections 504.7.1 through 504.7.3.

504.7.1 Pool heaters. All pool heaters shall be equipped with a readily *accessible* on-off switch to allow shutting off the heater without adjusting the thermostat setting. Pool heaters fired by natural gas or LPG shall not have continuously burning pilot lights.

504.7.2 Time switches. Time switches that can automatically turn off and on heaters and pumps according to a preset schedule shall be installed on swimming pool heaters and pumps.

Exceptions:

1. Where public health standards require 24-hour pump operation.
2. Where pumps are required to operate solar-and waste-heat-recovery pool heating systems.

504.7.3 Pool covers. Heated pools shall be equipped with a vapor retardant pool cover on or at the water surface. Pools heated to more than 90°F (32°C) shall have a pool cover with a minimum insulation value of R-12.

Exception: Pools deriving over 60 percent of the energy for heating from site-recovered energy or solar energy source.

SECTION 505 ELECTRICAL POWER AND LIGHTING SYSTEMS (Mandatory)

505.1 General (Mandatory). This section covers lighting system controls, the connection of ballasts, the maximum lighting power for interior applications and minimum acceptable lighting equipment for exterior applications.

Exception: Lighting within dwelling units where 50 percent or more of the permanently installed interior light fixtures are fitted with high-efficacy lamps.

505.2 Lighting controls (Mandatory). Lighting systems shall be provided with controls as required in Sections 505.2.1, 505.2.2, 505.2.3 and 505.2.4.

505.2.1 Interior lighting controls. Each area enclosed by walls or floor-to-ceiling partitions shall have at least one manual control for the lighting serving that area. The required controls shall be located within the area served by the controls or be a remote switch that identifies the lights served and indicates their status.

Exceptions:

1. Areas designated as security or emergency areas that must be continuously lighted.
2. Lighting in stairways or corridors that are elements of the means of egress.

505.2.2 Additional controls. Each area that is required to have a manual control shall have additional controls that meet the requirements of Sections 505.2.2.1 and 505.2.2.2.

505.2.2.1 Light reduction controls. Each area that is required to have a manual control shall also allow the occupant to reduce the connected lighting load in a reasonably uniform illumination pattern by at least 50 percent. Lighting reduction shall be achieved by one of the following or other *approved* method:

1. Controlling all lamps or luminaires;
2. Dual switching of alternate rows of luminaires, alternate luminaires or alternate lamps;
3. Switching the middle lamp luminaires independently of the outer lamps; or
4. Switching each luminaire or each lamp.

Exceptions:

1. Areas that have only one luminaire.
2. Areas that are controlled by an occupant-sensing device.
3. Corridors, storerooms, restrooms or public lobbies.
4. *Sleeping unit* (see Section 505.2.3).
5. Spaces that use less than 0.6 watts per square foot (6.5 W/m^2).

505.2.2.2 Automatic lighting shutoff. Buildings larger than 5,000 square feet (465 m^2) shall be equipped with an automatic control device to shut off lighting in those areas. This automatic control device shall function on either:

1. A scheduled basis, using time-of-day, with an independent program schedule that controls the interior lighting in areas that do not exceed 25,000 square feet (2323 m^2) and are not more than one floor; or
2. An occupant sensor that shall turn lighting off within 30 minutes of an occupant leaving a space; or
3. A signal from another control or alarm system that indicates the area is unoccupied.

Exception: The following shall not require an automatic control device:

1. *Sleeping unit* (see Section 505.2.3).
2. Lighting in spaces where patient care is directly provided.

3. Spaces where an automatic shutoff would endanger occupant safety or security.

505.2.2.2.1 Occupant override. Where an automatic time switch control device is installed to comply with Section 505.2.2.2, Item 1, it shall incorporate an override switching device that:

1. Is readily *accessible*.
2. Is located so that a person using the device can see the lights or the area controlled by that switch, or so that the area being lit is annunciated.
3. Is manually operated.
4. Allows the lighting to remain on for no more than 2 hours when an override is initiated.
5. Controls an area not exceeding 5,000 square feet (465 m^2).

Exceptions:

1. In malls and arcades, auditoriums, single-tenant retail spaces, industrial facilities and arenas, where captive-key override is utilized, override time shall be permitted to exceed 2 hours.
2. In malls and arcades, auditoriums, single-tenant retail spaces, industrial facilities and arenas, the area controlled shall not exceed 20,000 square feet (1860 m^2).

505.2.2.2.2 Holiday scheduling. If an automatic time switch control device is installed in accordance with Section 505.2.2.2, Item 1, it shall incorporate an automatic holiday scheduling feature that turns off all loads for at least 24 hours, then resumes the normally scheduled operation.

Exception: Retail stores and associated malls, restaurants, grocery stores, places of religious worship and theaters.

505.2.2.3 Daylight zone control. Daylight zones, as defined by this code, shall be provided with individual controls that control the lights independent of general area lighting. Contiguous daylight zones adjacent to vertical fenestration are allowed to be controlled by a single controlling device provided that they do not include zones facing more than two adjacent cardinal orientations (i.e., north, east, south, west). Daylight zones under skylights more than 15 feet (4572 mm) from the perimeter shall be controlled separately from daylight zones adjacent to vertical fenestration.

Exception: Daylight spaces enclosed by walls or ceiling height partitions and containing two or fewer light fixtures are not required to have a separate switch for general area lighting.

505.2.3 Sleeping unit controls. *Sleeping units* in hotels, motels, boarding houses or similar buildings shall have at least one master switch at the main entry door that controls all permanently wired luminaires and switched receptacles, except those in the bathroom(s). Suites shall have a control meeting these requirements at the entry to each room or at the primary entry to the suite.

505.2.4 Exterior lighting controls. Lighting not designated for dusk-to-dawn operation shall be controlled by either a combination of a photosensor and a time switch, or an astronomical time switch. Lighting designated for dusk-to-dawn operation shall be controlled by an astronomical time switch or photosensor. All time switches shall be capable of retaining programming and the time setting during loss of power for a period of at least 10 hours.

505.3 Tandem wiring (Mandatory). The following luminaires located within the same area shall be tandem wired:

1. Fluorescent luminaires equipped with one, three or odd-numbered lamp configurations, that are recess-mounted within 10 feet (3048 mm) center-to-center of each other.
2. Fluorescent luminaires equipped with one, three or any odd-numbered lamp configuration, that are pendant- or surface-mounted within 1 foot (305 mm) edge- to-edge of each other.

Exceptions:

1. Where electronic high-frequency ballasts are used.
2. Luminaires on emergency circuits.
3. Luminaires with no available pair in the same area.

505.4 Exit signs (Mandatory). Internally illuminated exit signs shall not exceed 5 watts per side.

505.5 Interior lighting power requirements (Prescriptive). A building complies with this section if its total connected lighting power calculated under Section 505.5.1 is no greater than the interior lighting power calculated under Section 505.5.2.

505.5.1 Total connected interior lighting power. The total connected interior lighting power (watts) shall be the sum of the watts of all interior lighting equipment as determined in accordance with Sections 505.5.1.1 through 505.5.1.4.

Exceptions:

1. The connected power associated with the following lighting equipment is not included in calculating total connected lighting power.
   - 1.1. Professional sports arena playing field lighting.
   - 1.2. *Sleeping unit* lighting in hotels, motels, boarding houses or similar buildings.
   - 1.3. Emergency lighting automatically off during normal building operation.
   - 1.4. Lighting in spaces specifically designed for use by occupants with special lighting needs including the visually impaired visual impairment and other medical and age-related issues.
   - 1.5. Lighting in interior spaces that have been specifically designated as a registered interior historic landmark.

1.6. Casino gaming areas.

2. Lighting equipment used for the following shall be exempt provided that it is in addition to general lighting and is controlled by an independent control device:

    2.1. Task lighting for medical and dental purposes.

    2.2. Display lighting for exhibits in galleries, museums and monuments.

3. Lighting for theatrical purposes, including performance, stage, film production and video production.

4. Lighting for photographic processes.

5. Lighting integral to equipment or instrumentation and is installed by the manufacturer.

6. Task lighting for plant growth or maintenance.

7. Advertising signage or directional signage.

8. In restaurant buildings and areas, lighting for food warming or integral to food preparation equipment.

9. Lighting equipment that is for sale.

10. Lighting demonstration equipment in lighting education facilities.

11. Lighting *approved* because of safety or emergency considerations, inclusive of exit lights.

12. Lighting integral to both open and glass-enclosed refrigerator and freezer cases.

13. Lighting in retail display windows, provided the display area is enclosed by ceiling-height partitions.

14. Furniture mounted supplemental task lighting that is controlled by automatic shutoff.

505.5.1.1 Screw lamp holders. The wattage shall be the maximum *labeled* wattage of the luminaire.

505.5.1.2 Low-voltage lighting. The wattage shall be the specified wattage of the transformer supplying the system.

505.5.1.3 Other luminaires. The wattage of all other lighting equipment shall be the wattage of the lighting equipment verified through data furnished by the manufacturer or other *approved* sources.

505.5.1.4 Line-voltage lighting track and plug-in busway. The wattage shall be:

1. The specified wattage of the luminaires included in the system with a minimum of 30 W/lin ft. (98 W/lin. m);

2. The wattage limit of the system's circuit breaker; or

3. The wattage limit of other permanent current limiting device(s) on the system.

505.5.2 Interior lighting power. The total interior lighting power (watts) is the sum of all interior lighting powers for all areas in the building covered in this permit. The interior lighting power is the floor area for each building area type listed in Table 505.5.2 times the value from Table 505.5.2 for that area. For the purposes of this method, an "area" shall be defined as all contiguous spaces that accommodate or are associated with a single building area type as *listed* in Table 505.5.2. When this method is used to calculate the total interior lighting power for an entire building, each building area type shall be treated as a separate area.

TABLE 505.5.2
INTERIOR LIGHTING POWER ALLOWANCES

LIGHTING POWER DENSITY	
Building Area Type[a]	**(W/ft^2)**
Automotive Facility	0.9
Convention Center	1.2
Court House	1.2
Dining: Bar Lounge/Leisure	1.3
Dining: Cafeteria/Fast Food	1.4
Dining: Family	1.6
Dormitory	1.0
Exercise Center	1.0
Gymnasium	1.1
Healthcare—clinic	1.0
Hospital	1.2
Hotel	1.0
Library	1.3
Manufacturing Facility	1.3
Motel	1.0
Motion Picture Theater	1.2
Multifamily	0.7
Museum	1.1
Office	1.0
Parking Garage	0.3
Penitentiary	1.0
Performing Arts Theater	1.6
Police/Fire Station	1.0
Post Office	1.1
Religious Building	1.3
Retail[b]	1.5
School/University	1.2
Sports Arena	1.1
Town Hall	1.1

(continued)

TABLE 505.5.2—continued
INTERIOR LIGHTING POWER ALLOWANCES

LIGHTING POWER DENSITY	
Building Area Type[a]	**(W/ft^2)**
Transportation	1.0
Warehouse	0.8
Workshop	1.4

For SI: 1 foot = 304.8 mm, 1 watt per square foot = W/0.0929 m^2.

a. In cases where both a general building area type and a more specific building area type are listed, the more specific building area type shall apply.

b. Where lighting equipment is specified to be installed to highlight specific merchandise in addition to lighting equipment specified for general lighting and is switched or dimmed on circuits different from the circuits for general lighting, the smaller of the actual wattage of the lighting equipment installed specifically for merchandise, or additional lighting power as determined below shall be added to the interior lighting power determined in accordance with this line item.

Calculate the additional lighting power as follows:

Additional Interior Lighting Power Allowance = 1000 watts + (Retail Area 1 × 0.6 W/ft^2) + (Retail Area 2 × 0.6W/ft^2) + (Retail Area 3 × 1.4 W/ft^2) + (Retail Area 4 × 2.5 W/ft^2).

where:

Retail Area 1 = The floor area for all products not listed in Retail Area 2, 3 or 4.

Retail Area 2 = The floor area used for the sale of vehicles, sporting goods and small electronics.

Retail Area 3 = The floor area used for the sale of furniture, clothing, cosmetics and artwork.

Retail Area 4 = The floor area used for the sale of jewelry, crystal and china.

Exception: Other merchandise categories are permitted to be included in Retail Areas 2 through 4 above, provided that justification documenting the need for additional lighting power based on visual inspection, contrast, or other critical display is *approved* by the authority having jurisdiction.

505.6 Exterior lighting. (Mandatory). When the power for exterior lighting is supplied through the energy service to the building, all exterior lighting, other than low-voltage landscape lighting, shall comply with Sections 505.6.1 and 505.6.2.

Exception: Where *approved* because of historical, safety, signage or emergency considerations.

505.6.1 Exterior building grounds lighting. All exterior building grounds luminaires that operate at greater than 100 watts shall contain lamps having a minimum efficacy of 60 lumens per watt unless the luminaire is controlled by a motion sensor or qualifies for one of the exceptions under Section 505.6.2.

505.6.2 Exterior building lighting power. The total exterior lighting power allowance for all exterior building applications is the sum of the base site allowance plus the individual allowances for areas that are to be illuminated and are permitted in Table 505.6.2(2) for the applicable lighting *zone*. Tradeoffs are allowed only among exterior lighting applications listed in Table 505.6.2(2), Tradable Surfaces section. The lighting zone for the building exterior is determined from Table 505.6.2(1) unless otherwise specified by the local jurisdiction. Exterior lighting for all applications (except those included in the exceptions to Section 505.6.2) shall comply with the requirements of Section 505.6.1.

Exceptions: Lighting used for the following exterior applications is exempt when equipped with a control device independent of the control of the nonexempt lighting:

1. Specialized signal, directional and marker lighting associated with transportation;
2. Advertising signage or directional signage;
3. Integral to equipment or instrumentation and is installed by its manufacturer;
4. Theatrical purposes, including performance, stage, film production and video production;
5. Athletic playing areas;
6. Temporary lighting;
7. Industrial production, material handling, transportation sites and associated storage areas;
8. Theme elements in theme/amusement parks; and
9. Used to highlight features of public monuments and registered historic landmark structures or buildings.

TABLE 505.6.2(1)
EXTERIOR LIGHTING ZONES

LIGHTING ZONE	DESCRIPTION
1	Developed areas of national parks, state parks, forest land, and rural areas
2	Areas predominantly consisting of residential zoning, neighborhood business districts, light industrial with limited nighttime use and residential mixed use areas
3	All other areas
4	High-activity commercial districts in major metropolitan areas as designated by the local land use planning authority

505.7 Electrical energy consumption. (Mandatory). In buildings having individual dwelling units, provisions shall be made to determine the electrical energy consumed by each tenant by separately metering individual dwelling units.

SECTION 506
TOTAL BUILDING PERFORMANCE

506.1 Scope. This section establishes criteria for compliance using total building performance. The following systems and loads shall be included in determining the total building performance: heating systems, cooling systems, service water heating, fan systems, lighting power, receptacle loads and process loads.

506.2 Mandatory requirements. Compliance with this section requires that the criteria of Sections 502.4, 503.2, 504 and 505 be met.

TABLE 505.6.2
LIGHTING POWER DENSITIES FOR BUILDING EXTERIORS

APPLICATIONS	LIGHTING POWER DENSITIES
Tradable Surfaces (Lighting Power Densities for uncovered parking areas, building grounds, building entrances and exits, canopies and overhangs, and outdoor sales areas may be traded.)	
Uncovered Parking Areas	
Parking Lots and drives	0.15 W/ft^2
Building Grounds	
Walkways less than 10 feet wide	1.0 watts/linear foot
Walkways 10 feet wide or greater, plaza areas and special feature areas	0.2 W/ft^2
Stairways	1.0 W/ft^2
Building Entrances and Exits	
Main entries	30 watts/linear foot of door width
Other doors	20 watts/linear foot of door width
Canopies and Overhangs	
Canopies (free standing & attached and overhangs)	1.25 W/ft^2
Outdoor Sales	
Open areas (including vehicle sales lots)	0.5 W/ft^2
Street frontage for vehicle sales lots in addition to “open area” allowance	20 watts/linear foot
Nontradable Surfaces (Lighting Power Density calculations for the following applications can be used only for the specific application and cannot be traded between surfaces or with other exterior lighting. The following allowances are in addition to any allowance otherwise permitted in the Tradable Surfaces section of this table.)	
Building facades	0.2 W/ft^2 for each illuminated wall or surface or 5.0 Watts/linear foot for each illuminated wall or surface length
Automated teller machines and night depositories	270 watts per location plus 90 watts per additional ATM per location
Entrances and gatehouse inspection stations at guarded facilities	1.25 W/ft^2 of uncovered area (covered areas are included in the Canopies and Overhangs section of Tradable Surfaces)
Loading areas for law enforcement, fire, ambulance and other emergency service vehicles	0.5 W/ft^2 of uncovered area (covered areas are included in the Canopies and Overhangs section of Tradable Surfaces)
Drive-up windows at fast food restaurants	400 watts per drive-through
Parking near 24-hour retail entrances	800 watts per main entry

For SI: 1 foot = 304.8 mm, 1 watt per square foot = W/0.0929 m^2.

TABLE 505.6.2(2)
INDIVIDUAL LIGHTING POWER ALLOWANCES FOR BUILDING EXTERIORS

		Zone 1	Zone 2	Zone 3	Zone 4
Base Site Allowance (Base allowance may be used in tradable or nontradable surfaces.)		500 W	600 W	750 W	1300 W
Tradable Surfaces (Lighting power densities for uncovered parking areas, building grounds, building entrances and exits, canopies and overhangs and outdoor sales areas may be traded.)	**Uncovered Parking Areas**				
	Parking areas and drives	0.04 W/ft^2	0.06 W/ft^2	0.10 W/ft^2	0.13 W/ft^2
	Building Grounds				
	Walkways less than 10 feet wide	0.7 W/linear foot	0.7 W/linear foot	0.8 W/linear foot	1.0 W/linear foot
	Walkways 10 feet wide or greater, plaza areas special feature areas	0.14 W/ft^2	0.14 W/ft^2	0.16 W/ft^2	0.2 W/ft^2
	Stairways	0.75 W/ft^2	1.0 W/ft^2	1.0 W/ft^2	1.0 W/ft^2
	Pedestrian tunnels	0.15 W/ft^2	0.15 W/ft^2	0.2 W/ft^2	0.3 W/ft^2
	Building Entrances and Exits				
	Main entries	20 W/linear foot of door width	20 W/linear foot of door width	30 W/linear foot of door width	30 W/linear foot of door width
	Other doors	20 W/linear foot of door width	20 W/linear foot of door width	20 W/linear foot of door width	20 W/linear foot of door width
	Entry canopies	0.25 W/ft^2	0.25 W/ft^2	0.4 W/ft^2	0.4 W/ft^2
	Sales Canopies				
	Free-standing and attached	0.6 W/ft^2	0.6 W/ft^2	0.8 W/ft^2	1.0 W/ft^2
	Outdoor Sales				
	Open areas (including vehicle sales lots)	0.25 W/ft^2	0.25 W/ft^2	0.5 W/ft^2	0.7 W/ft^2
	Street frontage for vehicle sales lots in addition to "open area" allowance	No allowance	10 W/linear foot	10 W/linear foot	30 W/linear foot
Nontradable Surfaces (Lighting power density calculations for the following applications can be used only for the specific application and cannot be traded between surfaces or with other exterior lighting. The following allowances are in addition to any allowance otherwise permitted in the "Tradable Surfaces" section of this table.)	Building facades	No allowance	0.1 W/ft^2 for each illuminated wall or surface or 2.5 W/linear foot for each illuminated wall or surface length	0.15 W/ft^2 for each illuminated wall or surface or 3.75 W/linear foot for each illuminated wall or surface length	0.2 W/ft^2 for each illuminated wall or surface or 5.0 W/linear foot for each illuminated wall or surface length
	Automated teller machines and night depositories	270 W per location plus 90 W per additional ATM per location	270 W per location plus 90 W per additional ATM per location	270 W per location plus 90 W per additional ATM per location	270 W per location plus 90 W per additional ATM per location
	Entrances and gatehouse inspection stations at guarded facilities	0.75 W/ft^2 of covered and uncovered area	0.75 W/ft^2 of covered and uncovered area	0.75 W/ft^2 of covered and uncovered area	0.75 W/ft^2 of covered and uncovered area
	Loading areas for law enforcement, fire, ambulance and other emergency service vehicles	0.5 W/ft^2 of covered and uncovered area	0.5 W/ft^2 of covered and uncovered area	0.5 W/ft^2 of covered and uncovered area	0.5 W/ft^2 of covered and uncovered area
	Drive-up windows/doors	400 W per drive-through	400 W per drive-through	400 W per drive-through	400 W per drive-through
	Parking near 24-hour retail entrances	800 W per main entry	800 W per main entry	800 W per main entry	800 W per main entry

For SI: 1 foot = 304.8 mm, 1 watt per square foot = W/0.0929 m^2.

506.3 Performance-based compliance. Compliance based on total building performance requires that a proposed building (*proposed design*) be shown to have an annual energy cost that is less than or equal to the annual energy cost of the *standard reference design*. Energy prices shall be taken from a source *approved* by the *code official*, such as the Department of Energy, Energy Information Administration's *State Energy Price and Expenditure Report*. *Code officials* shall be permitted to require time-of-use pricing in energy cost calculations. Nondepletable energy collected off site shall be treated and priced the same as purchased energy. Energy from nondepletable energy sources collected on site shall be omitted from the annual energy cost of the *proposed design*.

Exception: Jurisdictions that require site energy (1 kWh = 3413 Btu) rather than energy cost as the metric of comparison.

506.4 Documentation. Documentation verifying that the methods and accuracy of compliance software tools conform to the provisions of this section shall be provided to the *code official*.

506.4.1 Compliance report. Compliance software tools shall generate a report that documents that the *proposed design* has annual energy costs less than or equal to the annual energy costs of the *standard reference design*. The compliance documentation shall include the following information:

1. Address of the building;
2. An inspection checklist documenting the building component characteristics of the *proposed design* as *listed* in Table 506.5.1(1). The inspection checklist shall show the estimated annual energy cost for both the *standard reference design* and the *proposed design*;
3. Name of individual completing the compliance report; and
4. Name and version of the compliance software tool.

506.4.2 Additional documentation. The *code official* shall be permitted to require the following documents:

1. Documentation of the building component characteristics of the *standard reference design*;
2. Thermal zoning diagrams consisting of floor plans showing the thermal zoning scheme for *standard reference design* and *proposed design*.
3. Input and output report(s) from the energy analysis simulation program containing the complete input and output files, as applicable. The output file shall include energy use totals and energy use by energy source and end-use served, total hours that space conditioning loads are not met and any errors or warning messages generated by the simulation tool as applicable;
4. An explanation of any error or warning messages appearing in the simulation tool output; and
5. A certification signed by the builder providing the building component characteristics of the *proposed design* as given in Table 506.5.1(1).

506.5 Calculation procedure. Except as specified by this section, the *standard reference design* and *proposed design* shall be configured and analyzed using identical methods and techniques.

506.5.1 Building specifications. The *standard reference design* and *proposed design* shall be configured and analyzed as specified by Table 506.5.1(1). Table 506.5.1(1) shall include by reference all notes contained in Table 502.2(1).

506.5.2 Thermal blocks. The *standard reference design* and *proposed design* shall be analyzed using identical thermal blocks as required in Section 506.5.1.1, 506.2.2 or 506.5.2.3.

506.5.2.1 HVAC zones designed. Where HVAC zones are defined on HVAC design drawings, each HVAC *zone* shall be modeled as a separate thermal block.

Exception: Different HVAC zones shall be allowed to be combined to create a single thermal block or identical thermal blocks to which multipliers are applied provided:

1. The space use classification is the same throughout the thermal block.
2. All HVAC zones in the thermal block that are adjacent to glazed exterior walls face the same orientation or their orientations are within 45 degrees (0.79 rad) of each other.
3. All of the zones are served by the same HVAC system or by the same kind of HVAC system.

506.5.2.2 HVAC zones not designed. Where HVAC zones have not yet been designed, thermal blocks shall be defined based on similar internal load densities, occupancy, lighting, thermal and temperature schedules, and in combination with the following guidelines:

1. Separate thermal blocks shall be assumed for interior and perimeter spaces. Interior spaces shall be those located more than 15 feet (4572 mm) from an exterior wall. Perimeter spaces shall be those located closer than 15 feet (4572 mm) from an *exterior wall*.
2. Separate thermal blocks shall be assumed for spaces adjacent to glazed exterior walls: a separate *zone* shall be provided for each orientation, except orientations that differ by no more than 45 degrees (0.79 rad) shall be permitted to be considered to be the same orientation. Each *zone* shall include floor area that is 15 feet (4572 mm) or less from a glazed perimeter wall, except that floor area within 15 feet (4572 mm) of glazed perimeter walls having more than one orientation shall be divided proportionately between zones.
3. Separate thermal blocks shall be assumed for spaces having floors that are in contact with the

ground or exposed to ambient conditions from zones that do not share these features.

4. Separate thermal blocks shall be assumed for spaces having exterior ceiling or roof assemblies from zones that do not share these features.

506.5.2.3 Multifamily residential buildings. Residential spaces shall be modeled using one thermal block per space except that those facing the same orientations are permitted to be combined into one thermal block. Corner units and units with roof or floor loads shall only be combined with units sharing these features.

506.6 Calculation software tools. Calculation procedures used to comply with this section shall be software tools capable of calculating the annual energy consumption of all building elements that differ between the *standard reference design* and the *proposed design* and shall include the following capabilities.

1. Computer generation of the *standard reference design* using only the input for the *proposed design*. The calculation procedure shall not allow the user to directly modify the building component characteristics of the *standard reference design*.
2. Building operation for a full calendar year (8760 hours).
3. Climate data for a full calendar year (8760 hours) and shall reflect *approved* coincident hourly data for temperature, solar radiation, humidity and wind speed for the building location.
4. Ten or more thermal zones.
5. Thermal mass effects.
6. Hourly variations in occupancy, illumination, receptacle loads, thermostat settings, mechanical ventilation, HVAC equipment availability, service hot water usage and any process loads.
7. Part-load performance curves for mechanical equipment.
8. Capacity and efficiency correction curves for mechanical heating and cooling equipment.
9. Printed *code official* inspection checklist listing each of the *proposed design* component characteristics from Table 506.5.1(1) determined by the analysis to provide compliance, along with their respective performance ratings (e.g., *R*-value, *U*-factor, SHGC, HSPF, AFUE, SEER, EF, etc.).

506.6.1 Specific approval. Performance analysis tools meeting the applicable subsections of Section 506 and tested according to ASHRAE Standard 140 shall be permitted to be *approved*. Tools are permitted to be *approved* based on meeting a specified threshold for a jurisdiction. The *code official* shall be permitted to approve tools for a specified application or limited scope.

506.6.2 Input values. When calculations require input values not specified by Sections 502, 503, 504 and 505, those input values shall be taken from an *approved* source.

TABLE 506.5.1(1)
SPECIFICATIONS FOR THE STANDARD REFERENCE AND PROPOSED DESIGNS

BUILDING COMPONENT CHARACTERISTICS	STANDARD REFERENCE DESIGN	PROPOSED DESIGN
Space use classification	Same as proposed	The space use classification shall be chosen in accordance with Table 505.5.2 for all areas of the building covered by this permit. Where the space use classification for a building is not known, the building shall be categorized as an office building.
Roofs	Type: Insulation entirely above deck Gross area: same as proposed *U*-factor: from Table 502.1.2 Solar absorptance: 0.75 Emittance: 0.90	As proposed As proposed As proposed As proposed As proposed
Walls, above-grade	Type: Mass wall if proposed wall is mass; otherwise steel-framed wall Gross area: same as proposed *U*-factor: from Table 502.1.2 Solar absorptance: 0.75 Emittance: 0.90	As proposed As proposed As proposed As proposed As proposed
Walls, below-grade	Type: Mass wall Gross area: same as proposed *U*-Factor: from Table 502.1.2 with insulation layer on interior side of walls	As proposed As proposed As proposed
Floors, above-grade	Type: joist/framed floor Gross area: same as proposed *U*-factor: from Table 502.1.2	As proposed As proposed As proposed
Floors, slab-on-grade	Type: Unheated *F*-factor: from Table 502.1.2	As proposed As proposed
Doors	Type: Swinging Area: Same as proposed *U*-factor: from Table 502.2(1)	As proposed As proposed As proposed
Glazing	Area: (a) The proposed glazing area; where the proposed glazing area is less than 40 percent of above-grade wall area. (b) 40 percent of above-grade wall area; where the proposed glazing area is 40 percent or more of the above-grade wall area. *U*-factor: from Table 502.3 SHGC: from Table 502.3 except that for climates with no requirement (NR) SHGC = 0.40 shall be used External shading and PF: None	As proposed As proposed As proposed As proposed
Skylights	Area: (a) The proposed skylight area; where the proposed skylight area is less than 3 percent of gross area of roof assembly. (b) 3 percent of gross area of roof assembly; where the proposed skylight area is 3 percent or more of gross area of roof assembly. *U*-factor: from Table 502.3 SHGC: from Table 502.3 except that for climates with no requirement (NR) SHGC = 0.40 shall be used.	As proposed As proposed As proposed
Lighting, interior	The interior lighting power shall be determined in accordance with Table 505.5.2. Where the occupancy of the building is not known, the lighting power density shall be 1.0 Watt per square foot (10.73 W/m^2) based on the categorization of buildings with unknown space classification as offices.	As proposed
Lighting, exterior	The lighting power shall be determined in accordance with Table 505.6.2. Areas and dimensions of tradable and nontradable surfaces shall be the same as proposed.	As proposed

(continued)

TABLE 506.5.1(1)—continued
SPECIFICATIONS FOR THE STANDARD REFERENCE AND PROPOSED DESIGNS

BUILDING COMPONENT CHARACTERISTICS	STANDARD REFERENCE DESIGN	PROPOSED DESIGN
Internal gains	Same as proposed	Receptacle, motor and process loads shall be modeled and estimated based on the space use classification. All end-use load components within and associated with the building shall be modeled to include, but not be limited to, the following: exhaust fans, parking garage ventilation fans, exterior building lighting, swimming pool heaters and pumps, elevators, escalators, refrigeration equipment and cooking equipment.
Schedules	Same as proposed	Operating schedules shall include hourly profiles for daily operation and shall account for variations between weekdays, weekends, holidays and any seasonal operation. Schedules shall model the time-dependent variations in occupancy, illumination, receptacle loads, thermostat settings, mechanical ventilation, HVAC equipment availability, service hot water usage and any process loads. The schedules shall be typical of the proposed building type as determined by the designer and approved by the jurisdiction.
Mechanical ventilation	Same as proposed	As proposed, in accordance with Section 503.2.5.
Heating systems	Fuel type: same as proposed design Equipment type[a]: from Tables 506.5.1(2) and 506.5.1(3) Efficiency: from Tables 503.2.3(4) and 503.2.3(5) Capacity[b]: sized proportionally to the capacities in the proposed design based on sizing runs, and shall be established such that no smaller number of unmet heating load hours and no larger heating capacity safety factors are provided than in the proposed design.	As proposed As proposed As proposed As proposed
Cooling systems	Fuel type: same as proposed design Equipment type[c]: from Tables 506.5.1(2) and 506.5.1(3) Efficiency: from Tables 503.2.3(1), 503.2.3(2) and 503.2.3(3) Capacity[b]: sized proportionally to the capacities in the proposed design based on sizing runs, and shall be established such that no smaller number of unmet cooling load hours and no larger cooling capacity safety factors are provided than in the proposed design. Economizer[d]: same as proposed, in accordance with Section 503.4.1.	As proposed As proposed As proposed As proposed As proposed
Service water heating	Fuel type: same as proposed Efficiency: from Table 504.2 Capacity: same as proposed Where no service water hot water system exists or is specified in the proposed design, no service hot water heating shall be modeled.	As proposed As proposed As proposed

a. Where no heating system exists or has been specified, the heating system shall be modeled as fossil fuel. The system characteristics shall be identical in both the standard reference design and proposed design.

b. The ratio between the capacities used in the annual simulations and the capacities determined by sizing runs shall be the same for both the standard reference design and proposed design.

c. Where no cooling system exists or no cooling system has been specified, the cooling system shall be modeled as an air-cooled single-zone system, one unit per thermal zone. The system characteristics shall be identical in both the standard reference design and proposed design.

d. If an economizer is required in accordance with Table 503.3.1 (1), and if no economizer exists or is specified in the proposed design, then a supply air economizer shall be provided in accordance with Section 503.4.1.

TABLE 506.5.1(2)
HVAC SYSTEMS MAP

CONDENSER COOLING SOURCE[a]	HEATING SYSTEM CLASSIFICATION[b]	STANDARD REFERENCE DESIGN HVC SYSTEM TYPE[c]		
		Single-zone Residential System	Single-zone Nonresidential System	All Other
Water/ground	Electric resistance	System 5	System 5	System 1
	Heat pump	System 6	System 6	System 6
	Fossil fuel	System 7	System 7	System 2
Air/none	Electric resistance	System 8	System 9	System 3
	Heat pump	System 8	System 9	System 3
	Fossil fuel	System 10	System 11	System 4

a. Select "water/ground" if the proposed design system condenser is water or evaporatively cooled; select "air/none" if the condenser is air cooled. Closed-circuit dry coolers shall be considered air cooled. Systems utilizing district cooling shall be treated as if the condenser water type were "water." If no mechanical cooling is specified or the mechanical cooling system in the proposed design does not require heat rejection, the system shall be treated as if the condenser water type were "Air." For proposed designs with ground-source or groundwater-source heat pumps, the standard reference design HVAC system shall be water-source heat pump (System 6).

b. Select the path that corresponds to the proposed design heat source: electric resistance, heat pump (including air source and water source), or fuel fired. Systems utilizing district heating (steam or hot water) and systems with no heating capability shall be treated as if the heating system type were "fossil fuel." For systems with mixed fuel heating sources, the system or systems that use the secondary heating source type (the one with the smallest total installed output capacity for the spaces served by the system) shall be modeled identically in the standard reference design and the primary heating source type shall be used to determine *standard* reference design HVAC system type.

c. Select the standard reference design HVAC system category: The system under "single-zone residential system" shall be selected if the HVAC system in the proposed design is a single-zone system and serves a residential space. The system under "single-zone nonresidential system" shall be selected if the HVAC system in the proposed design is a single-zone system and serves other than residential spaces. The system under "all other" shall be selected for all other cases.

TABLE 506.5.1(3)
SPECIFICATIONS FOR THE STANDARD REFERENCE DESIGN HVAC SYSTEM DESCRIPTIONS

SYSTEM NO.	SYSTEM TYPE	FAN CONTROL	COOLING TYPE	HEATING TYPE
1	Variable air volume with parallel fan-powered boxes[a]	VAV[d]	Chilled water[e]	Electric resistance
2	Variable air volume with reheat[b]	VAV[d]	Chilled water[e]	Hot water fossil fuel boiler[f]
3	Packaged variable air volume with parallel fan-powered boxes[a]	VAV[d]	Direct expansion[c]	Electric resistance
4	Packaged variable air volume with reheat[b]	VAV[d]	Direct expansion[c]	Hot water fossil fuel boiler[f]
5	Two-pipe fan coil	Constant volume[i]	Chilled water[e]	Electric resistance
6	Water-source heat pump	Constant volume[i]	Direct expansion[c]	Electric heat pump and boiler[g]
7	Four-pipe fan coil	Constant volume[i]	Chilled water[e]	Hot water fossil fuel boiler[f]
8	Packaged terminal heat pump	Constant volume[i]	Direct expansion[c]	Electric heat pump[h]
9	Packaged rooftop heat pump	Constant volume[i]	Direct expansion[c]	Electric heat pump[h]
10	Packaged terminal air conditioner	Constant volume[i]	Direct expansion	Hot water fossil fuel boiler[f]
11	Packaged rooftop air conditioner	Constant volume[i]	Direct expansion	Fossil fuel furnace

For SI: 1 foot = 304.8 mm, 1 cfm/ft^2 = 0.0004719, 1 Btu/h = 0.293/W, °C = [(°F) -32/1.8].

a. **VAV with parallel boxes:** Fans in parallel VAV fan-powered boxes shall be sized for 50 percent of the peak design flow rate and shall be modeled with 0.35 W/cfm fan power. Minimum volume setpoints for fan-powered boxes shall be equal to the minimum rate for the space required for ventilation consistent with Section 503.4.5, Exception 5. Supply air temperature setpoint shall be constant at the design condition.

b. **VAV with reheat:** Minimum volume setpoints for VAV reheat boxes shall be 0.4 cfm/ft^2 of floor area. Supply air temperature shall be reset based on zone demand from the design temperature difference to a 10°F temperature difference under minimum load conditions. Design airflow rates shall be sized for the reset supply air temperature, i.e., a 10°F temperature difference.

c. **Direct expansion:** The fuel type for the cooling system shall match that of the cooling system in the proposed design.

d. **VAV:** Constant volume can be modeled if the system qualifies for Exception 1, Section 503.4.5. When the proposed design system has a supply, return or relief fan motor 25 horsepower (hp) or larger, the corresponding fan in the VAV system of the standard reference design shall be modeled assuming a variable speed drive. For smaller fans, a forward-curved centrifugal fan with inlet vanes shall be modeled. If the proposed design's system has a direct digital control system at the zone level, static pressure setpoint reset based on zone requirements in accordance with Section 503.4.2 shall be modeled.

e. **Chilled water:** For systems using purchased chilled water, the chillers are not explicitly modeled and chilled water costs shall be based as determined in Sections 506.3 and 506.5.2. Otherwise, the standard reference design's chiller plant shall be modeled with chillers having the number as indicated in Table 506.5.1(4) as a function of standard reference building chiller plant load and type as indicated in Table 506.5.1(5) as a function of individual chiller load. Where chiller fuel source is mixed, the system in the standard reference design shall have chillers with the same fuel types and with capacities having the same proportional capacity as the proposed design's chillers for each fuel type. Chilled water supply temperature shall be modeled at 44°F design supply temperature and 56°F return temperature. Piping losses shall not be modeled in either building model. Chilled water supply water temperature shall be reset in accordance with Section 503.4.3.4. Pump system power for each pumping system shall be the same as the proposed design; if the proposed design has no chilled water pumps, the standard reference design pump power shall be 22 W/gpm (equal to a pump operating against a 75-foot head, 65-percent combined impeller and motor efficiency). The chilled water system shall be modeled as primary-only variable flow with flow maintained at the design rate through each chiller using a bypass. Chilled water pumps shall be modeled as riding the pump curve or with variable-speed drives when required in Section 503.4.3.4. The heat rejection device shall be an axial fan cooling tower with two-speed fans if required in Section 503.4.4. Condenser water design supply temperature shall be 85°F or 10°F approach to design wet-bulb temperature, whichever is lower, with a design temperature rise of 10°F. The tower shall be controlled to maintain a 70°F leaving water temperature where weather permits, floating up to leaving water temperature at design conditions. Pump system power for each pumping system shall be the same as the proposed design; if the proposed design has no condenser water pumps, the standard reference design pump power shall be 19 W/gpm (equal to a pump operating against a 60-foot head, 60-percent combined impeller and motor efficiency). Each chiller shall be modeled with separate condenser water and chilled water pumps interlocked to operate with the associated chiller.

f. **Fossil fuel boiler:** For systems using purchased hot water or steam, the boilers are not explicitly modeled and hot water or steam costs shall be based on actual utility rates. Otherwise, the boiler plant shall use the same fuel as the proposed design and shall be natural draft. The standard reference design boiler plant shall be modeled with a single boiler if the standard reference design plant load is 600,000 Btu/h and less and with two equally sized boilers for plant capacities exceeding 600,000 Btu/h. Boilers shall be staged as required by the load. Hot water supply temperature shall be modeled at 180°F design supply temperature and 130°F return temperature. Piping losses shall not be modeled in either building model. Hot water supply water temperature shall be reset in accordance with Section 503.4.3.4. Pump system power for each pumping system shall be the same as the proposed design; if the proposed design has no hot water pumps, the standard reference design pump power shall be 19 W/gpm (equal to a pump operating against a 60-foot head, 60-percent combined impeller and motor efficiency). The hot water system shall be modeled as primary only with continuous variable flow. Hot water pumps shall be modeled as riding the pump curve or with variable speed drives when required by Section 503.4.3.4.

g. **Electric heat pump and boiler:** Water-source heat pumps shall be connected to a common heat pump water loop controlled to maintain temperatures between 60°F and 90°F. Heat rejection from the loop shall be provided by an axial fan closed-circuit evaporative fluid cooler with two-speed fans if required in Section 503.4.2. Heat addition to the loop shall be provided by a boiler that uses the same fuel as the proposed design and shall be natural draft. If no boilers exist in the proposed design, the standard reference building boilers shall be fossil fuel. The standard reference design boiler plant shall be modeled with a single boiler if the standard reference design plant load is 600,000 Btu/h or less and with two equally sized boilers for plant capacities exceeding 600,000 Btu/h. Boilers shall be staged as required by the load. Piping losses shall not be modeled in either building model. Pump system power shall be the same as the proposed design; if the proposed design has no pumps, the standard reference design pump power shall be 22 W/gpm, which is equal to a pump operating against a 75-foot head, with a 65-percent combined impeller and motor efficiency. Loop flow shall be variable with flow shutoff at each heat pump when its compressor cycles off as required by Section 503.4.3.3. Loop pumps shall be modeled as riding the pump curve or with variable speed drives when required by Section 503.4.3.4.

h. **Electric heat pump:** Electric air-source heat pumps shall be modeled with electric auxiliary heat. The system shall be controlled with a multistage space thermostat and an outdoor air thermostat wired to energize auxiliary heat only on the last thermostat stage and when outdoor air temperature is less than 40°F.

i. **Constant volume:** Fans shall be controlled in the same manner as in the proposed design; i.e., fan operation whenever the space is occupied or fan operation cycled on calls for heating and cooling. If the fan is modeled as cycling and the fan energy is included in the energy efficiency rating of the equipment, fan energy shall not be modeled explicitly.

TABLE 506.5.1(4)
NUMBER OF CHILLERS

TOTAL CHILLER PLANT CAPACITY	NUMBER OF CHILLERS
≤ 300 tons	1
> 300 tons, < 600 tons	2, sized equally
≥ 600 tons	2 minimum, with chillers added so that no chiller is larger than 800 tons, all sized equally

For SI: 1 ton = 3517 w.

TABLE 506.5.1(5)
WATER CHILLER TYPES

INDIVIDUAL CHILLER PLANT CAPACITY	ELECTRIC CHILLER TYPE	FOSSIL FUEL CHILLER TYPE
≤ 100 tons	Reciprocating	Single-effect absorption, direct fired
> 100 tons, < 300 tons	Screw	Double-effect absorption, direct fired
≥ 300 tons	Centrifugal	Double-effect absorption, direct fired

For SI: 1 ton = 3517 w.

CHAPTER 6

REFERENCED STANDARDS

This chapter lists the standards that are referenced in various sections of this document. The standards are listed herein by the promulgating agency of the standard, the standard identification, the effective date and title, and the section or sections of this document that reference the standard. The application of the referenced standards shall be as specified in Section 107.

AAMA

American Architectural Manufacturers Association
1827 Walden Office Square
Suite 550
Schaumburg, IL 60173-4268

Standard reference number	Title	Referenced in code section number
AAMA/WDMA/CSA 101/I.S.2/A c440—05	Specifications for Windows, Doors and Unit Skylights	402.4.4, 502.4.1

AHRI

Air Conditioning, Heating, and Refrigeration Institute
4100 North Fairfax Drive
Suite 200
Arlington, VA 22203

Standard reference number	Title	Referenced in code section number
210/240—03	Unitary Air-Conditioning and Air-Source Heat Pump Equipment	Table 503.2.3(1), Table 503.2.3(2)
310/380—93	Standard for Packaged Terminal Air-conditioners and Heat Pumps	Table 503.2.3(3)
340/360—2000	Commercial and Industrial Unitary Air-conditioning and Heat Pump Equipment	Table 503.2.3(1), Table 503.2.3(2)
365—02	Commercial and Industrial Unitary Air-conditioning Condensing Units	Table 503.2.3(6)
440—05	Room Fan-coil	503.2.8
550/590—98	Water Chilling Packages Using the Vapor Compression Cycle—with Addenda	Table 503.2.3(7)
560—00	Absorption Water Chilling and Water Heating Packages	Table 503.2.3(7)
840—1998	Unit Ventilators	503.2.8
13256-1 (2004)	Water-source Heat Pumps—Testing and Rating for Performance—Part 1: Water-to-air and Brine-to-air Heat Pumps	Table 503.2.3(2)
1160—2004	Performance Rating of Heat Pump Pool Heaters	Table 504.2

AMCA

Air Movement and Control Association International
30 West University Drive
Arlington Heights, IL 60004-1806

Standard reference number	Title	Referenced in code section number
500D—07	Laboratory Methods for Testing Dampers for Rating	502.4.5

ANSI

American National Standards Institute
25 West 43rd Street
Fourth Floor
New York, NY 10036

Standard reference number	Title	Referenced in code section number
Z21.10.3—01	Gas Water Heaters, Volume III - Storage Water Heaters with Input Ratings Above 75,000 Btu per Hour, Circulating Tank and Instantaneous—with Addenda Z21.10.3a-2003 and Z21.10.3b-2004	Table 504.2
Z21.13—04	Gas-fired Low Pressure Steam and Hot Water Boilers	Table 503.2.3(5)
Z21.47—03	Gas-fired Central Furnaces	Table 503.2.3(4)
Z83.8—02	Gas Unit Heaters and Gas-Fired Duct Furnaces—with Addendum Z83.8a-2003	Table 503.2.3(4)

ASHRAE

American Society of Heating, Refrigerating and Air-Conditioning Engineers, Inc.
1791 Tullie Circle, NE
Atlanta, GA 30329-2305

Standard reference number	Title	Referenced in code section number
119—88 (RA 2004)	Air Leakage Performance for Detached Single-family Residential Buildings	Table 405.5.2(1)
140—2007	Standard Method of Test for the Evaluation of Building Energy Analysis Computer Programs	506.6.1
146—1998	Testing and Rating Pool Heaters	Table 504.2
ANSI/ASHRAE/ACCA Standard 183—2007	Peak Cooling and Heating Load Calculations in Buildings Except Low-rise Residential Buildings	503.2.1
13256-1 (2005)	Water-source Heat Pumps—Testing and Rating for Performance—Part 1: Water-to-air and Brine-to-air Heat Pumps (ANSI/ASHRAE/IESNA 90.1-2004)	Table 503.2.3(2)
90.1—2007	Energy Standard for Buildings Except Low-rise Residential Buildings (ANSI/ASHRAE/IESNA 90.1-2007)	501.1, 501.2, 502.1.1, Table 502.2(2)
ASHRAE—2005	ASHRAE Handbook of Fundamentals	402.1.4, Table 405.5.2(1)
ASHRAE—2004	ASHRAE HVAC Systems and Equipment Handbook-2004	503.2.1

ASME

American Society of Mechanical Engineers
Three Park Avenue
New York, NY 10016-5990

Standard reference number	Title	Referenced in code section number
PTC 4.1 - 1964 (Reaffirmed 1991)	Steam Generating Units	Table 503.2.3(5)

ASTM

ASTM International
100 Barr Harbor Drive
West Conshohocken, PA 19428-2859

Standard reference number	Title	Referenced in code section number
C 90—06b	Specification for Load-bearing Concrete Masonry Units	Table 502.2(1)
E 283—04	Test Method for Determining the Rate of Air Leakage Through Exterior Windows, Curtain Walls and Doors Under Specified Pressure Differences Across the Specimen	402.4.5, 502.4.2, 502.4.8

CSA

Canadian Standards Association
5060 Spectrum Way
Mississauga, Ontario, Canada L4W 5N6

Standard reference number	Title	Referenced in code section number
101/I.S.2/A440—08	Specifications for Windows, Doors and Unit Skylights	402.4.4, 502.4.1

DOE

U.S. Department of Energy
c/o Superintendent of Documents
U.S. Government Printing Office
Washington, DC 20402-9325

Standard reference number	Title	Referenced in code section number
10 CFR Part 430, Subpart B, Appendix E (1998)	Uniform Test Method for Measuring the Energy Consumption of Water Heaters	Table 504.2
10 CFR Part 430, Subpart B, Appendix N (1998)	Uniform Test Method for Measuring the Energy Consumption of Furnaces and Boilers	Table 503.2.3(4), Table 503.2.3(5)

DOE—continued

10 CFR Part 431, Subpart E 2004	Test Procedures and Efficiency Standards for Commercial Packaged Boilers	Table 503.2.3(6)
DOE/EIA—0376 (Current Edition)	State Energy Prices and Expenditure Report	405.3, 506.2

ICC

International Code Council, Inc.
500 New Jersey Avenue, NW
6th Floor
Washington, DC 20001

Standard reference number	Title	Referenced in code section number
IBC—09	International Building Code®	201.3, 303.2, 402.2.9
IFC—09	International Fire Code®	201.3
IFGC—09	International Fuel Gas Code®	201.3
IMC—09	International Mechanical Code®	503.2.5, 503.2.6, 503.2.7.1, 503.2.7.1.1, 503.2.7.1.2, 503.2.9.1, 503.3.1, 503.4.5
IPC—09	International Plumbing Code®	201.3
IRC—09	International Residential Code®	201.3, 403.2.2, 403.6, 405.6.1, Table 405.5.2(1)

IESNA

Illuminating Engineering Society of North America
120 Wall Street, 17th Floor
New York, NY 10005-4001

Standard reference number	Title	Referenced in code section number
90.1—2007	Energy Standard for Buildings Except Low-rise Residential Buildings	501.1, 501.2, 502.1.1, Table 502.2(2)

NFRC

National Fenestration Rating Council, Inc.
6305 Ivy Lane, Suite 140
Greenbelt, MD 20770

Standard reference number	Title	Referenced in code section number
100—04	Procedure for Determining Fenestration Product U-factors—Second Edition	303.1.3
200—04	Procedure for Determining Fenestration Product Solar Heat Gain Coefficients and Visible Transmittance at Normal Incidence—Second Edition	303.1.3
400—04	Procedure for Determining Fenestration Product Air Leakage—Second Edition	402.4.2, 502.4.1

SMACNA

Sheet Metal and Air Conditioning Contractors National Association, Inc.
4021 Lafayette Center Drive
Chantilly, VA 20151-1209

Standard reference number	Title	Referenced in code section number
SMACNA—85	HVAC Air Duct Leakage Test Manual	503.2.7.1.3

UL

Underwriters Laboratories Inc.
333 Pfingsten Road
Northbrook, IL 60062-2096

Standard reference number	Title	Referenced in code section number
727—06	Oil-fired Central Furnaces	Table 503.2.3(4)
731—95	Oil-fired Unit Heaters—with Revisions through February 2006	Table 503.2.3(4)

US—FTC

United States - Federal Trade Commission
600 Pennsylvania Avenue NW
Washington, DC 20580

Standard reference number	Title	Referenced in code section number
CFR Title 16	R-value Rule	303.1.4

WDMA

Window and Door Manufacturers Association
1400 East Touhy Avenue, Suite 470
Des Plaines, IL 60018

Standard reference number	Title	Referenced in code section number
AAMA/WDMA/CSA 101/I.S.2/A440—08	Specifications for Windows, Doors and Unit Skylights	402.4.4, 502.4.1

INDEX

ANSI/ASHRAE/IESNA Standard 90.1-2007
(Supersedes ANSI/ASHRAE/IESNA Standard 90.1-2004)
Includes ANSI/ASHRAE/IESNA Addenda listed in Appendix F

ASHRAE STANDARD

Energy Standard for Buildings Except Low-Rise Residential Buildings

I-P Edition

See Appendix F for approval dates by the ASHRAE Standards Committee, the ASHRAE Board of Directors, the IESNA Board of Directors, and the American National Standards Institute.

This standard is under continuous maintenance by a Standing Standard Project Committee (SSPC) for which the Standards Committee has established a documented program for regular publication of addenda or revisions, including procedures for timely, documented, consensus action on requests for change to any part of the standard. The change submittal form, instructions, and deadlines may be obtained in electronic form from the ASHRAE Web site, http://www.ashrae.org, or in paper form from the Manager of Standards. The latest edition of an ASHRAE Standard may be purchased from ASHRAE Customer Service, 1791 Tullie Circle, NE, Atlanta, GA 30329-2305. E-mail: orders@ashrae.org. Fax: 404-321-5478. Telephone: 404-636-8400 (worldwide), or toll free 1-800-527-4723 (for orders in US and Canada).

ISSN 1041-2336

Jointly sponsored by

Illuminating Engineering Society of North America
www.iesna.org

www.ansi.org

American Society of Heating, Refrigerating and Air-Conditioning Engineers, Inc.
1791 Tullie Circle NE, Atlanta, GA 30329
www.ashrae.org

ASHRAE Standing Standard Project Committee 90.1
Cognizant TC: TC 7.6, Systems Energy Utilization
SPLS Liaisons: Hugh F. Crowther and Ross Montgomery
ASHRAE Staff Liaison: Steven C. Ferguson
IESNA Liaison: Rita M. Harrold

Jerry W. White, Jr., *Chair**
James M. Calm, *Vice-Chair**
Michael Schwedler, *Vice-Chair**
Carol E. Marriott, *Vice-Chair*
Wagdy A.Y. Anis*
Anthony Arbore
Karim Amrane*
William P. Bahnfleth*
Peter A. Baselici*
Van D. Baxter*
Denise M. Beach
Donald L. Beaty*
Valerie L. Block
Donald M. Brundage*
Ken Brendan*
Ernest A. Conrad
Charles C. Cottrell*
Roy Crane*
Joseph J. Deringer*
Keith I. Emerson*
Drake H. Erbe*
Thomas A. Farkas
Charles R. Foster, III
Allan Fraser*
James A. Garrigus*

Jason J. Glazer*
S. Pekka Hakkarainen*
Katherine G. Hammack*
Susanna S. Hanson
Richard V. Heinisch*
Ned B. Heminger
Randall T. Higa*
John F. Hogan*
William G. Holy*
Hyman M. Kaplan*
Larry Kouma*
Ronald D. Kurtz*
Samantha H. LaFleur
Michael D. Lane*
Dean E. Lewis
Richard Lord
Kenneth Luther*
Ronald Majette*
Itzhak H. Maor*
R. Christopher Mathis
Merle F. McBride
Michael W. Mehl
Harry P. Misuriello
Louis J. Molinini
John Montgomery*

Frank T. Morrison
Frank Myers*
Ronald G. Nickson
Jim A. Ranfone
Eric E. Richman*
Michael I. Rosenberg*
Steven Rosenstock*
Robert D. Ross*
David A. Schaaf, Jr.*
Leonard C. Sciarra*
Bipin Vadilal Shah*
Stephen V. Skalko*
Frank A. Stanonik*
Dennis Sczomak*
Joseph K. Ting
Cedric S. Trueman*
Martha G. VanGeem
Carl Wagus*
McHenry Wallace, Jr.*
Richard D. Watson*
David Weitz*
Robin Wilson
Michael W. Woodford
Donald R. Wulfinghoff
Stanley W. Zajac

**Denotes members of voting status when this standard was approved for publication.*

SPECIAL NOTE

This American National Standard (ANS) is a national voluntary consensus standard developed under the auspices of the American Society of Heating, Refrigerating and Air-Conditioning Engineers (ASHRAE). Consensus is defined by the American National Standards Institute (ANSI), of which ASHRAE is a member and which has approved this standard as an ANS, as "substantial agreement reached by directly and materially affected interest categories. This signifies the concurrence of more than a simple majority, but not necessarily unanimity. Consensus requires that all views and objections be considered, and that an effort be made toward their resolution." Compliance with this standard is voluntary until and unless a legal jurisdiction makes compliance mandatory through legislation.

ASHRAE obtains consensus through participation of its national and international members, associated societies, and public review.

ASHRAE Standards are prepared by a Project Committee appointed specifically for the purpose of writing the Standard. The Project Committee Chair and Vice-Chair must be members of ASHRAE; while other committee members may or may not be ASHRAE members, all must be technically qualified in the subject area of the Standard. Every effort is made to balance the concerned interests on all Project Committees.

The Manager of Standards of ASHRAE should be contacted for:

a. interpretation of the contents of this Standard,
b. participation in the next review of the Standard,
c. offering constructive criticism for improving the Standard, or
d. permission to reprint portions of the Standard.

DISCLAIMER

ASHRAE uses its best efforts to promulgate Standards and Guidelines for the benefit of the public in light of available information and accepted industry practices. However, ASHRAE does not guarantee, certify, or assure the safety or performance of any products, components, or systems tested, installed, or operated in accordance with ASHRAE's Standards or Guidelines or that any tests conducted under its Standards or Guidelines will be nonhazardous or free from risk.

ASHRAE INDUSTRIAL ADVERTISING POLICY ON STANDARDS

ASHRAE Standards and Guidelines are established to assist industry and the public by offering a uniform method of testing for rating purposes, by suggesting safe practices in designing and installing equipment, by providing proper definitions of this equipment, and by providing other information that may serve to guide the industry. The creation of ASHRAE Standards and Guidelines is determined by the need for them, and conformance to them is completely voluntary.

In referring to this Standard or Guideline and in marking of equipment and in advertising, no claim shall be made, either stated or implied, that the product has been approved by ASHRAE.

CONTENTS

ANSI/ASHRAE/IESNA Standard 90.1-2007
Energy Standard for Buildings Except Low-Rise Residential Buildings (I-P Edition)

NOTE

When addenda, interpretations, or errata to this standard have been approved, they can be downloaded free of charge from the ASHRAE Web site at http://www.ashrae.org.

1791 Tullie Circle NE
Atlanta, GA 30329
www.ashrae.org

(This foreword is not part of this standard. It is merely informative and does not contain requirements necessary for conformance to the standard. It has not been processed according to the ANSI requirements for a standard and may contain material that has not been subject to public review or a consensus process. Unresolved objectors on informative material are not offered the right to appeal at ASHRAE or ANSI.)

FOREWORD

The original Standard 90 was published in 1975 and revised editions were published in 1980, 1989, and 1999 using the ANSI and ASHRAE periodic maintenance procedures. Based upon these procedures, the entire standard was publicly reviewed and published in its entirety each time. As technology and energy prices began changing more rapidly, however, the ASHRAE Board of Directors voted in 1999 to place the standard on continuous maintenance, permitting the standard to be updated several times each year through the publication of approved addenda to the standard. Starting with the 2001 edition, the standard is now published in its entirety in the fall of every third year. This schedule allows the standard to be submitted and proposed by the deadline for inclusion or reference in model building and energy codes. All approved addenda and errata will be included in the new edition every three years. This procedure allows users to have some certainty about when new editions will be published.

This 2007 edition of the standard has several new features and includes changes resulting from the continuous maintenance proposals from the public. The committee welcomes suggestions for improving the standard. Users of the standard are encouraged and invited to use the continuous maintenance procedure to suggest changes. A form for submittal of a proposed change is included in the back of this standard. The committee will take formal action on every proposal received.

The project committee is continually considering changes and proposing addenda for public review. When addenda are approved, notices will be published on the ASHRAE and IESNA Web sites. Users are encouraged to sign up for the free ASHRAE and IESNA Internet Listserv for this standard to receive notice of all public reviews and approved and published addenda and errata.

This edition corrects all known typographical errors in the 2004 standard. It also includes the content of 31 addenda that were processed by the committee and approved by the ASHRAE and IESNA Boards of Directors. For brief descriptions and the publication dates of the addenda to 90.1-2004, see Appendix F.

1. PURPOSE

The purpose of this standard is to provide minimum requirements for the energy-efficient design of buildings except low-rise residential buildings.

2. SCOPE

2.1 This standard provides:

a. minimum energy-efficient requirements for the design and construction of:

   1. new buildings and their systems
   2. new portions of buildings and their systems
   3. new systems and equipment in existing buildings

b. criteria for determining compliance with these requirements.

2.2 The provisions of this standard apply to:

a. the envelope of buildings, provided that the enclosed spaces are

   1. heated by a heating system whose output capacity is greater than or equal to 3.4 Btu/h·ft^2 or
   2. cooled by a cooling system whose sensible output capacity is greater than or equal to 5 Btu/h·ft^2, and

b. the following systems and equipment used in conjunction with buildings:

   1. heating, ventilating, and air conditioning,
   2. service water heating,
   3. electric power distribution and metering provisions,
   4. electric motors and belt drives, and
   5. lighting.

2.3 The provisions of this standard do not apply to

a. single-family houses, multi-family structures of three stories or fewer above grade, manufactured houses (mobile homes), and manufactured houses (modular),
b. buildings that do not use either electricity or fossil fuel, or
c. equipment and portions of building systems that use energy primarily to provide for industrial, manufacturing, or commercial processes.

2.4 Where specifically noted in this standard, certain other buildings or elements of buildings shall be exempt.

2.5 This standard shall not be used to circumvent any safety, health, or environmental requirements.

3. DEFINITIONS, ABBREVIATIONS, AND ACRONYMS

3.1 General. Certain terms, abbreviations, and acronyms are defined in this section for the purposes of this standard. These definitions are applicable to all sections of this standard. Terms that are not defined shall have their ordinarily accepted meanings within the context in which they are used. Ordinarily accepted meanings shall be based upon American standard English language usage as documented in an unabridged dictionary accepted by the *adopting authority.*

3.2 Definitions

above-grade wall: see *wall.*

access hatch: see *door.*

addition: an extension or increase in floor area or height of a building outside of the existing building envelope.

adopting authority: the agency or agent that adopts this standard.

alteration: a replacement or addition to a building or its systems and equipment; routine maintenance, repair, and service or a change in the building's use classification or category shall not constitute an alteration.

annual fuel utilization efficiency (AFUE): an efficiency descriptor of the ratio of annual output energy to annual input energy as developed in accordance with the requirements of U.S. Department of Energy (DOE) 10 CFR Part 430.

astronomical time switch: a device that turns the lighting on at a time relative to sunset and off at a time relative to sunrise, accounting for geographic location and day of year.

attic and other roofs: see *roof.*

authority having jurisdiction: the agency or agent responsible for enforcing this standard.

automatic: self-acting, operating by its own mechanism when actuated by some nonmanual influence, such as a change in current strength, pressure, temperature, or mechanical configuration. (See *manual.*)

automatic control device: a device capable of automatically turning loads off and on without manual intervention.

balancing, air system: adjusting airflow rates through air distribution system devices, such as fans and diffusers, by manually adjusting the position of dampers, splitter vanes, extractors, etc., or by using automatic control devices, such as constant air volume or variable-air-volume (VAV) boxes.

balancing, hydronic system: adjusting water flow rates through hydronic distribution system devices, such as pumps and coils, by manually adjusting the position valves or by using automatic control devices, such as automatic flow control valves.

ballast: a device used in conjunction with an electric-discharge lamp to cause the lamp to start and operate under the proper circuit conditions of voltage, current, wave form, electrode heat, etc.

> ***ballast, electronic:*** a ballast constructed using electronic circuitry.
>
> ***ballast, hybrid:*** a ballast constructed using a combination of magnetic core and insulated wire winding and electronic circuitry.
>
> ***ballast, magnetic:*** a ballast constructed with magnetic core and a winding of insulated wire.

baseline building design: a computer representation of a hypothetical design based on the proposed building project. This representation is used as the basis for calculating the *baseline building performance* for rating above-standard design.

baseline building performance: the annual energy cost for a building design intended for use as a baseline for rating above-standard design.

below-grade wall: see *wall.*

boiler: a self-contained low-pressure appliance for supplying steam or hot water.

boiler, packaged: a boiler that is shipped complete with heating equipment, mechanical draft equipment, and automatic controls; usually shipped in one or more sections. A packaged boiler includes factory-built boilers manufactured as a unit or system, disassembled for shipment, and reassembled at the site.

branch circuit: the circuit conductors between the final overcurrent device protecting the circuit and the outlet(s); the final wiring run to the load.

budget building design: a computer representation of a hypothetical design based on the actual proposed building design. This representation is used as the basis for calculating the *energy cost budget.*

building: a structure wholly or partially enclosed within exterior walls, or within exterior and party walls, and a roof, affording shelter to persons, animals, or property.

building entrance: any doorway, set of doors, turnstile, vestibule, or other form of portal that is ordinarily used to gain access to the building by its users and occupants.

building envelope: the exterior plus the semi-exterior portions of a building. For the purposes of determining building envelope requirements, the classifications are defined as follows:

> ***building envelope, exterior:*** the elements of a building that separate conditioned spaces from the exterior.
>
> ***building envelope, semi-exterior:*** the elements of a building that separate conditioned space from unconditioned space or that enclose semiheated spaces through which thermal energy may be transferred to or from the exterior, or to or from unconditioned spaces, or to or from conditioned spaces.

building exit: any doorway, set of doors, or other form of portal that is ordinarily used only for emergency egress or convenience exit.

building grounds lighting: lighting provided through a building's electrical service for parking lot, site, roadway, pedestrian pathway, loading dock, or security applications.

building material: any element of the building envelope through which heat flows and that is included in the component U-factor calculations other than air films and insulation.

building official: the officer or other designated representative authorized to act on behalf of the *authority having jurisdiction.*

C-factor (thermal conductance): time rate of steady-state heat flow through unit area of a material or construction, induced by a unit temperature difference between the body surfaces. Units of *C* are Btu/h·ft^2·°F. Note that the C-factor does not include soil or air films.

circuit breaker: a device designed to open and close a circuit by nonautomatic means and to open the circuit automatically at a predetermined overcurrent without damage to itself when properly applied within its rating.

class of construction: for the building envelope, a subcategory of roof, above-grade wall, below-grade wall, floor, slab-on-grade floor, opaque door, vertical fenestration, or skylight. (See *roof*, *wall*, *floor*, *slab-on-grade floor*, *door*, and *fenestration*.)

clerestory: that part of a building that rises clear of the roofs or other parts and whose walls contain windows for lighting the interior.

code official: see *building official*.

coefficient of performance (COP)—cooling: the ratio of the rate of heat removal to the rate of energy input, in consistent units, for a complete refrigerating system or some specific portion of that system under designated operating conditions.

coefficient of performance (COP), heat pump—heating: the ratio of the rate of heat delivered to the rate of energy input, in consistent units, for a complete heat pump system, including the compressor and, if applicable, auxiliary heat, under designated operating conditions.

conditioned floor area: see *floor area*.

conditioned space: see *space*.

conductance: see *thermal conductance*.

continuous insulation (c.i.): insulation that is continuous across all structural members without thermal bridges other than fasteners and service openings. It is installed on the interior or exterior or is integral to any opaque surface of the building envelope.

control: to regulate the operation of equipment.

control device: a specialized device used to regulate the operation of equipment.

construction: the fabrication and erection of a new building or any addition to or alteration of an existing building.

construction documents: drawings and specifications used to construct a building, building systems, or portions thereof.

cool down: reduction of space temperature down to occupied setpoint after a period of shutdown or setup.

cooled space: see *space*.

cooling degree-day: see *degree-day*.

cooling design temperature: the outdoor dry-bulb temperature equal to the temperature that is exceeded by 1% of the number of hours during a typical weather year.

cooling design wet-bulb temperature: the outdoor wet-bulb temperature for sizing cooling systems and evaporative heat rejection systems such as cooling towers.

dead band: the range of values within which a sensed variable can vary without initiating a change in the controlled process.

decorative lighting: see *lighting, decorative*.

degree-day: the difference in temperature between the outdoor mean temperature over a 24-hour period and a given base temperature. For the purposes of determining building envelope requirements, the classifications are defined as follows:

> ***cooling degree-day base 50°F (CDD50):*** for any one day, when the mean temperature is more than 50°F, there are as many degree-days as degrees Fahrenheit temperature difference between the mean temperature for the day and 50°F. Annual cooling degree-days (CDDs) are the sum of the degree-days over a calendar year.
>
> ***heating degree-day base 65°F (HDD65):*** for any one day, when the mean temperature is less than 65°F, there are as many degree-days as degrees Fahrenheit temperature difference between the mean temperature for the day and 65°F. Annual heating degree-days (HDDs) are the sum of the degree-days over a calendar year.

demand: the highest amount of power (average Btu/h over an interval) recorded for a building or facility in a selected time frame.

demand control ventilation (DCV): a ventilation system capability that provides for the automatic reduction of outdoor air intake below design rates when the actual occupancy of spaces served by the system is less than design occupancy.

design capacity: output capacity of a system or piece of equipment at design conditions.

design conditions: specified environmental conditions, such as temperature and light intensity, required to be produced and maintained by a system and under which the system must operate.

design energy cost: the annual energy cost calculated for a proposed design.

design professional: an architect or engineer licensed to practice in accordance with applicable state licensing laws.

direct digital control (DDC): a type of control where controlled and monitored analog or binary data (e.g., temperature, contact closures) are converted to digital format for manipulation and calculations by a digital computer or microprocessor, then converted back to analog or binary form to control physical devices.

disconnect: a device or group of devices or other means by which the conductors of a circuit can be disconnected from their source of supply.

distribution system: conveying means, such as ducts, pipes, and wires, to bring substances or energy from a source to the point of use. The distribution system includes such auxiliary equipment as fans, pumps, and *transformers*.

door: all operable opening areas (which are not fenestration) in the building envelope, including swinging and roll-up doors, fire doors, and access hatches. Doors that are more than one-half glass are considered fenestration. (See *fenestration*.) For the purposes of determining building envelope requirements, the classifications are defined as follows:

nonswinging: roll-up, sliding, and all other doors that are not swinging doors.

swinging: all operable opaque panels with hinges on one side and opaque revolving doors.

door area: total area of the door measured using the rough opening and including the door slab and the frame. (See *fenestration area.*)

dwelling unit: a single unit providing complete independent living facilities for one or more persons, including permanent provisions for living, sleeping, eating, cooking, and sanitation.

economizer, air: a duct and damper arrangement and automatic control system that together allow a cooling system to supply *outdoor air* to reduce or eliminate the need for mechanical cooling during mild or cold weather.

economizer, water: a system by which the supply air of a cooling system is cooled indirectly with water that is itself cooled by heat or mass transfer to the environment without the use of mechanical cooling.

efficacy (of a lamp): the ratio of the total luminous output of a lamp to the total power input to the lamp; typically expressed in lumens per watt.

efficiency: performance at specified rating conditions.

emittance: the ratio of the radiant heat flux emitted by a specimen to that emitted by a blackbody at the same temperature and under the same conditions.

enclosed space: a volume substantially surrounded by solid surfaces such as walls, floors, roofs, and openable devices such as doors and operable windows.

energy: the capacity for doing work. It takes a number of forms that may be transformed from one into another such as thermal (heat), mechanical (work), electrical, and chemical. Customary measurement units are British thermal units (Btu).

energy cost budget: the annual energy cost for the budget building design intended for use in determining minimum compliance with this standard.

energy efficiency ratio (EER): the ratio of net cooling capacity in Btu/h to total rate of electric input in watts under designated operating conditions. (See *coefficient of performance [COP]—cooling.*)

energy factor (EF): a measure of water heater overall efficiency.

envelope performance factor: the trade-off value for the building envelope performance compliance option calculated using the procedures specified in Section 5. For the purposes of determining building envelope requirements, the classifications are defined as follows:

base envelope performance factor: the building envelope performance factor for the base design.

proposed envelope performance factor: the building envelope performance factor for the proposed design.

equipment: devices for comfort conditioning, electric power, lighting, transportation, or service water heating including, but not limited to, furnaces, boilers, air conditioners, heat pumps, chillers, water heaters, lamps, luminaires, ballasts, elevators, escalators, or other devices or installations.

existing building: a building or portion thereof that was previously occupied or approved for occupancy by the *authority having jurisdiction.*

existing equipment: equipment previously installed in an existing building.

existing system: a system or systems previously installed in an existing building.

exterior building envelope: see *building envelope.*

exterior lighting power allowance: see *lighting power allowance.*

eye adaptation: the process by which the retina becomes accustomed to more or less light than it was exposed to during an immediately preceding period. It results in a change in the sensitivity to light.

F-factor: the perimeter heat loss factor for slab-on-grade floors, expressed in Btu/h·ft·°F.

facade area: area of the facade, including overhanging soffits, cornices, and protruding columns, measured in elevation in a vertical plane parallel to the plane of the face of the building. Nonhorizontal roof surfaces shall be included in the calculation of vertical facade area by measuring the area in a plane parallel to the surface.

fan brake horsepower: the horsepower delivered to the fan's shaft. Brake horsepower (bhp) does not include the mechanical drive losses (belts, gears, etc.).

fan system bhp: the sum of the fan brake horsepower (bhp) of all fans that are required to operate at fan system design conditions to supply air from the heating or cooling source to the conditioned space(s) and return it to the source or exhaust it to the outdoors.

fan system design conditions: operating conditions that can be expected to occur during normal system operation that result in the highest supply airflow rate to conditioned spaces served by the system.

fan system motor nameplate horsepower: the sum of the motor nameplate horsepower (hp) of all fans that are required to operate at design conditions to supply air from the heating or cooling source to the conditioned space(s) and return it to the source or exhaust it to the outdoors.

feeder conductors: the wires that connect the service equipment to the branch circuit breaker panels.

fenestration: all areas (including the frames) in the building envelope that let in light, including windows, plastic panels, clerestories, skylights, doors that are more than one-half glass, and glass block walls. (See *building envelope* and *door.*)

skylight: a fenestration surface having a slope of less than 60 degrees from the horizontal plane. Other fenestration, even if mounted on the roof of a building, is considered *vertical fenestration*.

vertical fenestration: all fenestration other than *skylights*. Trombe wall assemblies, where glazing is installed within 12 in. of a mass wall, are considered walls, not fenestration.

fenestration area: total area of the fenestration measured using the rough opening and including the glazing, sash, and frame. For doors where the glazed vision area is less than 50% of the door area, the fenestration area is the glazed vision area. For all other doors, the fenestration area is the door area. (See *door area*.)

fenestration, vertical: see *fenestration* and *skylight*.

fixture: the component of a luminaire that houses the lamp or lamps, positions the lamp, shields it from view, and distributes the light. The fixture also provides for connection to the power supply, which may require the use of a ballast.

floor, envelope: that lower portion of the building envelope, including opaque area and fenestration, that has conditioned or semiheated space above and is horizontal or tilted at an angle of less than 60 degrees from horizontal but excluding slab-on-grade floors. For the purposes of determining building envelope requirements, the classifications are defined as follows:

mass floor: a floor with a heat capacity that exceeds (1) 7 $Btu/ft^2 \cdot °F$ or (2) 5 $Btu/ft^2 \cdot °F$ provided that the floor has a material unit mass not greater than 120 lb/ft^3.

steel-joist floor: a floor that (1) is not a mass floor and (2) that has steel joist members supported by structural members.

wood-framed and other floors: all other floor types, including wood joist floors.

(See *building envelope*, *fenestration*, *opaque area*, and *slab-on-grade floor*).

floor area, gross: the sum of the floor areas of the spaces within the building, including basements, mezzanine and intermediate-floored tiers, and penthouses with a headroom height of 7.5 ft or greater. It is measured from the exterior faces of exterior walls or from the centerline of walls separating buildings, but excluding covered walkways, open roofed-over areas, porches and similar spaces, pipe trenches, exterior terraces or steps, chimneys, roof overhangs, and similar features.

gross building envelope floor area: the gross floor area of the building envelope, but excluding slab-on-grade floors.

gross conditioned floor area: the gross floor area of conditioned spaces.

gross lighted floor area: the gross floor area of lighted spaces.

gross semiheated floor area: the gross floor area of semiheated spaces.

(See *building envelope*, *floor*, *slab-on-grade floor*, and *space*.)

flue damper: a device in the flue outlet or in the inlet of or upstream of the draft control device of an individual, automatically operated, fossil fuel-fired appliance that is designed to automatically open the flue outlet during appliance operation and to automatically close the flue outlet when the appliance is in a standby condition.

fossil fuel: fuel derived from a hydrocarbon deposit such as petroleum, coal, or natural gas derived from living matter of a previous geologic time.

fuel: a material that may be used to produce heat or generate power by combustion.

general lighting: see *lighting, general*.

generally accepted engineering standard: a specification, rule, guide, or procedure in the field of engineering, or related thereto, recognized and accepted as authoritative.

grade: the finished ground level adjoining a building at all exterior walls.

gross lighted area (GLA): see *floor area, gross*: *gross lighted floor area*.

gross roof area: see *roof area, gross*.

gross wall area: see *wall area, gross*.

heat capacity (HC): the amount of heat necessary to raise the temperature of a given mass 1°F. Numerically, the *HC* per unit area of surface ($Btu/ft^2 \cdot °F$) is the sum of the products of the mass per unit area of each individual material in the roof, wall, or floor surface multiplied by its individual specific heat.

heated space: see *space*.

heat trace: a heating system where the externally applied heat source follows (traces) the object to be heated, e.g., water piping.

heating design temperature: the outdoor dry-bulb temperature equal to the temperature that is exceeded at least 99.6% of the number of hours during a typical weather year.

heating degree-day: see *degree-day*.

heating seasonal performance factor (HSPF): the total heating output of a heat pump during its normal annual usage period for heating (in Btu) divided by the total electric energy input during the same period.

high-frequency electronic ballast: ballasts that operate at a frequency greater than 20 kHz.

historic: a building or space that has been specifically designated as historically significant by the adopting authority or is listed in The National Register of Historic Places or has been determined to be eligible for such listing by the US Secretary of the Interior.

hot-water supply boiler: a boiler used to heat water for purposes other than space heating.

humidistat: an automatic control device used to maintain humidity at a fixed or adjustable setpoint.

HVAC system: the equipment, distribution systems, and terminals that provide, either collectively or individually, the processes of heating, ventilating, or air conditioning to a building or portion of a building.

indirectly conditioned space: see *space.*

infiltration: the uncontrolled inward air leakage through cracks and crevices in any building element and around windows and doors of a building caused by pressure differences across these elements due to factors such as wind, inside and outside temperature differences (stack effect), and imbalance between supply and exhaust air systems.

installed interior lighting power: the power in watts of all permanently installed general, task, and furniture lighting systems and luminaires.

integrated part-load value (IPLV): a single-number figure of merit based on part-load EER, COP, or kW/ton expressing part-load efficiency for air-conditioning and heat pump equipment on the basis of weighted operation at various load capacities for the equipment.

interior lighting power allowance: see *lighting power allowance.*

isolation devices: devices that isolate HVAC zones so that they can be operated independently of one another. Isolation devices include, but are not limited to, separate systems, isolation dampers, and controls providing shutoff at terminal boxes.

joist, steel: any structural steel member of a building or structure made of hot-rolled or cold-rolled solid or open-web sections.

kilovolt-ampere (kVA): where the term *kilovolt-ampere* (kVA) is used in this standard, it is the product of the line current (amperes) times the nominal system voltage (kilovolts) times 1.732 for three-phase currents. For single-phase applications, kVA is the product of the line current (amperes) times the nominal system voltage (kilovolts).

kilowatt (kW): the basic unit of electric power, equal to 1000 W.

labeled: equipment or materials to which a symbol or other identifying mark has been attached by the manufacturer indicating compliance with specified standards or performance in a specified manner.

lamp: a generic term for a man-made light source often called a *bulb* or *tube*.

> ***compact fluorescent lamp:*** a fluorescent lamp of a small compact shape, with a single base that provides the entire mechanical support function.
>
> ***fluorescent lamp:*** a low-pressure electric discharge lamp in which a phosphor coating transforms some of the ultraviolet energy generated by the discharge into light.
>
> ***general service lamp:*** a class of incandescent lamps that provide light in virtually all directions. *General service lamps* are typically characterized by bulb shapes such as A, standard; S, straight side; F, flame; G, globe; and PS, pear straight.
>
> ***high-intensity discharge (HID) lamp:*** an electric discharge lamp in which light is produced when an electric arc is discharged through a vaporized metal such as mercury or sodium. Some HID lamps may also have a phosphor coating that contributes to the light produced or enhances the light color.
>
> ***incandescent lamp:*** a lamp in which light is produced by a filament heated to incandescence by an electric current.
>
> ***reflector lamp:*** a class of incandescent lamps that have an internal reflector to direct the light. Reflector lamps are typically characterized by reflective characteristics such as R, reflector; ER, ellipsoidal reflector; PAR, parabolic aluminized reflector; MR, mirrorized reflector; and others.

lighting, decorative: lighting that is purely ornamental and installed for aesthetic effect. Decorative lighting shall not include *general lighting.*

lighting, general: lighting that provides a substantially uniform level of illumination throughout an area. General lighting shall not include *decorative lighting* or lighting that provides a dissimilar level of illumination to serve a specialized application or feature within such area.

lighting system: a group of luminaires circuited or controlled to perform a specific function.

lighting power allowance:

> ***interior lighting power allowance:*** the maximum lighting power in watts allowed for the interior of a building.
>
> ***exterior lighting power allowance:*** the maximum lighting power in watts allowed for the exterior of a building.

lighting power density (LPD): the maximum lighting power per unit area of a building classification of space function.

low-rise residential buildings: single-family houses, multi-family structures of three stories or fewer above grade, manufactured houses (mobile homes), and manufactured houses (modular).

luminaire: a complete lighting unit consisting of a lamp or lamps together with the housing designed to distribute the light, position and protect the lamps, and connect the lamps to the power supply.

manual (nonautomatic): requiring personal intervention for control. Nonautomatic does not necessarily imply a manual controller, only that personal intervention is necessary. (See *automatic.*)

manufacturer: the company engaged in the original production and assembly of products or equipment or a company that purchases such products and equipment manufactured in accordance with company specifications.

mass floor: see *floor.*

mass wall: see *wall.*

mean temperature: one-half the sum of the minimum daily temperature and maximum daily temperature.

mechanical heating: raising the temperature of a gas or liquid by use of fossil fuel burners, electric resistance heaters, heat pumps, or other systems that require energy to operate.

mechanical cooling: reducing the temperature of a gas or liquid by using vapor compression, absorption, desiccant dehumidification combined with evaporative cooling, or another energy-driven thermodynamic cycle. Indirect or direct evaporative cooling alone is not considered mechanical cooling.

metal building: a complete integrated set of mutually dependent components and assemblies that form a building, which consists of a steel-framed superstructure and metal skin.

metal building roof: see *roof.*

metal building wall: see *wall.*

metering: instruments that measure electric voltage, current, power, etc.

motor power, rated: the rated output power from the motor.

nameplate horsepower: the nominal motor horsepower rating stamped on the motor nameplate.

nameplate rating: the design load operating conditions of a device as shown by the manufacturer on the nameplate or otherwise marked on the device.

nonautomatic: see *manual.*

nonrecirculating system: a domestic or service hot-water distribution system that is not a recirculating system.

nonrenewable energy: energy derived from a fossil fuel source.

nonresidential: all occupancies other than residential. (See *residential.*)

nonstandard part-load value (NPLV): a single-number part-load efficiency figure of merit calculated and referenced to conditions other than IPLV conditions, for units that are not designed to operate at ARI Standard Rating Conditions.

nonswinging door: see *door.*

north-oriented: facing within 45 degrees of true north (northern hemisphere).

occupant sensor: a device that detects the presence or absence of people within an area and causes lighting, equipment, or appliances to be regulated accordingly.

opaque: all areas in the building envelope, except fenestration and building service openings such as vents and grilles. (See *building envelope* and *fenestration.*)

optimum start controls: controls that are designed to automatically adjust the start time of an HVAC system each day with the intention of bringing the space to desired occupied temperature levels immediately before scheduled occupancy.

orientation: the direction an envelope element faces, i.e., the direction of a vector perpendicular to and pointing away from the surface outside of the element.

outdoor (outside) air: air that is outside the building envelope or is taken from outside the building that has not been previously circulated through the building.

overcurrent: any current in excess of the rated current of equipment or the ampacity of a conductor. It may result from overload, short circuit, or ground fault.

packaged terminal air conditioner (PTAC): a factory-selected wall sleeve and separate unencased combination of heating and cooling components, assemblies, or sections. It may include heating capability by hot water, steam, or electricity and is intended for mounting through the wall to serve a single room or zone.

packaged terminal heat pump (PTHP): a PTAC capable of using the refrigerating system in a reverse cycle or heat pump mode to provide heat.

party wall: a fire wall on an interior lot line used or adapted for joint service between two buildings.

Performance Rating Method: a calculation procedure that generates an index of merit for the performance of building designs that substantially exceeds the energy efficiency levels required by this standard.

permanently installed: equipment that is fixed in place and is not portable or movable.

photosensor: a device that detects the presence of visible light, infrared (IR) transmission, and/or ultraviolet (UV) energy.

plenum: a compartment or chamber to which one or more ducts are connected, that forms a part of the air distribution system, and that is not used for occupancy or storage. A plenum often is formed in part or in total by portions of the building.

pool: any structure, basin, or tank containing an artificial body of water for swimming, diving, or recreational bathing. The term includes, but is not limited to, swimming pool, whirlpool, spa, and hot tub.

process energy: energy consumed in support of a manufacturing, industrial, or commercial process other than conditioning spaces and maintaining comfort and amenities for the occupants of a building.

process load: the load on a building resulting from the consumption or release of process energy.

projection factor (PF): the ratio of the horizontal depth of the external shading projection divided by the sum of the height of the fenestration and the distance from the top of the fenestration to the bottom of the farthest point of the external shading projection, in consistent units.

proposed building performance: the annual energy cost calculated for a proposed design.

proposed design: a computer representation of the actual proposed building design or portion thereof used as the basis for calculating the design energy cost.

public facility restroom: a restroom used by the transient public.

pump system power: the sum of the nominal power demand (nameplate horsepower) of motors of all pumps that are required to operate at design conditions to supply fluid from the heating or cooling source to all heat transfer devices (e.g., coils, heat exchanger) and return it to the source.

purchased energy rates: costs for units of energy or power purchased at the building site. These costs may include energy costs as well as costs for power demand as determined by the *adopting authority*.

radiant heating system: a heating system that transfers heat to objects and surfaces within the heated space primarily (greater than 50%) by infrared radiation.

rated lamp wattage: see *lamp wattage, rated.*

rated motor power: see *motor power, rated.*

rated R-value of insulation: the thermal resistance of the insulation alone as specified by the manufacturer in units of $h \cdot ft^2 \cdot °F/Btu$ at a mean temperature of 75°F. Rated R-value refers to the thermal resistance of the added insulation in framing cavities or insulated sheathing only and does not include the thermal resistance of other building materials or air films. (See *thermal resistance*.)

rating authority: the organization or agency that adopts or sanctions use of this rating methodology.

readily accessible: capable of being reached quickly for operation, renewal, or inspection without requiring those to whom ready access is requisite to climb over or remove obstacles or to resort to portable ladders, chairs, etc. In public facilities, accessibility may be limited to certified personnel through locking covers or by placing equipment in locked rooms.

recirculating system: a domestic or service hot-water distribution system that includes a closed circulation circuit designed to maintain usage temperatures in hot-water pipes near terminal devices (e.g., lavatory faucets, shower heads) in order to reduce the time required to obtain hot water when the terminal device valve is opened. The motive force for circulation is either natural (due to water density variations with temperature) or mechanical (recirculation pump).

recooling: lowering the temperature of air that has been previously heated by a mechanical heating system.

record drawings: drawings that record the conditions of the project as constructed. These include any refinements of the construction or bid documents.

reflectance: the ratio of the light reflected by a surface to the light incident upon it.

reheating: raising the temperature of air that has been previously cooled either by mechanical refrigeration or an economizer system.

repair: the reconstruction or renewal of any part of an existing building for the purpose of its maintenance.

resistance, electric: the property of an electric circuit or of any object used as part of an electric circuit that determines for a given circuit the rate at which electric energy is converted into heat or radiant energy and that has a value such that the product of the resistance and the square of the current gives the rate of conversion of energy.

reset: automatic adjustment of the controller setpoint to a higher or lower value.

residential: spaces in buildings used primarily for living and sleeping. Residential spaces include, but are not limited to, dwelling units, hotel/motel guest rooms, dormitories, nursing homes, patient rooms in hospitals, lodging houses, fraternity/sorority houses, hostels, prisons, and fire stations.

roof: the upper portion of the building envelope, including opaque areas and fenestration, that is horizontal or tilted at an angle of less than 60° from horizontal. For the purposes of determining building envelope requirements, the classifications are defined as follows:

> ***attic and other roofs:*** all other roofs, including roofs with insulation entirely below (inside of) the roof structure (i.e., attics, cathedral ceilings, and single-rafter ceilings), roofs with insulation both above and below the roof structure, and roofs without insulation but excluding metal building roofs.
>
> ***metal building roof:*** a roof that is:
>
> 1. constructed with a metal, structural, weathering surface,
> 2. has no ventilated cavity, and
> 3. has the insulation entirely below deck (i.e., does not include composite concrete and metal deck construction nor a roof framing system that is separated from the superstructure by a wood substrate) and whose structure consists of one or more of the following configurations:
>
> a. metal roofing in direct contact with the steel framing members
> b. insulation between the metal roofing and the steel framing members
> c. insulated metal roofing panels installed as described in 1 or 2
>
> ***roof with insulation entirely above deck:*** a roof with all insulation
>
> 1. installed above (outside of) the roof structure and
> 2. continuous (i.e., uninterrupted by framing members).
>
> ***single-rafter roof:*** a subcategory of attic roofs where the roof above and the ceiling below are both attached to the same wood rafter and where insulation is located in the space between these wood rafters.

roof area, gross: the area of the roof measured from the exterior faces of walls or from the centerline of party walls. (See *roof* and *wall.*)

room air conditioner: an encased assembly designed as a unit to be mounted in a window or through a wall or as a console. It is designed primarily to provide direct delivery of conditioned air to an enclosed space, room, or zone. It includes a prime source of refrigeration for cooling and dehumidification and a means for circulating and cleaning air. It may also include a means for ventilating and heating.

room cavity ratio (RCR): a factor that characterizes room configuration as a ratio between the walls and ceiling and is based upon room dimensions.

seasonal coefficient of performance—cooling ($SCOP_C$): the total cooling output of an air conditioner during its normal annual usage period for cooling divided by the total electric energy input during the same period in consistent units (analogous to the SEER but in I-P or other consistent units).

seasonal coefficient of performance—heating ($SCOP_H$): the total heating output of a heat pump during its normal annual usage period for heating divided by the total electric energy input during the same period in consistent units (analogous to the HSPF but in I-P or other consistent units).

seasonal energy efficiency ratio (SEER): the total cooling output of an air conditioner during its normal annual usage period for cooling (in Btu) divided by the total electric energy input during the same period (in Wh).

semi-exterior building envelope: see *building envelope.*

semiheated floor area: see *floor area.*

semiheated space: see *space.*

service: the equipment for delivering energy from the supply or distribution system to the premises served.

service agency: an agency capable of providing calibration, testing, or manufacture of equipment, instrumentation, metering, or control apparatus, such as a contractor, laboratory, or manufacturer.

service equipment: the necessary equipment, usually consisting of a circuit breaker or switch and fuses and accessories, located near the point of entrance of supply conductors to a building or other structure (or an otherwise defined area) and intended to constitute the main control and means of cutoff of the supply. Service equipment may consist of circuit breakers or fused switches provided to disconnect all under-grounded conductors in a building or other structure from the service-entrance conductors.

service water heating: heating water for domestic or commercial purposes other than space heating and process requirements.

setback: reduction of heating (by reducing the setpoint) or cooling (by increasing the setpoint) during hours when a building is unoccupied or during periods when lesser demand is acceptable.

setpoint: point at which the desired temperature (°F) of the heated or cooled space is set.

shading coefficient (SC): the ratio of solar heat gain at normal incidence through glazing to that occurring through 1/8 in. thick clear, double-strength glass. SC, as used herein, does not include interior, exterior, or integral shading devices.

simulation program: a computer program that is capable of simulating the energy performance of building systems.

single-line diagram: a simplified schematic drawing that shows the connection between two or more items. Common multiple connections are shown as one line.

single-package vertical air conditioner (SPVAC): a type of air-cooled small or large commercial package air-conditioning and heating equipment; factory assembled as a single package having its major components arranged vertically, which is an encased combination of cooling and optional heating components; is intended for exterior mounting on, adjacent interior to, or through an outside wall; and is powered by single or three-phase current. It may contain separate indoor grille(s), outdoor louvers, various ventilation options, or indoor free air discharge, ductwork, wall plenum, or sleeve. Heating components may include electrical resistance, steam, hot water, gas, or no heat but may not include reverse cycle refrigeration as a heating means.

single-package vertical heat pump (SPVHP): an SPVAC that utilizes reverse cycle refrigeration as its primary heat source, with secondary supplemental heating by means of electrical resistance, steam, hot water, or gas.

single-rafter roof: see *roof.*

single-zone system: an HVAC system serving a single HVAC zone.

site-recovered energy: waste energy recovered at the building site that is used to offset consumption of purchased fuel or electrical energy supplies.

site-solar energy: thermal, chemical, or electrical energy derived from direct conversion of incident solar radiation at the building site and used to offset consumption of purchased fuel or electrical energy supplies. For the purposes of applying this standard, *site-solar energy* shall not include passive heat gain through fenestration systems.

skylight: see *fenestration.*

skylight well: the shaft from the skylight to the ceiling.

slab-on-grade floor: that portion of a slab floor of the building envelope that is in contact with the ground and that is either above grade or is less than or equal to 24 in. below the final elevation of the nearest exterior grade.

> ***heated slab-on-grade floor:*** a slab-on-grade floor with a heating source either within or below it.
>
> ***unheated slab-on-grade floor:*** a slab-on-grade floor that is not a heated slab-on-grade floor.

solar energy source: source of thermal, chemical, or electrical energy derived from direct conversion of incident solar radiation at the building site.

solar heat gain coefficient (SHGC): the ratio of the solar heat gain entering the space through the fenestration area to the incident solar radiation. Solar heat gain includes directly transmitted solar heat and absorbed solar radiation, which is then reradiated, conducted, or convected into the space. (See *fenestration area.*)

space: an enclosed space within a building. The classifications of spaces are as follows for the purpose of determining building envelope requirements:

conditioned space: a cooled space, heated space, or indirectly conditioned space defined as follows:

1. ***cooled space:*** an enclosed space within a building that is cooled by a cooling system whose sensible output capacity exceeds 5 Btu/h·ft^2 of floor area.
2. ***heated space:*** an enclosed space within a building that is heated by a heating system whose output capacity relative to the floor area is greater than or equal to the criteria in Table 3.1.
3. ***indirectly conditioned space:*** an enclosed space within a building that is not a heated space or a cooled space, which is heated or cooled indirectly by being connected to adjacent space(s) provided:
   a. the product of the U-factor(s) and surface area(s) of the space adjacent to connected space(s) exceeds the combined sum of the product of the U-factor(s) and surface area(s) of the space adjoining the outdoors, unconditioned spaces, and to or from semiheated spaces (e.g., corridors) or
   b. that air from heated or cooled spaces is intentionally transferred (naturally or mechanically) into the space at a rate exceeding 3 ach (e.g., atria).

semiheated space: an enclosed space within a building that is heated by a heating system whose output capacity is greater than or equal to 3.4 Btu/h·ft^2 of floor area but is not a conditioned space.

unconditioned space: an enclosed space within a building that is not a conditioned space or a semiheated space. Crawlspaces, attics, and parking garages with natural or mechanical ventilation are not considered enclosed spaces.

space-conditioning category:

nonresidential conditioned space,

residential conditioned space, and

nonresidential and residential semiheated space.

(See *nonresidential*, *residential*, and *space.*)

steel-framed wall: see *wall.*

steel-joist floor: see *floor.*

story: portion of a building that is between one finished floor level and the next higher finished floor level or the roof, provided, however, that a basement or cellar shall not be considered a story.

TABLE 3.1 Heated Space Criteria

Heating Output (Btu/h·ft^2)	Climate Zone
5	1 and 2
10	3
15	4 and 5
20	6 and 7
25	8

substantial contact: a condition where adjacent building materials are placed so that proximal surfaces are contiguous, being installed and supported so they eliminate voids between materials without compressing or degrading the thermal performance of either product.

swinging door: see *door.*

system: a combination of equipment and auxiliary devices (e.g., controls, accessories, interconnecting means, and terminal elements) by which energy is transformed so it performs a specific function such as HVAC, service water heating, or lighting.

system, existing: a system or systems previously installed in an existing building.

tandem wiring: pairs of luminaires operating with lamps in each luminaire powered from a single ballast contained in one of the luminaires.

task lighting: lighting directed to a specific surface or area that provides illumination for visual tasks.

terminal: a device by which energy from a system is finally delivered, e.g., registers, diffusers, lighting fixtures, faucets, etc.

thermal block: a collection of one or more HVAC zones grouped together for simulation purposes. Spaces need not be contiguous to be combined within a single thermal block.

thermal conductance: see *C-factor.*

thermal resistance (R-value): the reciprocal of the time rate of heat flow through a unit area induced by a unit temperature difference between two defined surfaces of material or construction under steady-state conditions. Units of R are h·ft^2·°F/Btu.

thermal transmittance: see *U-factor.*

thermostat: an automatic control device used to maintain temperature at a fixed or adjustable setpoint.

thermostatic control: an automatic control device or system used to maintain temperature at a fixed or adjustable setpoint.

tinted: (as applied to fenestration) bronze, green, blue, or gray coloring that is integral with the glazing material. Tinting does not include surface-applied films such as reflective coatings, applied either in the field or during the manufacturing process.

transformer: a piece of electrical equipment used to convert electric power from one voltage to another voltage.

dry-type transformer: a *transformer* in which the core and coils are in a gaseous or dry compound.

liquid-immersed transformer: a *transformer* in which the core and coils are immersed in an insulating liquid.

U-factor (thermal transmittance): heat transmission in unit time through unit area of a material or construction and the boundary air films, induced by unit temperature difference between the environments on each side. Units of *U* are Btu/h·ft^2·°F.

unmet load hour: an hour in which one or more zones is outside of the thermostat setpoint range.

unconditioned space: see *space*.

unenclosed space: a space that is not an enclosed space.

unitary cooling equipment: one or more factory-made assemblies that normally include an evaporator or cooling coil and a compressor and condenser combination. Units that perform a heating function are also included.

unitary heat pump: one or more factory-made assemblies that normally include an indoor conditioning coil, compressor(s), and an outdoor refrigerant-to-air coil or refrigerant-to-water heat exchanger. These units provide both heating and cooling functions.

variable-air-volume (VAV) system: HVAC system that controls the dry-bulb temperature within a space by varying the volumetric flow of heated or cooled supply air to the space.

vent damper: a device intended for installation in the venting system of an individual, automatically operated, fossil-fuel-fired appliance in the outlet or downstream of the appliance draft control device, which is designed to automatically open the venting system when the appliance is in operation and to automatically close off the venting system when the appliance is in a standby or shutdown condition.

ventilation: the process of supplying or removing air by natural or mechanical means to or from any space. Such air is not required to have been conditioned.

vertical fenestration: see *fenestration*.

voltage drop: a decrease in voltage caused by losses in the lines connecting the power source to the load.

wall: that portion of the building envelope, including opaque area and fenestration, that is vertical or tilted at an angle of 60 degrees from horizontal or greater. This includes above- and below-grade walls, between floor spandrels, peripheral edges of floors, and foundation walls. For the purposes of determining building envelope requirements, the classifications are defined as follows:

above-grade wall: a wall that is not a below-grade wall.

below-grade wall: that portion of a wall in the building envelope that is entirely below the finish grade and in contact with the ground.

mass wall: a wall with an *HC* exceeding (1) 7 Btu/ft^2·°F or (2) 5 Btu/ft^2·°F, provided that the wall has a material unit weight not greater than 120 lb/ft^3.

metal building wall: a wall whose structure consists of metal spanning members supported by steel structural members (i.e., does not include spandrel glass or metal panels in curtain wall systems).

steel-framed wall: a wall with a cavity (insulated or otherwise) whose exterior surfaces are separated by steel framing members (i.e., typical steel stud walls and curtain wall systems).

wood-framed and other walls: all other wall types, including wood stud walls.

wall area, gross: the area of the wall measured on the exterior face from the top of the floor to the bottom of the roof.

warm-up: increase in space temperature to occupied setpoint after a period of shutdown or setback.

water heater: vessel in which water is heated and is withdrawn for use external to the system.

wood-framed and other walls: see *wall*.

wood-framed and other floors: see *floor*.

zone, HVAC: a space or group of spaces within a building with heating and cooling requirements that are sufficiently similar so that desired conditions (e.g., temperature) can be maintained throughout using a single sensor (e.g., thermostat or temperature sensor).

3.3 Abbreviations and Acronyms

ac	alternating current
ach	air changes per hour
AFUE	annual fuel utilization efficiency
AHAM	Association of Home Appliance Manufacturers
ANSI	American National Standards Institute
ARI	Air-Conditioning and Refrigeration Institute
ASHRAE	American Society of Heating, Refrigerating and Air-Conditioning Engineers, Inc.
ASTM	American Society for Testing and Materials
bhp	brake horsepowerBSRBoard of Standards Review
Btu	British thermal unit
Btu/h	British thermal unit per hour
Btu/ft^2·°F	British thermal unit per square foot per degree FahrenheitBtu/h·ft^{2}British thermal unit per hour per square foot Btu/h·ft·°FBritish thermal unit per hour per linear foot per degree Fahrenheit
Btu/h·ft^2·°F	British thermal unit per hour per square foot per degree Fahrenheit
CDD	cooling degree-day
CDD50	cooling degree-days base 50°F
cfm	cubic feet per minute
c.i.	continuous insulation

COP	coefficient of performance
CTI	Cooling Technology Institute
DDC	direct digital control
DOE	U.S. Department of Energy
Ec	combustion efficiency
EER	energy efficiency ratio
EF	energy factor
ENVSTD	Envelope System Performance Compliance Program
Et	thermal efficiency
F	Fahrenheit
ft	foot
h	hour
HC	heat capacity
HDD	heating degree-day
HDD65	heating degree-days base 65°F
$h \cdot ft^2 \cdot °F/Btu$	hour per square foot per degree Fahrenheit per British thermal unit HIDhigh-intensity discharge
hp	horsepower
HSPF	heating seasonal performance factor
HVAC	heating, ventilating, and air conditioning
IESNA	Illuminating Engineering Society of North America
in.	inch
I-P	inch-pound
IPLV	integrated part-load value
K	kelvin
kVA	kilovolt-ampere
lin	linear
lin ft	linear foot
LPD	lighting power density
MICA	Midwest Insulation Contractors Association
NAECA	U.S. National Appliance Energy Conservation Act of 1987
NFPA	National Fire Protection Association
NFRC	National Fenestration Rating Council
NPLV	nonstandard part-load value
PF	projection factor
PTAC	packaged terminal air conditioner
PTHP	packaged terminal heat pump
R	R-value (thermal resistance)
R_c	thermal resistance of a material or construction from surface to surface
R_u	total thermal resistance of a material or construction including air film resistances
rpm	revolutions per minute
SC	shading coefficient
SEER	seasonal energy efficiency ratio
SHGC	solar heat gain coefficient
L	standby loss
SMACNA	Sheet Metal and Air Conditioning Contractors' National Association
T_{db}	dry-bulb temperature
T_{wb}	wet-bulb temperature
UL	Underwriters Laboratories Inc.
VAV	variable-air-volume
VLT	visible light transmittance
W	watt
W/ft^2	watts per square foot
Wh	watt-hour

4. ADMINISTRATION AND ENFORCEMENT

4.1 General

4.1.1 Scope

4.1.1.1 New Buildings. New buildings shall comply with the standard as described in Section 4.2.

4.1.1.2 Additions to Existing Buildings. An extension or increase in the floor area or height of a building outside of the *existing building* envelope shall be considered *additions* to *existing buildings* and shall comply with the standard as described in Section 4.2.

4.1.1.3 Alterations of Existing Buildings. *Alterations* of *existing buildings* shall comply with the standard as described in Section 4.2.

4.1.1.4 Replacement of Portions of Existing Buildings. Portions of a building envelope, heating, ventilating, air-conditioning, service water heating, power, lighting, and other systems and equipment that are being replaced shall be considered as alterations of existing buildings and shall comply with the standard as described in Section 4.2.

4.1.1.5 Changes in Space Conditioning. Whenever *unconditioned* or *semiheated* spaces in a building are converted to *conditioned spaces*, such *conditioned spaces* shall be brought into compliance with all the applicable requirements of this standard that would apply to the building envelope, heating, ventilating, air-conditioning, service water heating, power, lighting, and other systems and equipment of the space as if the building were new.

4.1.2 Administrative Requirements. Administrative requirements relating to permit requirements, enforcement by the *authority having jurisdiction*, locally adopted energy standards, interpretations, claims of exemption, and rights of appeal are specified by the *authority having jurisdiction*.

4.1.3 Alternative Materials, Methods of Construction, or Design. The provisions of this standard are not intended to prevent the use of any material, method of construction, design, equipment, or building system not specifically prescribed herein.

4.1.4 Validity. If any term, part, provision, section, paragraph, subdivision, table, chart, or referenced standard of this standard shall be held unconstitutional, invalid, or ineffective, in whole or in part, such determination shall not be deemed to invalidate any remaining term, part, provision, section, paragraph, subdivision, table, chart, or referenced standard of this standard.

4.1.5 Other Laws. The provisions of this standard shall not be deemed to nullify any provisions of local, state, or federal law. Where there is a conflict between a requirement of this standard and such other law affecting construction of the building, precedence shall be determined by the *authority having jurisdiction*.

4.1.6 Referenced Standards. The standards referenced in this standard and listed in Section 12 shall be considered part of the requirements of this standard to the prescribed extent of such reference. Where differences occur between the provision of this standard and referenced standards, the provisions of this standard shall apply. Informative references are cited to acknowledge sources and are not part of this standard. They are identified in Informative Appendix E.

4.1.7 Normative Appendices. The normative appendices to this standard are considered to be integral parts of the mandatory requirements of this standard, which, for reasons of convenience, are placed apart from all other normative elements.

4.1.8 Informative Appendices. The informative appendices to this standard and informative notes located within this standard contain additional information and are not mandatory or part of this standard.

4.2 Compliance

4.2.1 Compliance Paths

4.2.1.1 New Buildings. New Buildings shall comply with either the provisions of Sections 5, 6, 7, 8, 9, and 10 or Section 11.

4.2.1.2 Additions to Existing Buildings. *Additions* to *existing buildings* shall comply with either the provisions of Sections 5, 6, 7, 8, 9, and 10 or Section 11.

Exceptions: When an addition to an *existing building* cannot comply by itself, trade-offs will be allowed by modification to one or more of the existing components of the *existing building*. Modeling of the modified components of the *existing building* and addition shall employ the procedures of Section 11; the addition shall not increase the energy consumption of the *existing building* plus the addition beyond the energy that would be consumed by the *existing building* plus the addition if the addition alone did comply.

4.2.1.3 Alterations of Existing Buildings. *Alterations* of *existing buildings* shall comply with the provisions of Sections 5, 6, 7, 8, 9, and 10, provided, however, that nothing in this standard shall require compliance with any provision of this standard if such compliance will result in the increase of energy consumption of the building.

Exceptions:

a. A building that has been specifically designated as historically significant by the *adopting authority* or is listed in The National Register of Historic Places or has been determined to be eligible for listing by the US Secretary of the Interior need not comply with these requirements.

b. Where one or more components of an *existing building* or portions thereof are being replaced, the annual energy consumption of the comprehensive design shall not be greater than the annual energy consumption of a substantially identical design, using the same energy types, in which the applicable requirements of Sections 5, 6, 7, 8, 9, and 10, as provided in Section 4.2.1.3, and such compliance is verified by a *design professional*, by the use of any calculation methods acceptable to the *authority having jurisdiction*.

4.2.2 Compliance Documentation

4.2.2.1 Construction Details. Compliance documents shall show all the pertinent data and features of the building, equipment, and systems in sufficient detail to permit a determination of compliance by the *building official* and to indicate compliance with the requirements of this standard.

4.2.2.2 Supplemental Information. Supplemental information necessary to verify compliance with this standard, such as calculations, worksheets, compliance forms, vendor literature, or other data, shall be made available when required by the *building official*.

4.2.2.3 Manuals. Operating and maintenance information shall be provided to the building owner. This information shall include, but not be limited to, the information specified in Sections 6.7.2.2 and 8.7.2.

4.2.3 Labeling of Material and Equipment. Materials and equipment shall be labeled in a manner that will allow for a determination of their compliance with the applicable provisions of this standard.

4.2.4 Inspections. All building construction, *additions*, or *alterations* subject to the provisions of this standard shall be subject to inspection by the *building official*, and all such work shall remain accessible and exposed for inspection purposes until approved in accordance with the procedures specified by the *building official*. Items for inspection include at least the following:

a. wall insulation after the insulation and vapor retarder are in place but before concealment

b. roof/ceiling insulation after roof/insulation is in place but before concealment

c. slab/foundation wall after slab/foundation insulation is in place but before concealment

d. fenestration after all glazing materials are in place

e. mechanical systems and equipment and insulation after installation but before concealment

f. electrical equipment and systems after installation but before concealment

5. BUILDING ENVELOPE

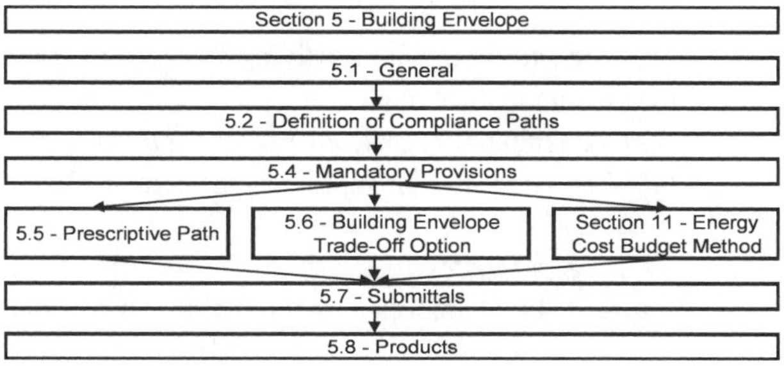

5.1 General

5.1.1 Scope. Section 5 specifies requirements for the *building envelope.*

5.1.2 Space-Conditioning Categories

5.1.2.1 Separate *exterior building envelope* requirements are specified for each of three categories of conditioned space: (a) *nonresidential conditioned* space, (b) *residential conditioned* space, and (c) *semiheated* space.

5.1.2.2 *Spaces* shall be assumed to be *conditioned spaces* and shall comply with the requirements for *conditioned space* at the time of construction, regardless of whether mechanical or electrical equipment is included in the building permit application or installed at that time.

5.1.2.3 In climate zones 3 through 8, a space may be designated as either *semiheated* or *unconditioned* only if approved by the *building official.*

5.1.3 Envelope Alterations. *Alterations* to the *building envelope* shall comply with the requirements of Section 5 for insulation, air leakage, and *fenestration* applicable to those specific portions of the building that are being altered.

Exceptions: The following *alterations* need not comply with these requirements, provided such *alterations* will not increase the energy usage of the building:

a. installation of storm windows over existing glazing
b. replacement of glazing in existing sash and frame provided the *U-factor* and *SHGC* will be equal to or lower than before the glass replacement
c. *alterations* to roof/ceiling, wall, or floor cavities, which are insulated to full depth with insulation having a minimum nominal value of R-3.0/in.
d. *alterations* to walls and floors, where the existing structure is without framing cavities and no new framing cavities are created
e. replacement of a roof membrane where either the roof sheathing or roof insulation is not exposed or, if there is existing roof insulation, below the roof deck
f. replacement of existing doors that separate conditioned space from the exterior shall not require the installation of a vestibule or revolving door, provided, however, that an existing vestibule that separates a conditioned space from the exterior shall not be removed
g. replacement of existing fenestration, provided, however, that the area of the replacement fenestration does not exceed 25% of the total fenestration area of an *existing building* and that the *U-factor* and *SHGC* will be equal to or lower than before the fenestration replacement

5.1.4 Climate. Determine the climate zone for the location. For US locations, follow the procedure in Section 5.1.4.1. For international locations, follow the procedure in Section 5.1.4.2.

5.1.4.1 United States Locations. Use Figure B-1 or Table B-1 in Appendix B to determine the required climate zone.

Exception: If there are recorded historical climatic data available for a construction site, they may be used to determine compliance if approved by the *building official.*

5.1.4.2 International Locations. For locations in Canada that are listed in Table B-2 in Appendix B, use this table to determine the required climate zone number and, when a climate zone letter is also required, use Table B-4 and the Major Climate Type Definitions in Appendix B to determine the letter (A, B, or C). For locations in other international countries that are listed in Table B-3, use this table to determine the required climate zone number and, when a climate zone letter is also required, use Table B-4 and the Major Climate Type Definitions in Appendix B to determine the letter (A, B, or C). For all international locations that are not listed either in Table B-2 or B-3, use Table B-4 and the Major Climate Type Definitions in Appendix B to determine both the climate zone letter and number.

5.2 Compliance Paths

5.2.1 Compliance. For the appropriate climate, *space-conditioning category*, and *class of construction*, the *building envelope* shall comply with Section 5.1, General; Section 5.4, Mandatory Provisions; Section 5.7, Submittals; and Section 5.8, Product Information and Installation Requirements; and either

a. 5.5, Prescriptive Building Envelope Option, provided that

   1. the *vertical fenestration area* does not exceed 40% of the *gross wall area* for each *space-conditioning category* and
   2. the *skylight fenestration area* does not exceed 5% of the *gross roof area* for each *space-conditioning category*, or

b. 5.6, Building Envelope Trade-Off Option.

5.2.2 Projects using the Energy Cost Budget Method (Section 11 of this standard) must comply with Section 5.4, the mandatory provisions of this section, as a portion of that compliance path.

5.3 Simplified Building (Not Used)

5.4 Mandatory Provisions

5.4.1 Insulation. Where insulation is required in Section 5.5 or 5.6, it shall comply with the requirements found in Sections 5.8.1.1 through 5.8.1.9.

5.4.2 Fenestration and Doors. Procedures for determining *fenestration* and door performance are described in Section 5.8.2. Product samples used for determining *fenestration* performance shall be production line units or representative of units purchased by the consumer or contractor.

5.4.3 Air Leakage

5.4.3.1 Building Envelope Sealing. The following areas of the *building envelope* shall be sealed, caulked, gasketed, or weather-stripped to minimize air leakage:

a. joints around *fenestration* and *door* frames
b. junctions between *walls* and foundations, between *walls* at building corners, between *walls* and structural *floors* or *roofs*, and between *walls* and *roof* or *wall* panels
c. openings at penetrations of utility services through *roofs*, *walls*, and *floors*
d. site-built *fenestration* and *doors*
e. building assemblies used as ducts or plenums
f. joints, seams, and penetrations of vapor retarders
g. all other openings in the *building envelope*

5.4.3.2 Fenestration and Doors. Air leakage for *fenestration* and *doors* shall be determined in accordance with NFRC 400. Air leakage shall be determined by a laboratory accredited by a nationally recognized accreditation organization, such as the National Fenestration Rating Council, and shall be *labeled* and certified by the *manufacturer*. Air leakage shall not exceed 1.0 cfm/ft^2 for glazed swinging entrance doors and for revolving doors and 0.4 cfm/ft^2 for all other products.

Exceptions:

a. Field-fabricated fenestration and doors.
b. For garage *doors*, air leakage determined by test at standard test conditions in accordance with ANSI/DASMA 105 shall be an acceptable alternate for compliance with air leakage requirements.

5.4.3.3 Loading Dock Weatherseals. In climate zones 4 through 8, cargo *doors* and loading dock *doors* shall be equipped with weatherseals to restrict *infiltration* when vehicles are parked in the doorway.

5.4.3.4 Vestibules. Building entrances that separate *conditioned space* from the exterior shall be protected with an enclosed vestibule, with all *doors* opening into and out of the vestibule equipped with self-closing devices. Vestibules shall be designed so that in passing through the vestibule it is not necessary for the interior and exterior *doors* to open at the same time. Interior and exterior *doors* shall have a minimum distance between them of not less than 7 ft when in the closed position.The exterior envelope of conditioned vestibules shall comply with the requirements for a conditioned space. The interior and exterior envelope of unconditioned vestibules shall comply with the requirements for a semiheated space.

Exceptions:

a. *Building entrances* with revolving *doors*.
b. *Doors* not intended to be used as a *building entrance*.
c. *Doors* opening directly from a *dwelling unit*.
d. *Building entrances* in buildings located in climate zone 1 or 2.
e. *Building entrances* in buildings located in climate zone 3 or 4 that are less than four stories above grade and less than 10,000 ft^2 in area.
f. *Building entrances* in buildings located in climate zone 5, 6, 7, or 8 that are less than 1000 ft^2 in area.
g. *Doors* that open directly from a *space* that is less than 3000 ft^2 in area and is separate from the *building entrance*.

5.5 Prescriptive Building Envelope Option

5.5.1 For a *conditioned space*, the *exterior building envelope* shall comply with either the "nonresidential" or "residential" requirements in Tables 5.5-1 through 5.5-8 for the appropriate climate.

5.5.2 If a building contains any *semiheated space* or *unconditioned space*, then the *semi-exterior building envelope* shall comply with the requirements for *semiheated space* in Tables 5.5-1 through 5.5-8 for the appropriate climate. (See Figure 5.5.)

5.5.3 Opaque Areas. For all opaque surfaces except doors, compliance shall be demonstrated by one of the following two methods:

1. Minimum *rated R-values of insulation* for the thermal resistance of the added insulation in framing cavities and *continuous insulation* only. Specifications listed in Normative Appendix A for each *class of construction* shall be used to determine compliance.
2. Maximum *U-factor, C-factor,* or *F-factor* for the entire assembly. The values for typical construction assemblies listed in Normative Appendix A shall be used to determine compliance.

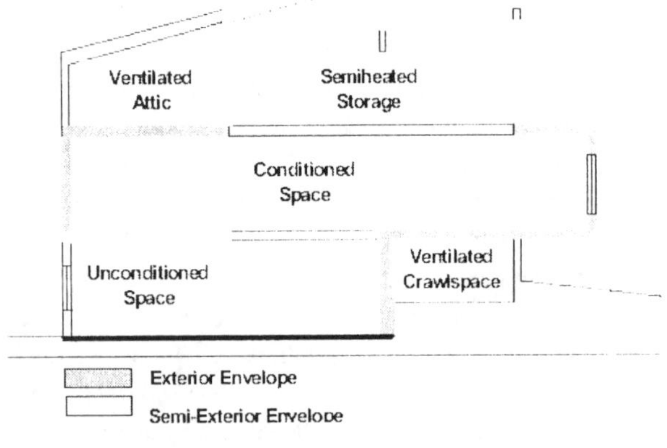

Figure 5.5 Exterior and semi-exterior building envelope.

TABLE 5.5-1 Building Envelope Requirements for Climate Zone 1 (A, B)*

Opaque Elements	Nonresidential		Residential		Semiheated	
	Assembly Maximum	Insulation Min. R-Value	Assembly Maximum	Insulation Min. R-Value	Assembly Maximum	Insulation Min. R-Value
Roofs						
Insulation Entirely above Deck	U-0.063	R-15.0 c.i.	U-0.048	R-20.0 c.i.	U-0.218	R-3.8 ci
Metal Building	U-0.065	R-19.0	U-0.065	R-19.0	U-1.280	NR
Attic and Other	U-0.034	R-30.0	U-0.027	R-38.0	U-0.081	R-13.0
Walls, Above-Grade						
Mass	U-0.580	NR	U-0.151[a]	R-5.7 c.i.[a]	U-0.580	NR
Metal Building	U-0.113	R-13.0	U-0.113	R-13.0	U-1.180	NR
Steel-Framed	U-0.124	R-13.0	U-0.124	R-13.0	U-0.352	NR
Wood-Framed and Other	U-0.089	R-13.0	U-0.089	R-13.0	U-0.292	NR
Walls, Below-Grade						
Below-Grade Wall	C-1.140	NR	C-1.140	NR	C-1.140	NR
Floors						
Mass	U-0.322	NR	U-0.322	NR	U-0.322	NR
Steel-Joist	U-0.350	NR	U-0.350	NR	U-0.350	NR
Wood-Framed and Other	U-0.282	NR	U-0.282	NR	U-0.282	NR
Slab-On-Grade Floors						
Unheated	F-0.730	NR	F-0.730	NR	F-0.730	NR
Heated	F-1.020	R-7.5 for 12 in.	F-1.020	R-7.5 for 12 in.	F-1.020	R-7.5 for 12 in.
Opaque Doors						
Swinging	U-0.700		U-0.700		U-0.700	
Nonswinging	U-1.450		U-1.450		U-1.450	

Fenestration	Assembly Max. U	Assembly Max. SHGC	Assembly Max. U	Assembly Max. SHGC	Assembly Max. U	Assembly Max. SHGC
Vertical Glazing, 0%–40% of Wall						
Nonmetal framing (all)[b]	U-1.20		U-1.20		U-1.20	
Metal framing (curtainwall/storefront)[c]	U-1.20	SHGC-0.25 all	U-1.20	SHGC-0.25 all	U-1.20	SHGC-NR all
Metal framing (entrance door)[c]	U-1.20		U-1.20		U-1.20	
Metal framing (all other)[c]	U-1.20		U-1.20		U-1.20	
Skylight with Curb, Glass, % of Roof						
0%–2.0%	U_{all}-1.98	$SHGC_{all}$-0.36	U_{all}-1.98	$SHGC_{all}$-0.19	U_{all}-1.98	$SHGC_{all}$-NR
2.1%–5.0%	U_{all}-1.98	$SHGC_{all}$-0.19	U_{all}-1.98	$SHGC_{all}$-0.16	U_{all}-1.98	$SHGC_{all}$-NR
Skylight with Curb, Plastic, % of Roof						
0%–2.0%	U_{all}-1.90	$SHGC_{all}$-0.34	U_{all}-1.90	$SHGC_{all}$-0.27	U_{all}-1.90	$SHGC_{all}$-NR
2.1%–5.0%	U_{all}-1.90	$SHGC_{all}$-0.27	U_{all}-1.90	$SHGC_{all}$-0.27	U_{all}-1.90	$SHGC_{all}$-NR
Skylight without Curb, All, % of Roof						
0%–2.0%	U_{all}-1.36	$SHGC_{all}$-0.36	U_{all}-1.36	$SHGC_{all}$-0.19	U_{all}-1.36	$SHGC_{all}$-NR
2.1%–5.0%	U_{all}-1.36	$SHGC_{all}$-0.19	U_{all}-1.36	$SHGC_{all}$-0.19	U_{all}-1.36	$SHGC_{all}$-NR

*The following definitions apply: c.i. = continuous insulation (see Section 3.2), NR = no (insulation) requirement.
[a]Exception to Section A3.1.3.1 applies.
[b]Nonmetal framing includes framing materials other than metal with or without metal reinforcing or cladding.
[c]Metal framing includes metal framing with or without thermal break. The "all other" subcategory includes operable windows, fixed windows, and non-entrance doors.

TABLE 5.5-2 Building Envelope Requirements for Climate Zone 2 (A, B)*

Opaque Elements	Nonresidential		Residential		Semiheated	
	Assembly Maximum	**Insulation Min. R-Value**	**Assembly Maximum**	**Insulation Min. R-Value**	**Assembly Maximum**	**Insulation Min. R-Value**
Roofs						
Insulation Entirely above Deck	U-0.048	R-20.0 c.i.	U-0.048	R-20.0 c.i.	U-0.218	R-3.8 c.i.
Metal Building	U-0.065	R-19.0	U-0.065	R-19.0	U-0.167	R-6.0
Attic and Other	U-0.027	R-38.0	U-0.027	R-38.0	U-0.081	R-13.0
Walls, Above-Grade						
Mass	U-0.151[a]	R-5.7 c.i.[a]	U-0.123	R-7.6 c.i.	U-0.580	NR
Metal Building	U-0.113	R-13.0	U-0.113	R-13.0	U-0.184	R-6.0
Steel-Framed	U-0.124	R-13.0	U-0.064	R-13.0 + R-7.5 c.i.	U-0.124	R-13.0
Wood-Framed and Other	U-0.089	R-13.0	U-0.089	R-13.0	U-0.089	R-13.0
Walls, Below-Grade						
Below-Grade Wall	C-1.140	NR	C-1.140	NR	C-1.140	NR
Floors						
Mass	U-0.107	R-6.3 c.i.	U-0.087	R-8.3 c.i.	U-0.322	NR
Steel-Joist	U-0.052	R-19.0	U-0.052	R-19.0	U-0.069	R-13.0
Wood-Framed and Other	U-0.051	R-19.0	U-0.033	R-30.0	U-0.066	R-13.0
Slab-On-Grade Floors						
Unheated	F-0.730	NR	F-0.730	NR	F-0.730	NR
Heated	F-1.020	R-7.5 for 12 in.	F-1.020	R-7.5 for 12 in.	F-1.020	R-7.5 for 12 in.
Opaque Doors						
Swinging	U-0.700		U-0.700		U-0.700	
Nonswinging	U-1.450		U-0.500		U-1.450	
Fenestration	**Assembly Max. U**	**Assembly Max. SHGC**	**Assembly Max. U**	**Assembly Max. SHGC**	**Assembly Max. U**	**Assembly Max. SHGC**
Vertical Glazing, 0%–40% of Wall						
Nonmetal framing (all)[b]	U-0.75	SHGC-0.25 all	U-0.75	SHGC-0.25 all	U-1.20	SHGC-NR all
Metal framing (curtainwall/storefront)[c]	U-0.70		U-0.70		U-1.20	
Metal framing (entrance door)[c]	U-1.10		U-1.10		U-1.20	
Metal framing (all other)[c]	U-0.75		U-0.75		U-1.20	
Skylight with Curb, Glass, % of Roof						
0%–2.0%	U_{all}-1.98	$SHGC_{all}$-0.36	U_{all}-1.98	$SHGC_{all}$-0.19	U_{all}-1.98	$SHGC_{all}$-NR
2.1%–5.0%	U_{all}-1.98	$SHGC_{all}$-0.19	U_{all}-1.98	$SHGC_{all}$-0.19	U_{all}-1.98	$SHGC_{all}$-NR
Skylight with Curb, Plastic, % of Roof						
0%–2.0%	U_{all}-1.90	$SHGC_{all}$-0.39	U_{all}-1.90	$SHGC_{all}$-0.27	U_{all}-1.90	$SHGC_{all}$-NR
2.1%–5.0%	U_{all}-1.90	$SHGC_{all}$-0.34	U_{all}-1.90	$SHGC_{all}$-0.27	U_{all}-1.90	$SHGC_{all}$-NR
Skylight without Curb, All, % of Roof						
0%–2.0%	U_{all}-1.36	$SHGC_{all}$-0.36	U_{all}-1.36	$SHGC_{all}$-0.19	U_{all}-1.36	$SHGC_{all}$-NR
2.1%–5.0%	U_{all}-1.36	$SHGC_{all}$-0.19	U_{all}-1.36	$SHGC_{all}$-0.19	U_{all}-1.36	$SHGC_{all}$-NR

*The following definitions apply: c.i. = continuous insulation (see Section 3.2), NR = no (insulation) requirement.

[a]Exception to Section A3.1.3.1 applies.

[b]Nonmetal framing includes framing materials other than metal with or without metal reinforcing or cladding.

[c]Metal framing includes metal framing with or without thermal break. The “all other” subcategory includes operable windows, fixed windows, and non-entrance doors.

TABLE 5.5-3 Building Envelope Requirements for Climate Zone 3 (A, B, C)*

Opaque Elements	Nonresidential Assembly Maximum	Nonresidential Insulation Min. R-Value	Residential Assembly Maximum	Residential Insulation Min. R-Value	Semiheated Assembly Maximum	Semiheated Insulation Min. R-Value
Roofs						
Insulation Entirely above Deck	U-0.048	R-20.0 c.i.	U-0.048	R-20.0 c.i.	U-0.173	R-5.0 c.i.
Metal Building	U-0.065	R-19.0	U-0.065	R-19.0	U-0.097	R-10.0
Attic and Other	U-0.027	R-38.0	U-0.027	R-38.0	U-0.053	R-19.0
Walls, Above-Grade						
Mass	U-0.123	R-7.6 c.i.	U-0.104	R-9.5 c.i.	U-0.580	NR
Metal Building	U-0.113	R-13.0	U-0.113	R-13.0	U-0.184	R-6.0
Steel-Framed	U-0.084	R-13.0 + R-3.8 c.i.	U-0.064	R-13.0 + R-7.5 c.i.	U-0.124	R-13.0
Wood-Framed and Other	U-0.089	R-13.0	U-0.089	R-13.0	U-0.089	R-13.0
Walls, Below-Grade						
Below-Grade Wall	C-1.140	NR	C-1.140	NR	C-1.140	NR
Floors						
Mass	U-0.107	R-6.3 c.i.	U-0.087	R-8.3 c.i.	U-0.322	NR
Steel-Joist	U-0.052	R-19.0	U-0.052	R-19.0	U-0.069	R-13.0
Wood-Framed and Other	U-0.051	R-19.0	U-0.033	R-30.0	U-0.066	R-13.0
Slab-On-Grade Floors						
Unheated	F-0.730	NR	F-0.730	NR	F-0.730	NR
Heated	F-0.900	R-10 for 24 in.	F-0.900	R-10 for 24 in.	F-1.020	R-7.5 for 12 in.
Opaque Doors						
Swinging	U-0.700		U-0.700		U-0.700	
Nonswinging	U-1.450		U-0.500		U-1.450	

Fenestration	Nonresidential Assembly Max. U	Nonresidential Assembly Max. SHGC	Residential Assembly Max. U	Residential Assembly Max. SHGC	Semiheated Assembly Max. U	Semiheated Assembly Max. SHGC
Vertical Glazing, 0%–40% of Wall						
Nonmetal framing (all)[b]	U-0.65	SHGC-0.25 all	U-0.65	SHGC-0.25 all	U-1.20	SHGC-NR all
Metal framing (curtainwall/storefront)[c]	U-0.60		U-0.60		U-1.20	
Metal framing (entrance door)[c]	U-0.90		U-0.90		U-1.20	
Metal framing (all other)[c]	U-0.65		U-0.65		U-1.20	
Skylight with Curb, Glass, % of Roof						
0%–2.0%	U_{all}-1.17	$SHGC_{all}$-0.39	U_{all}-1.17	$SHGC_{all}$-0.36	U_{all}-1.98	$SHGC_{all}$-NR
2.1%–5.0%	U_{all}-1.17	$SHGC_{all}$-0.19	U_{all}-1.17	$SHGC_{all}$-0.19	U_{all}-1.98	$SHGC_{all}$-NR
Skylight with Curb, Plastic, % of Roof						
0%–2.0%	U_{all}-1.30	$SHGC_{all}$-0.65	U_{all}-1.30	$SHGC_{all}$-0.27	U_{all}-1.90	$SHGC_{all}$-NR
2.1%–5.0%	U_{all}-1.30	$SHGC_{all}$-0.34	U_{all}-1.30	$SHGC_{all}$-0.27	U_{all}-1.90	$SHGC_{all}$-NR
Skylight without Curb, All, % of Roof						
0%–2.0%	U_{all}-0.69	$SHGC_{all}$-0.39	U_{all}-0.69	$SHGC_{all}$-0.36	U_{all}-1.36	$SHGC_{all}$-NR
2.1%–5.0%	U_{all}-0.69	$SHGC_{all}$-0.19	U_{all}-0.69	$SHGC_{all}$-0.19	U_{all}-1.36	$SHGC_{all}$-NR

*The following definitions apply: c.i. = continuous insulation (see Section 3.2), NR = no (insulation) requirement.
[b]Nonmetal framing includes framing materials other than metal with or without metal reinforcing or cladding.
[c]Metal framing includes metal framing with or without thermal break. The "all other" subcategory includes operable windows, fixed windows, and non-entrance doors.

TABLE 5.5-4 Building Envelope Requirements for Climate Zone 4 (A, B, C)*

Opaque Elements	Nonresidential Assembly Maximum	Nonresidential Insulation Min. R-Value	Residential Assembly Maximum	Residential Insulation Min. R-Value	Semiheated Assembly Maximum	Semiheated Insulation Min. R-Value
Roofs						
Insulation Entirely above Deck	U-0.048	R-20.0 c.i.	U-0.048	R-20.0 c.i.	U-0.173	R-5.0 c.i.
Metal Building	U-0.065	R-19.0	U-0.065	R-19.0	U-0.097	R-10.0
Attic and Other	U-0.027	R-38.0	U-0.027	R-38.0	U-0.053	R-19.0
Walls, Above-Grade						
Mass	U-0.104	R-9.5 c.i.	U-0.090	R-11.4 c.i.	U-0.580	NR
Metal Building	U-0.113	R-13.0	U-0.113	R-13.0	U-0.134	R-10.0
Steel-Framed	U-0.064	R-13.0 + R-7.5 c.i.	U-0.064	R-13.0 + R-7.5 c.i.	U-0.124	R-13.0
Wood-Framed and Other	U-0.089	R-13.0	U-0.064	R-13.0 + R-3.8 c.i.	U-0.089	R-13.0
Walls, Below-Grade						
Below-Grade Wall	C-1.140	NR	C-0.119	R-7.5 c.i.	C-1.140	NR
Floors						
Mass	U-0.087	R-8.3 c.i.	U-0.074	R-10.4 c.i.	U-0.137	R-4.2 c.i.
Steel-Joist	U-0.038	R-30.0	U-0.038	R-30.0	U-0.069	R-13.0
Wood-Framed and Other	U-0.033	R-30.0	U-0.033	R-30.0	U-0.066	R-13.0
Slab-On-Grade Floors						
Unheated	F-0.730	NR	F-0.540	R-10 for 24 in.	F-0.730	NR
Heated	F-0.860	R-15 for 24 in.	F-0.860	R-15 for 24in.	F-1.020	R-7.5 for 12 in.
Opaque Doors						
Swinging	U-0.700		U-0.700		U-0.700	
Nonswinging	U-1.500		U-0.500		U-1.450	

Fenestration	Nonresidential Assembly Max. U	Nonresidential Assembly Max. SHGC	Residential Assembly Max. U	Residential Assembly Max. SHGC	Semiheated Assembly Max. U	Semiheated Assembly Max. SHGC
Vertical Glazing, 0%–40% of Wall						
Nonmetal framing (all)[b]	U-0.40	SHGC-0.40 all	U-0.40	SHGC-0.40 all	U-1.20	SHGC-NR all
Metal framing (curtainwall/storefront)[c]	U-0.50		U-0.50		U-1.20	
Metal framing (entrance door)[c]	U-0.85		U-0.85		U-1.20	
Metal framing (all other)[c]	U-0.55		U-0.55		U-1.20	
Skylight with Curb, Glass, % of Roof						
0%–2.0%	U_{all}-1.17	$SHGC_{all}$-0.49	U_{all}-0.98	$SHGC_{all}$-0.36	U_{all}-1.98	$SHGC_{all}$-NR
2.1%–5.0%	U_{all}-1.17	$SHGC_{all}$-0.39	U_{all}-0.98	$SHGC_{all}$-0.19	U_{all}-1.98	$SHGC_{all}$-NR
Skylight with Curb, Plastic, % of Roof						
0%–2.0%	U_{all}-1.30	$SHGC_{all}$-0.65	U_{all}-1.30	$SHGC_{all}$-0.62	U_{all}-1.90	$SHGC_{all}$-NR
2.1%–5.0%	U_{all}-1.30	$SHGC_{all}$-0.34	U_{all}-1.30	$SHGC_{all}$-0.27	U_{all}-1.90	$SHGC_{all}$-NR
Skylight without Curb, All, % of Roof						
0%–2.0%	U_{all}-0.69	$SHGC_{all}$-0.49	U_{all}-0.58	$SHGC_{all}$-0.36	U_{all}-1.36	$SHGC_{all}$-NR
2.1%–5.0%	U_{all}-0.69	$SHGC_{all}$-0.39	U_{all}-0.58	$SHGC_{all}$-0.19	U_{all}-1.36	$SHGC_{all}$-NR

*The following definitions apply: c.i. = continuous insulation (see Section 3.2), NR = no (insulation) requirement.
[b]Nonmetal framing includes framing materials other than metal with or without metal reinforcing or cladding.
[c]Metal framing includes metal framing with or without thermal break. The "all other" subcategory includes operable windows, fixed windows, and non-entrance doors.

TABLE 5.5-5 Building Envelope Requirements for Climate Zone 5 (A, B, C)*

Opaque Elements	Nonresidential		Residential		Semiheated	
	Assembly Maximum	**Insulation Min. R-Value**	**Assembly Maximum**	**Insulation Min. R-Value**	**Assembly Maximum**	**Insulation Min. R-Value**
Roofs						
Insulation Entirely above Deck	U-0.048	R-20.0 c.i.	U-0.048	R-20.0 c.i.	U-0.119	R-7.6 c.i.
Metal Building	U-0.065	R-19.0	U-0.065	R-19.0	U-0.097	R-10.0
Attic and Other	U-0.027	R-38.0	U-0.027	R-38.0	U-0.053	R-19.0
Walls, Above-Grade						
Mass	U-0.090	R-11.4 c.i.	U-0.080	R-13.3 c.i.	U-0.151[a]	R-5.7 c.i.[a]
Metal Building	U-0.113	R-13.0	U-0.057	R-13.0 + R-13.0	U-0.123	R-11.0
Steel-Framed	U-0.064	R-13.0 + R-7.5 c.i.	U-0.064	R-13.0 + R-7.5 c.i.	U-0.124	R-13.0
Wood-Framed and Other	U-0.064	R-13.0 + R-3.8 c.i.	U-0.051	R-13.0 + R-7.5 c.i.	U-0.089	R-13.0
Walls, Below-Grade						
Below-Grade Wall	C-0.119	R-7.5 c.i.	C-0.119	R-7.5 c.i.	C-1.140	NR
Floors						
Mass	U-0.074	R-10.4 c.i.	U-0.064	R-12.5 c.i.	U-0.137	R-4.2 c.i.
Steel-Joist	U-0.038	R-30.0	U-0.038	R-30.0	U-0.052	R-19.0
Wood-Framed and Other	U-0.033	R-30.0	U-0.033	R-30.0	U-0.051	R-19.0
Slab-On-Grade Floors						
Unheated	F-0.730	NR	F-0.540	R-10 for 24 in.	F-0.730	NR
Heated	F-0.860	R-15 for 24 in.	F-0.860	R-15 for 24 in.	F-1.020	R-7.5 for 12 in.
Opaque Doors						
Swinging	U-0.700		U-0.500		U-0.700	
Nonswinging	U-0.500		U-0.500		U-1.450	

Fenestration	Assembly Max. U	Assembly Max. SHGC	Assembly Max. U	Assembly Max. SHGC	Assembly Max. U	Assembly Max. SHGC
Vertical Glazing, % of Wall						
Nonmetal framing (all)[b]	U-0.35	SHGC-0.40 all	U-0.35	SHGC-0.40 all	U-1.20	SHGC-NR all
Metal framing (curtainwall/storefront)[c]	U-0.45		U-0.45		U-1.20	
Metal framing (entrance door)[c]	U-0.80		U-0.80		U-1.20	
Metal framing (all other)[c]	U-0.55		U-0.55		U-1.20	
Skylight with Curb, Glass, % of Roof						
0%–2.0%	U_{all}-1.17	$SHGC_{all}$-0.49	U_{all}-1.17	$SHGC_{all}$-0.49	U_{all}-1.98	$SHGC_{all}$-NR
2.1%–5.0%	U_{all}-1.17	$SHGC_{all}$-0.39	U_{all}-1.17	$SHGC_{all}$-0.39	U_{all}-1.98	$SHGC_{all}$-NR
Skylight with Curb, Plastic, % of Roof						
0%–2.0%	U_{all}-1.10	$SHGC_{all}$-0.77	U_{all}-1.10	$SHGC_{all}$-0.77	U_{all}-1.90	$SHGC_{all}$-NR
2.1%–5.0%	U_{all}-1.10	$SHGC_{all}$-0.62	U_{all}-1.10	$SHGC_{all}$-0.62	U_{all}-1.90	$SHGC_{all}$-NR
Skylight without Curb, All, % of Roof						
0%–2.0%	U_{all}-0.69	$SHGC_{all}$-0.49	U_{all}-0.69	$SHGC_{all}$-0.49	U_{all}-1.36	$SHGC_{all}$-NR
2.1%–5.0%	U_{all}-0.69	$SHGC_{all}$-0.39	U_{all}-0.69	$SHGC_{all}$-0.39	U_{all}-1.36	$SHGC_{all}$-NR

*The following definitions apply: c.i. = continuous insulation (see Section 3.2), NR = no (insulation) requirement.
[a] Exception to Section A3.1.3.1 applies.
[b] Nonmetal framing includes framing materials other than metal with or without metal reinforcing or cladding.
[c] Metal framing includes metal framing with or without thermal break. The "all other" subcategory includes operable windows, fixed windows, and non-entrance doors.

TABLE 5.5-6 Building Envelope Requirements for Climate Zone 6 (A, B)*

Opaque Elements	Nonresidential Assembly Maximum	Nonresidential Insulation Min. R-Value	Residential Assembly Maximum	Residential Insulation Min. R-Value	Semiheated Assembly Maximum	Semiheated Insulation Min. R-Value
Roofs						
Insulation Entirely above Deck	U-0.048	R-20.0 c.i.	U-0.048	R-20.0 c.i.	U-0.093	R-10.0 c.i.
Metal Building	U-0.065	R-19.0	U-0.065	R-19.0	U-0.097	R-10.0
Attic and Other	U-0.027	R-38.0	U-0.027	R-38.0	U-0.034	R-30.0
Walls, Above-Grade						
Mass	U-0.080	R-13.3 c.i.	U-0.071	R-15.2 c.i.	U-0.151[a]	R-5.7 c.i.[a]
Metal Building	U-0.113	R-13.0	U-0.057	R-13.0 + R-13.0	U-0.113	R-13.0
Steel-Framed	U-0.064	R-13.0 + R-7.5 c.i.	U-0.064	R-13.0 + R-7.5 c.i.	U-0.124	R-13.0
Wood-Framed and Other	U-0.051	R-13.0 + R-7.5 c.i.	U-0.051	R-13.0 + R-7.5 c.i.	U-0.089	R-13.0
Walls, Below-Grade						
Below-Grade Wall	C-0.119	R-7.5 c.i.	C-0.119	R-7.5 c.i.	C-1.140	NR
Floors						
Mass	U-0.064	R-12.5 c.i.	U-0.057	R-14.6 c.i.	U-0.137	R-4.2 c.i.
Steel-Joist	U-0.038	R-30.0	U-0.032	R-38.0	U-0.052	R-19.0
Wood-Framed and Other	U-0.033	R-30.0	U-0.033	R-30.0	U-.0051	R-19.0
Slab-On-Grade Floors						
Unheated	F-0.540	R-10 for 24 in.	F-0.520	R-15 for 24 in.	F-0.730	NR
Heated	F-0.860	R-15 for 24 in.	F-0.688	R-20 for 48 in.	F-1.020	R-7.5 for 12 in.
Opaque Doors						
Swinging	U-0.700		U-0.500		U-0.700	
Nonswinging	U-0.500		U-0.500		U-1.450	

Fenestration	Nonresidential Assembly Max. U	Nonresidential Assembly Max. SHGC	Residential Assembly Max. U	Residential Assembly Max. SHGC	Semiheated Assembly Max. U	Semiheated Assembly Max. SHGC
Vertical Glazing, 0%–40% of Wall						
Nonmetal framing (all)[b]	U-0.35		U-0.35		U-0.65	
Metal framing (curtainwall/storefront)[c]	U-0.45	SHGC-0.40 all	U-0.45	SHGC-0.40 all	U-0.60	SHGC-NR all
Metal framing (entrance door)[c]	U-0.80		U-0.80		U-0.90	
Metal framing (all other)[c]	U-0.55		U-0.55		U-0.65	
Skylight with Curb, Glass, % of Roof						
0%–2.0%	U_{all}-1.17	$SHGC_{all}$-0.49	U_{all}-0.98	$SHGC_{all}$-0.46	U_{all}-1.98	$SHGC_{all}$-NR
2.1%–5.0%	U_{all}-1.17	$SHGC_{all}$-0.49	U_{all}-0.98	$SHGC_{all}$-0.36	U_{all}-1.98	$SHGC_{all}$-NR
Skylight with Curb, Plastic, % of Roof						
0%–2.0%	U_{all}-0.87	$SHGC_{all}$-0.71	U_{all}-0.74	$SHGC_{all}$-0.65	U_{all}-1.90	$SHGC_{all}$-NR
2.1%–5.0%	U_{all}-0.87	$SHGC_{all}$-0.58	U_{all}-0.74	$SHGC_{all}$-0.55	U_{all}-1.90	$SHGC_{all}$-NR
Skylight without Curb, All, % of Roof						
0%–2.0%	U_{all}-0.69	$SHGC_{all}$-0.49	U_{all}-0.58	$SHGC_{all}$-0.49	U_{all}-1.36	$SHGC_{all}$-NR
2.1%–5.0%	U_{all}-0.69	$SHGC_{all}$-0.49	U_{all}-0.58	$SHGC_{all}$-0.39	U_{all}-1.36	$SHGC_{all}$-NR

*The following definitions apply: c.i. = continuous insulation (see Section 3.2), NR = no (insulation) requirement.

[a] Exception to Section A3.1.3.1 applies.

[b] Nonmetal framing includes framing materials other than metal with or without metal reinforcing or cladding.

[c] Metal framing includes metal framing with or without thermal break. The "all other" subcategory includes operable windows, fixed windows, and non-entrance doors.

TABLE 5.5-7 Building Envelope Requirements for Climate Zone 7*

Opaque Elements	Nonresidential		Residential		Semiheated	
	Assembly Maximum	**Insulation Min. R-Value**	**Assembly Maximum**	**Insulation Min. R-Value**	**Assembly Maximum**	**Insulation Min. R-Value**
Roofs						
Insulation Entirely above Deck	U-0.048	R-20.0 c.i.	U-0.048	R-20.0 c.i.	U-0.093	R-10.0 c.i.
Metal Building	U-0.065	R-19.0	U-0.065	R-19.0	U-0.097	R-10.0
Attic and Other	U-0.027	R-38.0	U-0.027	R-38.0	U-0.034	R-30.0
Walls, Above-Grade						
Mass	U-0.071	R-15.2 c.i.	U-0.071	R-15.2 c.i.	U-0.123	R-7.6 c.i.
Metal Building	U-0.057	R-13.0 + R-13.0	U-0.057	R-13.0 + R-13.0	U-0.113	R-13.0
Steel-Framed	U-0.064	R-13.0 + R-7.5 c.i.	U-0.042	R-13.0 + R-15.6 c.i.	U-0.124	R-13.0
Wood-Framed and Other	U-0.051	R-13.0 + R-7.5 c.i.	U-0.051	R-13.0 + R-7.5 c.i.	U-0.089	R-13.0
Walls, Below-Grade						
Below-Grade Wall	C-0.119	R-7.5 c.i.	C-0.092	R-10.0 c.i.	C-1.140	NR
Floors						
Mass	U-0.064	R-12.5 c.i.	U-0.051	R-16.7 c.i.	U-0.107	R-6.3 c.i.
Steel-Joist	U-0.038	R-30.0	U-0.032	R-38.0	U-0.052	R-19.0
Wood-Framed and Other	U-0.033	R-30.0	U-0.033	R-30.0	U-0.051	R-19.0
Slab-On-Grade Floors						
Unheated	F-0.520	R-15 for 24 in.	F-0.520	R-15 for 24 in.	F-0.730	NR
Heated	F-0.843	R-20 for 24in.	F-0.688	R-20 for 48 in.	F-0.900	R-10 for 24 in.
Opaque Doors						
Swinging	U-0.500		U-0.500		U-0.700	
Nonswinging	U-0.500		U-0.500		U-1.450	
Fenestration	**Assembly Max. U**	**Assembly Max. SHGC**	**Assembly Max. U**	**Assembly Max. SHGC**	**Assembly Max. U**	**Assembly Max. SHGC**
Vertical Glazing, 0%–40% of Wall						
Nonmetal framing (all)[b]	U-0.35	SHGC-0.45 all	U-0.35	SHGC-NR all	U-0.65	SHGC-NR all
Metal framing (curtainwall/storefront)[c]	U-0.40		U-0.40		U-0.60	
Metal framing (entrance door)[c]	U-0.80		U-0.80		U-0.90	
Metal framing (all other)[c]	U-0.45		U-0.45		U-0.65	
Skylight with Curb, Glass, % of Roof						
0%–2.0%	U_{all}-1.17	$SHGC_{all}$-0.68	U_{all}-1.17	$SHGC_{all}$-0.64	U_{all}-1.98	$SHGC_{all}$-NR
2.1%–5.0%	U_{all}-1.17	$SHGC_{all}$-0.64	U_{all}-1.17	$SHGC_{all}$-0.64	U_{all}-1.98	$SHGC_{all}$-NR
Skylight with Curb, Plastic, % of Roof						
0%–2.0%	U_{all}-0.87	$SHGC_{all}$-0.77	U_{all}-0.61	$SHGC_{all}$-0.77	U_{all}-1.90	$SHGC_{all}$-NR
2.1%–5.0%	U_{all}-0.87	$SHGC_{all}$-0.71	U_{all}-0.61	$SHGC_{all}$-0.77	U_{all}-1.90	$SHGC_{all}$-NR
Skylight without Curb, All, % of Roof						
0%–2.0%	U_{all}-0.69	$SHGC_{all}$-0.68	U_{all}-0.69	$SHGC_{all}$-0.64	U_{all}-1.36	$SHGC_{all}$-NR
2.1%–5.0%	U_{all}-0.69	$SHGC_{all}$-0.64	U_{all}-0.69	$SHGC_{all}$-0.64	U_{all}-1.36	$SHGC_{all}$-NR

*The following definitions apply: c.i. = continuous insulation (see Section 3.2), NR = no (insulation) requirement.
[b]Nonmetal framing includes framing materials other than metal with or without metal reinforcing or cladding.
[c]Metal framing includes metal framing with or without thermal break. The "all other" subcategory includes operable windows, fixed windows, and non-entrance doors.

TABLE 5.5-8 Building Envelope Requirements for Climate Zone 8*

Opaque Elements	Nonresidential		Residential		Semiheated	
	Assembly Maximum	Insulation Min. R-Value	Assembly Maximum	Insulation Min. R-Value	Assembly Maximum	Insulation Min. R-Value
Roofs						
Insulation Entirely above Deck	U-0.048	R-20.0 c.i.	U-0.048	R-20.0 c.i.	U-0.063	R-15.0 c.i.
Metal Building	U-0.049	R-13.0 + R-19.0	U-0.049	R-13.0 + R-19.0	U-0.072	R-16.0
Attic and Other	U-0.021	R-49.0	U-0.021	R-49.0	U-0.034	R-30.0
Walls, Above-Grade						
Mass	U-0.071	R-15.2 c.i.	U-0.052	R-25.0 c.i.	U-0.104	R-9.5 c.i.
Metal Building	U-0.057	R-13.0 + R-13.0	U-0.057	R-13.0 + R-13.0	U-0.113	R-13.0
Steel-Framed	U-0.064	R-13.0 + R-7.5 c.i.	U-0.037	R-13.0 + R-18.8 c.i.	U-0.084	R-13.0 + R-3.8 c.i.
Wood-Framed and Other	U-0.036	R-13.0 + R-15.6 c.i.	U-0.036	R-13.0 + R-15.6 c.i.	U-0.089	R-13.0
Walls, Below-Grade						
Below-Grade Wall	C-0.119	R-7.5 c.i.	C-0.075	R-12.5 c.i.	C-1.140	NR
Floors						
Mass	U-0.057	R-14.6 c.i.	U-0.051	R-16.7 c.i.	U-0.087	R-8.3 c.i.
Steel-Joist	U-0.032	R-38.0	U-0.032	R-38.0	U-0.052	R-19.0
Wood-Framed and Other	U-0.033	R-30.0	U-0.033	R-30.0	U-0.033	R-30.0
Slab-On-Grade Floors						
Unheated	F-0.520	R-15 for 24 in.	F-0.510	R-20 for 24 in.	F-0.730	NR
Heated	F-0.688	R-20 for 48 in.	F-0.688	R-20 for 48 in.	F-0.900	R-10.0 for 24 in.
Opaque Doors						
Swinging	U-0.500		U-0.500		U-0.700	
Nonswinging	U-0.500		U-0.500		U-0.500	

Fenestration	Assembly Max. U	Assembly Max. SHGC	Assembly Max. U	Assembly Max. SHGC	Assembly Max. U	Assembly Max. SHGC
Vertical Glazing, 0%–40% of Wall						
Nonmetal framing (all)[b]	U-0.35		U-0.35		U-0.65	
Metal framing (curtainwall/storefront)[c]	U-0.40	SHGC-0.45 all	U-0.40	SHGC-NR all	U-0.60	SHGC-NR all
Metal framing (entrance door)[c]	U-0.80		U-0.80		U-0.90	
Metal framing (all other)[c]	U-0.45		U-0.45		U-0.65	
Skylight with Curb, Glass, % of Roof						
0%–2.0%	U_{all}-0.98	$SHGC_{all}$-NR	U_{all}-0.98	$SHGC_{all}$-NR	U_{all}-1.30	$SHGC_{all}$-NR
2.1%–5.0%	U_{all}-0.98	$SHGC_{all}$-NR	U_{all}-0.98	$SHGC_{all}$-NR	U_{all}-1.30	$SHGC_{all}$-NR
Skylight with Curb, Plastic, % of Roof						
0%–2.0%	U_{all}-0.61	$SHGC_{all}$-NR	U_{all}-0.61	$SHGC_{all}$-NR	U_{all}-1.10	$SHGC_{all}$-NR
2.1%–5.0%	U_{all}-0.61	$SHGC_{all}$-NR	U_{all}-0.61	$SHGC_{all}$-NR	U_{all}-1.10	$SHGC_{all}$-NR
Skylight without Curb, All, % of Roof						
0%–2.0%	U_{all}-0.58	$SHGC_{all}$-NR	U_{all}-0.58	$SHGC_{all}$-NR	U_{all}-0.81	$SHGC_{all}$-NR
2.1%–5.0%	U_{all}-0.58	$SHGC_{all}$-NR	U_{all}-0.58	$SHGC_{all}$-NR	U_{all}-0.81	$SHGC_{all}$-NR

*The following definitions apply: c.i. = continuous insulation (see Section 3.2), NR = no (insulation) requirement.

[b]Nonmetal framing includes framing materials other than metal with or without metal reinforcing or cladding.

[c]Metal framing includes metal framing with or without thermal break. The "all other" subcategory includes operable windows, fixed windows, and non-entrance doors.

Exceptions to Section 5.5.3:

a. For assemblies significantly different from those in Appendix A, calculations shall be performed in accordance with the procedures required in Appendix A.
b. For multiple assemblies within a single *class of construction* for a single *space-conditioning category*, compliance shall be shown for either (1) the most restrictive requirement or (2) an area-weighted average *U-factor, C-factor,* or *F-factor.*

5.5.3.1 Roof Insulation. All *roofs* shall comply with the insulation values specified in Tables 5.5-1 through 5.5-8 or shall comply with the insulation values specified in Section 5.5.3.1.1 and Table 5.5.3.1. Skylight curbs shall be insulated to the level of roofs with insulation entirely above deck or R-5, whichever is less.

5.5.3.1.1 High Albedo Roofs. For *roofs*, other than *roofs* over ventilated attics or *roofs* over *semi-heated spaces* or *roofs* over *conditioned spaces* that are not *cooled spaces*, where the exterior surface has

a. a solar reflectance of 0.70 when tested in accordance with ASTM C1549, ASTM E903, or ASTM E1918 and, in addition, a minimum thermal emittance of 0.75 when tested in accordance with ASTM C1371 or ASTM E408 or
b. a minimum Solar Reflective Index of 82 when determined in accordance with the Solar Reflectance Index method in ASTM E1980,

the insulation value for the roof shall comply with the values in Table 5.5.3.1. The values for solar reflectance and thermal emittance shall be determined by a laboratory accredited by a nationally recognized accreditation organization, such as the Cool Roof Rating Council CRRC-1 Product Rating Program, and shall be labeled and certified by the manufacturer.

5.5.3.2 Above-Grade Wall Insulation. All *above-grade walls* shall comply with the insulation values specified in Tables 5.5-1 through 5.5-8. When a *wall* consists of both *above-grade* and *below-grade* portions, the entire *wall* for that story shall be insulated on either the exterior or the interior or be integral.

a. If insulated on the interior, the *wall* shall be insulated to the *above-grade wall* requirements.
b. If insulated on the exterior or integral, the *below-grade wall* portion shall be insulated to the *below-grade wall* requirements, and the *above-grade wall* portion shall be insulated to the *above-grade wall* requirements.

5.5.3.3 Below-Grade Wall Insulation. *Below-grade walls* shall have a *rated R-value of insulation* not less that the insulation values specified in Tables 5.5-1 through 5.5-8.

Exception: Where framing, including metal and wood studs, is used, compliance shall be based on the maximum assembly *C-factor.*

5.5.3.4 Floor Insulation. All *floors* shall comply with the insulation values specified in Tables 5.5-1 through 5.5-8.

5.5.3.5 Slab-on-Grade Floor Insulation. All *slab-on-grade floors*, including *heated slab-on-grade floors* and *unheated slab-on-grade floors*, shall comply with the insulation values specified in Tables 5.5-1 through 5.5-8.

5.5.3.6 Opaque Doors. All *opaque doors* shall have a *U-factor* not greater than that specified in Tables 5.5-1 through 5.5-8.

TABLE 5.5.3.1 High Albedo Roof Insulation

Climate Zone	Opaque Elements (Roofs)	Nonresidential		Residential	
		Assembly Maximum	Insulation Min. R-Value	Assembly Maximum	Insulation Min. R-Value
1	Insulation entirely above deck	U-0.082	R-12.0 c.i.	U-0.081	R-12.0 c.i.
	Metal building	U-0.084	R-13	U-0.084	R-13.0
	Attic and other[a]	U-0.044	R-24.0	U-0.035	R-30.0
2	Insulation entirely above deck	U-0.076	R-13.0 c.i.	U-0.076	R-13.0 c.i.
	Metal building	U-0.078	R-13.0	U-0.078	R-13.0
	Attic and other[a]	U-0.041	R-25.0	U-0.032	R-30.0
3	Insulation entirely above deck	U-0.074	R-13.0 c.i.	U-0.074	R-13.0 c.i.
	Metal building	U-0.076	R-16	U-0.076	R-16.0
	Attic and other[a]	U-0.040	R-25.0	U-0.032	R-30.0
4, 5, 6, 7, 8	All roof opaque elements	NP	NP	NP	NP

NP = Not permitted.
[a]Excludes roofs over ventilated attics, or roofs over semiheated spaces, or roofs over conditioned spaces that are not cooled spaces.

5.5.4 Fenestration

5.5.4.1 General. Compliance with *U-factors* and *SHGC* shall be demonstrated for the overall fenestration product. Gross wall areas and gross roof areas shall be calculated separately for each *space-conditioning category* for the purposes of determining compliance.

Exception: If there are multiple assemblies within a single *class of construction* for a single *space-conditioning category*, compliance shall be based on an area-weighted average *U-factor* or *SHGC*. It is not acceptable to do an area-weighted average across multiple *classes of construction* or multiple *space-conditioning categories*.

5.5.4.2 Fenestration Area

5.5.4.2.1 Vertical Fenestration Area. The total *vertical fenestration area* shall be less than 40% of the *gross wall area*.

Exception: *Vertical fenestration* complying with Exception (b) to Section 5.5.4.4.1.

5.5.4.2.2 Skylight Fenestration Area. The total *skylight area* shall be less than 5% of the *gross roof area*.

5.5.4.3 Fenestration U-Factor. *Fenestration* shall have a *U-factor* not greater than that specified in Tables 5.5-1 through 5.5-8 for the appropriate *fenestration area*.

5.5.4.4 Fenestration Solar Heat aGain Coefficient (SHGC)

5.5.4.4.1 SHGC of Vertical Fenestration. *Vertical fenestration* shall have an *SHGC* not greater than that specified for "all" orientations in Tables 5.5-1 through 5.5-8 for the appropriate total *vertical fenestration area*.

Exceptions:

a. For demonstrating compliance for *vertical fenestration* shaded by opaque permanent projections that will last as long as the building itself, the *SHGC* in the proposed building shall be reduced by using the multipliers in Table 5.5.4.4.1. Permanent projections consisting of open louvers shall be considered to provide shading, provided that no sun penetrates the louvers during the peak sun angle on June 21.

b. For demonstrating compliance for *vertical fenestration* shaded by partially opaque permanent projections (e.g., framing with glass or perforated metal) that will last as long as the building itself, the *PF* shall be reduced by multiplying it by a factor of O_s, which is derived as follows:

$$O_s = (A_i \cdot O_i) + (A_f \cdot O_f)$$

where

O_s = percent opacity of the shading device

A_i = percent of the area of the shading device that is a partially opaque infill

O_i = percent opacity of the infill—for glass O_i = (100% – T_s), where T_s is the solar transmittance as determined in accordance with NFRC 300; for perforated or decorative metal panels O_i = percentage of solid material

A_f = percent of the area of the shading device that represents the framing members

O_f = percent opacity of the framing members; if solid, then 100%

And then the *SHGC* in the proposed building shall be reduced by using the multipliers in Table 5.5.4.4.1 for each *fenestration* product.

c. *Vertical fenestration* that is located on the street side of the street-level story only, provided that

1. the street side of the street-level story does not exceed 20 ft in height,
2. the *fenestration* has a continuous overhang with a weighted average *PF* greater than 0.5, and
3. the *fenestration area* for the street side of the street-level story is less than 75% of the *gross wall area* for the street side of the street-level story.

When this exception is utilized, separate calculations shall be performed for these sections of the *building envelope,* and these values shall not be averaged with any others for compliance purposes. No credit shall be given here or elsewhere in the building for not fully utilizing the *fenestration area* allowed.

TABLE 5.5.4.4.1 SHGC Multipliers for Permanent Projections

Projection Factor	SHGC Multiplier (All Other Orientations)	SHGC Multiplier (North-Oriented)
0–0.10	1.00	1.00
>0.10–0.20	0.91	0.95
>0.20–0.30	0.82	0.91
>0.30–0.40	0.74	0.87
>0.40–0.50	0.67	0.84
>0.50–0.60	0.61	0.81
>0.60–0.70	0.56	0.78
>0.70–0.80	0.51	0.76
>0.80–0.90	0.47	0.75
>0.90–1.00	0.44	0.73

5.5.4.4.2 SHGC of Skylights. *Skylights* shall have an *SHGC* not greater than that specified for "all" orientations in Tables 5.5-1 through 5.5-8 for the appropriate total *skylight area*.

5.6 Building Envelope Trade-Off Option

5.6.1 The *building envelope* complies with the standard if

a. the proposed building satisfies the provisions of Sections 5.1, 5.4, 5.7, and 5.8, and

b. the *envelope performance factor* of the proposed building is less than or equal to the *envelope performance factor* of the budget building.

5.6.1.1 The *envelope performance factor* considers only the *building envelope* components.

5.6.1.2 Schedules of operation, lighting power, equipment power, occupant density, and mechanical systems shall be the same for both the proposed building and the budget building.

5.6.1.3 *Envelope performance factor* shall be calculated using the procedures of Normative Appendix C.

5.7 Submittals

5.7.1 General. The *authority having jurisdiction* may require submittal of compliance documentation and supplemental information, in accordance with Section 4.2.2 of this standard.

5.7.2 Submittal Document Labeling of Space Conditioning Categories. For buildings that contain spaces that will be only semiheated or unconditioned, and compliance is sought using the "semiheated" envelope criteria, such spaces shall be clearly indicated on the floor plans that are submitted for review.

5.8 Product Information and Installation Requirements

5.8.1 Insulation

5.8.1.1 Labeling of Building Envelope Insulation. The *rated R-value* shall be clearly identified by an identification mark applied by the *manufacturer* to each piece of *building envelope* insulation.

Exception: When insulation does not have such an identification mark, the installer of such insulation shall provide a signed and dated certification for the installed insulation listing the type of insulation, the *manufacturer*, the *rated R-value*, and, where appropriate, the initial installed thickness, the settled thickness, and the coverage area.

5.8.1.2 Compliance with Manufacturers' Requirements. Insulation materials shall be installed in accordance with *manufacturers'* recommendations and in such a manner as to achieve *rated R-value of insulation*.

Exception: Where *metal building roof* and *metal building wall* insulation is compressed between the *roof* or *wall* skin and the structure.

5.8.1.3 Loose-Fill Insulation Limitation. Open-blown or poured loose-fill insulation shall not be used in *attic roof* spaces when the slope of the ceiling is more than three in twelve.

5.8.1.4 Baffles. When eave vents are installed, baffling of the vent openings shall be provided to deflect the incoming air above the surface of the insulation.

5.8.1.5 Substantial Contact. Insulation shall be installed in a permanent manner in *substantial contact* with the inside surface in accordance with *manufacturers'* recommendations for the framing system used. Flexible batt insulation installed in floor cavities shall be supported in a permanent manner by supports no greater than 24 in. on center.

Exception: Insulation materials that rely on air spaces adjacent to reflective surfaces for their rated performance.

5.8.1.6 Recessed Equipment. Lighting fixtures; heating, ventilating, and air-conditioning equipment, including wall heaters, ducts, and plenums; and other equipment shall not be recessed in such a manner as to affect the insulation thickness unless

a. the total combined area affected (including necessary clearances) is less than 1% of the opaque area of the assembly,
b. the entire *roof*, *wall*, or *floor* is covered with insulation to the full depth required, or
c. the effects of reduced insulation are included in calculations using an area-weighted average method and compressed insulation values obtained from Table A9.4.C.

In all cases, air leakage through or around the recessed equipment to the *conditioned space* shall be limited in accordance with Section 5.4.3.

5.8.1.7 Insulation Protection. Exterior insulation shall be covered with a protective material to prevent damage from sunlight, moisture, landscaping operations, equipment maintenance, and wind.

5.8.1.7.1 In *attics* and mechanical rooms, a way to access equipment that prevents damaging or compressing the insulation shall be provided.

5.8.1.7.2 Foundation vents shall not interfere with the insulation.

5.8.1.7.3 Insulation materials in ground contact shall have a water absorption rate no greater than 0.3% when tested in accordance with ASTM C272.

5.8.1.8 Location of Roof Insulation. The *roof* insulation shall not be installed on a suspended ceiling with removable ceiling panels.

5.8.1.9 Extent of Insulation. Insulation shall extend over the full component area to the required rated R-value of insulation, U-factor, C-factor, or F-factor, unless otherwise allowed in Section 5.8.1.

5.8.2 Fenestration and Doors

5.8.2.1 Rating of Fenestration Products. The U-factor, SHGC, and air leakage rate for all manufactured *fenestration* products shall be determined by a laboratory accredited by a nationally recognized accreditation organization, such as the National Fenestration Rating Council.

5.8.2.2 Labeling of Fenestration Products. All manufactured *fenestration* products shall have a permanent nameplate, installed by the *manufacturer*, listing the U-factor, SHGC, and air leakage rate.

Exception: When the *fenestration* product does not have such nameplate, the installer or supplier of such *fenestration* shall provide a signed and dated certification for the installed fenestration listing the U-factor, SHGC, and the air leakage rate.

5.8.2.3 Labeling of Doors. The *U-factor* and the air leakage rate for all manufactured *doors* installed between *conditioned space*, *semi-heated space*, *unconditioned space*, and exterior *space* shall be identified on a permanent nameplate installed on the product by the *manufacturer*.

Exception: When doors do not have such a nameplate, the installer or supplier of any such doors shall provide a signed and dated certification for the installed doors listing the *U-factor* and the air leakage rate.

5.8.2.4 U-factor. U-factors shall be determined in accordance with NFRC 100. U-factors for skylights shall be determined for a slope of 20 degrees above the horizontal.

Exceptions:

a. U-factors from Section A8.1 shall be an acceptable alternative for determining compliance with the U-factor criteria for *skylights*. Where credit is being taken for a low-emissivity coating, the emissivity of the coating shall be determined in accordance with NFRC 300. Emissivity shall be verified and certified by the *manufacturer*.

b. U-factors from Section A8.2 shall be an acceptable alternative for determining compliance with the U-factor criteria for *vertical fenestration*.

c. U-factors from Section A7 shall be an acceptable alternative for determining compliance with the U-factor criteria for *opaque doors*.

d. For garage doors, ANSI/DASMA105 shall be an acceptable alternative for determining *U-factors*.

5.8.2.5 Solar Heat Gain Coefficient. *SHGC* for the overall *fenestration area* shall be determined in accordance with NFRC 200.

Exceptions:

a. *SC* of the center-of-glass multiplied by 0.86 shall be an acceptable alternative for determining compliance with the *SHGC* requirements for the overall *fenestration area*. *SC* shall be determined using a spectral data file determined in accordance with NFRC 300. *SC* shall be verified and certified by the *manufacturer*.

b. *SHGC* of the center-of-glass shall be an acceptable alternative for determining compliance with the *SHGC* requirements for the overall *fenestration area*. *SHGC* shall be determined using a spectral data file determined in accordance with NFRC 300. SHGC shall be verified and certified by the *manufacturer*.

c. *SHGC* from Section A8.1 shall be an acceptable alternative for determining compliance with the *SHGC* criteria for *skylights*. Where credit is being taken for a low-emissivity coating, the emissivity of the coating shall be determined in accordance with NFRC 300. Emissivity shall be verified and certified by the *manufacturer*.

d. *SHGC* from Section A8.2 shall be an acceptable alternative for determining compliance with the *SHGC* criteria for *vertical fenestration*.

5.8.2.6 Visible Light Transmittance. VLT shall be determined in accordance with NFRC 200. VLT shall be verified and certified by the *manufacturer*.

6. HEATING, VENTILATING, AND AIR CONDITIONING

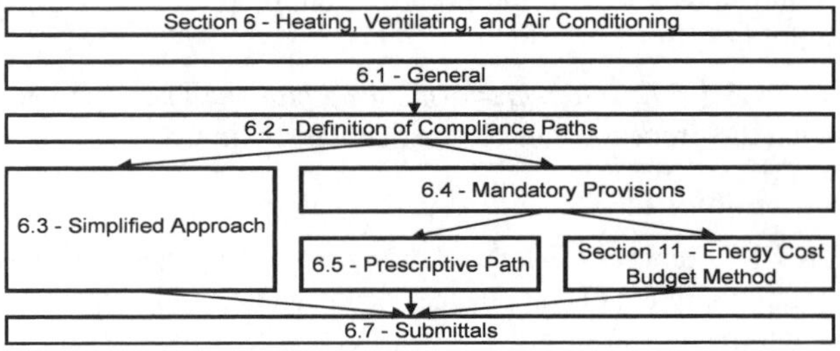

6.1 General

6.1.1 Scope

6.1.1.1 New Buildings. Mechanical equipment and systems serving the heating, cooling, or ventilating needs of new buildings shall comply with the requirements of this section as described in Section 6.2.

6.1.1.2 Additions to Existing Buildings. Mechanical equipment and systems serving the heating, cooling, or ventilating needs of *additions* to *existing buildings* shall comply with the requirements of this section as described in Section 6.2.

Exception: When HVAC to an *addition* is provided by existing *HVAC systems* and equipment, such existing *systems* and *equipment* shall not be required to comply with this standard. However, any new *systems* or *equipment* installed must comply with specific requirements applicable to those *systems* and *equipment*.

6.1.1.3 Alterations to Heating, Ventilating, and Air Conditioning in Existing Buildings

6.1.1.3.1 New HVAC equipment as a direct replacement of existing HVAC equipment shall comply with the specific minimum *efficiency* requirements applicable to that equipment.

6.1.1.3.2 New cooling systems installed to serve previously uncooled spaces shall comply with this section as described in Section 6.2.

6.1.1.3.3 *Alterations* to existing cooling systems shall not decrease economizer capability unless the system complies with Section 6.5.1.

6.1.1.3.4 New and replacement ductwork shall comply with Sections 6.4.4.1 and 6.4.4.2.

6.1.1.3.5 New and replacement piping shall comply with Section 6.4.4.1.

Exceptions: Compliance shall not be required

a. for *equipment* that is being modified or repaired but not replaced, provided that such modifications and/or repairs will not result in an increase in the annual energy consumption of the equipment using the same energy type;

b. where a replacement or *alteration* of *equipment* requires extensive revisions to other *systems*, *equip-*

ment, or elements of a *building*, and such replaced or altered equipment is a like-for-like replacement;

c. for a refrigerant change of existing *equipment*;
d. for the relocation of existing *equipment*; or
e. for ducts and pipes where there is insufficient space or access to meet these requirements.

6.2 Compliance Path(s)

6.2.1 Compliance with Section 6 shall be achieved by meeting all requirements for Section 6.1, General; Section 6.7, Submittals; Section 6.8, Minimum Equipment Efficiency; and either

a. Section 6.3, Simplified Approach Option for HVAC Systems; or
b. Section 6.4, Mandatory Provisions; and Section 6.5, Prescriptive Path.

6.2.2 Projects using the Energy Cost Budget Method (Section 11 of this standard) must comply with Section 6.4, the mandatory provisions of this section, as a portion of that compliance path.

6.3 Simplified Approach Option for HVAC Systems

6.3.1 Scope. The simplified approach is an optional path for compliance when the following conditions are met:

a. building is two stories or fewer in height
b. *gross floor area* is less than 25,000 ft^2
c. each HVAC *system* in the building complies with the requirements listed in Section 6.3.2

6.3.2 Criteria. The HVAC *system* must meet ALL of the following criteria:

a. The *system* serves a single *HVAC zone*.
b. Cooling (if any) shall be provided by a unitary packaged or split-system air conditioner that is either air-cooled or evaporatively cooled with *efficiency* meeting the requirements shown in Table 6.8.1A (air conditioners), Table 6.8.1B (heat pumps), or Table 6.8.1D (packaged terminal and room air conditioners and heat pumps) for the applicable equipment category.
c. The *system* shall have an air economizer where indicated in Table 6.5.1, with controls as indicated in Tables 6.5.1.1.3A and 6.5.1.1.3B and with either barometric or powered relief sized to prevent overpressurization of the building. Where the cooling *efficiency* meets or exceeds the *efficiency* requirement in Table 6.3.2, no economizer is required. *Outdoor air* dampers for economizer use shall be provided with blade and jamb seals.
d. Heating (if any) shall be provided by a unitary packaged or split-system heat pump that meets the applicable *efficiency* requirements shown in Table 6.8.1B (heat pumps) or Table 6.8.1D (packaged terminal and room air conditioners and heat pumps), a fuel-fired furnace that meets the applicable *efficiency* requirements shown in Table 6.8.1E (furnaces, duct furnaces, and unit heaters), an electric resistance heater, or a baseboard system connected to a boiler that meets the applicable *efficiency* requirements shown in Table 6.8.1F (boilers).
e. The *outdoor air* quantity supplied by the system shall be less than or equal to 3000 cfm and less than 70% of the supply air quantity at minimum *outdoor air* design conditions unless an energy recovery ventilation system is provided in accordance with the requirements in Section 6.5.6.
f. The *system* shall be controlled by a manual changeover or dual setpoint thermostat.

TABLE 6.3.2 Eliminate Required Economizer by Increasing Cooling Efficiency

Unitary Systems with Heat Pump Heating						
		Climate Zones				
System Size (kBtu/h)	Mandatory Minimum EER[a]	5 to 8	4	3	2	Test Procedure[c]
		Minimum Cooling Efficiency Required (EER)[a]				
≥65 and <135	10.1	N/A[b]	12.1	11.6	11.1	ARI 340/360
≥135 and <240	9.3	N/A[b]	11.3	10.8	10.4	
≥240 and <760	9.0	N/A[b]	10.9	10.5	10.0	

Other Unitary Systems						
		Climate Zones				
System Size (kBtu/h)	Mandatory Minimum EER[a]	5 to 8	4	3	2	Test Procedure[c]
		Minimum Cooling Efficiency Required (EER)[a]				
≥65 and <135	10.3	N/A[b]	12.5	12.0	11.5	ARI 340/360
≥135 and <240	9.7	N/A[b]	11.5	11.1	10.6	
≥240 and <760	9.5	N/A[b]	11.2	10.7	10.3	

[a] Each EER shown should be reduced by 0.2 for units with a heating section other than electric resistance heat.
[b] Elimination of required economizer is not allowed.
[c] Section 12 contains complete specification of the referenced test procedure, including the referenced year version of the test procedure.

g. If a heat pump equipped with auxiliary internal electric resistance heaters is installed, controls shall be provided that prevent supplemental heater operation when the heating load can be met by the heat pump alone during both steady-state operation and setback recovery. Supplemental heater operation is permitted during outdoor coil defrost cycles. Two means of meeting this requirement are (1) a digital or electronic thermostat designed for heat pump use that energizes auxiliary heat only when the heat pump has insufficient capacity to maintain setpoint or to warm up the space at a sufficient rate or (2) a multistage space thermostat and an outdoor air thermostat wired to energize auxiliary heat only on the last stage of the space thermostat and when outside air temperature is less than 40°F. Heat pumps whose minimum efficiency is regulated by NAECA and whose HSPF rating both meets the requirements shown in Table 6.8.1B and includes all usage of internal electric resistance heating are exempted from the control requirements of this part (Section 6.3.2[g]).

h. The *system* controls shall not permit reheat or any other form of simultaneous heating and cooling for humidity control.

i. *Systems* serving spaces other than hotel/motel guest rooms, and other than those requiring continuous operation, which have both a cooling or heating capacity greater than 15,000 Btu/h and a supply fan motor power greater than 0.75 hp, shall be provided with a time clock that (1) can start and stop the system under different schedules for seven different day-types per week, (2) is capable of retaining programming and time setting during a loss of power for a period of at least ten hours, (3) includes an accessible manual override that allows temporary operation of the system for up to two hours, (4) is capable of temperature setback down to 55°F during off hours, and (5) is capable of temperature setup to 90°F during off hours.

j. Except for piping within *manufacturers'* units, HVAC piping shall be insulated in accordance with Table 6.8.3. Insulation exposed to weather shall be suitable for outdoor service, e.g., protected by aluminum, sheet metal, painted canvas, or plastic cover. Cellular foam insulation shall be protected as above or painted with a coating that is water retardant and provides shielding from solar radiation.

k. Ductwork and plenums shall be insulated in accordance with Tables 6.8.2A and 6.8.2B and shall be sealed in accordance with Table 6.4.4.2A.

l. Construction documents shall require a ducted *system* to be air balanced in accordance with industry-accepted procedures.

m. Where separate heating and cooling equipment serves the same temperature zone, thermostats shall be interlocked to prevent simultaneous heating and cooling.

n. Exhausts with a design capacity of over 300 cfm on *systems* that do not operate continuously shall be equipped with gravity or motorized dampers that will automatically shut when the *systems* are not in use.

o. *Systems* with a design supply air capacity greater than 10,000 cfm shall have *optimum start controls*.

6.4 Mandatory Provisions

6.4.1 Equipment Efficiencies, Verification, and Labeling Requirements

6.4.1.1 Minimum Equipment Efficiencies—Listed Equipment—Standard Rating and Operating Conditions. Equipment shown in Tables 6.8.1A through 6.8.1G shall have a minimum performance at the specified rating conditions when tested in accordance with the specified test procedure. Where multiple rating conditions or performance requirements are provided, the equipment shall satisfy all stated requirements, unless otherwise exempted by footnotes in the table. Equipment covered under the Federal Energy Policy Act of 1992 (EPACT) shall have no minimum *efficiency* requirements for operation at minimum capacity or other than standard rating conditions. Equipment used to provide water heating functions as part of a combination system shall satisfy all stated requirements for the appropriate space heating or cooling category.

Tables are as follows:

a. Table 6.8.1A—Air Conditioners and Condensing Units
b. Table 6.8.1B—Heat Pumps
c. Table 6.8.1C—Water-Chilling Packages (see Section 6.4.1.2 for water-cooled centrifugal water-chilling packages that are designed to operate at nonstandard conditions)
d. Table 6.8.1D—Packaged Terminal and Room Air Conditioners and Heat Pumps
e. Table 6.8.1E—Furnaces, Duct Furnaces, and Unit Heaters
f. Table 6.8.1F—Boilers
g. Table 6.8.1G—Heat Rejection Equipment

All furnaces with input ratings of ≥225,000 Btu/h, including electric furnaces, that are not located within the conditioned space shall have jacket losses not exceeding 0.75% of the input rating.

6.4.1.2 Minimum Equipment Efficiencies—Listed Equipment—Nonstandard Conditions. Water-cooled centrifugal water-chilling packages that are not designed for operation at ARI Standard 550/590 test conditions (and, thus, cannot be tested to meet the requirements of Table 6.8.1C) of 44°F leaving chilled-water temperature and 85°F entering condenser-water temperature with 3 gpm/ton condenser-water flow shall have a minimum full-load COP and a minimum *NPLV* rating as shown in the tables referenced below.

a. Centrifugal chillers <150 tons shall meet the minimum full-load COP and IPLV/NPLV in Table 6.8.1H.
b. Centrifugal chillers ≥150 tons and <300 tons shall meet the minimum full-load COP and IPLV/NPLV in Table 6.8.1I.
c. Centrifugal chillers ≥300 tons shall meet the minimum full-load COP and IPLV/NPLV in Table 6.8.1J.

The table values are only applicable over the following full-load design ranges:

- Leaving Chiller-Water Temperature: 40°F to 48°F
- Entering Condenser-Water Temperature: 75°F to 85°F
- Condenser-Water Temperature Rise: 5°F to 15°F

Chillers designed to operate outside of these ranges or applications utilizing fluids or solutions with secondary coolants (e.g., glycol solutions or brines) with a freeze point of 27°F or lower for freeze protection are not covered by this standard.

6.4.1.3 Equipment Not Listed. Equipment not listed in the tables referenced in Sections 6.4.1.1 and 6.4.1.2 may be used.

6.4.1.4 Verification of Equipment Efficiencies. Equipment *efficiency* information supplied by *manufacturers* shall be verified as follows:

a. Equipment covered under EPACT shall comply with U.S. Department of Energy certification requirements.
b. If a certification program exists for a covered product, and it includes provisions for verification and challenge of equipment *efficiency* ratings, then the product shall be listed in the certification program, or
c. if a certification program exists for a covered product, and it includes provisions for verification and challenge of equipment *efficiency* ratings, but the product is not listed in the existing certification program, the ratings shall be verified by an independent laboratory test report, or
d. if no certification program exists for a covered product, the equipment *efficiency* ratings shall be supported by data furnished by the *manufacturer*, or
e. where components such as indoor or outdoor coils from different *manufacturers* are used, the system designer shall specify component efficiencies whose combined *efficiency* meets the minimum equipment *efficiency* requirements in Section 6.4.1.

6.4.1.5 Labeling

6.4.1.5.1 Mechanical Equipment. Mechanical equipment that is not covered by the U.S. National Appliance Energy Conservation Act (NAECA) of 1987 shall carry a permanent label installed by the *manufacturer* stating that the equipment complies with the requirements of Standard 90.1.

6.4.1.5.2 Packaged Terminal Air Conditioners. Packaged terminal air conditioners and heat pumps with sleeve sizes less than 16 in. high and 42 in. wide shall be factory labeled as follows: *Manufactured for replacement applications only: not to be installed in new construction projects.*

6.4.2 Load Calculations. Heating and cooling system design loads for the purpose of sizing systems and equipment shall be determined in accordance with generally accepted engineering standards and handbooks acceptable to the *adopting authority* (for example, *ASHRAE Handbook—Fundamentals*).

6.4.3 Controls

6.4.3.1 Zone Thermostatic Controls

6.4.3.1.1 General. The supply of heating and cooling energy to each *zone* shall be individually controlled by thermostatic controls responding to temperature within the *zone*. For the purposes of Section 6.4.3.1, a dwelling unit shall be permitted to be considered a single *zone*.

Exceptions: Independent perimeter systems that are designed to offset only *building envelope* loads shall be permitted to serve one or more *zones* also served by an interior system provided

a. the perimeter system includes at least one thermostatic control zone for each building exposure having exterior walls facing only one *orientation* for 50 contiguous feet or more, and
b. the perimeter system heating and cooling supply is controlled by a thermostatic control(s) located within the zones(s) served by the system.

Exterior walls are considered to have different *orientations* if the directions they face differ by more than 45 degrees.

6.4.3.1.2 Dead Band. Where used to control both heating and cooling, zone thermostatic controls shall be capable of providing a temperature range or dead band of at least 5°F within which the supply of heating and cooling energy to the zone is shut off or reduced to a minimum.

Exceptions:

a. Thermostats that require manual changeover between heating and cooling modes.
b. Special occupancy or special applications where wide temperature ranges are not acceptable (such as retirement homes, process applications, museums, some areas of hospitals) and are approved by the *authority having jurisdiction*.

6.4.3.2 Setpoint Overlap Restriction. Where heating and cooling to a zone are controlled by separate zone thermostatic controls located within the zone, means (such as limit switches, mechanical stops, or, for DDC systems, software programming) shall be provided to prevent the heating setpoint from exceeding the cooling setpoint minus any applicable proportional band.

6.4.3.3 Off-Hour Controls. HVAC systems shall have the off-hour controls required by Sections 6.4.3.3.1 through 6.4.3.3.4.

Exceptions:

a. *HVAC systems* intended to operate continuously.
b. *HVAC systems* having a design heating capacity and cooling capacity less than 15,000 Btu/h that are equipped with readily accessible manual ON/OFF controls.

6.4.3.3.1 Automatic Shutdown. *HVAC systems* shall be equipped with at least one of the following:

a. Controls that can start and stop the system under different time schedules for seven different day-types per week, are capable of retaining programming and time setting during loss of power for a period of at least ten hours, and include an accessible manual override, or equivalent function, that allows temporary operation of the system for up to two hours.

b. An *occupant sensor* that is capable of shutting the system off when no occupant is sensed for a period of up to 30 minutes.
c. A manually operated timer capable of being adjusted to operate the system for up to two hours.
d. An interlock to a security system that shuts the system off when the security system is activated.

Exception: Residential occupancies may use controls that can start and stop the system under two different time schedules per week.

6.4.3.3.2 Setback Controls. Heating systems located in climate zones 2–8 shall be equipped with controls that have the capability to automatically restart and temporarily operate the system as required to maintain *zone* temperatures above a heating setpoint adjustable down to 55°F or lower. Cooling systems located in climate zones 1b, 2b, and 3b shall be equipped with controls that have the capability to automatically restart and temporarily operate the system as required to maintain *zone* temperatures below a cooling setpoint adjustable up to 90°F or higher or to prevent high space humidity levels.

Exception: Radiant floor and ceiling heating *systems.*

6.4.3.3.3 Optimum Start Controls. Individual heating and cooling air distribution systems with a total design supply air capacity exceeding 10,000 cfm, served by one or more supply fans, shall have *optimum start controls*. The control algorithm shall, as a minimum, be a function of the difference between space temperature and occupied setpoint and the amount of time prior to scheduled occupancy.

6.4.3.3.4 Zone Isolation. *HVAC systems* serving *zones* that are intended to operate or be occupied nonsimultaneously shall be divided into isolation areas. Zones may be grouped into a single isolation area provided it does not exceed 25,000 ft^2 of conditioned floor area nor include more than one floor. Each isolation area shall be equipped with *isolation devices* capable of automatically shutting off the supply of conditioned air and *outdoor air* to and exhaust air from the area. Each isolation area shall be controlled independently by a device meeting the requirements of Section 6.4.3.3.1, Automatic Shutdown. For central systems and plants, controls and devices shall be provided to allow stable system and equipment operation for any length of time while serving only the smallest isolation area served by the system or plant.

Exceptions: Isolation devices and controls are not required for the following:

a. Exhaust air and *outdoor air* connections to isolation *zones* when the fan system to which they connect is 5000 cfm and smaller.
b. Exhaust airflow from a single isolation *zone* of less than 10% of the design airflow of the exhaust system to which it connects.
c. *Zones* intended to operate continuously or intended to be inoperative only when all other *zones* are inoperative.

6.4.3.4 Ventilation System Controls

6.4.3.4.1 Stair and Shaft Vents. Stair and elevator shaft vents shall be equipped with motorized dampers that are capable of being automatically closed during normal building operation and are interlocked to open as required by fire and smoke detection systems.

6.4.3.4.2 Gravity Hoods, Vents, and Ventilators. All *outdoor air* supply and exhaust hoods, vents, and ventilators shall be equipped with motorized dampers that will automatically shut when the spaces served are not in use.

Exceptions:

a. Gravity (nonmotorized) dampers are acceptable in buildings less than three stories in height above grade and for buildings of any height located in climate zones 1, 2, and 3.
b. Ventilation systems serving *unconditioned spaces*.

6.4.3.4.3 Shutoff Damper Controls. Both *outdoor air* supply and exhaust systems shall be equipped with motorized dampers that will automatically shut when the systems or spaces served are not in use. Ventilation *outdoor air* dampers shall be capable of automatically shutting off during preoccupancy building warm-up, cool down, and *setback,* except when *ventilation* reduces energy costs (e.g., night purge) or when ventilation must be supplied to meet code requirements.

Exceptions:

a. Gravity (nonmotorized) dampers are acceptable in buildings less than three stories in height and for buildings of any height located in climate zones 1, 2, and 3.
b. Gravity (nonmotorized) dampers are acceptable in systems with a design *outdoor air* intake or exhaust capacity of 300 cfm or less.

6.4.3.4.4 Dampers. Where *outdoor air* supply and exhaust air dampers are required by Section 6.4.3.4, they shall have a maximum leakage rate when tested in accordance with AMCA Standard 500 as indicated in Table 6.4.3.4.4.

6.4.3.4.5 Ventilation Fan Controls. Fans with motors greater than 0.75 hp shall have automatic controls complying with Section 6.4.3.3.1 that are capable of shutting off fans when not required.

Exception: *HVAC systems* intended to operate continuously.

6.4.3.5 Heat Pump Auxiliary Heat Control. Heat pumps equipped with internal electric resistance heaters shall

TABLE 6.4.3.4.4 Maximum Damper Leakage

Climate Zones	Maximum Damper Leakage at 1.0 in. w.g. (cfm per ft^2 of damper area)	
	Motorized	Nonmotorized
1, 2, 6, 7, 8	4	Not allowed
All others	10	20[a]

[a] Dampers smaller than 24 in. in either dimension may have leakage of 40 cfm/ft^2.

have controls that prevent supplemental heater operation when the heating load can be met by the heat pump alone during both steady-state operation and setback recovery. Supplemental heater operation is permitted during outdoor coil defrost cycles.

Exceptions: Heat pumps whose minimum *efficiency* is regulated by NAECA and whose HSPF rating both meets the requirements shown in Table 6.8.1B and includes all usage of internal electric resistance heating.

6.4.3.6 Humidifier Preheat. Humidifiers with preheating jackets mounted in the airstream shall be provided with an automatic valve to shut off preheat when humidification is not required.

6.4.3.7 Humidification and Dehumidification. Where a *zone* is served by a system or systems with both humidification and dehumidification capability, means (such as limit switches, mechanical stops, or, for DDC systems, software programming) shall be provided capable of preventing simultaneous operation of humidification and dehumidification equipment.

Exceptions:

a. Zones served by desiccant systems, used with direct evaporative cooling in series.
b. Systems serving zones where specific humidity levels are required, such as museums and hospitals, and approved by the *authority having jurisdiction*.

6.4.3.8 Freeze Protection and Snow/Ice Melting Systems. Freeze protection systems, such as heat tracing of outdoor piping and heat exchangers, including self-regulating heat tracing, shall include automatic controls capable of shutting off the systems when *outdoor air* temperatures are above 40°F or when the conditions of the protected fluid will prevent freezing. Snow- and ice-melting systems shall include automatic controls capable of shutting off the systems when the pavement temperature is above 50°F and no precipitation is falling and an automatic or manual control that will allow shutoff when the outdoor temperature is above 40°F so that the potential for snow or ice accumulation is negligible.

6.4.3.9 Ventilation Controls for High-Occupancy Areas. Demand control ventilation (DCV) is required for spaces larger than 500 ft^2 and with a design occupancy for ventilation of greater than 40 people per 1000 ft^2 of floor area and served by systems with one or more of the following:

a. an air-side economizer,
b. automatic modulating control of the outdoor air damper, or
c. a design outdoor airflow greater than 3000 cfm.

Exceptions:

a. Systems with energy recovery complying with Section 6.5.6.1.
b. Multiple-zone systems without DDC of individual zones communicating with a central control panel.
c. Systems with a design outdoor airflow less than 1200 cfm.
d. Spaces where the supply airflow rate minus any makeup or outgoing transfer air requirement is less than 1200 cfm.

6.4.4 HVAC System Construction and Insulation

6.4.4.1 Insulation

6.4.4.1.1 General. Insulation required by this section shall be installed in accordance with industry-accepted standards (see Informative Appendix E). These requirements do not apply to HVAC equipment. Insulation shall be protected from damage, including that due to sunlight, moisture, equipment maintenance and wind, but not limited to the following:

a. Insulation exposed to weather shall be suitable for outdoor service, e.g., protected by aluminum, sheet metal, painted canvas, or plastic cover. Cellular foam insulation shall be protected as above or painted with a coating that is water retardant and provides shielding from solar radiation that can cause degradation of the material.
b. Insulation covering chilled-water piping, refrigerant suction piping, or cooling ducts located outside the conditioned space shall include a vapor retardant located outside the insulation (unless the insulation is inherently vapor retardant), all penetrations and joints of which shall be sealed.

6.4.4.1.2 Duct and Plenum Insulation. All supply and return ducts and plenums installed as part of an HVAC air distribution system shall be thermally insulated in accordance with Tables 6.8.2A and 6.8.2B.

Exceptions:

a. Factory-installed plenums, casings, or ductwork furnished as a part of HVAC equipment tested and rated in accordance with Section 6.4.1.
b. Ducts or plenums located in heated spaces, *semiheated spaces*, or cooled spaces.
c. For runouts less than 10 ft in length to air terminals or air outlets, the rated R-value of insulation need not exceed R-3.5.
d. Backs of air outlets and outlet plenums exposed to unconditioned or indirectly *conditioned* spaces with face areas exceeding 5 ft^2 need not exceed R-2; those 5 ft^2 or smaller need not be insulated.

6.4.4.1.3 Piping Insulation. Piping shall be thermally insulated in accordance with Table 6.8.3.

Exceptions:

a. Factory-installed piping within HVAC equipment tested and rated in accordance with Section 6.4.1.
b. Piping that conveys fluids having a design operating temperature range between 60°F and 105°F, inclusive.
c. Piping that conveys fluids that have not been heated or cooled through the use of nonrenewable energy (such as roof and condensate drains, domestic cold water supply, natural gas piping, or refrigerant liquid piping) or where heat gain or heat loss will not increase energy usage.

d. Hot-water piping between the shutoff valve and the coil, not exceeding 4 ft in length, when located in *conditioned spaces*.
e. Pipe unions in heating systems (steam, steam condensate, and hot water).

6.4.4.2 Ducts and Plenum Leakage

6.4.4.2.1 Duct Sealing. Ductwork and plenums shall be sealed in accordance with Table 6.4.4.2A (Table 6.4.4.2B provides definitions of seal levels), as required to meet the requirements of Section 6.4.4.2.2 and with standard industry practice (see Informative Appendix E).

6.4.4.2.2 Duct Leakage Tests. Ductwork that is designed to operate at static pressures in excess of 3 in. w.c. shall be leak-tested according to industry-accepted test procedures (see Informative Appendix E). Representative sections totaling no less than 25% of the total installed duct area for the designated pressure class shall be tested. Duct systems with pressure ratings in excess of 3 in. w.c. shall be identified on the drawings. The maximum permitted duct leakage shall be

$$L_{max} = C_L P^{0.65},$$

TABLE 6.4.4.2A Minimum Duct Seal Level[a]

Duct Location	Duct Type			
	Supply		Exhaust	Return
	≤2 in. w.c.[b]	>2 in. w.c.[b]		
Outdoor	A	A	C	A
Unconditioned spaces	B	A	C	B
Conditioned spaces[c]	C	B	B	C

[a] See Table 6.4.4.2B description of seal level.
[b] Duct design static pressure classification.
[c] Includes indirectly conditioned spaces such as return air plenums.

TABLE 6.4.4.2B Duct Seal Levels

Seal Level	Sealing Requirements[a]
A	All transverse joints, longitudinal seams, and duct wall penetrations. Pressure-sensitive tape shall not be used as the primary sealant, unless it has been certified to comply with UL-181A or UL-181B by an independent testing laboratory and the tape is used in accordance with that certification.
B	All transverse joints, longitudinal seams. Pressure-sensitive tape shall not be used as the primary sealant, unless it has been certified to comply with UL-181A or UL-181B by an independent testing laboratory and the tape is used in accordance with that certification.
C	Transverse joints only.

[a] Longitudinal seams are joints oriented in the direction of airflow. Transverse joints are connections of two duct sections oriented perpendicular to airflow. Duct wall penetrations are openings made by any screw fastener, pipe, rod, or wire. Spiral lock seams in a round or flat oval duct need not be sealed. All other connections are considered transverse joints, including but not limited to spin-ins, taps, and other branch connections, access door frames and jambs, duct connections to equipment, etc.

where

L_{max} = maximum permitted leakage in cfm/100 ft^2 duct surface area;

C_L = duct leakage class, cfm/100 ft^2 at 1 in. w.c.,
6 for rectangular sheetmetal, rectangular fibrous, and round flexible ducts,
3 for round/flat oval sheetmetal or fibrous glass ducts; and

P = test pressure, which shall be equal to the design duct pressure class rating in in. w.c.

6.4.5 Completion Requirements. Completion requirements are as described in Section 6.7.2.

6.5 Prescriptive Path

6.5.1 Economizers. Each cooling system that has a fan shall include either an air or water economizer meeting the requirements of Sections 6.5.1.1 through 6.5.1.4.

Exceptions: Economizers are not required for the systems listed below.

a. Individual fan-cooling units with a supply capacity less than the minimum listed in Table 6.5.1.
b. Systems that include nonparticulate air treatment as required by Section 6.2.1 in Standard 62.1.
c. Where more than 25% of the air designed to be supplied by the system is to spaces that are designed to be humidified above 35°F dew-point temperature to satisfy process needs.
d. Systems that include a condenser heat recovery system required by Section 6.5.6.2.
e. Systems that serve *residential* spaces where the system capacity is less than five times the requirement listed in Table 6.5.1.
f. Systems that serve spaces whose sensible cooling load at design conditions, excluding transmission and infiltration loads, is less than or equal to transmission and infiltration losses at an outdoor temperature of 60°F.
g. Systems expected to operate less than 20 hours per week.
h. Where the use of *outdoor air* for cooling will affect supermarket open refrigerated casework systems.
i. Where the cooling *efficiency* meets or exceeds the *efficiency* requirements in Table 6.3.2.

6.5.1.1 Air Economizers

6.5.1.1.1 Design Capacity. Air economizer systems shall be capable of modulating *outdoor air* and return air

TABLE 6.5.1 Minimum System Size for Which an Economizer is Required

Climate Zones	Cooling Capacity for Which an Economizer is Required
1a, 1b, 2a, 3a, 4a	No economizer requirement
2b, 5a, 6a, 7, 8	≥135,000 Btu/h
3b, 3c, 4b, 4c, 5b, 5c, 6b	≥65,000 Btu/h

dampers to provide up to 100% of the design supply air quantity as *outdoor air* for cooling.

6.5.1.1.2 Control Signal. Economizer dampers shall be capable of being sequenced with the mechanical cooling equipment and shall not be controlled by only mixed air temperature.

Exception: The use of mixed air temperature limit control shall be permitted for systems controlled from space temperature (such as single-zone systems).

6.5.1.1.3 High-Limit Shutoff. All air economizers shall be capable of automatically reducing *outdoor air* intake to the design minimum *outdoor air* quantity when *outdoor air* intake will no longer reduce cooling energy usage. High-limit shutoff control types for specific climates shall be chosen from Table 6.5.1.1.3A. High-limit shutoff control settings for these control types shall be those listed in Table 6.5.1.1.3B.

6.5.1.1.4 Dampers. Both return air and *outdoor air* dampers shall meet the requirements of Section 6.4.3.3.4.

6.5.1.1.5 Relief of Excess *Outdoor Air*. Systems shall provide a means to relieve excess *outdoor air* during air economizer operation to prevent overpressurizing the building. The relief air outlet shall be located to avoid recirculation into the building.

6.5.1.2 Water Economizers

6.5.1.2.1 Design Capacity. Water economizer systems shall be capable of cooling supply air by indirect evaporation and providing up to 100% of the expected system cooling load at *outdoor air* temperatures of 50°F dry bulb/45°F wet bulb and below.

Exception: Systems in which a water economizer is used and where dehumidification requirements cannot be met using *outdoor air* temperatures of 50°F dry bulb/45°F

TABLE 6.5.1.1.3A High-Limit Shutoff Control Options for Air Economizers

Climate Zones	Allowed Control Types	Prohibited Control Types
1b, 2b, 3b, 3c, 4b, 4c, 5b, 5c, 6b, 7, 8	Fixed dry bulb Differential dry bulb Electronic enthalpy[a] Differential enthalpy Dew-point and dry-bulb temperatures	Fixed enthalpy
1a, 2a, 3a, 4a	Fixed dry bulb Fixed enthalpy Electronic enthalpy[a] Differential enthalpy Dew-point and dry-bulb temperatures	Differential dry bulb
All other climates	Fixed dry bulb Differential dry bulb Fixed enthalpy Electronic enthalpy[a] Differential enthalpy Dew-point and dry-bulb temperatures	

[a] Electronic enthalpy controllers are devices that use a combination of humidity and dry-bulb temperature in their switching algorithm.

TABLE 6.5.1.1.3B High-Limit Shutoff Control Settings for Air Economizers

Device Type	Climate	Required High Limit (Economizer Off When):	
		Equation	Description
Fixed dry bulb	1b, 2b, 3b, 3c, 4b, 4c, 5b, 5c, 6b, 7, 8 5a, 6a, 7a All other zones	$T_{OA} > 75°F$ $T_{OA} > 70°F$ $T_{OA} > 65°F$	Outdoor air temperature exceeds 75°F Outdoor air temperature exceeds 70°F Outdoor air temperature exceeds 65°F
Differential dry bulb	1b, 2b, 3b, 3c, 4b, 4c, 5a, 5b, 5c, 6a, 6b, 7, 8	$T_{OA} > T_{RA}$	Outdoor air temperature exceeds return air temperature
Fixed enthalpy	All	$h_{OA} > 28$ Btu/lb[a]	Outdoor air enthalpy exceeds 28 Btu/lb of dry air[a]
Electronic enthalpy	All	$(T_{OA}, RH_{OA}) > A$	Outdoor air temperature/RH exceeds the "A" setpoint curve[b]
Differential enthalpy	All	$h_{OA} > h_{RA}$	Outdoor air enthalpy exceeds return air enthalpy
Dew-point and dry-bulb temperatures	All	$DP_{oa} > 55°F$ or $T_{oa} > 75°F$	Outdoor air dry bulb exceeds 75°F or outside dew point exceeds 55°F (65 gr/lb)

[a] At altitudes substantially different than sea level, the Fixed Enthalpy limit shall be set to the enthalpy value at 75°F and 50% relative humidity. As an example, at approximately 6000 ft elevation the fixed enthalpy limit is approximately 30.7 Btu/lb.

[b] Setpoint "A" corresponds to a curve on the psychrometric chart that goes through a point at approximately 75°F and 40% relative humidity and is nearly parallel to dry-bulb lines at low humidity levels and nearly parallel to enthalpy lines at high humidity levels.

wet bulb must satisfy 100% of the expected system cooling load at 45°F dry bulb/40°F wet bulb.

6.5.1.2.2 Maximum Pressure Drop. Precooling coils and water-to-water heat exchangers used as part of a water economizer system shall either have a water-side pressure drop of less than 15 ft of water or a secondary loop shall be created so that the coil or heat exchanger pressure drop is not seen by the circulating pumps when the system is in the normal cooling (noneconomizer) mode.

6.5.1.3 Integrated Economizer Control. Economizer systems shall be integrated with the mechanical cooling system and be capable of providing partial cooling even when additional mechanical cooling is required to meet the remainder of the cooling load.

Exceptions:

a. Direct expansion systems that include controls that reduce the quantity of *outdoor air* required to prevent coil frosting at the lowest step of compressor unloading, provided this lowest step is no greater than 25% of the total system capacity.
b. Individual direct expansion units that have a rated cooling capacity less than 65,000 Btu/h and use non-integrated economizer controls that preclude simultaneous operation of the economizer and mechanical cooling.
c. Systems in climate zones 1, 2, 3a, 4a, 5a, 5b, 6, 7, and 8.

6.5.1.4 Economizer Heating System Impact. HVAC system design and economizer controls shall be such that economizer operation does not increase the building heating energy use during normal operation.

Exception: Economizers on VAV systems that cause zone level heating to increase due to a reduction in supply air temperature.

6.5.2 Simultaneous Heating and Cooling Limitation

6.5.2.1 Zone Controls. *Zone* thermostatic controls shall be capable of operating in sequence the supply of heating and cooling energy to the *zone*. Such controls shall prevent

1. *reheating*,
2. *recooling*,
3. mixing or simultaneously supplying air that has been previously mechanically heated and air that has been previously cooled, either by mechanical cooling or by economizer systems, and
4. other simultaneous operation of heating and cooling systems to the same *zone*.

Exceptions:

a. *Zones* for which the volume of air that is reheated, recooled, or mixed is no greater than the larger of the following:
    1. the volume of *outdoor air* required to meet the ventilation requirements of Section 6.2 of Standard 62.1 for the *zone*,
    2. 0.4 cfm/ft^2 of the *zone* conditioned floor area,
    3. 30% of the zone design peak supply rate,
    4. 300 cfm—this exception is for zones whose peak flow rate totals no more than 10% of the total fan system flow rate, or
    5. any higher rate that can be demonstrated, to the satisfaction of the *authority having jurisdiction*, to reduce overall system annual energy usage by offsetting reheat/recool energy losses through a reduction in *outdoor air* intake for the system.
b. *Zones* where special pressurization relationships, cross-contamination requirements, or code-required minimum circulation rates are such that VAV systems are impractical.
c. *Zones* where at least 75% of the energy for reheating or for providing warm air in mixing systems is provided from a *site-recovered* (including condenser heat) or *site-solar energy source.*

6.5.2.2 Hydronic System Controls. The heating of fluids in hydronic systems that have been previously mechanically cooled and the cooling of fluids that have been previously mechanically heated shall be limited in accordance with Sections 6.5.2.2.1 through 6.5.2.2.3.

6.5.2.2.1 Three-Pipe System. Hydronic systems that use a common return system for both hot water and chilled water shall not be used.

6.5.2.2.2 Two-Pipe Changeover System. Systems that use a common distribution system to supply both heated and chilled water are acceptable provided all of the following are met:

a. The system is designed to allow a dead band between changeover from one mode to the other of at least 15°F *outdoor air* temperature.
b. The system is designed to operate and is provided with controls that will allow operation in one mode for at least four hours before changing over to the other mode.
c. Reset controls are provided that allow heating and cooling supply temperatures at the changeover point to be no more than 30°F apart.

6.5.2.2.3 Hydronic (Water Loop) Heat Pump Systems. Hydronic heat pumps connected to a common heat pump water loop with central devices for heat rejection (e.g., cooling tower) and heat addition (e.g., boiler) shall have the following:

a. Controls that are capable of providing a heat pump water supply temperature dead band of at least 20°F between initiation of heat rejection and heat addition by the central devices (e.g., tower and boiler).
b. For climate zones 3 through 8, if a closed-circuit tower (fluid cooler) is used, either an automatic valve shall be installed to bypass all but a minimal flow of water around the tower (for freeze protection) or low-leakage positive closure dampers shall be provided. If an open-circuit tower is used directly in the heat pump loop, an automatic valve shall be installed to bypass all heat pump water flow

around the tower. If an open-circuit tower is used in conjunction with a separate heat exchanger to isolate the tower from the heat pump loop, then heat loss shall be controlled by shutting down the circulation pump on the cooling tower loop.

Exception: Where a system loop temperature optimization controller is used to determine the most efficient operating temperature based on real-time conditions of demand and capacity, dead bands of less than 20°F shall be allowed.

6.5.2.3 Dehumidification. Where humidistatic controls are provided, such controls shall prevent reheating, mixing of hot and cold airstreams, or other means of simultaneous heating and cooling of the same airstream.

Exceptions:

a. The system is capable of reducing supply air volume to 50% or less of the design airflow rate or the minimum rate specified in Section 6.2 of Standard 62.1, whichever is larger, before simultaneous heating and cooling takes place.
b. The individual fan cooling unit has a design cooling capacity of 80,000 Btu/h or less and is capable of unloading to 50% capacity before simultaneous heating and cooling takes place.
c. The individual mechanical cooling unit has a design cooling capacity of 40,000 Btu/h or less. An individual mechanical cooling unit is a single system composed of a fan or fans and a cooling coil capable of providing mechanical cooling.
d. Systems serving spaces where specific humidity levels are required to satisfy process needs, such as computer rooms, museums, surgical suites, and buildings with refrigerating systems, such as supermarkets, refrigerated warehouses, and ice arenas. This exception also applies to other applications for which fan volume controls in accordance with Exception (a) are proven to be impractical to the enforcement agency.
e. At least 75% of the energy for reheating or for providing warm air in mixing systems is provided from a *site-recovered* (including condenser heat) or *site-solar energy* source.
f. Systems where the heat added to the airstream is the result of the use of a desiccant system and 75% of the heat added by the desiccant system is removed by a heat exchanger, either before or after the desiccant system with energy recovery.

6.5.2.4 Humidification. Systems with hydronic cooling and humidification systems designed to maintain inside humidity at a dew-point temperature greater than 35°F shall use a water economizer if an economizer is required by Section 6.5.1.

6.5.3 Air System Design and Control. Each HVAC system having a total *fan system motor nameplate hp* exceeding 5 hp shall meet the provisions of Sections 6.5.3.1 through 6.5.3.2.

6.5.3.1 Fan System Power Limitation

6.5.3.1.1 Each HVAC system at fan system design conditions shall not exceed the allowable *fan system motor nameplate hp* (Option 1) or *fan system bhp* (Option 2) as shown in Table 6.5.3.1.1A. This includes supply fans, return/relief fans, exhaust fans, and fan-powered terminal units associated with systems providing heating or cooling capability.

Exceptions:

a. Hospital and laboratory systems that utilize flow control devices on exhaust and/or return to maintain space pressure relationships necessary for occupant health and safety or environmental control may use variable-volume fan power limitation.
b. Individual exhaust fans with motor nameplate horsepower of 1 hp or less.
c. Fans exhausting air from fume hoods. ***Note:*** If this exception is taken, no related exhaust side credits shall be taken from Table 6.5.3.1.1B and the Fume Hood Exhaust Exception Deduction must be taken from Table 6.5.3.1.1B.

6.5.3.1.2 Motor Nameplate Horsepower. For each fan, the selected fan motor shall be no larger than the first available motor size greater than the bhp. The fan bhp must be indicated on the design documents to allow for compliance verification by the code official.

TABLE 6.5.3.1.1A Fan Power Limitation[a]

	Limit	Constant Volume	Variable Volume
Option 1: Fan System Motor Nameplate hp	Allowable Nameplate Motor hp	hp $\le CFM_S \cdot 0.0011$	hp $\le CFM_S \cdot 0.0015$
Option 2: Fan System bhp	Allowable Fan System bhp	bhp $\le CFM_S \cdot 0.00094 + A$	bhp $\le CFM_S \cdot 0.0013 + A$

[a] where
CFM_S = the maximum design supply airflow rate to conditioned spaces served by the system in cubic feet per minute
hp = the maximum combined motor nameplate horsepower
bhp = the maximum combined fan brake horsepower
A = sum of $(PD \times CFM_D/4131)$
where
PD = each applicable pressure drop adjustment from Table 6.5.3.1.1B in in. w.c.
CFM_D = the design airflow through each applicable device from Table 6.5.3.1.1B in cubic feet per minute

TABLE 6.5.3.1.1B Fan Power Limitation Pressure Drop Adjustment

Device	Adjustment
Credits	
Fully ducted return and/or exhaust air systems	0.5 in. w.c.
Return and/or exhaust airflow control devices	0.5 in. w.c.
Exhaust filters, scrubbers, or other exhaust treatment	The pressure drop of device calculated at fan system design condition
Particulate Filtration Credit: MERV 9 through 12	0.5 in. w.c.
Particulate Filtration Credit: MERV 13 through 15	0.9 in. w.c.
Particulate Filtration Credit: MERV 16 and greater and electronically enhanced filters	Pressure drop calculated at 2× clean filter pressure drop at fan system design condition
Carbon and other gas-phase air cleaners	Clean filter pressure drop at fan system design condition
Heat recovery device	Pressure drop of device at fan system design condition
Evaporative humidifier/cooler in series with another cooling coil	Pressure drop of device at fan system design condition
Sound Attenuation Section	0.15 in. w.c.
Deductions	
Fume Hood Exhaust Exception (required if 6.5.3.1.1 Exception [c] is taken)	–1.0 in. w.c.

Exceptions:

a. For fans less than 6 bhp, where the first available motor larger than the bhp has a nameplate rating within 50% of the bhp, the next larger nameplate motor size may be selected.

b. For fans 6 bhp and larger, where the first available motor larger than the bhp has a nameplate rating with 30% of the bhp, the next larger nameplate motor size may be selected.

6.5.3.2 VAV Fan Control (Including Systems Using Series Fan Power Boxes)

6.5.3.2.1 Part-Load Fan Power Limitation. Individual VAV fans with motors 10 hp and larger shall meet one of the following:

a. The fan shall be driven by a mechanical or electrical variable-speed drive.

b. The fan shall be a vane-axial fan with variable-pitch blades.

c. The fan shall have other controls and devices that will result in fan motor demand of no more than 30% of design wattage at 50% of design air volume when static pressure setpoint equals one-third of the total design static pressure, based on *manufacturers'* certified fan data.

6.5.3.2.2 Static Pressure Sensor Location. Static pressure sensors used to control VAV fans shall be placed in a position such that the controller setpoint is no greater than one-third the total design fan static pressure, except for systems with zone reset control complying with Section 6.5.3.2.3. If this results in the sensor being located downstream of major duct splits, multiple sensors shall be installed in each major branch to ensure that static pressure can be maintained in each.

6.5.3.2.3 Setpoint Reset. For systems with DDC of individual zone boxes reporting to the central control panel, static pressure setpoint shall be reset based on the *zone* requiring the most pressure; i.e., the setpoint is reset lower until one *zone* damper is nearly wide open.

6.5.4 Hydronic System Design and Control. HVAC hydronic systems having a total *pump system power* exceeding 10 hp shall meet provisions of Sections 6.5.4.1 through 6.5.4.4.

6.5.4.1 Hydronic Variable Flow Systems. HVAC pumping systems that include control valves designed to modulate or step open and close as a function of load shall be designed for variable fluid flow and shall be capable of reducing pump flow rates to 50% or less of the design flow rate. Individual pumps serving variable flow systems having a pump head exceeding 100 ft and motor exceeding 50 hp shall have controls and/or devices (such as variable speed control) that will result in pump motor demand of no more than 30% of design wattage at 50% of design water flow. The controls or devices shall be controlled as a function of desired flow or to maintain a minimum required differential pressure. Differential pressure shall be measured at or near the most remote heat exchanger or the heat exchanger requiring the greatest differential pressure.

Exceptions:

a. Systems where the minimum flow is less than the minimum flow required by the equipment *manufacturer* for the proper operation of equipment served by the system, such as chillers, and where total pump system power is 75 hp or less.

b. Systems that include no more than three control valves.

6.5.4.2 Pump Isolation. When a chilled-water plant includes more than one chiller, provisions shall be made so that the flow in the chiller plant can be automatically reduced, correspondingly, when a chiller is shut down. Chillers referred to in this section, piped in series for the purpose of increased temperature differential, shall be considered as one chiller.

When a boiler plant includes more than one boiler, provisions shall be made so that the flow in the boiler plant can be automatically reduced, correspondingly, when a boiler is shut down.

6.5.4.3 Chilled- and Hot-Water Temperature Reset Controls. Chilled- and hot-water systems with a design capacity exceeding 300,000 Btu/h supplying chilled or heated water (or both) to comfort conditioning systems shall include controls that automatically reset supply water temperatures by representative building loads (including return water temperature) or by *outdoor air* temperature.

Exceptions:

a. Where the supply temperature reset controls cannot be implemented without causing improper operation of heating, cooling, humidifying, or dehumidifying systems.
b. Hydronic systems, such as those required by Section 6.5.4.1 that use variable flow to reduce pumping energy.

6.5.4.4 Hydronic (Water Loop) Heat Pump Systems. Each hydronic heat pump shall have a two-position automatic valve interlocked to shut off water flow when the compressor is off.

6.5.5 Heat Rejection Equipment

6.5.5.1 General. Section 6.5.5 applies to heat rejection equipment used in comfort cooling systems such as air-cooled condensers, open cooling towers, closed-circuit cooling towers, and evaporative condensers.

Exception: Heat rejection devices whose energy usage is included in the equipment *efficiency* ratings listed in Tables 6.8.1A through 6.8.1D.

6.5.5.2 Fan Speed Control. Each fan powered by a motor of 7.5 hp or larger shall have the capability to operate that fan at two-thirds of full speed or less and shall have controls that automatically change the fan speed to control the leaving fluid temperature or condensing temperature/pressure of the heat rejection device.

Exceptions:

a. Condenser fans serving multiple refrigerant circuits.
b. Condenser fans serving flooded condensers.
c. Installations located in climate zones 1 and 2.
d. Up to one-third of the fans on a condenser or tower with multiple fans, where the lead fans comply with the speed control requirement.

6.5.6 Energy Recovery

6.5.6.1 Exhaust Air Energy Recovery. Individual fan systems that have both a design supply air capacity of 5000 cfm or greater and have a minimum *outdoor air* supply of 70% or greater of the design supply air quantity shall have an energy recovery system with at least 50% recovery effectiveness. Fifty percent energy recovery effectiveness shall mean a change in the enthalpy of the *outdoor air* supply equal to 50% of the difference between the *outdoor air* and return air at design conditions. Provision shall be made to bypass or control the heat recovery system to permit air economizer operation as required by Section 6.5.1.1.

Exceptions:

a. Laboratory systems meeting Section 6.5.7.2.
b. Systems serving spaces that are not cooled and that are heated to less than 60°F.
c. Systems exhausting toxic, flammable, paint, or corrosive fumes or dust.
d. Commercial kitchen hoods used for collecting and removing grease vapors and smoke.
e. Where more than 60% of the *outdoor air* heating energy is provided from site-recovered or site-solar energy.
f. Heating systems in climate zones 1 through 3.
g. Cooling systems in climate zones 3c, 4c, 5b, 5c, 6b, 7, and 8.
h. Where the largest exhaust source is less than 75% of the design outdoor airflow.
i. Systems requiring dehumidification that employ energy recovery in series with the cooling coil.

6.5.6.2 Heat Recovery for Service Water Heating

6.5.6.2.1 Condenser heat recovery systems shall be installed for heating or preheating of service hot water provided all of the following are true:

a. The facility operates 24 hours a day.
b. The total installed heat rejection capacity of the water-cooled systems exceeds 6,000,000 Btu/h of heat rejection.
c. The design service water heating load exceeds 1,000,000 Btu/h.

6.5.6.2.2 The required heat recovery system shall have the capacity to provide the smaller of

a. 60% of the peak heat rejection load at design conditions or
b. preheat of the peak service hot water draw to 85°F.

Exceptions:

a. Facilities that employ condenser heat recovery for space heating with a heat recovery design exceeding 30% of the peak water-cooled condenser load at design conditions.
b. Facilities that provide 60% of their service water heating from *site-solar* or *site-recovered energy* or from other sources.

6.5.7 Exhaust Hoods

6.5.7.1 Kitchen Hoods. Individual kitchen exhaust hoods larger than 5000 cfm shall be provided with makeup air sized for at least 50% of exhaust air volume that is

a. unheated or heated to no more than 60°F and
b. uncooled or cooled without the use of mechanical cooling.

Exceptions:

a. Where hoods are used to exhaust ventilation air that would otherwise exfiltrate or be exhausted by other fan systems.

b. Certified grease extractor hoods that require a face velocity no greater than 60 fpm.

6.5.7.2 Fume Hoods. Buildings with fume hood systems having a total exhaust rate greater than 15,000 cfm shall include at least one of the following features:

a. VAV hood exhaust and room supply systems capable of reducing exhaust and makeup air volume to 50% or less of design values.
b. Direct makeup (auxiliary) air supply equal to at least 75% of the exhaust rate, heated no warmer than 2°F below room setpoint, cooled to no cooler than 3°F above room setpoint, no humidification added, and no simultaneous heating and cooling used for dehumidification control.
c. Heat recovery systems to precondition makeup air from fume hood exhaust in accordance with Section 6.5.6.1, Exhaust Air Energy Recovery, without using any exception.

6.5.8 Radiant Heating Systems

6.5.8.1 Heating Unenclosed Spaces. Radiant heating shall be used when heating is required for unenclosed spaces.

Exception: Loading docks equipped with air curtains.

6.5.8.2 Heating Enclosed Spaces. Radiant heating systems that are used as primary or supplemental enclosed space heating must be in conformance with the governing provisions of the standard, including, but not limited to, the following:

a. Radiant hydronic ceiling or floor panels (used for heating or cooling).
b. Combination or hybrid systems incorporating radiant heating (or cooling) panels.
c. Radiant heating (or cooling) panels used in conjunction with other systems such as VAV or thermal storage systems.

6.5.9 Hot Gas Bypass Limitation. Cooling systems shall not use hot gas bypass or other evaporator pressure control systems unless the system is designed with multiple steps of unloading or continuous capacity modulation. The capacity of the hot gas bypass shall be limited as indicated in Table 6.5.9.

Exception: Unitary packaged systems with cooling capacities not greater than 90,000 Btu/h.

6.6 Alternative Compliance Path (Not Used)

6.7 Submittals

6.7.1 General. The *Authority having jurisdiction* may require submittal of compliance documentation and supplemental information in accord with Section 4.2.2 of this standard.

6.7.2 Completion Requirements. The following requirements are mandatory provisions and are necessary for compliance with the standard.

6.7.2.1 Drawings. Construction documents shall require that, within 90 days after the date of system acceptance, record drawings of the actual installation be provided to the building owner or the designated representative of the building owner. Record drawings shall include, as a minimum, the location and performance data on each piece of equipment, general configuration of duct and pipe distribution system including sizes, and the terminal air or water design flow rates.

6.7.2.2 Manuals. Construction documents shall require that an operating manual and a maintenance manual be provided to the building owner or the designated representative of the building owner within 90 days after the date of system acceptance. These manuals shall be in accordance with industry-accepted standards (see Informative Appendix E) and shall include, at a minimum, the following:

a. Submittal data stating equipment size and selected options for each piece of equipment requiring maintenance.
b. Operation manuals and maintenance manuals for each piece of equipment requiring maintenance, except equipment not furnished as part of the project. Required routine maintenance actions shall be clearly identified.
c. Names and addresses of at least one *service agency.*
d. HVAC controls system maintenance and calibration information, including wiring diagrams, schematics, and control sequence descriptions. Desired or field-determined setpoints shall be permanently recorded on control drawings at control devices or, for digital control systems, in programming comments.
e. A complete narrative of how each system is intended to operate, including suggested setpoints.

6.7.2.3 System Balancing

6.7.2.3.1 General. Construction documents shall require that all HVAC systems be balanced in accordance with generally accepted engineering standards (see Informative Appendix E). Construction documents shall require that a written balance report be provided to the building owner or the designated representative of the building owner for HVAC systems serving *zones* with a total conditioned area exceeding 5000 ft^2.

6.7.2.3.2 Air System Balancing. Air systems shall be balanced in a manner to first minimize throttling losses. Then, for fans with *fan system power* greater than 1 hp, fan speed shall be adjusted to meet design flow conditions.

6.7.2.3.3 Hydronic System Balancing. Hydronic systems shall be proportionately balanced in a manner to first minimize throttling losses; then the pump impeller shall be trimmed or pump speed shall be adjusted to meet design flow conditions.

Exceptions: Impellers need not be trimmed nor pump speed adjusted

a. for pumps with pump motors of 10 hp or less or

TABLE 6.5.9 Hot Gas Bypass Limitation

Rated Capacity	Maximum Hot Gas Bypass Capacity (% of Total Capacity)
≤240,000 Btu/h	50%
>240,000 Btu/h	25%

b. when throttling results in no greater than 5% of the nameplate horsepower draw, or 3 hp , whichever is greater, above that required if the impeller was trimmed.

6.7.2.4 System Commissioning. HVAC control systems shall be tested to ensure that control elements are calibrated, adjusted, and in proper working condition. For projects larger than 50,000 ft^2 conditioned area, except warehouses and semiheated spaces, detailed instructions for commissioning HVAC systems (see Informative Appendix E) shall be provided by the designer in plans and specifications.

6.8 Minimum Equipment Efficiency Tables

6.8.1 Minimum Efficiency Requirement Listed Equipment—Standard Rating and Operating Conditions

TABLE 6.8.1A Electronically Operated Unitary Air Conditioners and Condensing Units—Minimum Efficiency Requirements

Equipment Type	Size Category	Heating Section Type	Subcategory or Rating Condition	Minimum Efficiency[a]	Test Procedure[b]
Air conditioners, air cooled	<65,000 Btu/h[c]	All	Split system	10.0 SEER (before 1/23/2006) 13.0 SEER (as of 1/23/2006)	ARI 210/240
			Single package	9.7 SEER (before 1/23/2006) 13.0 SEER (as of 1/23/2006)	
Through-the-wall, air cooled	≤30,000 Btu/h[c]	All	Split system	10.0 SEER (before 1/23/2006) 10.9 SEER(as of 1/23/2006) 12 SEER(as of 1/23/2010)	
			Single package	9.7 SEER (before 1/23/2006) 10.6 SEER(as of 1/23/2006) 12.0 SEER(as of 1/23/2010)	
Air conditioners, air cooled	≥65,000 Btu/h and <135,000 Btu/h	Electric resistance (or none)	Split system and single package	10.3 EER (before 1/1/2010) 11.2 EER (as of 1/1/2010)	ARI 340/360
		All other	Split system and single package	10.1 EER (before 1/1/2010) 11.0 EER (as of 1/1/2010)	
	≥135,000 Btu/h and <240,000 Btu/h	Electric resistance (or none)	Split system and single package	9.7 EER (before 1/1/2010) 11.0 EER (as of 1/1/2010)	
		All other	Split system and single package	9.5 EER (before 1/1/2010) 10.8 EER (as of 1/1/2010)	
	≥240,000 Btu/h and <760,000 Btu/h	Electric resistance (or none)	Split system and single package	9.5 EER (before 1/1/2010) 10.0 EER (as of 1/1/2010) 9.7 IPLV	
		All other	Split system and single package	9.3 EER (before 1/1/2010) 9.8 EER (as of 1/1/2010) 9.5 IPLV	
	≥760,000 Btu/h	Electric resistance (or none)	Split system and single package	9.2 EER (before 1/1/2010) 9.7 EER (as of 1/1/2010) 9.4 IPLV	
		All other	Split system and single package	9.0 EER (as of 1/1/2010) 9.5 EER (as of 1/1/2010) 9.2 IPLV	

TABLE 6.8.1A Electronically Operated Unitary Air Conditioners and Condensing Units—Minimum Efficiency Requirements *(continued)*

Equipment Type	Size Category	Heating Section Type	Sub-Category or Rating Condition	Minimum Efficiency[a]	Test Procedure[b]
Air conditioners, water and evaporatively cooled	<65,000 Btu/h	All	Split system and single package	12.1 EER	ARI 210/240
	≥65,000 Btu/h and <135,000 Btu/h	Electric resistance (or none)	Split system and single package	11.5 EER	ARI 340/360
		All other	Split system and single package	11.3 EER	
	≥135,000 Btu/h and <240,000 Btu/h	Electric resistance (or none)	Split system and single package	11.0 EER	
		All other	Split system and single package	10.8 EER	
	≥240,000 Btu/h	Electric resistance (or none)	Split system and single package	11.0 EER 10.3 IPLV	
		All other	Split system and single package	10.8 EER 10.1 IPLV	
Condensing units, air cooled	≥135,000 Btu/h	–		10.1 EER 11.2 IPLV	ARI 365
Condensing units, water or evaporatively cooled	≥135,000 Btu/h	–		13.1 EER 13.1 IPLV	

[a] IPLVs and part-load rating conditions are only applicable to equipment with capacity modulation.
[b] Section 12 contains a complete specification of the referenced test procedure, including the referenced year version of the test procedure.
[c] Single-phase, air-cooled air conditioners <65,000 Btu/h are regulated by NAECA. SEER values are those set by NAECA.

TABLE 6.8.1B Electrically Operated Unitary and Applied Heat Pumps—Minimum Efficiency Requirements

Equipment Type	Size Category	Heating Section Type	Subcategory or Rating Condition	Minimum Efficiency[a]	Test Procedure[b]
Air cooled (cooling mode)	<65,000 Btu/h[c]	All	Split system	10.0 SEER (before 1/23/2006) 13.0 SEER (as of 1/23/2006)	ARI 210/240
			Single package	9.7 SEER (before 1/23/2006) 13.0 SEER (as of 1/23/2006)	
Through-the-wall (air cooled, cooling mode)	≤30,000 Btu/h[c]	All	Split system	10.0 SEER (before 1/23/2006) 10.9 SEER (as of 1/23/2006) 12 SEER (as of 1/23/2010)	
			Single package	9.7 SEER (before 1/23/2006) 10.6 SEER (as of 1/23/2006) 12.0 SEER (as of 1/23/2010)	
Air cooled (cooling mode)	≥65,000 Btu/h and <135,000 Btu/h	Electric resistance (or none)	Split system and single package	10.1 EER (before 1/1/2010) 11.0 EER (as of 1/1/2010)	ARI 340/360
		All other	Split system and single package	9.9 EER (before 1/1/2010) 10.8 EER (as of 1/1/2010)	
	≥135,000 Btu/h and <240,000 Btu/h	Electric resistance (or none)	Split system and single package	9.3 EER (before 1/1/2010) 10.6 EER (as of 1/1/2010)	
		All other	Split system and single package	9.1 EER (before 1/1/2010) 10.4 EER (as of 1/1/2010)	
	≥240,000 Btu/h	Electric resistance (or none)	Split system and single package	9.0 EER (before 1/1/2010) 9.5 EER (as of 1/1/2010) 9.2 IPLV	
		All other	Split system and single package	8.8 EER (before 1/1/2010) 9.3 EER (as of 1/1/2010) 9.0 IPLV	
Water source (cooling mode)	<17,000 Btu/h	All	86°F entering water	11.2 EER	ISO-13256-1
	≥17,000 Btu/h and <65,000 Btu/h	All	86°F entering water	12.0 EER	ISO-13256-1
	≥65,000 Btu/h and <135,000 Btu/h	All	86°F entering water	12.0 EER	ISO-13256-1
Groundwater source (cooling mode)	<135,000 Btu/h	All	59°F entering water	16.2 EER	ISO-13256-1
Ground source (cooling mode)	<135,000 Btu/h	All	77°F entering water	13.4 EER	ISO-13256-1
Air cooled (heating mode)	<65,000 Btu/h[c] (cooling capacity)	—	Split system	6.8 HSPF (before 1/23/2006) 7.7 HSPF (as of 1/23/2006)	ARI 210/240
			Single package	6.6 HSPF (before 1/23/2006) 7.7 HSPF (as of 1/23/2006)	
Through-the-wall, (air cooled, heating mode)	≤30,000 Btu/h[c] (cooling capacity)	—	Split system	6.8 HSPF (before 1/23/2006) 7.1 HSPF (as of 1/23/2006) 7.4 HSPF (as of 1/23/2010)	
			Single package	6.6 HSPF (before 1/23/2006) 7.0 HSPF (as of 1/23/2006) 7.4 HSPF (as of 1/23/2010)	

TABLE 6.8.1B Electrically Operated Unitary and Applied Heat Pumps— Minimum Efficiency Requirements *(continued)*

Equipment Type	Size Category	Heating Section Type	Subcategory or Rating Condition	Minimum Efficiency[a]	Test Procedure[b]
Air cooled (heating mode)	≥65,000 Btu/h and <135,000 Btu/h (cooling capacity)	—	47°F db/43°F wb outdoor air	3.2 COP (before 1/1/2010) 3.3 COP (as of 1/1/2010)	ARI 340/360
			17°F db/15°F wb outdoor air	2.2 COP	
	≥135,000 Btu/h (cooling capacity)	—	47°F db/43°F wb outdoor air	3.1 COP (before 1/1/2010) 3.2 COP (as of 1/1/2010)	
			17°F db/15°F wb outdoor air	2.0 COP	
Water source (heating mode)	<135,000 Btu/h (cooling capacity)	—	68°F entering water	4.2 COP	ISO-13256-1
Groundwater source (heating mode)	<135,000 Btu/h (cooling capacity)	—	50°F entering water	3.6 COP	ISO-13256-1
Ground source (heating mode)	<135,000 Btu/h (cooling capacity)	—	32°F entering water	3.1 COP	ISO-13256-1

[a] IPLVs and part-load rating conditions are only applicable to equipment with capacity modulation.
[b] Section 12 contains a complete specification of the referenced test procedure, including the referenced year version of the test procedure.
[c] Single-phase, air-cooled heat pumps <65,000 Btu/h are regulated by NAECA. SEER and HSPF values are those set by NAECA

TABLE 6.8.1C Water Chilling Packages–Minimum Efficiency Requirements

Equipment Type	Size Category	Subcategory or Rating Condition	Minimum Efficiency[a]	Test Procedure[b]
Air cooled, with condenser, electrically operated	All capacities	—	2.80 COP 3.05 IPLV	ARI 550/590
Air cooled, without condenser, electrically operated	All capacities	—	3.10 COP 3.45 IPLV	
Water cooled, electrically operated, positive displacement (reciprocating)	All capacities	—	4.20 COP 5.05 IPLV	ARI 550/590
Water cooled, electrically operated, positive displacement (rotary screw and scroll)	<150 tons	—	4.45 COP 5.20 IPLV	ARI 550/590
	≥150 tons and <300 tons	—	4.90 COP 5.60 IPLV	
	≥300 tons	—	5.50 COP 6.15 IPLV	
Water cooled, electrically operated, centrifugal	<150 tons	—	5.00 COP 5.25 IPLV	ARI 550/590
	≥150 tons and <300 tons	—	5.55 COP 5.90 IPLV	
	≥300 tons	—	6.10 COP 6.40 IPLV	
Air-cooled absorption single effect	All capacities	—	0.60 COP	ARI 560
Water-cooled absorption single effect	All capacities	—	0.70 COP	
Absorption double effect, indirect-fired	All capacities	—	1.00 COP 1.05 IPLV	
Absorption double effect, direct-fired	All capacities	—	1.00 COP 1.00 IPLV	

[a] The chiller equipment requirements do not apply for chillers used in low-temperature applications where the design leaving fluid temperature is <40°F.
[b] Section 12 contains a complete specification of the referenced test procedure, including the referenced year version of the test procedure.

TABLE 6.8.1D Electrically Operated Packaged Terminal Air Conditioners, Packaged Terminal Heat Pumps, Single-Package Vertical Air Conditioners, Single-Package Vertical Heat Pumps, Room Air Conditioners, and Room Air-Conditioner Heat Pumps—Minimum Efficiency Requirements

Equipment Type	Size Category (Input)	Subcategory or Rating Condition	Minimum Efficiency	Test Procedure[a]
PTAC (cooling mode) new construction	All capacities	95°F db outdoor air	12.5 – (0.213 × Cap/1000)[c] EER	ARI 310/380
PTAC (cooling mode) replacements[b]	All capacities	95°F db outdoor air	10.9 – (0.213 × Cap/1000)[c] EER	
PTHP (cooling mode) new construction	All capacities	95°F db outdoor air	12.3 – (0.213 × Cap/1000)[c] EER	
PTHP (cooling mode) replacements[b]	All capacities	95°F db outdoor air	10.8 – (0.213 × Cap/1000)[c] EER	
PTHP (heating mode) new construction	All capacities		3.2 – (0.026 × Cap/1000)[c] COP	
PTHP (heating mode) replacements[b]	All capacities		2.9 – (0.026 × Cap/1000)[c] COP	
SPVAC (cooling mode)	<65,000 Btu/h	95°F db/75°F wb outdoor air	9.0 EER	ARI 390
	≥65,000 Btu/h and <135,000 Btu/h	95°F db/75°F wb outdoor air	8.9EER	
	≥135,000 Btu/h and <240,000 Btu/h	95°F db/75°F wb outdoor air	8.6 EER	
SPVHP (cooling mode)	<65,000 Btu/h	95°F db/75°F wb outdoor air	9.0 EER	
	≥65,000 Btu/h and <135,000 Btu/h	95°F db/75°F wb outdoor air	8.9EER	
	≥135,000 Btu/h and <240,000 Btu/h	95°F db/75°F wb outdoor air	8.6 EER	
SPVHP (heating mode)	<65,000 Btu/h	47°F db/43°F wb outdoor air	3.0 COP	
	≥65,000 Btu/h and <135,000 Btu/h	47°F db/43°F wb outdoor air	3.0 COP	
	≥135,000 Btu/h and <240,000 Btu/h	47°F db/43°F wb outdoor air	2.9 COP	

Equipment Type	Size Category (Input)	Subcategory or Rating Condition	Minimum Efficiency	Test Procedure[a]
Room air conditioners, with louvered sides	<6000 Btu/h	—	9.7 SEER	ANSI/AHAM RAC-1
	≥6000 Btu/h and <8000 Btu/h		9.7 SEER	
	≥8000 Btu/h and <14,000 Btu/h		9.8 EER	
	≥14,000 Btu/h and <20,000 Btu/h		9.7 SEER	
	≥20,000 Btu/h		8.5 EER	
Room air conditioners, without louvered sides	<8000 Btu/h	—	9.0 EER	
	≥8000 Btu/h and <20,000 Btu/h		8.5 EER	
	≥20,000 Btu/h		8.5 EER	
Room air-conditioner heat pumps with louvered sides	<20,000 Btu/h	—	9.0 EER	
	≥20,000 Btu/h		8.5 EER	
Room air-conditioner heat pumps without louvered sides	<14,000 Btu/h	—	8.5 EER	
	≥14,000 Btu/h		8.0 EER	
Room air conditioner, casement only	All capacities	—	8.7 EER	
Room air conditioner, casement–slider	All capacities	—	9.5 EER	

[a] Section 12 contains a complete specification of the referenced test procedure, including the referenced year version of the test procedure.
[b] Replacement units must be factory labeled as follows: "MANUFACTURED FOR REPLACEMENT APPLICATIONS ONLY; NOT TO BE INSTALLED IN NEW CONSTRUCTION PROJECTS." Replacement efficiencies apply only to units with existing sleeves less than 16 in. high and less than 42 in. wide.
[c] *Cap* means the rated cooling capacity of the product in Btu/h. If the unit's capacity is less than 7000 Btu/h, use 7000 Btu/h in the calculation. If the unit's capacity is greater than 15,000 Btu/h, use 15,000 Btu/h in the calculation.

TABLE 6.8.1E Warm Air Furnaces and Combination Warm Air Furnaces/Air-Conditioning Units, Warm Air Duct Furnaces and Unit Heaters

Equipment Type	Size Category (Input)	Subcategory or Rating Condition	Minimum Efficiency[a]	Test Procedure[b]
Warm air furnace, gas-fired	<225,000 Btu/h	Maximum capacity[d]	78% AFUE or 80% E_t[f]	DOE 10 CFR Part 430 or ANSI Z21.47
	≥225,000 Btu/h		80% E_c[f]	ANSI Z21.47
Warm air furnace, oil-fired	<225,000 Btu/h	Maximum capacity[c]	78% AFUE or 80% E_t[d]	DOE 10 CFR Part 430 or UL 727
	≥225,000 Btu/h		81% E_t[e]	UL 727
Warm air duct furnaces, gas-fired	All capacities	Maximum capacity[d]	80% E_c[g,h]	ANSI Z83.8
Warm air unit heaters, gas-fired	All capacities	Maximum capacity[d]	80% E_c[g,h]	ANSI Z83.8
Warm air unit heaters, oil-fired	All capacities	Maximum capacity[d]	80% E_c[f,h]	UL 731

[a] E_t = thermal *efficiency.* See test procedure for detailed discussion.
[b] Section 12 contains a complete specification of the referenced test procedure, including the referenced year version of the test procedure.
[c] Minimum and maximum ratings as provided for and allowed by the unit's controls.
[d] Combination units not covered by NAECA (three-phase power or cooling capacity greater than or equal to 65,000 Btu/h) may comply with either rating.
[e] E_t = thermal *efficiency.* Units must also include an interrupted or intermittent ignition device (IID), have jacket losses not exceeding 0.75% of the input rating, and have either power venting or a flue damper. A vent damper is an acceptable alternative to a flue damper for those furnaces where combustion air is drawn from the conditioned space.
[f] E_c = combustion *efficiency* (100% less flue losses). See test procedure for detailed discussion.
[h] As of August 8, 2008, according to the Energy Policy Act of 2005, units must also include an interrupted or intermittent ignition device (IID) and have either power venting or an automatic flue damper. A vent damper is an acceptable alternative to a flue damper for those unit heaters where combustion air is drawn from the conditioned space.

TABLE 6.8.1F Gas- and Oil-Fired Boilers, Minimum Efficiency Requirements

Equipment Type[a]	Subcategory or Rating Condition	Size Category (Input)	Minimum Efficiency[b,c]	Efficiency as of 3/2/2010 (Date 3 yrs after ASHRAE Board Approval)	Efficiency as of 3/2/2020 (Date 13 yrs after ASHRAE Board Approval)	Test Procedure
Boilers, hot water	Gas-fired	<300,000 Btu/h	80% AFUE	80% AFUE	80% AFUE	10 CFR Part 430
		≥300,000 Btu/h and ≤2,500,000 Btu/h[d]	75% E_t	80% E_t	80% E_t	10 CFR Part 431
		>2,500,000 Btu/h[a]	80% E_c	82% E_c	82% E_c	
	Oil-fired[e]	<300,000 Btu/h	80% AFUE	80% AFUE	80% AFUE	10 CFR Part 430
		≥300,000 Btu/h and ≤2,500,000 Btu/h[d]	78% E_t	82% E_t	82% E_t	10 CFR Part 431
		>2,500,000 Btu/h[a]	83% E_c	84% E_c	84% E_c	
Boilers, steam	Gas-fired	<300,000 Btu/h	75% AFUE	75% AFUE	75% AFUE	10 CFR Part 430
	Gas-fired—all, except natural draft	≥300,000 Btu/h and ≤2,500,000 Btu/h[d]	75% E_t	79% E_t	79% E_t	10 CFR Part 431
		>2,500,000 Btu/h[a]	80% E_c	79% E_t	79% E_t	
	Gas-fired—natural draft	≥300,000 Btu/h and ≤2,500,000 Btu/h[d]	75% E_t	77% E_t	79% E_t	
		>2,500,000 Btu/h[a]	80% E_c	77% E_t	79% E_t	
	Oil-fired[e]	<300,000 Btu/h	80% AFUE	80% AFUE	80% AFUE	10 CFR Part 430
		≥300,000 Btu/h and ≤2,500,000 Btu/h[d]	78% E_t	81% E_t	81% E_t	10 CFR Part 431
		>2,500,000 Btu/h[a]	83% E_c	81% E_t	81% E_t	

[a] These requirements apply to boilers with rated input of 8,000,000 Btu/h or less that are not packaged boilers and to all packaged boilers. Minimum efficiency requirements for boilers cover all capacities of packaged boilers.
[b] E_c = combustion efficiency (100% less flue losses). See reference document for detailed information.
[c] E_t = thermal efficiency. See reference document for detailed information.
[d] Maximum capacity – minimum and maximum ratings as provided for and allowed by the unit's controls.
[e] Includes oil-fired (residual).

TABLE 6.8.1G Performance Requirements for Heat Rejection Equipment

Equipment Type	Total System Heat Rejection Capacity at Rated Conditions	Subcategory or Rating Condition	Performance Required[a,b]	Test Procedure[c]
Propeller or axial fan cooling towers	All	95°F entering water 85°F leaving water 75°F wb *outdoor air*	≥38.2 gpm/hp	CTI ATC-105 and CTI STD-201
Centrifugal fan cooling towers	All	95°F entering water 85°F leaving water 75°F wb *outdoor air*	≥20.0 gpm/hp	CTI ATC-105 and CTI STD-201
Air-cooled condensers	All	125°F condensing temperature R-22 test fluid 190°F entering gas temperature 15°F subcooling 95°F entering db	≥176,000 Btu/h·hp	ARI 460

[a] For purposes of this table, cooling tower performance is defined as the maximum flow rating of the tower divided by the fan nameplate rated motor power.
[b] For purposes of this table, air-cooled condenser performance is defined as the heat rejected from the refrigerant divided by the fan nameplate rated motor power.
[c] Section 12 contains a complete specification of the referenced test procedure, including the referenced year version of the test procedure.

TABLE 6.8.1H Minimum Efficiencies for Centrifugal Chillers <150 tons

Centrifugal Chillers <150 tons

COP_{std} = 5.00; $IPLV_{std}$ = 5.25

Condenser Flow Rate

Leaving Chilled-Water Temperature (°F)	Entering Condenser-Water Temperature (°F)	LIFT[a] (°F)	2 gpm/ton		2.5 gpm/ton		3 gpm/ton		4 gpm/ton		5 gpm/ton		6 gpm/ton	
			COP	NPLV[c]	COP	NPLV[c]	COP	NPLV[c]	COP	NPLV[c]	COP	NPLV[c]	COP	NPLV[c]
40	75	35	5.11	5.35	5.33	5.58	5.48	5.73	5.67	5.93	5.79	6.06	5.88	6.15
40	80	40	4.62	4.83	4.92	5.14	5.09	5.32	5.27	5.52	5.38	5.63	5.45	5.70
40	85	45	3.84	4.01	4.32	4.52	4.58	4.79	4.84	5.06	4.98	5.20	5.06	5.29
41	75	34	5.19	5.43	5.41	5.66	5.56	5.81	5.75	6.02	5.89	6.16	5.99	6.26
41	80	39	4.73	4.95	5.01	5.24	5.17	5.41	5.35	5.60	5.46	5.71	5.53	5.78
41	85	44	4.02	4.21	4.46	4.67	4.70	4.91	4.94	5.17	5.06	5.30	5.14	5.38
42	75	33	5.27	5.51	5.49	5.74	5.64	5.90	5.85	6.12	6.00	6.27	6.11	6.39
42	80	38	4.84	5.06	5.10	5.33	5.25	5.49	5.43	5.67	5.53	5.79	5.61	5.87
42	85	43	4.19	4.38	4.59	4.80	4.81	5.03	5.03	5.26	5.15	5.38	5.22	5.46
43	75	32	5.35	5.59	5.57	5.82	5.72	5.99	5.95	6.23	6.11	6.39	6.23	6.52
43	80	37	4.94	5.16	5.18	5.42	5.32	5.57	5.50	5.76	5.62	5.87	5.70	5.96
43	85	42	4.35	4.55	4.71	4.93	4.91	5.13	5.12	5.35	5.23	5.47	5.30	5.54
44	75	31	5.42	5.67	5.65	5.91	5.82	6.08	6.07	6.34	6.24	6.53	6.37	6.67
44	80	36	5.03	5.26	5.26	5.50	5.40	5.65	5.58	5.84	5.70	5.96	5.79	6.05
44	85	41	4.49	4.69	4.82	5.04	5.00	5.25	5.20	5.43	5.30	5.55	5.38	5.62
45	75	30	5.50	5.75	5.74	6.00	5.92	6.19	6.19	6.47	6.38	6.68	6.53	6.83
45	80	35	5.11	5.35	5.33	5.58	5.48	5.73	5.67	5.93	5.79	6.06	5.88	6.15
45	85	40	4.62	4.83	4.92	5.14	5.09	5.32	5.27	5.52	5.38	5.63	5.45	5.70
46	75	29	5.58	5.84	5.83	6.10	6.03	6.30	6.32	6.61	6.54	6.84	6.70	7.00
46	80	34	5.19	5.43	5.41	5.66	5.56	5.81	5.75	6.02	5.89	6.16	5.99	6.26
46	85	39	4.73	4.95	5.01	5.24	5.17	5.41	5.35	5.60	5.46	5.71	5.53	5.78
47	75	28	5.66	5.92	5.93	6.20	6.15	6.43	6.47	6.77	6.71	7.02	6.88	7.20
47	80	33	5.27	5.51	5.49	5.74	5.64	5.90	5.85	6.12	6.00	6.27	6.11	6.39
47	85	38	4.84	5.06	5.10	5.33	5.25	5.49	5.43	5.67	5.53	5.79	5.61	5.87
48	75	27	5.75	6.02	6.04	6.32	6.28	6.56	6.64	6.94	6.89	7.21	7.09	7.41
48	80	32	5.35	5.59	5.57	5.82	5.72	5.99	5.95	6.23	6.11	6.39	6.23	6.52
48	85	37	4.94	5.16	5.18	5.42	5.32	5.57	5.50	5.76	5.62	5.87	5.70	5.96
Condenser ΔT[b]			14.04		11.23		9.36		7.02		5.62		4.68	

[a] LIFT = entering condenser water temperature – leaving chilled-water temperature (°F)

[b] Condenser ΔT = leaving condenser-water temperature (°F) – entering condenser-water temperature (°F)

[c] All NPLV values shown are NPLV except at conditions of 3 gpm/ton condenser flow rate with 44°F leaving chilled-water temperature and 85°F entering condenser-water temperature, which is IPLV

$K_{adj} = 6.1507 - 0.30244(X) + 0.0062692(X)^2 - 0.000045595(X)^3$

where X = Condenser ΔT + LIFT

$COP_{adj} = K_{adj} \cdot COP_{std}$

TABLE 6.8.1I Minimum Efficiencies for Centrifugal Chillers ≥150 tons, < 300 tons

Centrifugal Chillers ≥150 tons, <300 tons

COP_{std} = 5.55; $IPLV_{std}$ = 5.90

			Condenser Flow Rate											
			2 gpm/ton		2.5 gpm/ton		3 gpm/ton		4 gpm/ton		5 gpm/ton		6 gpm/ton	
Leaving Chilled-Water Temperature (°F)	Entering Condenser-Water Temperature (°F)	LIFT[a] (°F)	COP	NPLV[c]	COP	NPLV[c]	COP	NPLV[c]	COP	NPLV[c]	COP	NPLV[c]	COP	NPLV[c]
40	75	35	5.65	6.03	5.90	6.29	6.05	6.46	6.26	6.68	6.40	6.83	6.51	6.94
40	80	40	5.10	5.44	5.44	5.80	5.62	6.00	5.83	6.22	5.95	6.35	6.03	6.43
40	85	45	4.24	4.52	4.77	5.09	5.06	5.40	5.35	5.71	5.50	5.87	5.59	5.97
41	75	34	5.74	6.13	5.80	6.38	6.14	6.55	6.36	6.79	6.51	6.95	6.62	7.06
41	80	39	5.23	5.58	5.54	5.91	5.71	6.10	5.91	6.31	6.03	6.44	6.11	6.52
41	85	44	4.45	4.74	4.93	5.26	5.19	5.54	5.46	5.82	5.60	5.97	5.69	6.07
42	75	33	5.83	6.22	6.07	6.47	6.23	6.65	6.47	6.90	6.63	7.07	6.75	7.20
42	80	38	5.35	5.71	5.64	6.01	5.80	6.19	6.00	6.40	6.12	6.53	6.20	6.62
42	85	43	4.63	4.94	5.08	5.41	5.31	5.67	5.56	5.93	5.69	6.07	5.77	6.16
43	75	32	5.91	6.31	6.15	6.56	6.33	6.75	6.58	7.02	6.76	7.21	6.89	7.35
43	80	37	5.46	5.82	5.73	6.11	5.89	6.28	6.08	6.49	6.21	6.62	6.30	6.72
43	85	42	4.81	5.13	5.21	5.55	5.42	5.79	5.66	6.03	5.78	6.16	5.86	6.25
44	75	31	6.00	6.40	6.24	6.66	6.43	6.86	6.71	7.15	6.90	7.36	7.05	7.52
44	80	36	5.56	5.93	5.81	6.20	5.97	6.37	6.17	6.58	6.30	6.72	6.40	6.82
44	85	41	4.96	5.29	5.33	5.68	5.55	5.90	5.74	6.13	5.86	6.26	5.94	6.34
45	75	30	6.08	6.49	6.34	6.76	6.54	6.98	6.84	7.30	7.06	7.53	7.22	7.70
45	80	35	5.65	6.03	5.90	6.29	6.05	6.46	6.26	6.68	6.40	6.83	6.51	6.94
45	85	40	5.10	5.44	5.44	5.80	5.62	6.00	5.83	6.22	5.95	6.35	6.03	6.43
46	75	29	6.17	6.58	6.44	6.87	6.66	7.11	6.99	7.46	7.23	7.71	7.40	7.90
46	80	34	5.74	6.13	5.80	6.38	6.14	6.55	6.36	6.79	6.51	6.95	6.62	7.06
46	85	39	5.23	5.58	5.54	5.91	5.71	6.10	5.91	6.31	6.03	6.44	6.11	6.52
47	75	28	6.26	6.68	6.56	6.99	6.79	7.24	7.16	7.63	7.42	7.91	7.61	8.11
47	80	33	5.83	6.21	6.07	6.47	6.23	6.64	6.47	6.90	6.63	7.07	6.75	7.20
47	85	38	5.35	5.70	5.64	6.01	5.80	6.19	6.00	6.40	6.12	6.52	6.20	6.61
48	75	27	6.36	6.78	6.68	7.12	6.94	7.40	7.34	7.82	7.62	8.13	7.83	8.35
48	80	32	5.91	6.30	6.15	6.56	6.33	6.75	6.58	7.02	6.76	7.21	6.89	7.35
48	85	37	5.46	5.82	5.73	6.10	5.89	6.28	6.08	6.49	6.21	6.62	6.30	6.71
Condenser ΔT[b]			14.04		11.23		9.36		7.02		5.62		4.68	

[a] LIFT = entering condenser-water temperature – leaving chilled-water temperature (°F)

[b] Condenser ΔT = leaving condenser-water temperature (°F) – entering condenser-water temperature (°F)

[c] All NPLV values shown are NPLV except at conditions of 3 gpm/ton condenser flow rate with 44°F leaving chilled-water temperature and 85°F entering condenser-water temperature, which is IPLV

$K_{adj} = 6.1507 - 0.30244(X) + 0.0062692(X)^2 - 0.000045595(X)^3$

where X = Condenser ΔT + LIFT

$COP_{adj} = K_{adj} \cdot COP_{std}$

TABLE 6.8.1J Minimum Efficiencies for Centrifugal Chillers ≥300 tons

Centrifugal Chillers ≥300 tons

COP_{std} = 6.10; $IPLV_{std}$ = 6.40

			Condenser Flow Rate											
			2 gpm/ton		2.5 gpm/ton		3 gpm/ton		4 gpm/ton		5 gpm/ton		6 gpm/ton	
Leaving Chilled-Water Temperature (°F)	Entering Condenser-Water Temperature (°F)	LIFT[a] (°F)	COP	NPLV[c]	COP	NPLV[c]	COP	NPLV[c]	COP	NPLV[c]	COP	NPLV[c]	COP	NPLV[c]
40	75	35	6.23	6.55	6.50	6.83	6.68	7.01	6.91	7.26	7.06	7.42	7.17	7.54
40	80	40	5.63	5.91	6.00	6.30	6.20	6.52	6.43	6.76	6.56	6.89	6.65	6.98
40	85	45	4.68	4.91	5.26	5.53	5.58	5.86	5.90	6.20	6.07	6.37	6.17	6.48
41	75	34	6.33	6.65	6.60	6.93	6.77	7.12	7.02	7.37	7.18	7.55	7.30	7.67
41	80	39	5.77	6.06	6.11	6.42	6.30	6.62	6.52	6.85	6.65	6.99	6.74	7.08
41	85	44	4.90	5.15	5.44	5.71	5.72	6.01	6.02	6.33	6.17	6.49	6.27	6.59
42	75	33	6.43	6.75	6.69	7.03	6.87	7.22	7.13	7.49	7.31	7.68	7.44	7.82
42	80	38	5.90	6.20	6.21	6.53	6.40	6.72	6.61	6.95	6.75	7.09	6.84	7.19
42	85	43	5.11	5.37	5.60	5.88	5.86	6.16	6.13	6.44	6.28	6.59	6.37	6.69
43	75	32	6.52	6.85	6.79	7.13	6.98	7.33	7.26	7.63	7.45	7.83	7.60	7.98
43	80	37	6.02	6.32	6.31	6.63	6.49	6.82	6.71	7.05	6.85	7.19	6.94	7.30
43	85	42	5.30	5.57	5.74	6.03	5.98	6.28	6.24	6.55	6.37	6.70	6.46	6.79
44	75	31	6.61	6.95	6.89	7.23	7.09	7.45	7.40	7.77	7.61	8.00	7.77	8.16
44	80	36	6.13	6.44	6.41	6.73	6.58	6.92	6.81	7.15	6.95	7.30	7.05	7.41
44	85	41	5.47	5.75	5.87	6.17	6.10	6.40	6.33	6.66	6.47	6.79	6.55	6.89
45	75	30	6.71	7.05	6.99	7.35	7.21	7.58	7.55	7.93	7.78	8.18	7.96	8.36
45	80	35	6.23	6.55	6.50	6.83	6.68	7.01	6.91	7.26	7.06	7.42	7.17	7.54
45	85	40	5.63	5.91	6.00	6.30	6.20	6.52	6.43	6.76	6.56	6.89	6.65	6.98
46	75	29	6.80	7.15	7.11	7.47	7.35	7.72	7.71	8.10	7.97	8.37	8.16	8.58
46	80	34	6.33	6.65	6.60	6.93	6.77	7.12	7.02	7.37	7.18	7.55	7.30	7.67
46	85	39	5.77	6.06	6.11	6.42	6.30	6.62	6.52	6.85	6.65	6.99	6.74	7.08
47	75	28	6.91	7.26	7.23	7.60	7.49	7.87	7.89	8.29	8.18	8.59	8.39	8.82
47	80	33	6.43	6.75	6.69	7.03	6.87	7.22	7.13	7.49	7.31	7.68	7.44	7.82
47	85	38	5.90	6.20	6.21	6.53	6.40	6.72	6.61	6.95	6.75	7.09	6.84	7.19
48	75	27	7.01	7.37	7.36	7.74	7.65	8.04	8.09	8.50	8.41	8.83	8.64	9.08
48	80	32	6.52	6.85	6.79	7.13	6.98	7.33	7.26	7.63	7.45	7.83	7.60	7.98
48	85	37	6.02	6.32	6.31	6.63	6.49	6.82	6.71	7.05	6.85	7.19	6.94	7.30
Condenser ΔT[b]			14.04		11.23		9.36		7.02		5.62		4.68	

[a] LIFT = entering condenser-water temperature – leaving chilled-water temperature (°F)

[b] Condenser ΔT = leaving condenser-water temperature (°F) – entering condenser-water temperature (°F)

[c] All NPLV values shown are NPLV except at conditions of 3 gpm/ton condenser flow rate with 44°F leaving chilled-water temperature and 85°F entering condenser-water temperature, which is IPLV

$K_{adj} = 6.1507 - 0.30244(X) + 0.0062692(X)^2 - 0.000045595(X)^3$

where X = Condenser ΔT + LIFT

$COP_{adj} = K_{adj} \cdot COP_{std}$

6.8.2 Duct Insulation Tables

TABLE 6.8.2A Minimum Duct Insulation R-Value,[a] Cooling and Heating Only Supply Ducts and Return Ducts

Climate Zone	Duct Location						
	Exterior	Ventilated Attic	Unvented Attic Above Insulated Ceiling	Unvented Attic with Roof Insulation[a]	Unconditioned Space[b]	Indirectly Conditioned Space[c]	Buried
				Heating-Only Ducts			
1, 2	none	none	none	none	none	none	none
3	R-3.5	none	none	none	none	none	none
4	R-3.5	none	none	none	none	none	none
5	R-6	R-3.5	none	none	none	none	R-3.5
6	R-6	R-6	R-3.5	none	none	none	R-3.5
7	R-8	R-6	R-6	none	R-3.5	none	R-3.5
8	R-8	R-8	R-6	none	R-6	none	R-6
				Cooling-Only Ducts			
1	R-6	R-6	R-8	R-3.5	R-3.5	none	R-3.5
2	R-6	R-6	R-6	R-3.5	R-3.5	none	R-3.5
3	R-6	R-6	R-6	R-3.5	R-1.9	none	none
4	R-3.5	R-3.5	R-6	R-1.9	R-1.9	none	none
5, 6	R-3.5	R-1.9	R-3.5	R-1.9	R-1.9	none	none
7, 8	R-1.9	R-1.9	R-1.9	R-1.9	R-1.9	none	none
				Return Ducts			
1 to 8	R-3.5	R-3.5	R-3.5	none	none	none	none

[a] Insulation R-values, measured in $(h \cdot ft^2 \cdot °F)/Btu$, are for the insulation as installed and do not include film resistance. The required minimum thicknesses do not consider water vapor transmission and possible surface condensation. Where exterior walls are used as plenum walls, wall insulation shall be as required by the most restrictive condition of Section 6.4.4.2 or Section 5. Insulation resistance measured on a horizontal plane in accordance with ASTM C518 at a mean temperature of 75°F at the installed thickness.
[b] Includes crawlspaces, both ventilated and nonventilated.
[c] Includes return air plenums with or without exposed roofs above.

TABLE 6.8.2B Minimum Duct Insulation R-Value,[a] Combined Heating and Cooling Supply Ducts and Return Ducts

Climate Zone	Duct Location						
	Exterior	Ventilated Attic	Unvented Attic Above Insulated Ceiling	Unvented Attic with Roof Insulation[a]	Unconditioned Space[b]	Indirectly Conditioned Space[c]	Buried
				Supply Ducts			
1	R-6	R-6	R-8	R-3.5	R-3.5	none	R-3.5
2	R-6	R-6	R-6	R-3.5	R-3.5	none	R-3.5
3	R-6	R-6	R-6	R-3.5	R-3.5	none	R-3.5
4	R-6	R-6	R-6	R-3.5	R-3.5	none	R-3.5
5	R-6	R-6	R-6	R-1.9	R-3.5	none	R-3.5
6	R-8	R-6	R-6	R-1.9	R-3.5	none	R-3.5
7	R-8	R-6	R-6	R-1.9	R-3.5	none	R-3.5
8	R-8	R-8	R-8	R-1.9	R-6	none	R-6
				Return Ducts			
1 to 8	R-3.5	R-3.5	R-3.5	none	none	none	none

[a] Insulation R-values, measured in $(h \cdot ft^2 \cdot °F)/Btu$, are for the insulation as installed and do not include film resistance. The required minimum thicknesses do not consider water vapor transmission and possible surface condensation. Where exterior walls are used as plenum walls, wall insulation shall be as required by the most restrictive condition of Section 6.4.4.2 or Section 5. Insulation resistance measured on a horizontal plane in accordance with ASTM C518 at a mean temperature of 75°F at the installed thickness.
[b] Includes crawlspaces, both ventilated and nonventilated.
[c] Includes return air plenums with or without exposed roofs above.

TABLE 6.8.3 Minimum Pipe Insulation Thickness[a]

Fluid Design Operating Temp. Range (°F)	Insulation Conductivity		Nominal Pipe or Tube Size (in.)				
	Conductivity Btu·in./(h·ft²·°F)	Mean Rating Temp. °F	<1	1 to <1-1/2	1-1/2 to <4	4 to <8	≥8
Heating Systems (Steam, Steam Condensate, and Hot Water)[b,c]							
>350	0.32 – 0.34	250	2.5	3.0	3.0	4.0	4.0
251 – 350	0.29 – 0.32	200	1.5	2.5	3.0	3.0	3.0
201 – 250	0.27 – 0.30	150	1.5	1.5	2.0	2.0	2.0
141 – 200	0.25 – 0.29	125	1.0	1.0	1.0	1.5	1.5
105 – 140	0.22 – 0.28	100	0.5	0.5	1.0	1.0	1.0
Domestic and Service Hot-Water Systems							
105+	0.22 – 0.28	100	0.5	0.5	1.0	1.0	1.0
Cooling Systems (Chilled Water, Brine, and Refrigerant)[d]							
40 – 60	0.22 – 0.28	100	0.5	0.5	1.0	1.0	1.0
<40	0.22 – 0.28	100	0.5	1.0	1.0	1.0	1.5

[a] For insulation outside the stated conductivity range, the minimum thickness (T) shall be determined as follows:
$T = r\{(1 + t/r)^{K/k} - 1\}$
where T = minimum insulation thickness (in.), r = actual outside radius of pipe (in.), t = insulation thickness listed in this table for applicable fluid temperature and pipe size, K = conductivity of alternate material at mean rating temperature indicated for the applicable fluid temperature (Btu·in.[h·ft²·°F]); and k = the upper value of the conductivity range listed in this table for the applicable fluid temperature.

[b] These thicknesses are based on energy *efficiency* considerations only. Additional insulation is sometimes required relative to safety issues/surface temperature.

[c] Piping insulation is not required between the control valve and coil on run-outs when the control valve is located within 4 ftin. of the coil and the pipe size is 1 in. or less.

[d] These thicknesses are based on energy *efficiency* considerations only. Issues such as water vapor permeability or surface condensation sometimes require vapor retarders or additional insulation.

7. SERVICE WATER HEATING

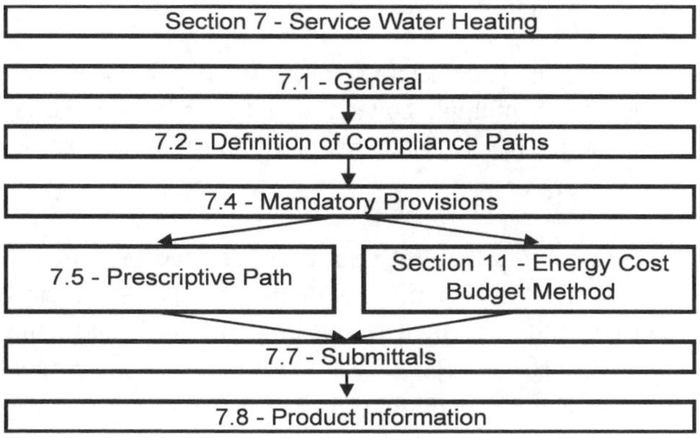

7.1 General

7.1.1 Service Water Heating Scope

7.1.1.1 New Buildings. Service water heating *systems* and *equipment* shall comply with the requirements of this section as described in Section 7.2.

7.1.1.2 Additions to Existing Buildings. Service water heating *systems* and *equipment* shall comply with the requirements of this section.

Exception: When the service water heating to an *addition* is provided by existing service water heating systems and equipment, such systems and equipment shall not be required to comply with this standard. However, any new systems or equipment installed must comply with specific requirements applicable to those systems and equipment.

7.1.1.3 Alterations to Existing Buildings. Building service water heating equipment installed as a direct replacement for *existing building* service water heating equipment shall comply with the requirements of Section 7 applicable to the equipment being replaced. New and replacement piping shall comply with Section 7.4.3.

Exception: Compliance shall not be required where there is insufficient space or access to meet these requirements.

7.2 Compliance Path(s)

7.2.1 Compliance shall be achieved by meeting the requirements of Section 7.1, General; Section 7.4, Mandatory Provisions; Section 7.5, Prescriptive Path; Section 7.7, Submittals; and Section 7.8, Product Information.

7.2.2 Projects using the Energy Cost Budget Method (Section 11) for demonstrating compliance with the standard shall meet the requirements of Section 7.4, Mandatory Provisions, in conjunction with Section 11, Energy Cost Budget Method.

7.3 Simplified/Small Building Option (Not Used)

7.4 Mandatory Provisions

7.4.1 Load Calculations. Service water heating *system* design loads for the purpose of sizing *systems* and *equipment* shall be determined in accordance with *manufacturers'* published sizing guidelines or generally accepted engineering standards and handbooks acceptable to the *adopting authority* (e.g., *ASHRAE Handbook—HVAC Applications*).

7.4.2 Equipment Efficiency. All water heating *equipment*, hot-water supply boilers used solely for heating potable water, pool heaters, and hot-water storage tanks shall meet the criteria listed in Table 7.8. Where multiple criteria are listed, all criteria shall be met. Omission of minimum performance requirements for certain classes of *equipment* does not preclude use of such *equipment* where appropriate. Equipment not listed in Table 7.8 has no minimum performance requirements.

Exception: All water heaters and hot-water supply boilers having more than 140 gal of storage capacity are not required to meet the *standby loss* (SL) requirements of Table 7.8 when

a. the tank surface is thermally insulated to R-12.5,
b. a standing pilot light is not installed, and
c. gas- or oil-fired storage water heaters have a flue damper or fan-assisted combustion.

7.4.3 Service Hot-Water Piping Insulation. The following piping shall be insulated to levels shown in Section 6, Table 6.8.3:

a. recirculating system piping, including the supply and return piping of a circulating tank type water heater
b. the first 8 ft of outlet piping for a constant temperature nonrecirculating storage *system*
c. the inlet pipe between the storage tank and a heat trap in a nonrecirculating storage *system*
d. pipes that are externally heated (such as heat trace or impedance heating)

7.4.4 Service Water Heating System Controls

7.4.4.1 Temperature Controls. Temperature controls shall be provided that allow for storage temperature adjustment from 120°F or lower to a maximum temperature compatible with the intended use.

Exception: When the *manufacturers'* installation instructions specify a higher minimum thermostat setting to minimize condensation and resulting corrosion.

7.4.4.2 Temperature Maintenance Controls. Systems designed to maintain usage temperatures in hot-water pipes, such as recirculating hot-water systems or heat trace, shall be equipped with automatic time switches or other controls that can be set to switch off the usage temperature maintenance system during extended periods when hot water is not required.

7.4.4.3 Outlet Temperature Controls. Temperature controlling means shall be provided to limit the maximum temperature of water delivered from lavatory faucets in public facility restrooms to 110°F.

7.4.4.4 Circulating Pump Controls. When used to maintain storage tank water temperature, recirculating pumps shall be equipped with controls limiting operation to a period from the start of the heating cycle to a maximum of five minutes after the end of the heating cycle.

7.4.5 Pools

7.4.5.1 Pool Heaters. Pool heaters shall be equipped with a readily accessible ON/OFF switch to allow shutting off the heater without adjusting the thermostat setting. Pool heaters fired by natural gas shall not have continuously burning pilot lights.

7.4.5.2 Pool Covers. Heated pools shall be equipped with a vapor retardant pool cover on or at the water surface. Pools heated to more than 90°F shall have a pool cover with a minimum insulation value of R-12.

Exception: Pools deriving over 60% of the energy for heating from *site-recovered energy* or *solar energy source.*

7.4.5.3 Time Switches. Time switches shall be installed on swimming pool heaters and pumps.

Exceptions:

a. Where public health standards require 24-hour pump operation.
b. Where pumps are required to operate solar and waste heat recovery pool heating *systems*.

7.4.6 Heat Traps.

Vertical pipe risers serving storage water heaters and storage tanks not having integral heat traps and serving a nonrecirculating system shall have heat traps on both the inlet and outlet piping as close as practical to the storage tank. A heat trap is a means to counteract the natural convection of heated water in a vertical pipe run. The means is either a device specifically designed for the purpose or an arrangement of tubing that forms a loop of 360 degrees or piping that from the point of connection to the water heater (inlet or outlet) includes a length of piping directed downward before connection to the vertical piping of the supply water or hot-water distribution system, as applicable.

7.5 Prescriptive Path

7.5.1 Space Heating and Water Heating. The use of a gas-fired or oil-fired space-heating boiler system otherwise complying with Section 6 to provide the total space heating and water heating for a building is allowed when one of the following conditions is met:

a. The single space-heating boiler, or the component of a modular or multiple boiler system that is heating the service water, has a standby loss in Btu/h not exceeding

$$(13.3 \times pmd + 400) / n,$$

where pmd is the probable maximum demand in gal/h, determined in accordance with the procedures described in generally accepted engineering standards and handbooks, and n is the fraction of the year when the outdoor daily mean temperature is greater than 64.9°F.

The standby loss is to be determined for a test period of 24 hours duration while maintaining a boiler water temperature of at least 90°F above ambient, with an ambient temperature between 60°F and 90°F. For a boiler with a modulating burner, this test shall be conducted at the lowest input.

b. It is demonstrated to the satisfaction of the *authority having jurisdiction* that the use of a single heat source will consume less energy than separate units.
c. The energy input of the combined boiler and water heater system is less than 150,000 Btu/h.

7.5.2 Service Water Heating Equipment. Service water heating *equipment* used to provide the additional function of space heating as part of a combination (integrated) *system* shall satisfy all stated requirements for the service water heating *equipment.*

7.6 Alternative Compliance Path (Not Used)

7.7 Submittals

7.7.1 General. The *authority having jurisdiction* may require submittal of compliance documentation and supplemental information, in accord with Section 4.2.2 of this standard.

7.8 Product Information

TABLE 7.8 Performance Requirements for Water Heating Equipment

Equipment Type	Size Category (Input)	Subcategory or Rating Condition	Performance Required [a]	Test Procedure[b]
Electric water heaters	≤12 kW	Resistance ≥20 gal	0.93–0.00132V EF	DOE 10 CFR Part 430
	>12 kW	Resistance ≥20 gal	$20 + 35\sqrt{V}$ SL, Btu/h	ANSI Z21.10.3
	≤24 Amps and ≤250 Volts	Heat Pump	0.93–0.00132V EF	DOE 10 CFR Part 430
Gas storage water heaters	≤75,000 Btu/h	≥20 gal	0.62–0.0019V EF	DOE 10 CFR Part 430
	>75,000 Btu/h	<4000 (Btu/h)/gal	80% E_t $(Q/800 + 110\sqrt{V})$ SL, Btu/h	ANSI Z21.10.3
Gas instantaneous water heaters	>50,000 Btu/h and <200,000 Btu/h	≥4000 (Btu/h)/gal and <2 gal	0.62–0.0019V EF	DOE 10 CFR Part 430
	≥200,000 Btu/h[c]	≥4000 (Btu/h)/gal and <10 gal	80% E_t	ANSI Z21.10.3
	≥200,000 Btu/h	≥4000 (Btu/h)/gal and ≥10 gal	80% E_t $(Q/800 + 110\sqrt{V})$ SL, Btu/h	
Oil storage water heaters	≤105,000 Btu/h	≥20 gal	0.59–0.0019V EF	DOE 10 CFR Part 430
	>105,000 Btu/h	<4000 (Btu/h)/gal	78% E_t $(Q/800 + 110\sqrt{V})$ SL, Btu/h	ANSI Z21.10.3
Oil instantaneous water heaters	≤210,000 Btu/h	≥4000 (Btu/h)/gal and <2 gal	0.59–0.0019V EF	DOE 10 CFR Part 430
	>210,000 Btu/h	≥4000 (Btu/h)/gal and <10 gal	80% E_t	ANSI Z21.10.3
	>210,000 Btu/h	≥4000 (Btu/h)/gal and ≥10 gal	78% E_t $(Q/800 + 110\sqrt{V})$ SL, Btu/h	
Hot-water supply boilers, gas and oil	≥300,000 Btu/h and <12,500,000 Btu/h	≥4000 (Btu/h)/gal and <10 gal	80% E_t	ANSI Z21.10.3
Hot-water supply boilers, gas		≥4000 (Btu/h)/gal and ≥10 gal	80% E_t $(Q/800 + 110\sqrt{V})$ SL, Btu/h	
Hot-water supply boilers, oil		≥4000 (Btu/h)/gal and ≥10 gal	78% E_t $(Q/800 + 110\sqrt{V})$ SL, Btu/h	
Pool heaters, oil and gas	All		78% E_t	ASHRAE 146
Heat pump pool heaters	All		4.0 COP	ASHRAE 146
Unfired storage tanks	All		R-12.5	(none)

[a] Energy factor (EF) and thermal *efficiency* (E_t) are minimum requirements, while standby loss (SL) is maximum Btu/h based on a 70°F temperature difference between stored water and ambient requirements. In the EF equation, V is the rated volume in gallons. In the SL equation, V is the rated volume in gallons and Q is the nameplate input rate in Btu/h.

[b] Section 12 contains a complete specification, including the year version, of the referenced test procedure.

[c] Instantaneous water heaters with input rates below 200,000 Btu/h must comply with these requirements if the water heater is designed to heat water to temperatures of 180°F or higher.

8. POWER

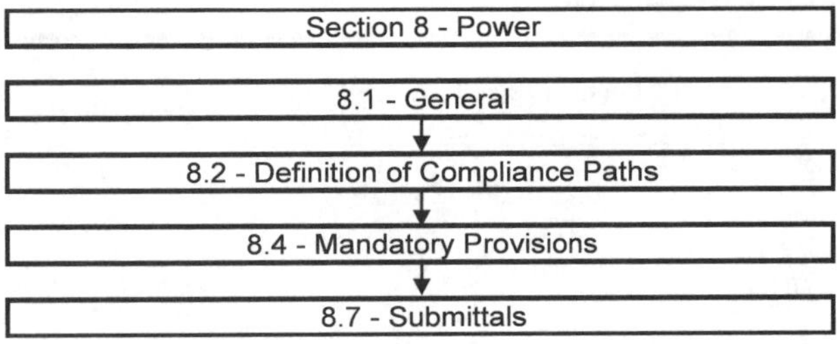

8.1 General. This section applies to all building power distribution *systems.*

8.2 Compliance Path(s)

8.2.1 Power distribution systems in all projects shall comply with the requirements of Section 8.1, General; Section 8.4, Mandatory Provisions; and Section 8.7, Submittals.

8.3 Simplified/Small Building Option (Not Used)

8.4 Mandatory Provisions

8.4.1 Voltage Drop

8.4.1.1 Feeders. *Feeder conductors* shall be sized for a maximum *voltage drop* of 2% at design load.

8.4.1.2 Branch Circuits. *Branch circuit* conductors shall be sized for a maximum *voltage drop* of 3% at design load.

8.5 Prescriptive Path (Not Used)

8.6 Alternative Compliance Path (Not Used)

8.7 Submittals

8.7.1 Drawings. Construction documents shall require that within 30 days after the date of system acceptance, record drawings of the actual installation shall be provided to the building owner, including

a. a single-line diagram of the building electrical distribution system and

b. floor plans indicating location and area served for all distribution.

8.7.2 Manuals. Construction documents shall require that an operating manual and maintenance manual be provided to the building owner. The manuals shall include, at a minimum, the following:

a. Submittal data stating *equipment* rating and selected options for each piece of *equipment* requiring maintenance.

b. Operation manuals and maintenance manuals for each piece of *equipment* requiring maintenance. Required routine maintenance actions shall be clearly identified.

c. Names and addresses of at least one qualified *service agency.*

d. A complete narrative of how each system is intended to operate.

(Enforcement agencies should only check to be sure that the construction documents require this information to be transmitted to the owner and should not expect copies of any of the materials.)

8.8 Product Information (Not Used)

9. LIGHTING

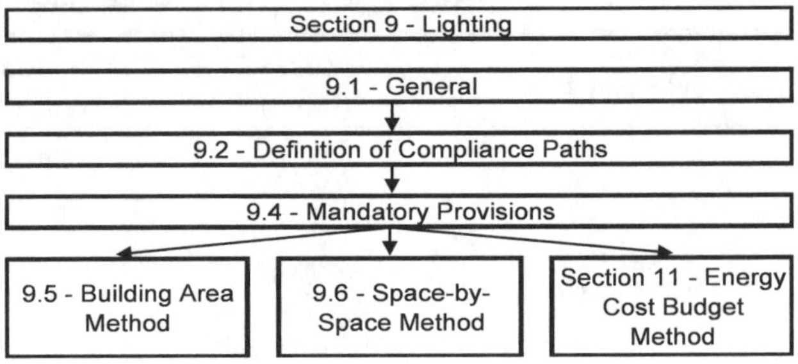

9.1 General

9.1.1 Scope. This section shall apply to the following:

a. interior spaces of *buildings*

b. exterior building features, including facades, illuminated roofs, architectural features, entrances, exits, loading docks, and illuminated canopies

c. exterior building grounds lighting provided through the *building's* electrical *service*

Exceptions:

a. emergency lighting that is automatically off during normal *building* operation

b. lighting within *dwelling units*

c. lighting that is specifically designated as required by a health or life safety statute, ordinance, or regulation

d. decorative gas lighting systems

9.1.2 Lighting Alterations. The replacement of lighting *systems* in any building space shall comply with the *LPD* requirements of Section 9 applicable to that space. New lighting *systems* shall comply with the applicable *LPD* requirements of Section 9. Any new *control devices* as a direct replacement of existing *control devices* shall comply with the specific requirements of Section 9.4.1.2(b).

Exception: *Alterations* that replace less than 50% of the *luminaires* in a *space* need not comply with these requirements provided that such *alterations* do not increase the installed interior lighting power.

9.1.3 Installed Interior Lighting Power. The *installed interior lighting power* shall include all power used by the *luminaires*, including *lamps*, *ballasts*, *transformers*, and *control devices* except as specifically exempted in Section 9.2.2.3.

Exception: If two or more independently operating lighting systems in a space are capable of being controlled to prevent simultaneous user operation, the installed interior lighting power shall be based solely on the lighting system with the highest wattage.

9.1.4 Luminaire Wattage. Luminaire wattage incorporated into the installed interior lighting power shall be determined in accordance with the following criteria:

a. The wattage of incandescent or tungsten-halogen luminaires with medium screw base sockets and not containing permanently installed ballasts shall be the maximum labeled wattage of the luminaire.
b. The wattage of luminaires with permanently installed or remote ballasts or *transformers* shall be the operating input wattage of the maximum lamp/auxiliary combination based on values from the auxiliary *manufacturers'* literature or recognized testing laboratories or shall be the maximum labeled wattage of the luminaire.
c. For line-voltage lighting track and plug-in busway, designed to allow the addition and/or relocation of luminaires without altering the wiring of the system, the wattage shall be

   1. the specified wattage of the luminaires included in the system with a minimum of 30 W/lin ft or
   2. the wattage limit of the system's circuit breaker or
   3. the wattage limit of other permanent current-limiting device(s) on the system.

d. The wattage of low-voltage lighting track, cable conductor, rail conductor, and other flexible lighting systems that allow the addition and/or relocation of luminaires without altering the wiring of the system shall be the specified wattage of the transformer supplying the system.
e. The wattage of all other miscellaneous lighting equipment shall be the specified wattage of the lighting equipment.

9.2 Compliance Path(s)

9.2.1 Lighting systems and equipment shall comply with Section 9.1, General; Section 9.4, Mandatory Provisions; and the prescriptive requirements of either

a. Section 9.5, Building Area Method; or
b. Section 9.6, Space-by-Space Method.

9.2.2 Prescriptive Requirements

9.2.2.1 The Building Area Method for determining the *interior lighting power allowance*, described in Section 9.5, is a simplified approach for demonstrating compliance.

9.2.2.2 The Space-by-Space Method, described in Section 9.6, is an alternative approach that allows greater flexibility.

9.2.2.3 Interior Lighting Power. The *interior lighting power allowance* for a *building* or a separately metered or permitted portion of a *building* shall be determined by either the *Building* Area Method described in Section 9.5 or the Space-by-Space Method described in Section 9.6. Trade-offs of *interior lighting power allowance* among portions of the *building* for which a different method of calculation has been used are not permitted. The *installed interior lighting power* identified in accordance with Section 9.1.3 shall not exceed the *interior lighting power allowance* developed in accordance with Section 9.5 or 9.6.

Exceptions: The following *lighting equipment* and applications shall not be considered when determining the *interior lighting power allowance* developed in accordance with Section 9.5 or 9.6, nor shall the wattage for such lighting be included in the *installed interior lighting power* identified in accordance with Section 9.1.3. However, any such lighting shall not be exempt unless it is an addition to general lighting and is controlled by an independent *control device*.

a. Display or accent lighting that is an essential element for the function performed in galleries, museums, and monuments.
b. Lighting that is integral to *equipment* or instrumentation and is installed by its *manufacturer.*
c. Lighting specifically designed for use only during medical or dental procedures and lighting integral to medical *equipment*.
d. Lighting integral to both open and glass-enclosed refrigerator and freezer cases.
e. Lighting integral to food warming and food preparation *equipment*.
f. Lighting for plant growth or maintenance.
g. Lighting in spaces specifically designed for use by occupants with special lighting needs including visual impairment and other medical and age-related issues.
h. Lighting in *retail* display windows, provided the display area is enclosed by ceiling-height partitions.
i. Lighting in interior spaces that have been specifically designated as a registered interior *historic* landmark.
j. Lighting that is an integral part of advertising or directional signage.
k. Exit signs.
l. Lighting that is for sale or lighting educational demonstration systems.
m. Lighting for theatrical purposes, including performance, stage, and film and video production.
n. Lighting for television broadcasting in sporting activity areas.
o. Casino gaming areas.
p. Furniture-mounted supplemental task lighting that is controlled by automatic shutoff and complies with Section 9.4.1.4(d).

9.3 (Not Used)

9.4 Mandatory Provisions

9.4.1 Lighting Control

9.4.1.1 Automatic Lighting Shutoff. Interior lighting in *buildings* larger than 5000 ft^2 shall be controlled with an *automatic control device* to shut off *building* lighting in all spaces. This *automatic control device* shall function on either

a. a scheduled basis using a time-of-day operated control device that turns lighting off at specific programmed times—an independent program schedule shall be provided for areas of no more than 25,000 ft^2 but not more than one floor—or
b. an *occupant sensor* that shall turn lighting off within 30 minutes of an occupant leaving a space or
c. a signal from another control or alarm system that indicates the area is unoccupied.

Exceptions: The following shall not require an *automatic control device:*

a. Lighting intended for 24-hour operation.
b. Lighting in spaces where patient care is rendered.
c. Lighting in spaces where an automatic shutoff would endanger the safety or security of the room or building occupant(s).

9.4.1.2 Space Control. Each space enclosed by ceiling-height partitions shall have at least one *control device* to independently *control* the *general lighting* within the space. Each manual device shall be readily accessible and located so the occupants can see the controlled lighting.

a. A control device shall be installed that automatically turns lighting off within 30 minutes of all occupants leaving a space, except spaces with multi-scene control, in

   1. classrooms (not including shop classrooms, laboratory classrooms, and preschool through 12th grade classrooms),
   2. conference/meeting rooms, and
   3. employee lunch and break rooms.

   These spaces are not required to be connected to other automatic lighting shutoff controls.
b. For all other spaces, each *control device* shall be activated either manually by an occupant or automatically by sensing an occupant. Each *control device* shall *control* a maximum of 2500 ft^2 area for a space 10,000 ft^2 or less and a maximum of 10,000 ft^2 area for a space greater than 10,000 ft^2 and be capable of overriding any time-of-day scheduled shutoff *control* for no more than four hours.

Exception: Remote location shall be permitted for reasons of safety or security when the remote control device has an indicator pilot light as part of or next to the control device and the light is clearly labeled to identify the controlled lighting.

9.4.1.3 Exterior Lighting Control. Lighting for all exterior applications not exempted in Section 9.1 shall have automatic controls capable of turning off exterior lighting when sufficient daylight is available or when the lighting is not required during nighttime hours. Lighting not designated for dusk-to-dawn operation shall be controlled by either

a. a combination of a photosensor and a time switch or
b. an astronomical time switch.

Lighting designated for dusk-to-dawn operation shall be controlled by an astronomical time switch or photosensor. All time switches shall be capable of retaining programming and the time setting during loss of power for a period of at least ten hours.

Exception: Lighting for covered vehicle entrances or exits from buildings or parking structures where required for safety, security, or eye adaptation.

9.4.1.4 Additional Control

a. *Display/Accent Lighting*—display or accent lighting shall have a separate *control device*.
b. *Case Lighting*—lighting in cases used for display purposes shall have a separate *control device*.
c. *Hotel and Motel Guest Room Lighting*—hotel and motel guest rooms and guest suites shall have a master *control device* at the main room entry that *controls* all *permanently installed luminaires* and switched receptacles.
d. *Task Lighting*—supplemental task lighting, including *permanently installed* undershelf or undercabinet lighting, shall have a *control device* integral to the *luminaires* or be controlled by a wall-mounted *control device* provided the *control device* is readily accessible and located so that the occupant can see the controlled lighting.
e. *Nonvisual Lighting*—lighting for nonvisual applications, such as plant growth and food warming, shall have a separate *control device*.
f. *Demonstration Lighting*—*lighting equipment* that is for sale or for demonstrations in lighting education shall have a separate *control device*.

9.4.2 Tandem Wiring. Luminaires designed for use with one or three linear fluorescent lamps greater than 30 W each shall use two-lamp tandem-wired ballasts in place of single-lamp ballasts when two or more luminaires are in the same space and on the same control device.

Exceptions:

a. Recessed luminaires more than 10 ft apart measured center to center.
b. Surface-mounted or pendant luminaires that are not continuous.
c. Luminaires using single-lamp high-frequency electronic ballasts.
d. Luminaires using three-lamp high-frequency electronic or three-lamp electromagnetic ballasts.
e. Luminaires on emergency circuits.
f. Luminaires with no available pair.

9.4.3 Exit Signs. Internally illuminated exit signs shall not exceed 5 W per face.

9.4.4 Exterior Building Grounds Lighting. All exterior building grounds luminaires that operate at greater than 100 W shall contain lamps having a minimum efficacy of 60 lm/W unless the luminaire is controlled by a motion sensor or qualifies for one of the exceptions under Section 9.1.1 or 9.4.5.

9.4.5 Exterior Building Lighting Power. The total *exterior lighting power allowance* for all exterior building applications is the sum of the individual lighting power densities permitted in Table 9.4.5 for these applications plus an additional unrestricted allowance of 5% of that sum. Trade-offs are allowed only among exterior lighting applications listed in the Table 9.4.5 "Tradable Surfaces" section.

Exceptions: Lighting used for the following exterior applications is exempt when equipped with a *control device* independent of the control of the nonexempt lighting:

a. Specialized signal, directional, and marker lighting associated with transportation.
b. Advertising signage or directional signage.

TABLE 9.4.5 Lighting Power Densities for Building Exteriors

Tradable Surfaces (*LPD*s for uncovered parking areas, building grounds, building entrances and exits, canopies and overhangs, and outdoor sales areas may be traded.)	**Uncovered parking areas**	
	Parking lots and drives	**0.15** W/ft^2
	Building grounds	
	Walkways less than 10 ft wide	**1.0** W/linear foot
	Walkways 10 ft wide or greater Plaza areas Special feature areas	**0.2** W/ft^2
	Stairways	**1.0** W/ft^2
	Building entrances and exits	
	Main entries	**30** W/linear foot of door width
	Other doors	**20** W/linear foot of door width
	Canopies and overhangs	
	Canopies (free standing and attached and overhangs)	**1.25** W/ft^2
	Outdoor sales	
	Open areas (including vehicle sales lots)	**0.5** W/ft^2
	Street frontage for vehicle sales lots in addition to "open area" allowance	**20** W/linear foot
Nontradable Surfaces (*LPD* calculations for the following applications can be used only for the specific application and cannot be traded between surfaces or with other exterior lighting. The following allowances are in addition to any allowance otherwise permitted in the "Tradable Surfaces" section of this table.)	**Building facades**	**0.2** W/ft^2 for each illuminated wall or surface or **5.0** W/linear foot for each illuminated wall or surface length
	Automated teller machines and night depositories	**270** W per location plus **90** W per additional ATM per location
	Entrances and gatehouse inspection stations at guarded facilities	**1.25** W/ft^2 of uncovered area (covered areas are included in the "Canopies and Overhangs" section of "Tradable Surfaces")
	Loading areas for law enforcement, fire, ambulance, and other emergency service vehicles	**0.5** W/ft^2 of uncovered area (covered areas are included in the "Canopies and Overhangs" section of "Tradable Surfaces")
	Drive-through windows at fast food restaurants	**400** W per drive-through
	Parking near 24-hour retail entrances	**800** W per main entry

c. Lighting integral to *equipment* or instrumentation and installed by its *manufacturer.*

d. Lighting for theatrical purposes, including performance, stage, film production, and video production.

e. Lighting for athletic playing areas.

f. Temporary lighting.

g. Lighting for industrial production, material handling, transportation sites, and associated storage areas.

h. Theme elements in theme/amusement parks.

i. Lighting used to highlight features of public monuments and registered *historic* landmark structures or *buildings.*

9.5 Building Area Method Compliance Path

9.5.1 Building Area Method of Calculating Interior Lighting Power Allowance. Use the following steps to determine the interior lighting power allowance by the Building Area Method:

a. Determine the appropriate building area type from Table 9.5.1 and the allowed *LPD* (watts per unit area) from the "Building Area Method" column. For building area types not listed, selection of a reasonably equivalent type shall be permitted.

b. Determine the gross lighted floor area (square feet) of the building area type.

c. Multiply the gross lighted floor areas of the building area type(s) times the *LPD.*

d. The *interior lighting power allowance* for the building is the sum of the *lighting power allowances* of all building area types. Trade-offs among building area types are permitted provided that the total *installed interior lighting power* does not exceed the *interior lighting power allowance.*

TABLE 9.5.1 Lighting Power Densities Using the Building Area Method

Building Area Type[a]	*LPD* (W/ft^2)
Automotive facility	0.9
Convention center	1.2
Courthouse	1.2
Dining: bar lounge/leisure	1.3
Dining: cafeteria/fast food	1.4
Dining: family	1.6
Dormitory	1.0
Exercise center	1.0
Gymnasium	1.1
Health-care clinic	1.0
Hospital	1.2
Hotel	1.0
Library	1.3
Manufacturing facility	1.3
Motel	1.0
Motion picture theater	1.2
Multifamily	0.7
Museum	1.1
Office	1.0
Parking garage	0.3
Penitentiary	1.0
Performing arts theater	1.6
Police/fire station	1.0
Post office	1.1
Religious building	1.3
Retail	1.5
School/university	1.2
Sports arena	1.1
Town hall	1.1
Transportation	1.0
Warehouse	0.8
Workshop	1.4

[a] In cases where both a general building area type and a specific building area type are listed, the specific building area type shall apply.

9.6 Alternative Compliance Path: Space-by-Space Method

9.6.1 Space-by-Space Method of Calculating Interior Lighting Power Allowance. Use the following steps to determine the interior lighting power allowance by the Space-by-Space Method:

a. Determine the appropriate building type from Table 9.6.1. For building types not listed, selection of a reasonably equivalent type shall be permitted.
b. For each space enclosed by partitions 80% or greater than ceiling height, determine the gross interior floor area by measuring to the center of the partition wall. Include the floor area of balconies or other projections. Retail spaces do not have to comply with the 80% partition height requirements.
c. Determine the *interior lighting power allowance* by using the columns designated Space-by-Space Method in Table 9.6.1. Multiply the floor area(s) of the space(s) times the allowed *LPD* for the space type that most closely represents the proposed use of the space(s). The product is the *lighting power allowance* for the space(s). For space types not listed, selection of a reasonable equivalent category shall be permitted.
d. The *interior lighting power allowance* is the sum of *lighting power allowances* of all spaces. Trade-offs among spaces are permitted provided that the total *installed interior lighting power* does not exceed the *interior lighting power allowance.*

9.6.2 Additional Interior Lighting Power. When using the Space-by-Space Method, an increase in the *interior lighting power allowance* is allowed for specific lighting functions. Additional power shall be allowed only if the specified lighting is installed and automatically controlled, separately from the general lighting, to be turned off during nonbusiness hours. This additional power shall be used only for the specified *luminaires* and shall not be used for any other purpose.

An increase in the *interior lighting power allowance* is permitted in the following cases:

a. For spaces in which lighting is specified to be installed in addition to the general lighting for the purpose of decorative appearance, such as chandelier-type luminaries or sconces or for highlighting art or exhibits, provided that the additional lighting power shall not exceed 1.0 W/ft^2 of such spaces.
b. For lighting equipment installed in sales areas and specifically designed and directed to highlight merchandise, calculate the additional lighting power as follows:

$$\text{Additional Interior Lighting Power Allowance} = 1000 \text{ watts} + (\text{Retail Area 1} \times 1.0 \text{ W/ft}^2) + (\text{Retail Area 2} \times 1.7 \text{ W/ft}^2) + (\text{Retail Area 3} \times 2.6 \text{ W/ft}^2) + (\text{Retail Area 4} \times 4.2 \text{ W/ft}^2),$$

where

Retail Area 1 = the floor area for all products not listed in Retail Areas 2, 3, or 4;

Retail Area 2 = the floor area used for the sale of vehicles, sporting goods, and small electronics;

Retail Area 3 = the floor area used for the sale of furniture, clothing, cosmetics, and artwork; and

Retail Area 4 = the floor area used for the sale of jewelry, crystal, and china.

Exception: Other merchandise categories may be included in Retail Areas 2 through 4 above, provided that justification documenting the need for additional lighting power based on visual inspection, contrast, or other critical display is approved by the *authority having jurisdiction.*

9.7 Submittals (Not Used)

9.8 Product Information (Not Used)

TABLE 9.6.1 Lighting Power Densities Using the Space-by-Space Method

Common Space Types[a]	*LPD*, W/ft^2	Building-Specific Space Types	*LPD*, W/ft^2
Office—Enclosed	1.1	Gymnasium/Exercise Center	
Office—Open Plan	1.1	Playing Area	1.4
Conference/Meeting/Multipurpose	1.3	Exercise Area	0.9
Classroom/Lecture/Training	1.4	Courthouse/Police Station/Penitentiary	
For Penitentiary	1.3	Courtroom	1.9
Lobby	1.3	Confinement Cells	0.9
For Hotel	1.1	Judges' Chambers	1.3
For Performing Arts Theater	3.3	Fire Stations	
For Motion Picture Theater	1.1	Engine Room	0.8
Audience/Seating Area	0.9	Sleeping Quarters	0.3
For Gymnasium	0.4	Post Office—Sorting Area	1.2
For Exercise Center	0.3	Convention Center—Exhibit Space	1.3
For Convention Center	0.7	Library	
For Penitentiary	0.7	Card File and Cataloging	1.1
For Religious Buildings	1.7	Stacks	1.7
For Sports Arena	0.4	Reading Area	1.2
For Performing Arts Theater	2.6	Hospital	
For Motion Picture Theater	1.2	Emergency	2.7
For Transportation	0.5	Recovery	0.8
Atrium—First Three Floors	0.6	Nurses' Station	1.0
Atrium—Each Additional Floor	0.2	Exam/Treatment	1.5
Lounge/Recreation	1.2	Pharmacy	1.2
For Hospital	0.8	Patient Room	0.7
Dining Area	0.9	Operating Room	2.2
For Penitentiary	1.3	Nursery	0.6
For Hotel	1.3	Medical Supply	1.4
For Motel	1.2	Physical Therapy	0.9
For Bar Lounge/Leisure Dining	1.4	Radiology	0.4
For Family Dining	2.1	Laundry—Washing	0.6
Food Preparation	1.2	Automotive—Service/Repair	0.7
Laboratory	1.4	Manufacturing	
Restrooms	0.9	Low Bay (<25 ft Floor to Ceiling Height)	1.2
Dressing/Locker/Fitting Room	0.6	High Bay (≥25 ft Floor to Ceiling Height)	1.7
Corridor/Transition	0.5	Detailed Manufacturing	2.1
For Hospital	1.0	Equipment Room	1.2
For Manufacturing Facility	0.5	Control Room	0.5
Stairs—Active	0.6	Hotel/Motel Guest Rooms	1.1
Active Storage	0.8	Dormitory—Living Quarters	1.1
For Hospital	0.9	Museum	
Inactive Storage	0.3	General Exhibition	1.0
For Museum	0.8	Restoration	1.7
Electrical/Mechanical	1.5	Bank/Office—Banking Activity Area	1.5

TABLE 9.6.1 Lighting Power Densities Using the Space-by-Space Method *(continued)*

Common Space Types[a]	*LPD*, W/ft²	Building-Specific Space Types	*LPD*, W/ft²
Workshop	1.9	Religious Buildings	
Sales Area [for accent lighting, see Section 9.6.2(b)]	1.7	Worship Pulpit, Choir	2.4
		Fellowship Hall	0.9
		Retail	
		Sales Area [for accent lighting, see Section 9.6.3(c)]	1.7
		Mall Concourse	1.7
		Sports Arena	
		Ring Sports Area	2.7
		Court Sports Area	2.3
		Indoor Playing Field Area	1.4
		Warehouse	
		Fine Material Storage	1.4
		Medium/Bulky Material Storage	0.9
		Parking Garage—Garage Area	0.2
		Transportation	
		Airport—Concourse	0.6
		Air/Train/Bus—Baggage Area	1.0
		Terminal—Ticket Counter	1.5

[a] In cases where both a common space type and a building-specific type are listed, the building specific space type shall apply.

10. OTHER EQUIPMENT

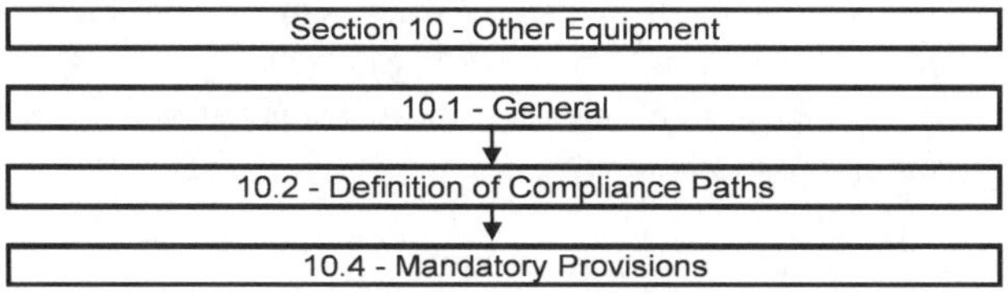

10.1 General

10.1.1 Scope. This section applies only to the equipment described below.

10.1.1.1 New Buildings. Other equipment installed in new buildings shall comply with the requirements of this section.

10.1.1.2 Additions to Existing Buildings. Other equipment installed in *additions* to *existing buildings* shall comply with the requirements of this section.

10.1.1.3 Alterations to Existing Buildings

10.1.1.3.1 *Alterations* to other building service equipment or systems shall comply with the requirements of this section applicable to those specific portions of the building and its systems that are being altered.

10.1.1.3.2 Any new equipment subject to the requirements of this section that is installed in conjunction with the *alterations*, as a direct replacement of existing equipment or control devices, shall comply with the specific requirements applicable to that equipment or control devices.

Exception: Compliance shall not be required for the relocation or reuse of existing equipment.

10.2 Compliance Path(s)

10.2.1 Compliance with Section 10 shall be achieved by meeting all requirements of Section 10.1, General; Section 10.4, Mandatory Provisions; and Section 10.8, Product Information.

10.2.2 Projects using the Energy Cost Budget Method (Section 11 of this standard) must comply with Section 10.4, the mandatory provisions of this section, as a portion of that compliance path.

10.3 Simplified/Small Building Option (Not Used)

10.4 Mandatory Provisions

10.4.1 Electric Motors. Electric motors shall comply with the requirements of the Energy Policy Act of 1992 where applicable, as shown in Table 10.8. Motors that are not included in the scope of the Energy Policy Act of 1992 have no performance requirements in this section.

10.5 Prescriptive Compliance Path (Not Used)

10.6 Alternative Compliance Path (Not Used)

10.7 Submittals (Not Used)

10.8 Product Information

TABLE 10.8 Minimum Nominal Efficiency for General Purpose Design A and Design B Motors[a]

	Minimum Nominal Full-Load Efficiency (%)					
	Open Motors			Enclosed Motors		
Number of Poles ⇒	**2**	**4**	**6**	**2**	**4**	**6**
Synchronous Speed (RPM) ⇒	**3600**	**1800**	**1200**	**3600**	**1800**	**1200**
Motor Horsepower						
1	—	82.5	80.0	75.5	82.5	80.0
1.5	82.5	84.0	84.0	82.5	84.0	85.5
2	84.0	84.0	85.5	84.0	84.0	86.5
3	84.0	86.5	86.5	85.5	87.5	87.5
5	85.5	87.5	87.5	87.5	87.5	87.5
7.5	87.5	88.5	88.5	88.5	89.5	89.5
10	88.5	89.5	90.2	89.5	89.5	89.5
15	89.5	91.0	90.2	90.2	91.0	90.2
20	90.2	91.0	91.0	90.2	91.0	90.2
25	91.0	91.7	91.7	91.0	92.4	91.7
30	91.0	92.4	92.4	91.0	92.4	91.7
40	91.7	93.0	93.0	91.7	93.0	93.0
50	92.4	93.0	93.0	92.4	93.0	93.0
60	93.0	93.6	93.6	93.0	93.6	93.6
75	93.0	94.1	93.6	93.0	94.1	93.6
100	93.0	94.1	94.1	93.6	94.5	94.1
125	93.6	94.5	94.1	94.5	94.5	94.1
150	93.6	95.0	94.5	94.5	95.0	95.0
200	94.5	95.0	94.5	95.0	95.0	95.0

[a] Nominal efficiencies shall be established in accordance with NEMA Standard MG1. Design A and Design B are National Electric Manufacturers Association (NEMA) design class designations for fixed-frequency small and medium AC squirrel-cage induction motors.

11. ENERGY COST BUDGET METHOD

11.1 General

11.1.1 Energy Cost Budget Method Scope. The building Energy Cost Budget Method is an alternative to the prescriptive provisions of this standard. It may be employed for evaluating the compliance of all proposed designs except designs with no mechanical system.

11.1.2 Trade-Offs Limited to Building Permit. When the building permit being sought applies to less than the whole building, only the calculation parameters related to the systems to which the permit applies shall be allowed to vary. Parameters relating to unmodified existing conditions or to future building components shall be identical for both the *energy cost budget* and the *design energy cost* calculations. Future building components shall meet the prescriptive requirements of Sections 5.5, 6.5, 7.5, and either 9.5 or 9.6.

11.1.3 Envelope Limitation. For new buildings or *additions*, the building Energy Cost Budget Method results shall not be submitted for building permit approval to the *authority having jurisdiction* prior to submittal for approval of the building envelope design.

11.1.4 Compliance. Compliance with Section 11 will be achieved if

a. all requirements of Sections 5.4, 6.4, 7.4, 8.4, 9.4, and 10.4 are met;
b. the *design energy cost*, as calculated in Section 11.3, does not exceed the *energy cost budget*, as calculated by the simulation program described in Section 11.2; and
c. the energy *efficiency* level of components specified in the building design meet or exceed the *efficiency* levels used to calculate the *design energy cost.*

Informative Note: *The energy cost budget and the design energy cost calculations are applicable only for determining compliance with this standard. They are not predictions of actual energy consumption or costs of the proposed design after construction. Actual experience will differ from these calculations due to variations such as occupancy, building operation and maintenance, weather, energy use not covered by this standard, changes in energy rates between design of the building and occupancy, and precision of the calculation tool.*

11.1.5 Documentation Requirements. Compliance shall be documented and submitted to the *authority having jurisdiction*. The information submitted shall include the following:

a. The *energy cost budget* for the *budget building design* and the *design energy cost* for the *proposed design.*
b. A list of the energy-related features that are included in the design and on which compliance with the provisions of Section 11 is based. This list shall document all energy features that differ between the models used in the *energy cost budget* and the *design energy cost* calculations.
c. The input and output report(s) from the *simulation program*, including a breakdown of energy usage by at least the following components: lights, internal equipment loads, service water heating equipment, space heating equipment, space cooling and heat rejection equipment, fans, and other HVAC equipment (such as pumps). The output reports shall also show the amount of time any loads are not met by the HVAC system for both the *proposed design* and *budget building design*.
d. An explanation of any error messages noted in the *simulation program* output.

11.2 Simulation General Requirements

11.2.1 Simulation Program. The *simulation program* shall be a computer-based program for the analysis of energy consumption in buildings (a program such as, but not limited to, DOE-2 or BLAST). The *simulation program* shall include calculation methodologies for the building components being modeled.

Note to Adopting Authority: *ASHRAE Standing Standard Project Committee 90.1 recommends that a compliance shell implementing the rules of a compliance supplement that controls inputs to and reports outputs from the required computer analysis program be adopted for the purposes of easier use and simpler compliance.*

11.2.1.1 The *simulation program* shall be approved by the *adopting authority* and shall, at a minimum, have the ability to explicitly model all of the following:

a. a minimum of 1400 hours per year
b. hourly variations in occupancy, lighting power, miscellaneous equipment power, thermostat setpoints, and HVAC system operation, defined separately for each day of the week and holidays
c. thermal mass effects
d. ten or more thermal zones
e. part-load performance curves for mechanical equipment
f. capacity and *efficiency* correction curves for mechanical heating and cooling equipment
g. air-side and water-side economizers with integrated control
h. the *budget building design* characteristics specified in Section 11.2.5

11.2.1.2 The *simulation program* shall have the ability to either

a. directly determine the *design energy cost* and *energy cost budget* or
b. produce hourly reports of energy use by energy source suitable for determining *the design energy cost* and *energy cost budget* using a separate calculation engine.

11.2.1.3 The *simulation program* shall be capable of performing design load calculations to determine required HVAC equipment capacities and air and water flow rates in accordance with Section 6.4.2 for both the *proposed design* and the *budget building design*.

11.2.1.4 The simulation program shall be tested according to Standard 140, and the results shall be furnished by the software provider.

11.2.2 Climatic Data. The *simulation program* shall perform the simulation using hourly values of climatic data, such as temperature and humidity from representative climatic data, for the city in which the *proposed design* is to be located. For cities or urban regions with several climatic data entries, and for locations where weather data are not available, the designer shall select available weather data that best represent the climate at the construction site. Such selected weather data shall be approved by the *authority having jurisdiction*.

11.2.3 Purchased Energy Rates. Annual energy costs shall be determined using rates for purchased energy, such as electricity, gas, oil, propane, steam, and chilled water, and approved by the *adopting authority*.

Exception: On-site renewable energy sources or site-recovered energy shall not be considered to be purchased energy and shall not be included in the *design energy cost*. Where on-site renewable or site-recovered sources are used, the *budget building design* shall be based on the energy source used as the backup energy source or electricity if no backup energy source has been specified.

11.2.4 Compliance Calculations. The *design energy cost* and *energy cost budget* shall be calculated using

a. the same *simulation program*,
b. the same weather data, and
c. the same *purchased energy rates*.

11.2.5 Exceptional Calculation Methods. Where no *simulation program* is available that adequately models a design, material, or device, the *authority having jurisdiction* may approve an exceptional calculation method to be used to demonstrate compliance with Section 11. Applications for approval of an exceptional method to include theoretical and empirical information verifying the method's accuracy shall include the following documentation to demonstrate that the exceptional calculation method and results

a. make no change in any input parameter values specified by this standard and the *adopting authority*;
b. provide input and output documentation that facilitates the enforcement agency's review and meets the formatting and content required by the *adopting authority*; and
c. are supported with instructions for using the method to demonstrate that the *energy cost budget* and *design energy cost* required by Section 11 are met.

11.3 Calculation of Design Energy Cost and Energy Cost Budget

11.3.1 The simulation model for calculating the design energy cost and the *energy cost budget* shall be developed in accordance with the requirements in Table 11.3.1.

TABLE 11.3.1 Modeling Requirements for Calculating Design Energy Cost and Energy Cost Budget

No.	Proposed Building Design (Column A) Design Energy Cost (DEC)	Budget Building Design (Column B) Energy Cost Budget (ECB)
1. Design Model		
a.	The simulation model of the *proposed building design* shall be consistent with the design documents, including proper accounting of fenestration and opaque envelope types and area; interior lighting power and controls; HVAC system types, sizes, and controls; and service water heating systems and controls.	The *budget building design* shall be developed by modifying the *proposed design* as described in this table. Except as specifically instructed in this table, all building systems and equipment shall be modeled identically in the *budget building design* and *proposed building design*.
b.	All conditioned spaces in the *proposed building design* shall be simulated as being both heated and cooled even if no cooling or heating system is being installed.	
c.	When the *energy cost budget* method is applied to buildings in which energy-related features have not yet been designed (e.g., a lighting system), those yet-to-be-designed features shall be described in the *proposed building design* so that they minimally comply with applicable mandatory and prescriptive requirements from Sections 5 through 10. Where the space classification for a building is not known, the building shall be categorized as an office building.	
2. Additions and Alterations		
	It is acceptable to demonstrate compliance using building models that exclude parts of the *existing building* provided all of the following conditions are met:	Same as *proposed building design*
a.	Work to be performed under the current permit application in excluded parts of the building shall meet the requirements of Sections 5 through 10.	
b.	Excluded parts of the building are served by HVAC systems that are entirely separate from those serving parts of the building that are included in the building model.	
c.	Design space temperature and HVAC system operating setpoints and schedules, on either side of the boundary between included and excluded parts of the building, are identical.	
d.	If a declining block or similar utility rate is being used in the analysis and the excluded and included parts of the building are on the same utility meter, the rate shall reflect the utility block or rate for the building plus the addition.	

No.	Proposed Building Design (Column A) Design Energy Cost (DEC)	Budget Building Design (Column B) Energy Cost Budget (ECB)
3. Space Use Classification		
	The building type or space type classifications shall be chosen in accordance with Section 9.5.1 or 9.6.1. The user or designer shall specify the space use classifications using either the building type or space type categories but shall not combine the two types of categories within a single permit application. More than one building type category may be used for a building if it is a mixed-use facility.	Same as *proposed building design*
4. Schedules		
	The schedule types listed in Section 11.2.1.1(b) shall be required input. The schedules shall be typical of the proposed building type as determined by the designer and approved by the *authority having jurisdiction*. Required schedules shall be identical for the *proposed building design* and *budget building design*.	Same as *proposed building design*
5. Building Envelope		
	All components of the building envelope in the *proposed building design* shall be modeled as shown on architectural drawings or as installed for *existing building* envelopes. **Exceptions:** The following building elements are permitted to differ from architectural drawings. a. Any envelope assembly that covers less than 5% of the total area of that assembly type (e.g., exterior walls) need not be separately described. If not separately described, the area of an envelope assembly must be added to the area of the adjacent assembly of that same type. b. Exterior surfaces whose azimuth orientation and tilt differ by no more than 45 degrees and are otherwise the same may be described as either a single surface or by using multipliers. c. For exterior roofs other than roofs with ventilated attics, the roof surface may be modeled with a reflectance of 0.45 if the reflectance of the proposed design roof is greater than 0.70 and its emittance is greater than 0.75. The reflectance and emittance shall be tested in accordance with the Exception to Section 5.5.3.1. All other roof surfaces shall be modeled with a reflectance of 0.3. d. Manually operated fenestration shading devices such as blinds or shades shall not be modeled. Permanent shading devices such as fins, overhangs, and lightshelves shall be modeled.	The *budget building design* shall have identical *conditioned floor area* and identical exterior dimensions and orientations as the *proposed building design*, except as noted in (a), (b), and (c) in this clause. a. Opaque assemblies such as roof, floors, doors, and walls shall be modeled as having the same *heat capacity* as the *proposed building design* but with the minimum U-factor required in Section 5.5 for new buildings or *additions* and Section 5.1.3 for *alterations*. b. Roof albedo—All roof surfaces shall be modeled with a reflectivity of 0.3. c. Fenestration—No shading projections are to be modeled; fenestration shall be assumed to be flush with the exterior wall or roof. If the fenestration area for new buildings or *additions* exceeds the maximum allowed by Section 5.5.4.2, the area shall be reduced proportionally along each exposure until the limit set in Section 5.5.4.2 is met. Fenestration U-factor shall be the minimum required for the climate, and the SHGC shall be the maximum allowed for the climate and orientation. The fenestration model for envelope *alterations* shall reflect the limitations on area, U-factor, and SHGC as described in Section 5.1.3. **Exception:** When trade-offs are made between an *addition* and an *existing building* as described in the Exception to Section 4.2.1.2, the envelope assumptions for the *existing building* in the *budget building design* shall reflect existing conditions prior to any revisions that are part of this permit.

TABLE 11.3.1 Modeling Requirements for Calculating Design Energy Cost and Energy Cost Budget *(continued)*

No.	Proposed Building Design (Column A) Design Energy Cost (DEC)	Budget Building Design (Column B) Energy Cost Budget (ECB)
6. Lighting		
	Lighting power in the *proposed building design* shall be determined as follows: a. Where a complete lighting system exists, the actual lighting power for each thermal block shall be used in the model. b. Where a lighting system has been designed, lighting power shall be determined in accordance with Sections 9.1.3 and 9.1.4. c. Where no lighting exists or is specified, lighting power shall be determined in accordance with the Building Area Method for the appropriate building type. d. Lighting system power shall include all lighting system components shown or provided for on plans (including lamps, ballasts, task fixtures, and furniture-mounted fixtures).	Lighting power in the *budget building design* shall be determined using the same categorization procedure (building area or space function) and categories as the *proposed building design* with lighting power set equal to the maximum allowed for the corresponding method and category in either Section 9.5 or 9.6. Power for fixtures not included in the *LPD* calculation shall be modeled identically in the *proposed building design* and *budget building design*. Lighting controls shall be the minimum required.
7. Thermal Blocks—HVAC Zones Designed		
	Where HVAC zones are defined on HVAC design drawings, each HVAC zone shall be modeled as a separate *thermal block.* **Exception:** Different HVAC zones may be combined to create a single *thermal block* or identical *thermal blocks* to which multipliers are applied provided all of the following conditions are met: a. The space use classification is the same throughout the *thermal block.* b. All HVAC zones in the *thermal block* that are adjacent to glazed exterior walls face the same orientation or their orientations are within 45 degrees of each other. c. All of the zones are served by the same HVAC system or by the same kind of HVAC system.	Same as *proposed building design*
8. Thermal Blocks—HVAC Zones Not Designed		
	Where the HVAC zones and systems have not yet been designed, *thermal blocks* shall be defined based on similar internal load densities, occupancy, lighting, thermal and space temperature schedules, and in combination with the following guidelines: a. Separate *thermal blocks* shall be assumed for interior and perimeter spaces. Interior spaces shall be those located more than 15 ft from an exterior wall. Perimeter spaces shall be those located closer than 15 ft from an exterior wall. b. Separate *thermal blocks* shall be assumed for spaces adjacent to glazed exterior walls; a separate zone shall be provided for each orientation, except orientations that differ by no more than 45 degrees may be considered to be the same orientation. Each zone shall include all floor area that is 15 ft or less from a glazed perimeter wall, except that floor area within 15 ft of glazed perimeter walls having more than one orientation shall be divided proportionately between zones. c. Separate *thermal blocks* shall be assumed for spaces having floors that are in contact with the ground or exposed to ambient conditions from zones that do not share these features. d. Separate *thermal blocks* shall be assumed for spaces having exterior ceiling or roof assemblies from zones that do not share these features.	Same as *proposed building design*
9. Thermal Blocks—Multifamily Residential Buildings		
	Residential spaces shall be modeled using one *thermal block* per space except that those facing the same orientations may be combined into one *thermal block.* Corner units and units with roof or floor loads shall only be combined with units sharing these features.	Same as Proposed Design

No.	Proposed Building Design (Column A) Design Energy Cost (DEC)	Budget Building Design (Column B) Energy Cost Budget (ECB)
10. HVAC Systems		
	The HVAC system type and all related performance parameters, such as equipment capacities and efficiencies, in the *proposed building design* shall be determined as follows: a. Where a complete HVAC system exists, the model shall reflect the actual system type using actual component capacities and efficiencies. b. Where an HVAC system has been designed, the HVAC model shall be consistent with design documents. Mechanical equipment efficiencies shall be adjusted from actual design conditions to the standard rating conditions specified in Section 6.4.1, if required by the simulation model. c. Where no heating system exists or no heating system has been specified, the heating system shall be modeled as fossil fuel. The system characteristics shall be identical to the system modeled in the *budget building design*. d. Where no cooling system exists or no cooling system has been specified, the cooling system shall be modeled as an air-cooled single-zone system, one unit per *thermal block*. The system characteristics shall be identical to the system modeled in the *budget building design*.	The HVAC system type and related performance parameters for the *budget building design* shall be determined from Figure 11.3.2, the system descriptions in Table 11.3.2A and accompanying notes, and in accord with rules specified in Section 11.3.2 (a)–(j).
11. Service Hot-Water Systems		
	The service hot-water system type and all related performance parameters, such as equipment capacities and efficiencies, in the *proposed building design* shall be determined as follows: a. Where a complete service hot-water system exists, the model shall reflect the actual system type using actual component capacities and efficiencies. b. Where a service hot-water system has been designed, the service hot-water model shall be consistent with design documents. c. Where no service hot-water system exists or is specified, no service hot-water heating shall be modeled.	The service hot-water system type and related performance in the *budget building design* shall be identical to the *proposed building design.* **Exceptions:** a. Where Section 7.5 applies, the boiler shall be split into a separate space heating boiler and hot-water heater with *efficiency* requirements set to the least efficient allowed. b. For 24-hour-per-day facilities that meet the prescriptive criteria for use of condenser heat recovery systems described in Section 6.5.6.2, a system meeting the requirements of that section shall be included in the *baseline building design* regardless of the exceptions to Section 6.5.6.2. If a condenser heat recovery system meeting the requirements described in Section 6.5.6.2 cannot be modeled, the requirement for including such a system in the actual building shall be met as a prescriptive requirement in accordance with Section 6.5.6.2 and no heat-recovery system shall be included in the *proposed* or *budget building design*.
12. Miscellaneous Loads		
	Receptacle, motor, and process loads shall be modeled and estimated based on the building type or space type category and shall be assumed to be identical in the *proposed* and *budget building designs.* These loads shall be included in simulations of the building and shall be included when calculating the *energy cost budget* and *design energy cost.* All end-use load components within and associated with the building shall be modeled, unless specifically excluded by Sections 13 and 14 of Table 11.3.1: including, but not limited to, exhaust fans, parking garage ventilation fans, exterior building lighting, swimming pool heaters and pumps, elevators and escalators, refrigeration equipment, and cooking equipment.	Receptacle, motor, and process loads shall be modeled and estimated based on the building type or space type category and shall be assumed to be identical in the *proposed* and *budget building designs.* These loads shall be included in simulations of the building and shall be included when calculating the *energy cost budget* and *design energy cost.* All end-use load components within and associated with the building shall be modeled, unless specifically excluded by Sections 13 and 14 of Table 11.3.1: including, but not limited to, exhaust fans, parking garage ventilation fans, exterior building lighting, swimming pool heaters and pumps, elevators and escalators, refrigeration equipment, and cooking equipment.

TABLE 11.3.1 Modeling Requirements for Calculating Design Energy Cost and Energy Cost Budget *(continued)*

No.	Proposed Building Design (Column A) Design Energy Cost (DEC)	Budget Building Design (Column B) Energy Cost Budget (ECB)
13.	**Modeling Exceptions**	
	All elements of the *proposed building design* envelope, HVAC, service water heating, lighting, and electrical systems shall be modeled in the *proposed building design* in accordance with the requirements of Sections 1 through 12 of Table 11.3.1. **Exception:** Components and systems in the *proposed building design* may be excluded from the simulation model provided: a. component energy usage does not affect the energy usage of systems and components that are being considered for trade-off; b. the applicable prescriptive requirements of Sections 5.5, 6.5, 7.5, and either 9.5 or 9.6 applying to the excluded components are met.	None
14.	**Modeling Limitations to the Simulation Program**	
	If the simulation program cannot model a component or system included in the *proposed building design*, one of the following methods shall be used with the approval of the *authority having jurisdiction*: a. Ignore the component if the energy impact on the trade-offs being considered is not significant. b. Model the component substituting a thermodynamically similar component model. c. Model the HVAC system components or systems using the *budget building design's* HVAC system in accordance with Section 10 of Table 11.3.1. Whichever method is selected, the component shall be modeled identically for both the *proposed building design* and *budget building design* models.	Same as *proposed building design*

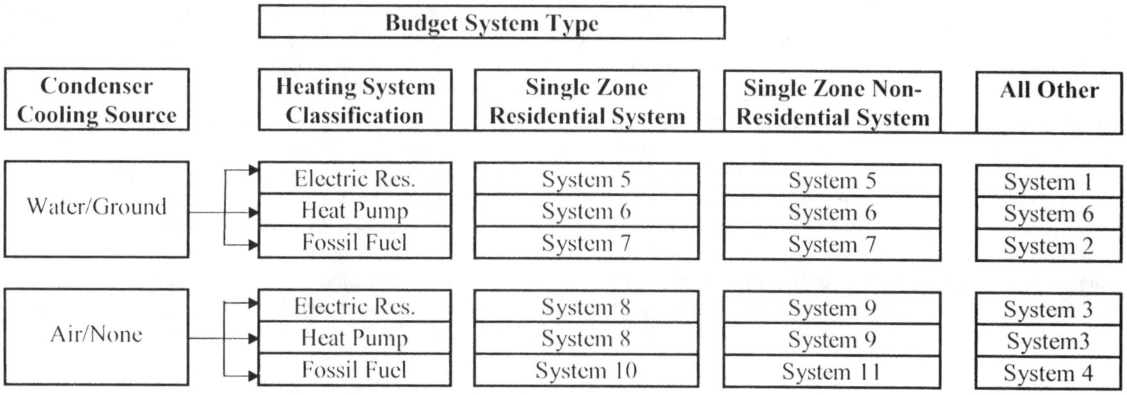

Figure 11.3.2 HVAC systems map.

11.3.2 HVAC Systems. The *HVAC system* type and related performance parameters for the *budget building design* shall be determined from Figure 11.3.2, the system descriptions in Table 11.3.2A and accompanying notes, and the following rules:

a. Components and parameters not listed in Figure 11.3.2 and Table 11.3.2A or otherwise specifically addressed in this subsection shall be identical to those in the *proposed building design.*

   Exception: Where there are specific requirements in Sections 6.4 and 6.5, the component *efficiency* in the *budget building design* shall be adjusted to the lowest *efficiency* level allowed by the requirement for that component type.

b. All HVAC and service water heating equipment in the *budget building* shall be modeled at the minimum *efficiency* levels, both part load and full load, in accordance with Sections 6.4 and 7.4.

c. Where *efficiency* ratings, such as EER and COP, include fan energy, the descriptor shall be broken down into its components so that supply fan energy can be modeled separately. Supply and return/relief system fans shall be modeled as operating at least whenever the spaces served are occupied except as specifically noted in Table 11.3.2A.

d. Minimum *outdoor air* ventilation rates shall be the same for both the *budget building design* and *proposed building design*. Heat recovery shall be modeled for the *budget building design* in accordance with Section 6.5.6.1.

TABLE 11.3.2A Budget System Descriptions

System No.	System Type	Fan Control	Cooling Type	Heating Type
1	VAV with parallel fan-powered boxes[a]	VAV[d]	Chilled water[e]	Electric resistance
2	VAV with reheat[b]	VAV[d]	Chilled water[e]	Hot-water fossil fuel boiler[f]
3	Packaged VAV with parallel fan-powered boxes[a]	VAV[d]	Direct expansion[c]	Electric resistance
4	Packaged VAV with reheat[b]	VAV[d]	Direct expansion[c]	Hot-water fossil fuel boiler[f]
5	Two-pipe fan-coil	Constant volume[i]	Chilled water[e]	Electric resistance
6	Water-source heat pump	Constant volume[i]	Direct expansion[c]	Electric heat pump and boiler[g]
7	Four-pipe fan-coil	Constant volume[i]	Chilled water[e]	Hot-water fossil fuel boiler[f]
8	Packaged terminal heat pump	Constant volume[i]	Direct expansion[c]	Electric heat pump[h]
9	Packaged rooftop heat pump	Constant volume[i]	Direct expansion[c]	Electric heat pump[h]
10	Packaged terminal air conditioner	Constant volume[i]	Direct expansion	Hot-water fossil fuel boiler[f]
11	Packaged rooftop air conditioner	Constant volume[i]	Direct expansion	Fossil fuel furnace

[a] **VAV with parallel boxes:** Fans in parallel VAV fan-powered boxes shall be sized for 50% of the peak design flow rate and shall be modeled with 0.35 W/cfm fan power. Minimum volume setpoints for fan-powered boxes shall be equal to the minimum rate for the space required for ventilation consistent with Section 6.5.2.1 Exception (a)1. Supply air temperature setpoint shall be constant at the design condition [see Section 11.3.2 (h)].

[b] **VAV with reheat:** Minimum volume setpoints for VAV reheat boxes shall be 0.4 cfm/ft^2 of floor area, or the minimum ventilation rate, whichever is larger, consistent with Section 6.5.2.1 Exception (a)2. Supply air temperature shall be reset based on zone demand from the design temperature difference to a 10°F temperature difference under minimum load conditions. Design airflow rates shall be sized for the reset supply air temperature, i.e., a 10°F temperature difference.

[c] **Direct expansion:** The fuel type for the cooling system shall match that of the cooling system in the *proposed building design*.

[d] **VAV:** Constant volume can be modeled if the system qualifies for Exception (b) to Section 6.5.2.1. When the *proposed building design* system has a supply, return, or relief fan motor 25 hp or larger, the corresponding fan in the VAV system of the *budget building design* shall be modeled assuming a variable speed drive. For smaller fans, a forward-curved centrifugal fan with inlet vanes shall be modeled. If the *proposed building design's* system has a DDC system at the zone level, static pressure setpoint reset based on zone requirements in accordance with Section 6.5.3.2.3 shall be modeled.

[e] **Chilled water:** For systems using purchased chilled water, the chillers are not explicitly modeled and chilled water costs shall be based as determined in Section 11.2.3. Otherwise, the *budget building design's* chiller plant shall be modeled with chillers having the number as indicated in Table 11.3.2B as a function of *budget building design* chiller plant load and type as indicated in Table 11.3.2C as a function of individual chiller load. Where chiller fuel source is mixed, the system in the *budget building design* shall have chillers with the same fuel types and with capacities having the same proportional capacity as the *proposed building design's* chillers for each fuel type. Chilled-water supply temperature shall be modeled at 44°F design supply temperature and 56°F return temperature. Piping losses shall not be modeled in either building model. Chilled-water supply water temperature shall be reset in accordance with Section 6.5.4.3. Pump system power for each pumping system shall be the same as the *proposed building design*; if the *proposed building design* has no chilled-water pumps, the *budget building design* pump power shall be 22 W/gpm (equal to a pump operating against a 75 ft head, 65% combined impeller and motor *efficiency*). The chilled-water system shall be modeled as primary-only variable flow with flow maintained at the design rate through each chiller using a bypass. Chilled-water pumps shall be modeled as riding the pump curve or with variable-speed drives when required in Section 6.5.4.1. The heat rejection device shall be an axial fan cooling tower with two-speed fans if required in Section 6.5.5. Condenser water design supply temperature shall be 85°F or 10°F approach to design wet-bulb temperature, whichever is lower, with a design temperature rise of 10°F. The tower shall be controlled to maintain a 70°F leaving water temperature where weather permits, floating up to leaving water temperature at design conditions. Pump system power for each pumping system shall be the same as the *proposed building design*; if the *proposed building design* has no condenser water pumps, the *budget building design* pump power shall be 19 W/gpm (equal to a pump operating against a 60 ft head, 60% combined impeller and motor *efficiency*). Each chiller shall be modeled with separate condenser water and chilled-water pumps interlocked to operate with the associated chiller.

[f] **Fossil fuel boiler:** For systems using purchased hot water or steam, the boilers are not explicitly modeled and hot water or steam costs shall be based on actual utility rates. Otherwise, the boiler plant shall use the same fuel as the *proposed building design* and shall be natural draft. The *budget building design* boiler plant shall be modeled with a single boiler if the *budget building design* plant load is 600,000 Btu/h and less and with two equally sized boilers for plant capacities exceeding 600,000 Btu/h. Boilers shall be staged as required by the load. Hot-water supply temperature shall be modeled at 180°F design supply temperature and 130°F return temperature. Piping losses shall not be modeled in either building model. Hot-water supply water temperature shall be reset in accordance with Section 6.5.4.3. Pump system power for each pumping system shall be the same as the *proposed building design*; if the *proposed building design* has no hot-water pumps, the *budget building design* pump power shall be 19 W/gpm (equal to a pump operating against a 60 ft head, 60% combined impeller and motor *efficiency*). The hot-water system shall be modeled as primary-only with continuous variable flow. Hot-water pumps shall be modeled as riding the pump curve or with variable speed drives when required by Section 6.5.4.1.

[g] **Electric heat pump and boiler:** Water-source heat pumps shall be connected to a common heat pump water loop controlled to maintain temperatures between 60°F and 90°F. Heat rejection from the loop shall be provided by an axial fan closed-circuit evaporative fluid cooler with two-speed fans if required in Section 6.5.5.2. Heat addition to the loop shall be provided by a boiler that uses the same fuel as the *proposed building design* and shall be natural draft. If no boilers exist in the *proposed building design*, the budget building boilers shall be fossil fuel. The *budget building design* boiler plant shall be modeled with a single boiler if the *budget building design* plant load is 600,000 Btu/h or less and with two equally sized boilers for plant capacities exceeding 600,000 Btu/h. Boilers shall be staged as required by the load. Piping losses shall not be modeled in either building model. Pump system power shall be the same as the *proposed building design*; if the *proposed building design* has no pumps, the *budget building design* pump power shall be 22 W/gpm, which is equal to a pump operating against a 75 ft head, with a 65% combined impeller and motor *efficiency*. Loop flow shall be variable with flow shutoff at each heat pump when its compressor cycles off as required by Section 6.5.4.4. Loop pumps shall be modeled as riding the pump curve or with variable speed drives when required by Section 6.5.4.1.

[h] **Electric heat pump:** Electric air-source heat pumps shall be modeled with electric auxiliary heat. The system shall be controlled with a multi-stage space thermostat and an *outdoor air* thermostat wired to energize auxiliary heat only on the last thermostat stage and when *outdoor air* temperature is less than 40°F.

[i] **Constant volume:** Fans shall be controlled in the same manner as in the *proposed building design*; i.e., fan operation whenever the space is occupied or fan operation cycled on calls for heating and cooling. If the fan is modeled as cycling and the fan energy is included in the energy *efficiency* rating of the equipment, fan energy shall not be modeled explicitly.

e. *Budget building* systems as listed in Table 11.3.2A shall have *outdoor air* economizers or water economizers, the same as in the proposed building, in accordance with Section 6.5.1. The high-limit shutoff shall be in accordance with Table 11.3.2D.
f. If the *proposed building design* system has a preheat coil, the *budget building design's* system shall be modeled with a preheat coil controlled in the same manner.
g. System design supply air rates for the *budget building design* shall be based on a supply-air-to-room-air temperature difference of 20°F. If return or relief fans are specified in the *proposed building design*, the *budget building design* shall also be modeled with the same fan type sized for the budget system supply fan air quantity less the minimum *outdoor air*, or 90% of the supply fan air quantity, whichever is larger.
h. Fan system *efficiency* (bhp per cfm of supply air including the effect of belt losses but excluding motor and motor drive losses) shall be the same as the *proposed building design* or up to the limit prescribed in Section 6.5.3.1, whichever is smaller. If this limit is reached, each fan shall be proportionally reduced in brake horsepower until the limit is met. Fan electrical power shall then be determined by adjusting the calculated fan hp by the minimum motor *efficiency* prescribed by Section 10.4 for the appropriate motor size for each fan.
i. The equipment capacities for the *budget building design* shall be sized proportionally to the capacities in the *proposed building design* based on sizing runs, i.e., the ratio between the capacities used in the annual simulations and the capacities determined by the sizing runs shall be the same for both the *proposed building design* and *budget building design*. Unmet load hours for the *proposed building design* shall not differ from unmet load hours for the *budget building design* by more than 50 hours.
j. Each *HVAC system* in a *proposed building design* is mapped on a one-to-one correspondence with one of eleven *HVAC systems* in the *budget building design*. To determine the budget building system:

    1. Enter Figure 11.3.2 at "Water" if the *proposed building design* system condenser is water or evaporatively cooled; enter at "Air" if the condenser is air-cooled. Closed-circuit dry-coolers shall be considered air-cooled. Systems utilizing district cooling shall be treated as if the condenser water type were "water." If no mechanical cooling is specified or the mechanical cooling system in the *proposed building design* does not require heat rejection, the system shall be treated as if the condenser water type were "Air." For *proposed building designs* with ground-source or groundwater-source heat pumps, the budget system shall be water-source heat pump (System 6).
    2. Select the path that corresponds to the *proposed building design* heat source: electric resistance, heat pump (including air-source and water-source), or fuel-fired. Systems utilizing district heating (steam or hot water) shall be treated as if the heating system type were "Fossil Fuel." Systems with no heating capability shall be treated as if the heating system type were "Fossil Fuel." For systems with mixed fuel heating sources, the system or systems that use the secondary heating source type (the one with the smallest total installed output capacity for the spaces served by the system) shall be modeled identically in the *budget building design* and the primary heating source type shall be used in Figure 11.3.2 to determine budget system type.
    3. Select the *budget building design* system category: The system under "Single Zone Residential System" shall be selected if the HVAC system in the proposed design is a single-zone system and serves a residential space. The system under "Single Zone Nonresidential System" shall be selected if the HVAC system in the proposed design is a single-zone system and serves other than residential spaces. The system under "All Other" shall be selected for all other cases.

TABLE 11.3.2B Number of Chillers

Total Chiller Plant Capacity	Number of Chillers
≤300 tons	One
>300 tons, <600 tons	Two sized equally
≥600 tons	Two minimum with chillers added so that no chiller is larger than 800 tons, all sized equally

TABLE 11.3.2C Water Chiller Types

Individual Chiller Plant Capacity	Electric Chiller Type	Fossil Fuel Chiller Type
≤100 tons	Reciprocating	Single-effect absorption, direct fired
>100 tons, <300 tons	Screw	Double-effect absorption, direct fired
≥300 tons	Centrifugal	Double-effect absorption, direct fired

TABLE 11.3.2D Economizer High-Limit Shutoff

Economizer Type	High-Limit Shutoff
Air	Table 6.5.1.1.3B
Water (integrated)	When its operation will no longer reduce HVAC system energy
Water (nonintegrated)	When its operation can no longer provide the cooling load

12. NORMATIVE REFERENCES

Reference	Title
Air Movement and Control Association International, 30 West University Drive, Arlington Heights, IL 60004-1806	
AMCA 500-D-98	Test Methods for Louvers, Dampers, and Shutters
American National Standards Institute, 11 West 42nd Street, New York, NY 10036	
ANSI Z21.10.3-1998	Gas Water Heater, Volume 3, Storage, with Input Ratings above 75,000 Btu/h, Circulating and Instantaneous Water Heaters
ANSI Z21.47-2001	Gas-Fired Central Furnaces (Except Direct Vent and Separated Combustion System Furnaces)
ANSI Z83.8-2002	Gas Unit Heaters and Duct Furnaces
Association of Home Appliance Manufacturers, 20 North Wacker Drive, Chicago, IL 60606	
ANSI/AHAM RAC-1-87	Room Air Conditioners
Air-Conditioning and Refrigeration Institute, 4100 North Fairfax Drive, Suite 200, Arlington, VA 22203	
ARI 210/240-2003	Unitary Air Conditioning and Air-Source Heat Pump Equipment
ARI 310/380-2004	Packaged Terminal Air-Conditioners and Heat Pumps
ARI 340/360-2004	Commercial and Industrial Unitary Air-Conditioning and Heat Pump Equipment
ARI 365-2002	Commercial and Industrial Unitary Air-Conditioning Condensing Units
ARI 390-2003	Performance Rating of Single Packaged Vertical Air-Conditioners and Heat Pumps
ARI 460-2000	Remote Mechanical Draft Air Cooled Refrigerant Condensers
ARI 550/590-98 with Addenda through July 2002	Water-Chilling Packages Using the Vapor Compression Cycle
ARI 560-2000	Absorption Water Chilling and Water Heating Packages
American Society of Heating, Refrigerating and Air-Conditioning Engineers, 1791 Tullie Circle, NE, Atlanta, GA 30329	
ANSI/ASHRAE Standard 62.1-2004	Ventilation for Acceptable Indoor Air Quality
ANSI/ASHRAE Standard 140-2004	Standard Method of Test for the Evaluation of Building Energy Analysis Computer Programs
ANSI/ASHRAE 146-1998	Method of Testing for Rating Pool Heaters
American Society for Testing and Materials, 100 Barr Harbor Dr., West Conshohocken, PA 19428-2959	
ASTM C90-03	Standard Specification for Loadbearing Concrete Masonry Units
ASTM C177-97	Standard Test Method for Steady-State Heat Flux Measurements and Thermal Transmittance Properties by Means of the Guarded-Hot-Plate Apparatus
ASTM C272-01	Test Method for Water Absorption of Core Materials for Structural Sandwich Constructions
ASTM C518-04	Standard Test Method for Steady-State Thermal Transmittance Properties by Means of the Heat Flow Meter Apparatus
ASTM C835-01	Standard Test Method for Total Hemispherical Emittance of Surfaces From 20°C to 1400°C

Reference	Title
ASTM C1363-97	Standard Test Method for the Thermal Performance of Building Assemblies by Means of a Hot Box Apparatus
ASTM C1371-04	Standard Test Method for Determination of Emittance of Materials Near Room Temperature Using Portable Emissometers
ASTM C1549-04	Standard Test Method for Determination of Solar Reflectance Near Ambient Temperature Using a Portable Solar Reflectometer
ASTM E1980 (2001)	Standard Practice for Calculating Solar Reflectance Index of Horizontal and Low Sloped Opaque Surfaces
ASTM E408-71 (2002)	Test Methods for Total Normal Emittance of Surfaces Using Inspection-Meter Techniques
ASTM E903-96	Test Method for Solar Absorptance, Reflectance, and Transmittance of Materials Using Integrating Spheres
ASTM E1175-87 (2003)	Standard Test Method for Determining Solar or Photopic Reflectance, Transmittance, and Absorptance of Materials Using a Large Diameter Integrating Sphere
ASTM E1918-97	Standard Test Method for Measuring Solar Reflectance of Horizontal or Low-Sloped Surfaces in the Field
Cooling Technology Institute, 2611 FM 1960 West, Suite A-101, Houston, TX 77068-3730; P.O. Box 73383, Houston, TX 77273-3383	
CTI ATC-105 (00)	Acceptance Test Code for Water Cooling Towers
CTI STD-201 (04)	Standard for Certification of Water Cooling Tower Thermal Performance
Hydronics Institute, Division of Gama, 35 Russo Place, P.O. Box 218, Berkeley Heights, NJ 07922	
BTS 2000.	Testing Standard Method to Determine Efficiency of Commercial Space Heating Boilers
International Organization for Standardization, 1, rue de Varembe, Case postale 56, CH-1211 Geneve 20, Switzerland	
ISO 13256-1 (1998)	Water-Source Heat Pumps—Testing and Rating for Performance—Part 1: Water-to-Air and Brine-to-Air Heat Pumps
Door and Access Systems Manufacturers Association (DASMA), 1300 Sumner Avenue, Cleveland, OH 44115-2851	
ANSI/DASMA 105-92 (R 1998)	Test Method for Thermal Transmittance and Air Infiltration of Garage Doors
National Electrical Manufacturers Association, 1300 N. 17th Street, Suite 1847, Rosslyn, VA 22209	
ANSI/NEMA MG 1-1993	Motors and Generators
National Fire Protection Association, 1 Battery March Park, P.O. Box 9101, Quincy, MA 02269-9101	
NFPA 96-94	Ventilation Control and Fire Protection of Commercial Cooking Operations
National Fenestration Rating Council, 1300 Spring Street, Suite 500, Silver Springs, MD 20910	
NFRC 100-2004	Procedure for Determining Fenestration Product U-Factors
NFRC 200-2004	Procedure for Determining Fenestration Product Solar Heat Gain Coefficients and Visible Transmittance at Normal Incidence
NFRC 300-2004	Standard Test Method for Determining the Solar Optical Properties of Glazing Materials and Systems

Reference	Title
NFRC 400-2004	Procedure for Determining Fenestration Product Air Leakage
Underwriters Laboratories, Inc., 333 Pfingsten Rd., Northbrook, IL 60062	
UL 181A-94	Closure Systems for Use with Rigid Air Ducts and Air Connectors
UL 181B-95	Closure Systems for Use with Flexible Air Ducts and Air Connectors
UL 727-94	UL Standard for Safety—Oil Fired Central Furnaces
UL 731-95	UL Standard for Safety—Oil-Fired Unit Heaters
U.S. Department of Energy 1000 Independence Avenue, SW, Washington, DC 20585	
10 CFR Part 430, App N	Uniform Test Method for Measuring the Energy Consumption of Furnaces
42 USC 6831, et seq., Public Law 102-486	Energy Policy Act of 1992

(This is a normative appendix and is part of this standard.)

NORMATIVE APPENDIX A

RATED R-VALUE OF INSULATION AND ASSEMBLY U-FACTOR, C-FACTOR, AND F-FACTOR DETERMINATIONS

A1. GENERAL

A1.1 Pre-Calculated Assembly U-Factors, C-Factors, F-Factors, or Heat Capacities. The *U-factors*, *C-factors, F-factors,* and *heat capacities* for typical construction assemblies are included in Sections A2 through A8. These values shall be used for all calculations unless otherwise allowed by Section A1.2. Interpolation between values in a particular table in Normative Appendix A is allowed for *rated R-values of insulation*, including insulated sheathing. Extrapolation beyond values in a table in Normative Appendix A is not allowed.

A1.2 Applicant-Determined Assembly U-Factors, C-Factors, F-Factors, or Heat Capacities. If the *building official* determines that the proposed construction assembly is not adequately represented in Sections A2 through A8, the applicant shall determine appropriate values for the assembly using the assumptions in Section A9. An assembly is deemed to be adequately represented if

a. the interior structure, hereafter referred to as the *base assembly*, for the *class of construction* is the same as described in Sections A2 through A8 and
b. changes in exterior or interior surface *building materials* added to the base assembly do not increase or decrease the R-value by more than 2 from that indicated in the descriptions in Sections A2 through A8.

Insulation, including insulated sheathing, is not considered a *building material.*

A2. ROOFS

A2.1 General. The buffering effect of suspended ceilings or attic spaces shall not be included in *U-factor* calculations.

A2.2 Roofs with Insulation Entirely Above Deck

A2.2.1 General. For the purpose of Section A1.2, the base assembly is *continuous insulation* over a structural deck. The *U-factor* includes R-0.17 for exterior air film, R-0 for metal deck, and R-0.61 for interior air film heat flow up. Added insulation is continuous and uninterrupted by framing. The framing factor is zero.

A2.2.2 Rated R-Value of Insulation. For *roofs with insulation entirely above deck*, the *rated R-value of insulation* is for *continuous insulation*.

Exception: Interruptions for framing and pads for mechanical equipment are permitted with a combined total area not exceeding one percent of the total opaque assembly area.

A2.2.3 U-Factor. *U-factors* for *roofs with insulation entirely above deck* shall be taken from Table A2.2. It is not acceptable to use these *U-factors* if the insulation is not entirely above deck or not continuous.

TABLE A2.2 Assembly U-Factors for Roofs with Insulation Entirely Above Deck

Rated R-Value of Insulation Alone	Overall U-Factor for Entire Assembly
R-0	U-1.282
R-1	U-0.562
R-2	U-0.360
R-3	U-0.265
R-4	U-0.209
R-5	U-0.173
R-6	U-0.147
R-7	U-0.129
R-8	U-0.114
R-9	U-0.102
R-10	U-0.093
R-11	U-0.085
R-12	U-0.078
R-13	U-0.073
R-14	U-0.068
R-15	U-0.063
R-16	U-0.060
R-17	U-0.056
R-18	U-0.053
R-19	U-0.051
R-20	U-0.048
R-21	U-0.046
R-22	U-0.044
R-23	U-0.042
R-24	U-0.040
R-25	U-0.039
R-26	U-0.037
R-27	U-0.036
R-28	U-0.035
R-29	U-0.034
R-30	U-0.032
R-35	U-0.028
R-40	U-0.025
R-45	U-0.022
R-50	U-0.020
R-55	U-0.018
R-60	U-0.016

A2.3 Metal Building Roofs

A2.3.1 General. For the purpose of Section A1.2, the base assembly is a *roof* where the insulation is draped over the steel structure (purlins) and then compressed when the metal roof panels are attached to the steel structure (purlins). Additional assemblies include *continuous insulation*, uncompressed and uninterrupted by framing.

A2.3.2 Rated R-Value of Insulation

A2.3.2.1 The first *rated R-value of insulation* is for insulation draped over purlins and then compressed when the metal roof panels are attached, or for insulation hung between the purlins. A minimum 1 in. thermal spacer block between the purlins and the metal roof panels is required when specified in Table A2.3.

A2.3.2.2 For double-layer installations, the second *rated R-value of insulation* is for insulation installed parallel to the purlins.

A2.3.2.3 For continuous insulation (e.g., insulation boards or blankets), it is assumed that the insulation is installed below the purlins and is uninterrupted by framing members. Insulation exposed to the *conditioned space* or *semiheated space* shall have a facing, and all insulation seams shall be continuously sealed to provide a continuous air barrier.

TABLE A2.3 Assembly U-Factors for Metal Building Roofs

Insulation System	Rated R-Value of Insulation	Total Rated R-Value of Insulation	Overall U-Factor for Entire Base Roof Assembly	Overall U-Factor for Assembly of Base Roof Plus Continuous Insulation (Uninterrupted by Framing) — Rated R-Value of Continuous Insulation: R-5.6	R-11.2	R-16.8	R-22.4	R-28.0	R-33.6
Standing Seam Roofs with Thermal Spacer Blocks									
Single Layer	None	0	**U-1.280**	0.162	0.087	0.059	0.045	0.036	0.030
	R-6	6	**U-0.167**	0.086	0.058	0.044	0.035	0.029	0.025
	R-10	10	**U-0.097**	0.063	0.046	0.037	0.031	0.026	0.023
	R-11	11	**U-0.092**	0.061	0.045	0.036	0.030	0.026	0.022
	R-13	13	**U-0.083**	0.057	0.043	0.035	0.029	0.025	0.022
	R-16	16	**U-0.072**	0.051	0.040	0.033	0.028	0.024	0.021
	R-19	19	**U-0.065**	0.048	0.038	0.031	0.026	0.023	0.020
Double Layer	R-10 + R-10	20	**U-0.063**	0.047	0.037	0.031	0.026	0.023	0.020
	R-10 + R-11	21	**U-0.061**	0.045	0.036	0.030	0.026	0.023	0.020
	R-11 + R-11	22	**U-0.060**	0.045	0.036	0.030	0.026	0.022	0.020
	R-10 + R-13	23	**U-0.058**	0.044	0.035	0.029	0.025	0.022	0.020
	R-11 + R-13	24	**U-0.057**	0.043	0.035	0.029	0.025	0.022	0.020
	R-13 + R-13	26	**U-0.055**	0.042	0.034	0.029	0.025	0.022	0.019
	R-10 + R-19	29	**U-0.052**	0.040	0.033	0.028	0.024	0.021	0.019
	R-11 + R-19	30	**U-0.051**	0.040	0.032	0.027	0.024	0.021	0.019
	R-13 + R-19	32	**U-0.049**	0.038	0.032	0.027	0.023	0.021	0.019
	R-16 + R-19	35	**U-0.047**	0.037	0.031	0.026	0.023	0.020	0.018
	R-19 + R-19	38	**U-0.046**	0.037	0.030	0.026	0.023	0.020	0.018
Thru-Fastened without Thermal Spacer Blocks									
	R-10	10	**U-0.153**						
	R-11	11	**U-0.139**						
	R-13	13	**U-0.130**						
	R-16	16	**U-0.106**						
	R-19	19	**U-0.098**						
Filled Cavity with Thermal Spacer Blocks									
	R-19 + R-10	29	**U-0.041**	0.033	0.028	0.024	0.021	0.020	0.017

(Multiple R-values are listed in order from inside to outside.)

A2.3.3 U-factor. *U-factors* for *metal building roofs* shall be taken from Table A2.3 It is not acceptable to use these *U-factors* if additional insulated sheathing is not continuous.

A2.4 Attic Roofs with Wood Joists

A2.4.1 General. For the purpose of Section A1.2, the base *attic roof* assembly is a *roof* with nominal 4 in. deep wood as the lower chord of a roof truss or ceiling joist. The ceiling is attached directly to the lower chord of the truss and the attic space above is ventilated. Insulation is located directly on top of the ceiling, first filling the cavities between the wood and then later covering both the wood and cavity areas. No credit is given for roofing materials. The *single-rafter roof* is similar to the base *attic roof*, with the key difference being that there is a single, deep rafter to which both the *roof* and the ceiling are attached. The heat flow path through the rafter is calculated to be the same depth as the insulation. The *U-factors* include R-0.46 for semi-exterior air film, R-0.56 for 0.625 in. gypsum board, and R-0.61 for interior air film heat flow up. *U-factors* are provided for the following configurations:

a. *Attic roof, standard framing:* insulation is tapered around the perimeter with a resultant decrease in thermal resistance. Weighting factors are 85% full-depth insulation, 5% half-depth insulation, and 10% joists.
b. *Attic roof, advanced framing:* full and even depth of insulation extending to the outside edge of exterior walls. Weighting factors are 90% full-depth insulation and 10% joists.
c. *Single-rafter roof:* an *attic roof* where the roof sheathing and ceiling are attached to the same rafter. Weighting factors are 90% full-depth insulation and 10% joists.

A2.4.2 Rated R-Value of Insulation

A2.4.2.1 For *attics and other roofs*, the *rated R-value of insulation* is for insulation installed both inside and outside the roof or entirely inside the roof cavity.

A2.4.2.2 Occasional interruption by framing members is allowed but requires that the framing members be covered with insulation when the depth of the insulation exceeds the depth of the framing cavity.

A2.4.2.3 Insulation in such roofs shall be permitted to be tapered at the eaves where the building structure does not allow full depth.

A2.4.2.4 For *single-rafter roofs*, the requirement is the lesser of the values for *attics and other roofs* and those listed in Table A2.4.2.

A2.4.3 U-factors for Attic Roofs with Wood Joists. *U-factors* for *attic roofs* with wood joists shall be taken from Table A2.4. It is not acceptable to use these *U-factors* if the framing is not wood. For *attic roofs* with steel joists, see Section A2.5.

A2.5 Attic Roofs with Steel Joists

A2.5.1 General. For the purpose of Section A1.2, the base assembly is a roof supported by steel joists with insula-

TABLE A2.4 Assembly U-Factors for Attic Roofs with Wood Joists

Rated R-Value of Insulation Alone	Overall U-Factor for Entire Assembly
Wood-Framed Attic, Standard Framing	
None	U-0.613
R-11	U-0.091
R-13	U-0.081
R-19	U-0.053
R-30	U-0.034
R-38	U-0.027
R-49	U-0.021
R-60	U-0.017
R-71	U-0.015
R-82	U-0.013
R-93	U-0.011
R-104	U-0.010
R-115	U-0.009
R-126	U-0.008
Wood-Framed Attic, Advanced Framing	
None	U-0.613
R-11	U-0.088
R-13	U-0.078
R-19	U-0.051
R-30	U-0.032
R-38	U-0.026
R-49	U-0.020
R-60	U-0.016
R-71	U-0.014
R-82	U-0.012
R-93	U-0.011
R-104	U-0.010
R-115	U-0.009
R-126	U-0.008
Wood Joists, Single-rafter Roof	
None	U-0.417
R-11	U-0.088
R-13	U-0.078
R-15	U-0.071
R-19	U-0.055
R-21	U-0.052
R-25	U-0.043
R-30	U-0.036
R-38	U-0.028

TABLE A2.4.2 Single-Rafter Roofs

Climate Zone	Minimum Insulation R-Value or Maximum Assembly U-Factor — Wood Rafter Depth, *d* (Actual)		
	$d \le 8$ in.	$8 < d \le 10$ in.	$10 < d \le 12$ in.
1–7	R-19 U-0.055	R-30 U-0.036	R-38 U-0.028
8	R-21 U-0.052	R-30 U-0.036	R-38 U-0.028

tion between the joists. The assembly represents a *roof* in many ways similar to a *roof with insulation entirely above deck* and a *metal building roof.* It is distinguished from the *metal building roof* category in that there is no metal exposed to the exterior. It is distinguished from the *roof with insulation entirely above deck* in that the insulation is located below the deck and is interrupted by metal trusses that provide thermal bypasses to the insulation. The *U-factors* include R-0.17 for exterior air film, R-0 for metal deck, and R-0.61 for interior air film heat flow up. The performance of the insulation/framing layer is calculated using the values in Table A9.2A.

A2.5.2 *U-factors* for *attic roofs* with steel joists shall be taken from Table A2.5. It is acceptable to use these *U-factors* for any *attic roof* with steel joists.

A3. ABOVE-GRADE WALLS

A3.1 Mass Wall

A3.1.1 General. For the purpose of Section A1.2, the base assembly is a masonry or concrete *wall. Continuous insulation* is installed on the interior or exterior or within the masonry units, or it is installed on the interior or exterior of the concrete. The *U-factors* include R-0.17 for exterior air film and R-0.68 for interior air film, vertical surfaces. For insulated walls, the U-factor also includes R-0.45 for 0.5 in. gypsum board. *U-factors* are provided for the following configurations:

a. Concrete *wall*: 8 in. normal weight concrete wall with a density of 145 lb/ft^3.
b. Solid grouted concrete block *wall*: 8 in. medium weight ASTM C90 concrete block with a density of 115 lb/ft^3 and solid grouted cores.
c. Partially grouted concrete block *wall*: 8 in. medium weight ASTM C90 concrete block with a density of 115 lb/ft^3 having reinforcing steel every 32 in. vertically and every 48 in. horizontally, with cores grouted in those areas only. Other cores are filled with insulating material only if there is no other insulation.

A3.1.2 Mass Wall Rated R-Value of Insulation

A3.1.2.1 Mass wall *HC* shall be determined from Table A3.1B or A3.1C.

A3.1.2.2 The *rated R-value of insulation* is for *continuous insulation* uninterrupted by framing other than 20 gauge 1 in. metal clips spaced no closer than 24 in. on center horizontally and 16 in. on center vertically.

TABLE A2.5 Assembly U-Factors for Attic Roofs with Steel Joists (4.0 ft on Center)

Rated R-Value of Insulation Alone	Overall U-Factor for Entire Assembly
R-0	U-1.282
R-4	U-0.215
R-5	U-0.179
R-8	U-0.120
R-10	U-0.100
R-11	U-0.093
R-12	U-0.086
R-13	U-0.080
R-15	U-0.072
R-16	U-0.068
R-19	U-0.058
R-20	U-0.056
R-21	U-0.054
R-24	U-0.049
R-25	U-0.048
R-30	U-0.041
R-35	U-0.037
R-38	U-0.035
R-40	U-0.033
R-45	U-0.031
R-50	U-0.028
R-55	U-0.027

A3.1.2.3 Where other framing, including metal and wood studs, is used, compliance shall be based on the maximum assembly *U-factor.*

A3.1.2.4 Where *rated R-value of insulation* is used for concrete sandwich panels, the insulation shall be continuous throughout the entire panel.

A3.1.3 Mass Wall U-Factor

A3.1.3.1 *U-factors* for *mass walls* shall be taken from Table A3.1A or determined by the procedure in this subsection. It is acceptable to use the *U-factors* in Table A3.1A for all *mass walls*, provided that the grouting is equal to or less than that specified. *HC* for *mass walls* shall be taken from Table A3.1B or A3.1C.

Exception: For *mass walls*, where the requirement in Tables 5.5-1 through 5.5-8 is for a maximum assembly U-0.151 followed by footnote "a," ASTM C90 concrete block walls, ungrouted or partially grouted at 32 in. or less on center vertically and 48 in. or less on center horizontally, shall have ungrouted cores filled with material having a maximum thermal conductivity

TABLE A3.1A Assembly U-Factors for Above-Grade Concrete Walls and Masonry Walls

Framing Type and Depth	Rated R-Value of Insulation Alone	Assembly U-Factors for 8 in. Normal Weight 145 lb/ft^3 Solid Concrete Walls	Assembly U-Factors for 8 in. Medium Weight 115 lb/ft^3 Concrete Block Walls: Solid Grouted	Assembly U-Factors for 8 in. Medium Weight 115 lb/ft^3 Concrete Block Walls: Partially Grouted (Cores Uninsulated Except Where Specified)
No Framing	**R-0**	**U-0.740**	**U-0.580**	**U-0.480**
	Ungrouted Cores Filled with Loose-Fill Insulation	**N/A**	**N/A**	**U-0.350**
Continuous Metal Framing at 24 in. on Center Horizontally				
3.5 in.	R-11.0	U-0.168	U-0.158	U-0.149
3.5 in.	R-13.0	U-0.161	U-0.152	U-0.144
3.5 in.	R-15.0	U-0.155	U-0.147	U-0.140
4.5 in.	R-17.1	U-0.133	U-0.126	U-0.121
4.5 in.	R-22.5	U-0.124	U-0.119	U-0.114
4.5 in.	R-25.2	U-0.122	U-0.116	U-0.112
5.0 in.	R-19.0	U-0.122	U-0.117	U-0.112
5.0 in.	R-25.0	U-0.115	U-0.110	U-0.106
5.0 in.	R-28.0	U-0.112	U-0.107	U-0.103
5.5 in.	R-19.0	U-0.118	U-0.113	U-0.109
5.5 in.	R-20.9	U-0.114	U-0.109	U-0.105
5.5 in.	R-21.0	U-0.113	U-0.109	U-0.105
5.5 in.	R-27.5	U-0.106	U-0.102	U-0.099
5.5 in.	R-30.8	U-0.104	U-0.100	U-0.096
6.0 in.	R-22.8	U-0.106	U-0.102	U-0.098
6.0 in.	R-30.0	U-0.099	U-0.095	U-0.092
6.0 in.	R-33.6	U-0.096	U-0.093	U-0.090
6.5 in.	R-24.7	U-0.099	U-0.096	U-0.092
7.0 in.	R-26.6	U-0.093	U-0.090	U-0.087
7.5 in.	R-28.5	U-0.088	U-0.085	U-0.083
8.0 in.	R-30.4	U-0.083	U-0.081	U-0.079
1 in. Metal Clips at 24 in. on Center Horizontally and 16 in. Vertically				
1.0 in.	R-3.8	U-0.210	U-0.195	U-0.182
1.0 in.	R-5.0	U-0.184	U-0.172	U-0.162
1.0 in.	R-5.6	U-0.174	U-0.163	U-0.154
1.5 in.	R-5.7	U-0.160	U-0.151	U-0.143
1.5 in.	R-7.5	U-0.138	U-0.131	U-0.125
1.5 in.	R-8.4	U-0.129	U-0.123	U-0.118
2.0 in.	R-7.6	U-0.129	U-0.123	U-0.118
2.0 in.	R-10.0	U-0.110	U-0.106	U-0.102
2.0 in.	R-11.2	U-0.103	U-0.099	U-0.096
2.5 in.	R-9.5	U-0.109	U-0.104	U-0.101
2.5 in.	R-12.5	U-0.092	U-0.089	U-0.086
2.5 in.	R-14.0	U-0.086	U-0.083	U-0.080
3.0 in.	R-11.4	U-0.094	U-0.090	U-0.088
3.0 in.	R-15.0	U-0.078	U-0.076	U-0.074
3.0 in.	R-16.8	U-0.073	U-0.071	U-0.069
3.5 in.	R-13.3	U-0.082	U-0.080	U-0.077
3.5 in.	R-17.5	U-0.069	U-0.067	U-0.065
3.5 in.	R-19.6	U-0.064	U-0.062	U-0.061
4.0 in.	R-15.2	U-0.073	U-0.071	U-0.070
4.0 in.	R-20.0	U-0.061	U-0.060	U-0.058
4.0 in.	R-22.4	U-0.057	U-0.056	U-0.054

TABLE A3.1A Assembly U-Factors for Above-Grade Concrete Walls and Masonry Walls *(continued)*

Framing Type and Depth	Rated R-Value of Insulation Alone	Assembly U-Factors for 8 in. Normal Weight 145 lb/ft³ Solid Concrete Walls	Assembly U-Factors for 8 in. Medium Weight 115 lb/ft³ Concrete Block Walls: Solid Grouted	Assembly U-Factors for 8 in. Medium Weight 115 lb/ft³ Concrete Block Walls: Partially Grouted (Cores Uninsulated Except Where Specified)
No Framing	**R-0**	**U-0.740**	**U-0.580**	**U-0.480**
	Ungrouted Cores Filled with Loose-Fill Insulation	**N/A**	**N/A**	**U-0.350**
1 in. Metal Clips at 24 in. on Center Horizontally and 16 in. Vertically *(continued)*				
5.0 in.	R-28.0	U-0.046	U-0.046	U-0.045
6.0 in.	R-33.6	U-0.039	U-0.039	U-0.038
7.0 in.	R-39.2	U-0.034	U-0.034	U-0.033
8.0 in.	R-44.8	U-0.030	U-0.030	U-0.029
9.0 in.	R-50.4	U-0.027	U-0.027	U-0.026
10.0 in.	R-56.0	U-0.024	U-0.024	U-0.024
11.0 in.	R-61.6	U-0.022	U-0.022	U-0.022
Continuous Insulation Uninterrupted by Framing				
No framing	R-1.0	U-0.425	U-0.367	U-0.324
No framing	R-2.0	U-0.298	U-0.269	U-0.245
No framing	R-3.0	U-0.230	U-0.212	U-0.197
No framing	R-4.0	U-0.187	U-0.175	U-0.164
No framing	R-5.0	U-0.157	U-0.149	U-0.141
No framing	R-6.0	U-0.136	U-0.129	U-0.124
No framing	R-7.0	U-0.120	U-0.115	U-0.110
No framing	R-8.0	U-0.107	U-0.103	U-0.099
No framing	R-9.0	U-0.097	U-0.093	U-0.090
No framing	R-10.0	U-0.088	U-0.085	U-0.083
No framing	R-11.0	U-0.081	U-0.079	U-0.076
No framing	R-12.0	U-0.075	U-0.073	U-0.071
No framing	R-13.0	U-0.070	U-0.068	U-0.066
No framing	R-14.0	U-0.065	U-0.064	U-0.062
No framing	R-15.0	U-0.061	U-0.060	U-0.059
No framing	R-16.0	U-0.058	U-0.056	U-0.055
No framing	R-17.0	U-0.054	U-0.053	U-0.052
No framing	R-18.0	U-0.052	U-0.051	U-0.050
No framing	R-19.0	U-0.049	U-0.048	U-0.047
No framing	R-20.0	U-0.047	U-0.046	U-0.045
No framing	R-21.0	U-0.045	U-0.044	U-0.043
No framing	R-22.0	U-0.043	U-0.042	U-0.042
No framing	R-23.0	U-0.041	U-0.040	U-0.040
No framing	R-24.0	U-0.039	U-0.039	U-0.038
No framing	R-25.0	U-0.038	U-0.037	U-0.037
No framing	R-30.0	U-0.032	U-0.032	U-0.031
No framing	R-35.0	U-0.028	U-0.027	U-0.027
No framing	R-40.0	U-0.024	U-0.024	U-0.024
No framing	R-45.0	U-0.022	U-0.021	U-0.021
No framing	R-50.0	U-0.019	U-0.019	U-0.019
No framing	R-55.0	U-0.018	U-0.018	U-0.018
No framing	R-60.0	U-0.016	U-0.016	U-0.016

TABLE A3.1B Assembly U-Factors, C-Factors, R_u, R_c, and *HC* for Concrete

Density, lb/ft³	Properties	Thickness, in. 3	4	5	6	7	8	9	10	11	12
	U-factor	0.22	0.17	0.14	0.12	0.10	0.09	0.08	0.07	0.07	0.06
	C-factor	0.27	0.20	0.16	0.13	0.11	0.10	0.09	0.08	0.07	0.07
20	R_u	4.60	5.85	7.10	8.35	9.60	10.85	12.10	13.35	14.60	15.85
	R_c	3.75	5.00	6.25	7.50	8.75	10.00	11.25	12.50	13.75	15.00
	HC	1.0	1.3	1.7	2.0	2.3	2.7	3.0	3.3	3.7	4.0
	U-factor	0.28	0.22	0.19	0.16	0.14	0.12	0.11	0.10	0.09	0.09
	C-factor	0.37	0.28	0.22	0.18	0.16	0.14	0.12	0.11	0.10	0.09
30	R_u	3.58	4.49	5.40	6.30	7.21	8.12	9.03	9.94	10.85	11.76
	R_c	2.73	3.64	4.55	5.45	6.36	7.27	8.18	9.09	10.00	10.91
	HC	1.5	2.0	2.5	3.0	3.5	4.0	4.5	5.0	5.5	6.0
	U-factor	0.33	0.27	0.23	0.19	0.17	0.15	0.14	0.13	0.11	0.11
	C-factor	0.47	0.35	0.28	0.23	0.20	0.18	0.16	0.14	0.13	0.12
40	R_u	2.99	3.71	4.42	5.14	5.85	6.56	7.28	7.99	8.71	9.42
	R_c	2.14	2.86	3.57	4.29	5.00	5.71	6.43	7.14	7.86	8.57
	HC	2.0	2.7	3.3	4.0	4.7	5.3	6.0	6.7	7.3	8.0
	U-factor	0.38	0.31	0.26	0.23	0.20	0.18	0.16	0.15	0.14	0.13
	C-factor	0.57	0.43	0.34	0.28	0.24	0.21	0.19	0.17	0.15	0.14
50	R_u	2.61	3.20	3.79	4.38	4.97	5.56	6.14	6.73	7.32	7.91
	R_c	1.76	2.35	2.94	3.53	4.12	4.71	5.29	5.88	6.47	7.06
	HC	2.5	3.3	4.2	5.0	5.8	6.7	7.5	8.3	9.2	10.0
	U-factor	0.65	0.56	0.50	0.44	0.40	0.37	0.34	0.31	0.29	0.27
	C-factor	1.43	1.08	0.86	0.71	0.61	0.54	0.48	0.43	0.39	0.36
85	R_u	1.55	1.78	2.01	2.25	2.48	2.71	2.94	3.18	3.41	3.64
	R_c	0.70	0.93	1.16	1.40	1.63	1.86	2.09	2.33	2.56	2.79
	HC	4.3	5.7	7.1	8.5	9.9	11.3	12.8	14.2	15.6	17.0
	U-factor	0.72	0.64	0.57	0.52	0.48	0.44	0.41	0.38	0.36	0.33
	C-factor	1.85	1.41	1.12	0.93	0.80	0.70	0.62	0.56	0.51	0.47
95	R_u	1.39	1.56	1.74	1.92	2.10	2.28	2.46	2.64	2.81	2.99
	R_c	0.54	0.71	0.89	1.07	1.25	1.43	1.61	1.79	1.96	2.14
	HC	4.8	6.3	7.9	9.5	11.1	12.7	14.3	15.8	17.4	19.0
	U-factor	0.79	0.71	0.65	0.59	0.54	0.51	0.47	0.44	0.42	0.39
	C-factor	2.38	1.79	1.43	1.18	1.01	0.88	0.79	0.71	0.65	0.59
105	R_u	1.27	1.41	1.56	1.70	1.84	1.98	2.12	2.26	2.40	2.54
	R_c	0.42	0.56	0.70	0.85	0.99	1.13	1.27	1.41	1.55	1.69
	HC	5.3	7.0	8.8	10.5	12.3	14.0	15.8	17.5	19.3	21.0
	U-factor	0.84	0.77	0.70	0.65	0.61	0.57	0.53	0.50	0.48	0.45
	C-factor	2.94	2.22	1.75	1.47	1.25	1.10	0.98	0.88	0.80	0.74
115	R_u	1.19	1.30	1.42	1.53	1.65	1.76	1.87	1.99	2.10	2.21
	R_c	0.34	0.45	0.57	0.68	0.80	0.91	1.02	1.14	1.25	1.36
	HC	5.8	7.7	9.6	11.5	13.4	15.3	17.3	19.2	21.1	23.0
	U-factor	0.88	0.82	0.76	0.71	0.67	0.63	0.60	0.56	0.53	0.51
	C-factor	3.57	2.70	2.17	1.79	1.54	1.35	1.20	1.03	0.98	0.90
125	R_u	1.13	1.22	1.31	1.41	1.50	1.59	1.68	1.78	1.87	1.96
	R_c	0.28	0.37	0.46	0.56	0.65	0.74	0.83	0.93	1.02	1.11
	HC	6.3	8.3	10.4	12.5	14.6	16.7	18.8	20.8	22.9	25.0
	U-factor	0.93	0.87	0.82	0.77	0.73	0.69	0.66	0.63	0.60	0.57
	C-factor	4.55	3.33	2.70	2.22	1.92	1.67	1.49	1.33	1.22	1.11
135	R_u	1.07	1.15	1.22	1.30	1.37	1.45	1.52	1.60	1.67	1.75
	R_c	0.22	0.30	0.37	0.45	0.52	0.60	0.67	0.75	0.82	0.90
	HC	6.8	9.0	11.3	13.5	15.8	18.0	20.3	22.5	24.8	27.0
	U-factor	0.96	0.91	0.86	0.81	0.78	0.74	0.71	0.68	0.65	0.63
	C-factor	5.26	4.00	3.23	2.63	2.27	2.00	1.79	1.59	1.45	1.33
144	R_u	1.04	1.10	1.16	1.23	1.29	1.35	1.41	1.48	1.54	1.60
	R_c	0.19	0.25	0.31	0.38	0.44	0.50	0.56	0.63	0.69	0.75
	HC	7.2	9.6	12.0	14.4	16.8	19.2	21.6	24.0	26.4	28.8

The U-factors and R_u include standard air film resistances.
The C-factors and R_c are for the same assembly without air film resistances.
Note that the following assemblies do not qualify as a mass wall or mass floor:
3 in. thick concrete with densities of 85, 95, 125, and 135 lb/ft³.

TABLE A3.1C Assembly U-Factors, C-Factors, R_u, R_c, and *HC* for Concrete Block Walls

Product Size, in.	Density, lb/ft^3	Properties	Concrete Block Grouting and Cell Treatment				
			Solid Grouted	Partly Grouted, Cells Empty	Partly Grouted, Cells Insulated	Unreinforced, Cells Empty	Unreinforced, Cells Insulated
6 in. block	85	U-factor	0.57	0.46	0.34	0.40	0.20
		C-factor	1.11	0.75	0.47	0.60	0.23
		R_u	1.75	2.18	2.97	2.52	5.13
		R_c	0.90	1.33	2.12	1.67	4.28
		HC	10.9	6.7	7.0	4.2	4.6
	95	U-factor	0.61	0.49	0.36	0.42	0.22
		C-factor	1.25	0.83	0.53	0.65	0.27
		R_u	1.65	2.06	2.75	2.38	4.61
		R_c	0.80	1.21	1.90	1.53	3.76
		HC	11.4	7.2	7.5	4.7	5.1
	105	U-factor	0.64	0.51	0.39	0.44	0.24
		C-factor	1.38	0.91	0.58	0.71	0.30
		R_u	1.57	1.95	2.56	2.26	4.17
		R_c	0.72	1.10	1.71	1.41	3.32
		HC	11.9	7.7	7.9	5.1	5.6
	115	U-factor	0.66	0.54	0.41	0.46	0.26
		C-factor	1.52	0.98	0.64	0.76	0.34
		R_u	1.51	1.87	2.41	2.16	3.79
		R_c	0.66	1.02	1.56	1.31	2.94
		HC	12.3	8.1	8.4	5.6	6.0
	125	U-factor	0.70	0.56	0.45	0.49	0.30
		C-factor	1.70	1.08	0.73	0.84	0.40
		R_u	1.44	1.78	2.23	2.04	3.38
		R_c	0.59	0.93	1.38	1.19	2.53
		HC	12.8	8.6	8.8	6.0	6.5
	135	U-factor	0.73	0.60	0.49	0.53	0.35
		C-factor	1.94	1.23	0.85	0.95	0.49
		R_u	1.36	1.67	2.02	1.90	2.89
		R_c	0.51	0.82	1.17	1.05	2.04
		HC	13.2	9.0	9.3	6.5	6.9
8 in. block	85	U-factor	0.49	0.41	0.28	0.37	0.15
		C-factor	0.85	0.63	0.37	0.53	0.17
		R_u	2.03	2.43	3.55	2.72	6.62
		R_c	1.18	1.58	2.70	1.87	5.77
		HC	15.0	9.0	9.4	5.4	6.0
	95	U-factor	0.53	0.44	0.31	0.39	0.17
		C-factor	0.95	0.70	0.41	0.58	0.20
		R_u	1.90	2.29	3.27	2.57	5.92
		R_c	1.05	1.44	2.42	1.72	5.07
		HC	15.5	9.6	10.0	6.0	6.6
	105	U-factor	0.55	0.46	0.33	0.41	0.19
		C-factor	1.05	0.76	0.46	0.63	0.22
		R_u	1.81	2.17	3.04	2.44	5.32
		R_c	0.96	1.32	2.19	1.59	4.47
		HC	16.1	10.2	10.6	6.6	7.2
	115	U-factor	0.58	0.48	0.35	0.43	0.21
		C-factor	1.14	0.82	0.50	0.68	0.25
		R_u	1.72	2.07	2.84	2.33	4.78
		R_c	0.87	1.22	1.99	1.48	3.93
		HC	16.7	10.8	11.2	7.2	7.8
	125	U-factor	0.61	0.51	0.38	0.45	0.24
		C-factor	1.27	0.90	0.57	0.74	0.30
		R_u	1.64	1.96	2.62	2.20	4.20
		R_c	0.79	1.11	1.77	1.35	3.35
		HC	17.3	11.4	11.8	7.8	8.4
	135	U-factor	0.65	0.55	0.42	0.49	0.28
		C-factor	1.44	1.02	0.67	0.83	0.37
		R_u	1.54	1.83	2.35	2.05	3.55
		R_c	0.69	0.98	1.50	1.20	2.70
		HC	17.9	12.0	12.4	8.4	9.0

TABLE A3.1C Assembly U-Factors, C-Factors, R_u, R_c, and *HC* for Concrete Block Walls *(continued)*

Product Size, in.	Density, lb/ft³	Properties	Concrete Block Grouting and Cell Treatment				
			Solid Grouted	Partly Grouted, Cells Empty	Partly Grouted, Cells Insulated	Unreinforced, Cells Empty	Unreinforced, Cells Insulated
10 in. block	85	U-factor	0.44	0.38	0.25	0.35	0.13
		C-factor	0.70	0.57	0.31	0.50	0.14
		R_u	2.29	2.61	4.05	2.84	7.87
		R_c	1.44	1.76	3.20	1.99	7.02
		HC	19.0	11.2	11.7	6.5	7.3
	95	U-factor	0.47	0.41	0.27	0.37	0.14
		C-factor	0.77	0.62	0.35	0.55	0.16
		R_u	2.15	2.46	3.73	2.67	6.94
		R_c	1.30	1.61	2.88	1.82	6.09
		HC	19.7	11.9	12.4	7.3	8.1
	105	U-factor	0.49	0.43	0.29	0.39	0.16
		C-factor	0.85	0.68	0.39	0.59	0.19
		R_u	2.03	2.33	3.45	2.54	6.17
		R_c	1.18	1.48	2.60	1.69	5.32
		HC	20.4	12.6	13.1	8.0	8.8
	115	U-factor	0.52	0.45	0.31	0.41	0.18
		C-factor	0.92	0.73	0.42	0.64	0.21
		R_u	1.94	2.22	3.21	2.42	5.52
		R_c	1.09	1.37	2.36	1.57	4.67
		HC	21.1	13.4	13.9	8.7	9.5
	125	U-factor	0.54	0.48	0.34	0.44	0.21
		C-factor	1.01	0.80	0.48	0.70	0.25
		R_u	1.84	2.10	2.95	2.28	4.81
		R_c	0.99	1.25	2.10	1.43	3.96
		HC	21.8	14.1	14.6	9.4	10.2
	135	U-factor	0.58	0.51	0.38	0.47	0.25
		C-factor	1.14	0.90	0.56	0.79	0.32
		R_u	1.72	1.96	2.64	2.12	4.00
		R_c	0.87	1.11	1.79	1.27	3.15
		HC	22.6	14.8	15.3	10.2	11.0
12 in. block	85	U-factor	0.40	0.36	0.22	0.34	0.11
		C-factor	0.59	0.52	0.27	0.48	0.12
		R_u	2.53	2.77	4.59	2.93	9.43
		R_c	1.68	1.92	3.74	2.08	8.58
		HC	23.1	13.3	14.0	7.5	8.5
	95	U-factor	0.42	0.38	0.24	0.36	0.12
		C-factor	0.66	0.57	0.30	0.52	0.13
		R_u	2.30	2.60	4.22	2.76	8.33
		R_c	1.53	1.75	3.37	1.91	7.48
		HC	23.9	14.2	14.8	8.3	9.3
	105	U-factor	0.44	0.41	0.26	0.38	0.14
		C-factor	0.71	0.62	0.33	0.57	0.15
		R_u	2.25	2.47	3.90	2.62	7.35
		R_c	1.40	1.62	3.05	1.77	6.50
		HC	24.7	15.0	15.6	9.1	10.2
	115	U-factor	0.47	0.42	0.28	0.40	0.15
		C-factor	0.77	0.66	0.36	0.61	0.18
		R_u	2.15	2.36	3.63	2.49	6.54
		R_c	1.30	1.51	2.78	1.64	5.69
		HC	25.6	15.8	16.4	10.0	11.0
	125	U-factor	0.49	0.45	0.30	0.42	0.18
		C-factor	0.84	0.72	0.40	0.66	0.21
		R_u	2.04	2.23	3.34	2.36	5.68
		R_c	1.19	1.38	2.49	1.51	4.83
		HC	26.4	16.6	17.3	10.8	11.8
	135	U-factor	0.52	0.48	0.34	0.46	0.21
		C-factor	0.94	0.81	0.47	0.74	0.26
		R_u	1.91	2.08	2.98	2.19	4.67
		R_c	1.06	1.23	2.13	1.34	3.82
		HC	27.2	17.5	18.1	11.6	12.6

of 0.44 Btu·in./h·ft^2·°F. Other *mass walls* with integral insulation shall meet the criteria when their *U-factors* are equal to or less than those for the appropriate thickness and density in the "Partly Grouted Cells Insulated" column of Table A3.1C.

A3.1.3.2 Determination of Mass Wall U-Factors. If not taken from Table A3.1A, *mass wall U-factors* shall be determined from Tables A3.1B, A3.1C, or A3.1D using the following procedure:

1. If the *mass wall* is uninsulated or only the cells are insulated:

    a. For concrete *walls*, determine the *U-factor* from Table A3.1B based on the concrete density and *wall* thickness.

    b. For concrete block *walls*, determine the *U-factor* from Table A3.1C based on the block size, concrete density, degree of grouting in the cells, and whether the cells are insulated.

2. If the *mass wall* has additional insulation:

    a. For concrete *walls*, determine the R_u from Table A3.1B based on the concrete density and *wall* thickness. Next, determine the effective R-value for the insulation/framing layer from Table A3.1D based on the *rated R-value of insulation* installed, the thickness of the insulation, and whether it is installed between wood or metal framing or with no framing. Then, determine the *U-factor* by adding the R_u and the effective R-value together and taking the inverse of the total.

    b. For concrete block *walls*, determine the R_u from Table A3.1C based on the block size, concrete density, degree of grouting in the cells, and whether the cells are insulated. Next, determine the effective R-value for the insulation/framing layer from Table A3.1D based on the *rated R-value of insulation* installed, the thickness of the insulation, and whether it is installed between wood or metal framing or with no framing. Then, determine the *U-factor* by adding the R_u and the effective R-value together and taking the inverse of the total.

A3.2 Metal Building Walls

A3.2.1 General. For the purpose of Section A1.2, the base assembly is a *wall* where the insulation is compressed between metal wall panels and the metal structure. Additional assemblies include *continuous insulation*, uncompressed and uninterrupted by framing.

A3.2.2 Rated R-Value of Insulation for Metal Building Walls

A3.2.2.1 The first *rated R-Value of insulation* is for insulation compressed between metal wall panels and the steel structure.

A3.2.2.2 For double-layer installations, the second *rated R-value of insulation* is for insulation installed from the inside, covering the girts.

A3.2.2.3 For continuous insulation (e.g., insulation boards) it is assumed that the insulation boards are installed on the inside of the girts and uninterrupted by the framing members.

A3.2.2.4 Insulation exposed to the *conditioned space* or *semiheated space* shall have a facing, and all insulation seams shall be continuously sealed to provide a continuous air barrier.

A3.2.3 *U-Factors* for *Metal Building Walls*. U-factors for metal building walls shall be taken from Table A3.2. It is not acceptable to use these *U-factors* if additional insulation is not continuous.

A3.3 Steel-Framed Walls

A3.3.1 General. For the purpose of Section A1.2, the base assembly is a *wall* where the insulation is installed within the cavity of the steel stud framing but where there is not a metal exterior surface spanning member. The steel stud framing is a minimum uncoated thickness of 0.043 in. for 18 gauge or 0.054 in. for 16 gauge. The *U-factors* include R-0.17 for exterior air film, R-0.08 for stucco, R-0.56 for 0.625 in. gypsum board on the exterior, R-0.56 for 0.625 in. gypsum board on the interior, and R-0.68 for interior vertical surfaces air film. The performance of the insulation/framing layer is calculated using the values in Table A-9.2B. Additional assemblies include *continuous insulation*, uncompressed and uninterrupted by framing. *U-factors* are provided for the following configurations:

a. *Standard framing*: steel stud framing at 16 in. on center with cavities filled with 16 in. wide insulation for both 3.5 in. deep and 6.0 in. deep wall cavities.

b. *Advanced framing*: steel stud framing at 24 in. on center with cavities filled with 24 in. wide insulation for both 3.5 in. deep and 6.0 in. deep wall cavities.

A3.3.2 *Rated R-Value of Insulation* for Steel-Framed Walls

A3.3.2.1 The first *rated R-value of insulation* is for uncompressed insulation installed in the cavity between steel studs. It is acceptable for this insulation to also be *continuous insulation* uninterrupted by framing.

A3.3.2.2 If there are two values, the second *rated R-value of insulation* is for *continuous insulation* uninterrupted by framing, etc., to be installed in addition to the first insulation.

A3.3.2.3 Opaque mullions in spandrel glass shall be covered with insulation complying with the steel-framed wall requirements.

A3.3.3 U-Factors for Steel-Framed Walls

A3.3.3.1 U-factors for steel-framed walls shall be taken from Table A3.3.

A3.3.3.2 For *steel-framed walls* with framing at less than 24 in. on center, use the standard framing values as described in Section A3.3.1(a).

A3.3.3.3 For *steel-framed walls* with framing from 24 to 32 in. on center, use the advanced framing values as described in Section A3.3.1(b).

TABLE A3.1D Effective R-Values for Insulation/Framing Layers Added to Above-Grade Mass Walls and Below-Grade Walls

Depth, in.	Framing Type	Rated R-Value of Insulation																									
		0	1	2	3	4	5	6	7	8	9	10	11	12	13	14	15	16	17	18	19	20	21	22	23	24	25
Effective R-value if continuous insulation uninterrupted by framing (includes gypsum board)																											
	None	0.5	1.5	2.5	3.5	4.5	5.5	6.5	7.5	8.5	9.5	10.5	11.5	12.5	13.5	14.5	15.5	16.5	17.5	18.5	19.5	20.5	21.5	22.5	23.5	24.5	25.5
Effective R-value if insulation is installed in cavity between framing (includes gypsum board)																											
0.5	Wood	1.3	1.3	1.9	2.4	2.7	na	na	na	na	na	na	na	na	na	na	na	na	na	na	na	na	na	na	na	na	na
	Metal	0.9	0.9	1.1	1.1	1.2	na	na	na	na	na	na	na	na	na	na	na	na	na	na	na	na	na	na	na	na	na
0.75	Wood	1.4	1.4	2.1	2.7	3.1	3.5	3.8	na	na	na	na	na	na	na	na	na	na	na	na	na	na	na	na	na	na	na
	Metal	1.0	1.0	1.3	1.4	1.5	1.5	1.6	na	na	na	na	na	na	na	na	na	na	na	na	na	na	na	na	na	na	na
1.0	Wood	1.3	1.5	2.2	2.9	3.4	3.9	4.3	4.6	4.9	na	na	na	na	na	na	na	na	na	na	na	na	na	na	na	na	na
	Metal	1.0	1.1	1.4	1.6	1.7	1.8	1.8	1.9	1.9	na	na	na	na	na	na	na	na	na	na	na	na	na	na	na	na	na
1.5	Wood	1.3	1.5	2.4	3.1	3.8	4.4	4.9	5.4	5.8	6.2	6.5	6.8	7.1	na	na	na	na	na	na	na	na	na	na	na	na	na
	Metal	1.1	1.2	1.6	1.9	2.1	2.2	2.3	2.4	2.5	2.5	2.6	2.6	2.7	na	na	na	na	na	na	na	na	na	na	na	na	na
2.0	Wood	1.4	1.5	2.5	3.3	4.0	4.7	5.3	5.9	6.4	6.9	7.3	7.7	8.1	8.4	8.7	9.0	9.3	na	na	na	na	na	na	na	na	na
	Metal	1.1	1.2	1.7	2.1	2.3	2.5	2.7	2.8	2.9	3.0	3.1	3.2	3.2	3.3	3.3	3.4	3.4	na	na	na	na	na	na	na	na	na
2.5	Wood	1.4	1.5	2.5	3.4	4.2	4.9	5.6	6.3	6.8	7.4	7.9	8.4	8.8	9.2	9.6	10.0	10.3	10.6	10.9	11.2	11.5	na	na	na	na	na
	Metal	1.2	1.3	1.8	2.3	2.6	2.8	3.0	3.2	3.3	3.5	3.6	3.6	3.7	3.8	3.9	3.9	4.0	4.0	4.1	4.1	4.1	na	na	na	na	na
3.0	Wood	1.4	1.5	2.5	3.5	4.3	5.1	5.8	6.5	7.2	7.8	8.3	8.9	9.4	9.9	10.3	10.7	11.1	11.5	11.9	12.2	12.5	12.9	na	na	na	na
	Metal	1.2	1.3	1.9	2.4	2.8	3.1	3.3	3.5	3.7	3.8	4.0	4.1	4.2	4.3	4.4	4.4	4.5	4.6	4.6	4.7	4.7	4.8	na	na	na	na
3.5	Wood	1.4	1.5	2.6	3.5	4.4	5.2	6.0	6.7	7.4	8.1	8.7	9.3	9.8	10.4	10.9	11.3	11.8	12.2	12.6	13.0	13.4	13.8	14.1	14.5	14.8	15.1
	Metal	1.2	1.3	2.0	2.5	2.9	3.2	3.5	3.8	4.0	4.2	4.3	4.5	4.6	4.7	4.8	4.9	5.0	5.1	5.1	5.2	5.2	5.3	5.4	5.4	5.4	5.5
4.0	Wood	1.4	1.6	2.6	3.6	4.5	5.3	6.1	6.9	7.6	8.3	9.0	9.6	10.2	10.8	11.3	11.9	12.4	12.8	13.3	13.7	14.2	14.6	14.9	15.3	15.7	16.0
	Metal	1.2	1.3	2.0	2.6	3.0	3.4	3.7	4.0	4.2	4.5	4.6	4.8	5.0	5.1	5.2	5.3	5.4	5.5	5.6	5.7	5.8	5.8	5.9	5.9	6.0	6.0
4.5	Wood	1.4	1.6	2.6	3.6	4.5	5.4	6.2	7.1	7.8	8.5	9.2	9.9	10.5	11.2	11.7	12.3	12.8	13.3	13.8	14.3	14.8	15.2	15.7	16.1	16.5	16.9
	Metal	1.2	1.3	2.1	2.6	3.1	3.5	3.9	4.2	4.5	4.7	4.9	5.1	5.3	5.4	5.6	5.7	5.8	5.9	6.0	6.1	6.2	6.3	6.4	6.4	6.5	6.6
5.0	Wood	1.4	1.6	2.6	3.6	4.6	5.5	6.3	7.2	8.0	8.7	9.4	10.1	10.8	11.5	12.1	12.7	13.2	13.8	14.3	14.8	15.3	15.8	16.3	16.7	17.2	17.6
	Metal	1.2	1.4	2.1	2.7	3.2	3.7	4.1	4.4	4.7	5.0	5.2	5.4	5.6	5.8	5.9	6.1	6.2	6.3	6.5	6.6	6.7	6.8	6.8	6.9	7.0	7.1
5.5	Wood	1.4	1.6	2.6	3.6	4.6	5.5	6.4	7.3	8.1	8.9	9.6	10.3	11.0	11.7	12.4	13.0	13.6	14.2	14.7	15.3	15.8	16.3	16.8	17.3	17.8	18.2
	Metal	1.3	1.4	2.1	2.8	3.3	3.8	4.2	4.6	4.9	5.2	5.4	5.7	5.9	6.1	6.3	6.4	6.6	6.7	6.8	7.0	7.1	7.2	7.3	7.4	7.5	7.6

TABLE A3.2 Assembly U-Factors for Metal Building Walls

Insulation System	Rated R-Value of Insulation	Total Rated R-Value of Insulation	Overall U-Factor for Entire Base Wall Assembly	Overall U-Factor for Assembly of Base Wall Plus Continuous Insulation (Uninterrupted by Framing)					
				Rated R-Value of Continuous Insulation					
				R-5.6	R-11.2	R-16.8	R-22.4	R-28.0	R-33.6
Single Layer of Mineral Fiber									
	None	0	**1.180**	0.161	0.086	0.059	0.045	0.036	0.030
	R-6	6	**0.184**	0.091	0.060	0.045	0.036	0.030	0.026
	R-10	10	**0.134**	0.077	0.054	0.051	0.033	0.028	0.024
	R-11	11	**0.123**	0.073	0.052	0.040	0.033	0.028	0.024
	R-13	13	**0.113**	0.069	0.050	0.039	0.032	0.027	0.024
Double Layer of Mineral Fiber									
(Second layer inside of girts)									
(Multiple layers are listed in order from inside to outside)									
	R-6 + R-13	19	**0.070**	N/A	N/A	N/A	N/A	N/A	N/A
	R-10 + R-13	23	**0.061**	N/A	N/A	N/A	N/A	N/A	N/A
	R-13 + R-13	26	**0.057**	N/A	N/A	N/A	N/A	N/A	N/A
	R-19 + R-13	32	**0.048**	N/A	N/A	N/A	N/A	N/A	N/A

A3.3.3.4 For *steel-framed walls* with framing greater than 32 in. on center, use the *metal building wall* values in Table A3.2.

A3.4 Wood-Framed Walls

A3.4.1 General. For the purpose of Section A1.2, the base assembly is a *wall* where the insulation is installed between 2 in. nominal wood framing. Cavity insulation is full depth, but values are taken from Table A9.4C for R-19 insulation, which is compressed when installed in a 5.5 in. cavity. Headers are double 2 in. nominal wood framing. The *U-factors* include R-0.17 for exterior air film, R-0.08 for stucco, R-0.56 for 0.625 in. gypsum board on the exterior, R-0.56 for 0.625 in. gypsum board on the interior, and R-0.68 for interior air film, vertical surfaces. Additional assemblies include *continuous insulation*, uncompressed and uninterrupted by framing. *U-factors* are provided for the following configurations:

a. *Standard framing:* wood framing at 16 in. on center with cavities filled with 14.5 in. wide insulation for both 3.5 in. deep and 5.5 in. deep wall cavities. Double headers leave no cavity. Weighting factors are 75% insulated cavity, 21% studs, plates, and sills, and 4% headers.
b. *Advanced framing:* wood framing at 24 in. on center with cavities filled with 22.5 in. wide insulation for both 3.5 in. deep and 5.5 in. deep wall cavities. Double headers leave uninsulated cavities. Weighting factors are 78% insulated cavity, 18% studs, plates, and sills, and 4% headers.
c. *Advanced framing with insulated headers:* wood framing at 24 in. on center with cavities filled with 22.5 in. wide insulation for both 3.5 in. deep and 5.5 in. deep wall cavities. Double header cavities are insulated. Weighting factors are 78% insulated cavity, 18% studs, plates, and sills, and 4% headers.

A3.4.2 Rated R-value of Insulation for Wood-Framed and Other Walls

A3.4.2.1 The first *rated R-value of insulation* is for uncompressed insulation installed in the cavity between wood studs. It is acceptable for this insulation to also be *continuous insulation* uninterrupted by framing.

A3.4.2.2 If there are two values, the second *rated R-value of insulation* is for *continuous insulation* uninterrupted by framing, etc., to be installed in addition to the first insulation.

A3.4.3 U-Factors for Wood-Framed Walls

A3.4.3.1 U-factors for wood-framed walls shall be taken from Table A3.4.

A3.4.3.2 For *wood-framed walls* with framing at less than 24 in. on center, use the standard framing values as described in Section A3.4.1(a).

A3.4.3.3 For *wood-framed walls* with framing from 24 to 32 in. on center, use the advanced framing values as described in Section A3.4.1(b) if the headers are uninsulated or the advanced framing with insulated header values as described in Section A3.4.1(c) if the headers are insulated.

A3.4.3.4 For *wood-framed walls* with framing greater than 32 in. on center, U-factors shall be determined in accordance with Section A9.

TABLE A3.3 Assembly U-Factors for Steel-Frame Walls

Framing Type and Spacing Width (Actual Depth)	Cavity Insulation R-Value: Rated (Effective Installed [see Table A9.2B])	Overall U-Factor for Entire Base Wall Assembly	Overall U-Factor for Assembly of Base Wall Plus Continuous Insulation (Uninterrupted by Framing), Rated R-Value of Continuous Insulation																			
			R-1.00	R-2.00	R-3.00	R-4.00	R-5.00	R-6.00	R-7.00	R-8.00	R-9.00	R-10.00	R-11.00	R-12.00	R-13.00	R-14.00	R-15.00	R-20.00	R-25.00	R-30.00	R-35.00	R-40.00
Steel Framing at 16 in. on center																						
	None (0.0)	**0.352**	0.260	0.207	0.171	0.146	0.128	0.113	0.102	0.092	0.084	0.078	0.072	0.067	0.063	0.059	0.056	0.044	0.036	0.030	0.026	0.023
3.5 in. depth	R-11 (5.5)	**0.132**	0.117	0.105	0.095	0.087	0.080	0.074	0.069	0.064	0.060	0.057	0.054	0.051	0.049	0.046	0.044	0.036	0.031	0.027	0.024	0.021
	R-13 (6.0)	**0.124**	0.111	0.100	0.091	0.083	0.077	0.071	0.066	0.062	0.059	0.055	0.052	0.050	0.048	0.045	0.043	0.036	0.030	0.026	0.023	0.021
	R-15 (6.4)	**0.118**	0.106	0.096	0.087	0.080	0.074	0.069	0.065	0.061	0.057	0.054	0.051	0.049	0.047	0.045	0.043	0.035	0.030	0.026	0.023	0.021
6.0 in. depth	R-19 (7.1)	**0.109**	0.099	0.090	0.082	0.076	0.071	0.066	0.062	0.058	0.055	0.052	0.050	0.047	0.045	0.043	0.041	0.034	0.029	0.026	0.023	0.020
	R-21 (7.4)	**0.106**	0.096	0.087	0.080	0.074	0.069	0.065	0.061	0.057	0.054	0.051	0.049	0.047	0.045	0.043	0.041	0.034	0.029	0.025	0.022	0.020
Steel Framing at 24 in. on center																						
	None (0.0)	**0.338**	0.253	0.202	0.168	0.144	0.126	0.112	0.100	0.091	0.084	0.077	0.072	0.067	0.063	0.059	0.056	0.044	0.036	0.030	0.026	0.023
3.5 in. depth	R-11 (6.6)	**0.116**	0.104	0.094	0.086	0.079	0.073	0.068	0.064	0.060	0.057	0.054	0.051	0.048	0.046	0.044	0.042	0.035	0.030	0.026	0.023	0.021
	R-13 (7.2)	**0.108**	0.098	0.089	0.082	0.075	0.070	0.066	0.062	0.058	0.055	0.052	0.049	0.047	0.045	0.043	0.041	0.034	0.029	0.025	0.023	0.020
	R-15 (7.8)	**0.102**	0.092	0.084	0.078	0.072	0.067	0.063	0.059	0.056	0.053	0.050	0.048	0.046	0.044	0.042	0.040	0.034	0.029	0.025	0.022	0.020
6.0 in. depth	R-19 (8.6)	**0.094**	0.086	0.079	0.073	0.068	0.064	0.060	0.057	0.054	0.051	0.048	0.046	0.044	0.042	0.041	0.039	0.033	0.028	0.025	0.022	0.020
	R-21 (9.0)	**0.090**	0.083	0.077	0.071	0.066	0.062	0.059	0.055	0.052	0.050	0.048	0.045	0.043	0.042	0.040	0.038	0.032	0.028	0.024	0.022	0.020

TABLE A3.4 Assembly U-Factors for Wood-Frame Walls

Framing Type and Spacing Width (Actual Depth)	Cavity Insulation R-Value: Rated (Effective Installed [see Table A9.4C])	Overall U-Factor for Entire Base Wall Assembly	Overall U-Factor for Assembly of Base Wall Plus Continuous Insulation (Uninterrupted by Framing) — Rated R-Value of Continuous Insulation																			
			R-1.00	R-2.00	R-3.00	R-4.00	R-5.00	R-6.00	R-7.00	R-8.00	R-9.00	R-10.00	R-11.00	R-12.00	R-13.00	R-14.00	R-15.00	R-20.00	R-25.00	R-30.00	R-35.00	R-40.00
Wood Studs at 16 in. on center																						
3.5 in. depth	None (0.0)	**0.292**	0.223	0.181	0.152	0.132	0.116	0.104	0.094	0.086	0.079	0.073	0.068	0.064	0.060	0.056	0.053	0.042	0.035	0.030	0.026	0.023
	R-11 (11.0)	**0.096**	0.087	0.079	0.073	0.068	0.063	0.059	0.056	0.053	0.050	0.048	0.046	0.044	0.042	0.040	0.038	0.032	0.028	0.024	0.022	0.020
	R-13 (13.0)	**0.089**	0.080	0.074	0.068	0.063	0.059	0.056	0.053	0.050	0.047	0.045	0.043	0.041	0.040	0.038	0.037	0.031	0.027	0.024	0.021	0.019
	R-15 (15.0)	**0.083**	0.075	0.069	0.064	0.060	0.056	0.053	0.050	0.047	0.045	0.043	0.041	0.039	0.038	0.036	0.035	0.030	0.026	0.023	0.020	0.019
5.5 in. depth	R-19 (18.0)	**0.067**	0.062	0.058	0.054	0.051	0.048	0.046	0.044	0.042	0.040	0.038	0.037	0.036	0.034	0.033	0.032	0.027	0.024	0.021	0.019	0.018
	R-21 (21.0)	**0.063**	0.058	0.054	0.051	0.048	0.045	0.043	0.041	0.039	0.038	0.036	0.035	0.034	0.032	0.031	0.030	0.026	0.023	0.021	0.019	0.017
+ R-10 headers	R-19 (18.0)	**0.063**	0.059	0.055	0.052	0.049	0.047	0.045	0.043	0.041	0.039	0.038	0.036	0.035	0.034	0.033	0.031	0.027	0.024	0.021	0.019	0.017
	R-21 (21.0)	**0.059**	0.055	0.051	0.049	0.046	0.044	0.042	0.040	0.038	0.037	0.035	0.034	0.033	0.032	0.031	0.030	0.026	0.023	0.020	0.018	0.017
Wood Studs at 24 in. on center																						
3.5 in. depth	None (0.0)	**0.298**	0.227	0.183	0.154	0.133	0.117	0.105	0.095	0.086	0.079	0.074	0.068	0.064	0.060	0.057	0.054	0.042	0.035	0.030	0.026	0.023
	R-11 (11.0)	**0.094**	0.085	0.078	0.072	0.067	0.062	0.059	0.055	0.052	0.050	0.047	0.045	0.043	0.041	0.040	0.038	0.032	0.027	0.024	0.022	0.019
	R-13 (13.0)	**0.086**	0.078	0.072	0.067	0.062	0.058	0.055	0.052	0.049	0.047	0.045	0.043	0.041	0.039	0.038	0.036	0.031	0.026	0.023	0.021	0.019
	R-15 (15.0)	**0.080**	0.073	0.067	0.062	0.058	0.055	0.052	0.049	0.046	0.044	0.042	0.040	0.039	0.037	0.036	0.035	0.029	0.026	0.023	0.020	0.018
5.5 in. depth	R-19 (18.0)	**0.065**	0.060	0.056	0.053	0.050	0.047	0.045	0.043	0.041	0.039	0.038	0.036	0.035	0.034	0.033	0.032	0.027	0.024	0.021	0.019	0.018
	R-21 (21.0)	**0.060**	0.056	0.052	0.049	0.046	0.044	0.042	0.040	0.038	0.037	0.036	0.034	0.033	0.032	0.031	0.030	0.026	0.023	0.020	0.018	0.017
+ R-10 headers	R-19 (18.0)	**0.062**	0.058	0.054	0.051	0.048	0.046	0.044	0.042	0.040	0.039	0.037	0.036	0.034	0.033	0.032	0.031	0.027	0.024	0.021	0.019	0.017
	R-21 (21.0)	**0.057**	0.053	0.050	0.047	0.045	0.043	0.041	0.039	0.037	0.036	0.035	0.033	0.032	0.031	0.030	0.029	0.025	0.023	0.020	0.018	0.017

A4. BELOW-GRADE WALLS

A4.1 General. For the purpose of Section A1.2, The base assembly is 8 in. medium-weight concrete block with a density of 115 lb/ft^3 and solid grouted cores. *Continuous insulation* is installed on the interior or exterior. In contrast to the *U-factor* for *above-grade walls*, the *C-factor* for *below-grade walls* does not include R-values for exterior or interior air films or for soil. For insulated walls, the *C-factor* does include R-0.45 for 0.5 in. gypsum board.

A4.2 C-Factors for Below-Grade Walls

A4.2.1 C-factors for below-grade walls shall be taken from Table A4.2 or determined by the procedure described in this subsection.

A4.2.2 It is acceptable to use the *C-factors* in Table A4.2 for all *below-grade walls*.

A4.2.3 If not taken from Table A4.2, *below-grade wall C-factors* shall be determined from Tables A3.1B, A3.1C, or A3.1D using the following procedure:

a. If the *below-grade wall* is uninsulated or only the cells are insulated:

   1. For concrete *walls*, determine the *C-factor* from Table A3.1B based on the concrete density and *wall* thickness.
   2. For concrete block *walls*, determine the *C-factor* from Table A3.1C based on the block size, concrete density, degree of grouting in the cells, and whether the cells are insulated.

b. If the *mass wall* has additional insulation:

   1. For concrete *walls*, determine the R_c from Table A3.1B based on the concrete density and *wall* thickness. Next, determine the effective R-value for the insulation/framing layer from Table A3.1D based on the *rated R-value of insulation* installed, the thickness of the insulation, and whether it is installed between wood or metal framing or with no framing. Then, determine the *C-factor* by adding the R_c and the effective R-value together and taking the inverse of the total.
   2. For concrete block *walls*, determine the R_c from Table A3.1C based on the block size, concrete density, degree of grouting in the cells, and whether the cells are insulated. Next, determine the effective R-value for the insulation/framing layer from Table A3.1D based on the *rated R-value of insulation* installed, the thickness of the insulation, and whether it is installed between wood or metal framing or with no framing. Then, determine the *C-factor* by adding the R_c and the effective R-value together and taking the inverse of the total.

A5. FLOORS

A5.1 General. The buffering effect of crawlspaces or parking garages shall not be included in *U-factor* calculations. See Section A6 for *slab-on-grade floors*.

A5.2 Mass Floors

A5.2.1 General. For the purpose of Section A1.2, the base assembly is *continuous insulation* over or under a solid concrete *floor*. The *U-factors* include R-0.92 for interior air film-heat flow down, R-1.23 for carpet and rubber pad, R-0.50 for 8 in. concrete, and R-0.46 for semi-exterior air film. Added insulation is continuous and uninterrupted by framing. Framing factor is zero.

A5.2.2 Rated R-Value of Insulation for Mass Floors

A5.2.2.1 The *rated R-value of insulation* is for *continuous insulation* uninterrupted by framing.

A5.2.2.2 Where framing, including metal and wood joists, is used, compliance shall be based on the maximum assembly *U-factor* rather than the minimum *rated R-value of insulation*.

A5.2.2.3 For waffle-slab *floors*, the *floor* shall be insulated either on the interior above the slab or on all exposed surfaces of the waffle.

A5.2.2.4 For *floors* with beams that extend below the floor slab, the *floor* shall be insulated either on the interior above the slab or on the exposed floor and all exposed surfaces of the beams that extend 24 in. and less below the exposed floor.

A5.2.3 U-Factors for Mass Floors

A5.2.3.1 The *U-factors* for mass walls shall be taken from Table A5.2.

A5.2.3.2 It is not acceptable to use the *U-factors* in Table A5.2 if the insulation is not continuous.

A5.3 Steel-Joist Floors

A5.3.1 General. For the purpose of Section A1.2, the base assembly is a *floor* where the insulation is either placed between the steel joists or is sprayed on the underside of the *floor* and the joists. In both cases, the steel provides a thermal bypass to the insulation. The *U-factors* include R-0.92 for interior air film—heat flow down, R-1.23 for carpet and pad, R-0.25 for 4 in. concrete, R-0 for metal deck, and R-0.46 for semi-exterior air film. The performance of the insulation/framing layer is calculated using the values in Table A9.2A.

A5.3.2 Rated R-Value of Insulation for Steel-Joist Floors

A5.3.2.1 The first *rated R-value of insulation* is for uncompressed insulation installed in the cavity between steel joists or for spray-on insulation.

A5.3.2.2 It is acceptable for this insulation to also be *continuous insulation* uninterrupted by framing. All *continuous insulation* shall be installed either on the interior above the floor structure or below a framing cavity completely filled with insulation.

A5.3.3 U-Factors for Steel-Joist Floors

A5.3.3.1 The *U-factors* for steel-joist floors shall be taken from Table A5.3.

A5.3.3.2 It is acceptable to use these *U-factors* for any *steel-joist floor*.

A5.4 Wood-Framed and Other Floors

A5.4.1 General. For the purpose of Section A1.2, the base assembly is a *floor* attached directly to the top of the wood joist

TABLE A4.2 Assembly C-Factors for Below-Grade Walls

Framing Type and Depth	Rated R-Value of Insulation Alone	Specified C-Factors (Wall Only, without Soil and Air Films)
No Framing	R-0	C-1.140
Exterior Insulation, Continuous and Uninterrupted by Framing		
No Framing	R-5.0	C-0.170
No Framing	R-7.5	C-0.119
No Framing	R-10.0	C-0.092
No Framing	R-12.5	C-0.075
No Framing	R-15.0	C-0.063
No Framing	R-17.5	C-0.054
No Framing	R-20.0	C-0.048
No Framing	R-25.0	C-0.039
No Framing	R-30.0	C-0.032
No Framing	R-35.0	C-0.028
No Framing	R-40.0	C-0.025
No Framing	R-45.0	C-0.022
No Framing	R-50.0	C-0.020
Continuous Metal Framing at 24 in. on Center Horizontally		
3.5 in.	R-11.0	C-0.182
3.5 in.	R-13.0	C-0.174
3.5 in.	R-15.0	C-0.168
5.5 in.	R-19.0	C-0.125
5.5 in.	R-21.0	C-0.120
1 in. Metal Clips at 24 in. on Center Horizontally and 16 in. Vertically		
1.0 in.	R-3.8	C-0.233
1.0 in.	R-5.0	C-0.201
1.0 in.	R-5.6	C-0.189
1.5 in.	R-5.7	C-0.173
1.5 in.	R-7.5	C-0.147
1.5 in.	R-8.4	C-0.138
2.0 in.	R-7.6	C-0.138
2.0 in.	R-10.0	C-0.116
2.0 in.	R-11.2	C-0.108
2.5 in.	R-9.5	C-0.114
2.5 in.	R-12.5	C-0.096
2.5 in.	R-14.0	C-0.089
3.0 in.	R-11.4	C-0.098
3.0 in.	R-15.0	C-0.082
3.0 in.	R-16.8	C-0.076
3.5 in.	R-13.3	C-0.085
3.5 in.	R-17.5	C-0.071
3.5 in.	R-19.6	C-0.066
4.0 in.	R-15.2	C-0.076
4.0 in.	R-20.0	C-0.063
4.0 in.	R-22.4	C-0.058

TABLE A5.2 Assembly U-Factors for Mass Floors

Framing Type and Spacing Width (Actual Depth)	Cavity Insulation R-Value: Rated (Effective Installed)	Overall U-Factor for Entire Base Floor Assembly	Overall U-Factor for Assembly of Base Floor Plus Continuous Insulation (Uninterrupted by Framing) — Rated R-Value of Continuous Insulation																			
			R-1.00	R-2.00	R-3.00	R-4.00	R-5.00	R-6.00	R-7.00	R-8.00	R-9.00	R-10.00	R-11.00	R-12.00	R-13.00	R-14.00	R-15.00	R-20.00	R-25.00	R-30.00	R-35.00	R-40.00
Concrete Floor with Rigid Foam																						
	None (0.0)	**0.322**	0.243	0.196	0.164	0.141	0.123	0.110	0.099	0.090	0.083	0.076	0.071	0.066	0.062	0.058	0.055	0.043	0.036	0.030	0.026	0.023
Concrete Floor with Pinned Boards																						
	R-4.2 (4.2)	**0.137**	0.121	0.108	0.097	0.089	0.081	0.075	0.070	0.065	0.061	0.058	0.055	0.052	0.049	0.047	0.045	0.037	0.031	0.027	0.024	0.021
	R-6.3 (6.3)	**0.107**	0.096	0.088	0.081	0.075	0.070	0.065	0.061	0.058	0.054	0.052	0.049	0.047	0.045	0.043	0.041	0.034	0.029	0.025	0.023	0.020
	R-8.3 (8.3)	**0.087**	0.080	0.074	0.069	0.065	0.061	0.057	0.054	0.051	0.049	0.047	0.045	0.043	0.041	0.039	0.038	0.032	0.027	0.024	0.022	0.019
	R-10.4 (10.4)	**0.074**	0.069	0.064	0.060	0.057	0.054	0.051	0.049	0.046	0.044	0.042	0.041	0.039	0.038	0.036	0.035	0.030	0.026	0.023	0.021	0.019
	R-12.5 (12.5)	**0.064**	0.060	0.057	0.054	0.051	0.048	0.046	0.044	0.042	0.041	0.039	0.038	0.036	0.035	0.034	0.033	0.028	0.025	0.022	0.020	0.018
	R-14.6 (14.6)	**0.056**	0.053	0.051	0.048	0.046	0.044	0.042	0.040	0.039	0.037	0.036	0.035	0.034	0.033	0.032	0.031	0.027	0.023	0.021	0.019	0.017
	R-16.7 (16.7)	**0.051**	0.048	0.046	0.044	0.042	0.040	0.039	0.037	0.036	0.035	0.034	0.032	0.031	0.030	0.030	0.029	0.025	0.022	0.020	0.018	0.017
Concrete Floor with Spray-On Insulation																						
1 in.	R-4 (4.0)	**0.141**	0.123	0.110	0.099	0.090	0.083	0.076	0.071	0.066	0.062	0.058	0.055	0.052	0.050	0.047	0.045	0.037	0.031	0.027	0.024	0.021
2 in.	R-8 (8.0)	**0.090**	0.083	0.076	0.071	0.066	0.062	0.058	0.055	0.052	0.050	0.047	0.045	0.043	0.041	0.040	0.038	0.032	0.028	0.024	0.022	0.020
3 in.	R-12 (12.0)	**0.066**	0.062	0.058	0.055	0.052	0.050	0.047	0.045	0.043	0.041	0.040	0.038	0.037	0.036	0.034	0.033	0.028	0.025	0.022	0.020	0.018
4 in.	R-16 (16.0)	**0.052**	0.050	0.047	0.045	0.043	0.041	0.040	0.038	0.037	0.036	0.034	0.033	0.032	0.031	0.030	0.029	0.026	0.023	0.020	0.018	0.017
5 in.	R-20 (20.0)	**0.043**	0.041	0.040	0.038	0.037	0.036	0.034	0.033	0.032	0.031	0.030	0.029	0.028	0.028	0.027	0.026	0.023	0.021	0.019	0.017	0.016
6 in.	R-24 (24.0)	**0.037**	0.036	0.034	0.033	0.032	0.031	0.030	0.029	0.028	0.028	0.027	0.026	0.026	0.025	0.024	0.024	0.021	0.019	0.018	0.016	0.015

TABLE A5.3 Assembly U-Factors for Steel-Joist Floors

Framing Type and Spacing Width (Actual Depth)	Cavity Insulation R-Value: Rated (Effective Installed [See Table A9.2A])	Overall U-Factor for Entire Base Floor Assembly	Overall U-Factor for Assembly of Base Floor Plus Continuous Insulation (Uninterrupted by Framing) — Rated R-Value of Continuous Insulation																			
			R-1.00	R-2.00	R-3.00	R-4.00	R-5.00	R-6.00	R-7.00	R-8.00	R-9.00	R-10.00	R-11.00	R-12.00	R-13.00	R-14.00	R-15.00	R-20.00	R-25.00	R-30.00	R-35.00	R-40.00
Steel Joist Floor with Rigid Foam																						
	None (0.0)	**0.350**	0.259	0.206	0.171	0.146	0.127	0.113	0.101	0.092	0.084	0.078	0.072	0.067	0.063	0.059	0.056	0.044	0.036	0.030	0.026	0.023
Steel Joist Floor with Spray-on Insulation																						
1 in.	R-4 (3.88)	**0.148**	0.129	0.114	0.103	0.093	0.085	0.078	0.073	0.068	0.064	0.060	0.056	0.053	0.051	0.048	0.046	0.037	0.032	0.027	0.024	0.021
2 in.	R-8 (7.52)	**0.096**	0.088	0.081	0.075	0.070	0.065	0.061	0.058	0.054	0.052	0.049	0.047	0.045	0.043	0.041	0.039	0.033	0.028	0.025	0.022	0.020
3 in.	R-12 (10.80)	**0.073**	0.068	0.064	0.060	0.057	0.054	0.051	0.048	0.046	0.044	0.042	0.041	0.039	0.038	0.036	0.035	0.030	0.026	0.023	0.021	0.019
4 in.	R-16 (13.92)	**0.060**	0.056	0.053	0.051	0.048	0.046	0.044	0.042	0.040	0.039	0.037	0.036	0.035	0.034	0.032	0.031	0.027	0.024	0.021	0.019	0.018
5 in.	R-20 (17.00)	**0.050**	0.048	0.046	0.044	0.042	0.040	0.039	0.037	0.036	0.035	0.033	0.032	0.031	0.030	0.030	0.029	0.025	0.022	0.020	0.018	0.017
6 in.	R-24 (19.68)	**0.044**	0.042	0.041	0.039	0.038	0.036	0.035	0.034	0.033	0.032	0.031	0.030	0.029	0.028	0.027	0.027	0.024	0.021	0.019	0.017	0.016
Steel Joist Floor with Batt Insulation																						
	None (0.0)	**0.350**	0.259	0.206	0.171	0.146	0.127	0.113	0.101	0.092	0.084	0.078	0.072	0.067	0.063	0.059	0.056	0.044	0.036	0.030	0.026	0.023
	R-11 (10.01)	**0.078**	0.072	0.067	0.063	0.059	0.056	0.053	0.050	0.048	0.046	0.044	0.042	0.040	0.039	0.037	0.036	0.030	0.026	0.023	0.021	0.019
	R-13 (11.70)	**0.069**	0.064	0.060	0.057	0.054	0.051	0.049	0.046	0.044	0.042	0.041	0.039	0.038	0.036	0.035	0.034	0.029	0.025	0.022	0.020	0.018
	R-15 (13.20)	**0.062**	0.059	0.055	0.052	0.050	0.047	0.045	0.043	0.042	0.040	0.038	0.037	0.036	0.034	0.033	0.032	0.028	0.024	0.022	0.020	0.018
	R-19 (16.34)	**0.052**	0.050	0.047	0.045	0.043	0.041	0.040	0.038	0.037	0.035	0.034	0.033	0.032	0.031	0.030	0.029	0.026	0.023	0.020	0.018	0.017
	R-21 (17.64)	**0.049**	0.047	0.044	0.043	0.041	0.039	0.038	0.036	0.035	0.034	0.033	0.032	0.031	0.030	0.029	0.028	0.025	0.022	0.020	0.018	0.017
	R-25 (20.25)	**0.043**	0.041	0.040	0.038	0.037	0.036	0.034	0.033	0.032	0.031	0.030	0.029	0.028	0.028	0.027	0.026	0.023	0.021	0.019	0.017	0.016
	R-30C (23.70)	**0.038**	0.036	0.035	0.034	0.033	0.032	0.031	0.030	0.029	0.028	0.027	0.027	0.026	0.025	0.025	0.024	0.021	0.019	0.018	0.016	0.015
	R-30 (23.70)	**0.038**	0.036	0.035	0.034	0.033	0.032	0.031	0.030	0.029	0.028	0.027	0.027	0.026	0.025	0.025	0.024	0.021	0.019	0.018	0.016	0.015
	R-38C (28.12)	**0.032**	0.031	0.030	0.029	0.029	0.028	0.027	0.026	0.026	0.025	0.024	0.024	0.023	0.023	0.022	0.022	0.020	0.018	0.016	0.015	0.014
	R-38 (28.12)	**0.032**	0.031	0.030	0.029	0.029	0.028	0.027	0.026	0.026	0.025	0.024	0.024	0.023	0.023	0.022	0.022	0.020	0.018	0.016	0.015	0.014

with insulation located directly below the *floor* and ventilated airspace below the insulation. The heat flow path through the joist is calculated to be the same depth as the insulation. The *U-factors* include R-0.92 for interior air film—heat flow down, R-1.23 for carpet and pad, R-0.94 for 0.75 in. wood subfloor, and R-0.46 for semi-exterior air film. The weighting factors are 91% insulated cavity and 9% framing.

A5.4.2 Rated R-Value of Insulation for Wood-Framed and Other Floors

A5.4.2.1 The first *rated R-value of insulation* is for uncompressed insulation installed in the cavity between wood joists.

A5.4.2.2 It is acceptable for this insulation to also be *continuous insulation* uninterrupted by framing. All *continuous insulation* shall be installed either on the interior above the floor structure or below a framing cavity completely filled with insulation.

A5.4.3 *U-Factors* for *Wood-Framed Floors*

A5.4.3.1 The *U-factors* for wood-framed floors shall be taken from Table A5.4.

A5.4.3.2 It is not acceptable to use these *U-factors* if the framing is not wood.

A6. SLAB-ON-GRADE FLOORS

A6.1 General. For the purpose of Section A1.2, the base assembly is a slab floor of 6 in. concrete poured directly on to the earth, the bottom of the slab is at grade line, and soil conductivity is 0.75 Btu/h·ft·°F. In contrast to the *U-factor* for *floors*, the *F-factor* for *slab-on-grade floors* is expressed per linear foot of building perimeter. *F-factors* are provided for unheated slabs and for heated slabs. *Unheated slab-on-grade floors* do not have heating elements, and *heated slab-on-grade floors* do have heating elements within or beneath the slab. *F-factors* are provided for three insulation configurations:

1. *Horizontal Insulation: continuous insulation* is applied directly to the underside of the slab and extends inward horizontally from the perimeter for the distance specified or *continuous insulation* is applied downward from the top of the slab and then extends horizontally to the interior or the exterior from the perimeter for the distance specified.
2. *Vertical Insulation: continuous insulation* is applied directly to the slab exterior, extending downward from the top of the slab for the distance specified.
3. *Fully Insulated Slab: continuous insulation* extends downward from the top of the slab and along the entire perimeter and completely covers the entire area under the slab.

A6.2 Rated R-Value of Insulation for Slab-on-Grade Floors

A6.2.1 The *rated R-value of insulation* shall be installed around the perimeter of the *slab-on-grade floor* to the distance specified.

Exception: For a monolithic *slab-on-grade floor*, the insulation shall extend from the top of the slab-on-grade to the bottom of the footing.

A6.2.2 Insulation installed inside the foundation wall shall extend downward from the top of the slab a minimum of the distance specified or to the top of the footing, whichever is less.

A6.2.3 Insulation installed outside the foundation wall shall extend from the top of the slab or downward to at least the bottom of the slab and then horizontally to a minimum of the distance specified. In all climates, the horizontal insulation extending outside of the foundation shall be covered by pavement or by soil a minimum of 10 in. thick.

A6.3 F-Factors for Slab-on-Grade Floors

A6.3.1 *F-factors* for slab-on-grade floors shall be taken from Table A6.3.

A6.3.2 These *F-factors* are acceptable for all *slab-on-grade floors*.

A7. OPAQUE DOORS

All *opaque doors* with *U-factors* determined, certified, and labeled in accordance with NFRC 100 shall be assigned those *U-factors*.

A7.1 Unlabeled Opaque Doors. Unlabeled *opaque doors* shall be assigned the following *U-factors*:

a. Uninsulated single-layer metal *swinging doors* or *nonswinging doors*, including single-layer uninsulated *access hatches* and uninsulated smoke vents: 1.45
b. Uninsulated double-layer metal *swinging doors* or *nonswinging doors*, including double-layer uninsulated *access hatches* and uninsulated smoke vents: 0.70
c. Insulated metal *swinging doors*, including fire-rated *doors*, insulated *access hatches*, and insulated smoke vents: 0.50
d. Wood *doors*, minimum nominal thickness of 1.75 in., including panel *doors* with minimum panel thickness of 1.125 in., solid core flush *doors*, and hollow core flush *doors*: 0.50
e. Any other wood *door*: 0.60

A8. FENESTRATION

All *fenestration* with *U-factors, SHGC,* or visible light transmittance determined, certified, and labeled in accordance with NFRC 100, 200, and 300, respectively, shall be assigned those values.

A8.1 Unlabeled Skylights. Unlabeled *skylights* shall be assigned the *U-factors* in Table A8.1A and are allowed to use the *SHGC*s and VLTs in Table A8.1B. The metal with thermal break frame category shall not be used unless all frame members have a thermal break equal to or greater than 0.25 in.

A8.2 Unlabeled Vertical Fenestration. Unlabeled *vertical fenestration*, both operable and fixed, shall be assigned the *U-factors, SHGC*s, and VLTs in Table A8.2.

TABLE A5.4 Assembly U-Factors for Wood-Joist Floors

Framing Type and Spacing Width (Actual Depth)	Cavity Insulation R-Value: Rated (Effective Installed)	Overall U-Factor for Entire Base Floor Assembly	Overall U-Factor for Assembly of Base Floor Plus Continuous Insulation (Uninterrupted by Framing) — Rated R-Value of Continuous Insulation																			
			R-1.00	R-2.00	R-3.00	R-4.00	R-5.00	R-6.00	R-7.00	R-8.00	R-9.00	R-10.00	R-11.00	R-12.00	R-13.00	R-14.00	R-15.00	R-20.00	R-25.00	R-30.00	R-35.00	R-40.00
Wood Joists																						
5.5 in.	None (0.0)	**0.282**	0.220	0.180	0.153	0.132	0.117	0.105	0.095	0.087	0.080	0.074	0.069	0.064	0.060	0.057	0.054	0.042	0.035	0.030	0.026	0.023
	R-11 (11.0)	**0.074**	0.069	0.064	0.060	0.057	0.054	0.051	0.048	0.046	0.044	0.042	0.040	0.039	0.037	0.036	0.035	0.030	0.026	0.023	0.020	0.019
	R-13 (13.0)	**0.066**	0.062	0.058	0.055	0.052	0.049	0.047	0.045	0.043	0.041	0.039	0.038	0.036	0.035	0.034	0.033	0.028	0.025	0.022	0.020	0.018
	R-15 (15.0)	**0.060**	0.057	0.053	0.050	0.048	0.046	0.044	0.042	0.040	0.038	0.037	0.036	0.034	0.033	0.032	0.031	0.027	0.024	0.021	0.019	0.017
	R-19 (18.0)	**0.051**	0.048	0.046	0.044	0.042	0.040	0.038	0.037	0.036	0.034	0.033	0.032	0.031	0.030	0.029	0.028	0.025	0.022	0.020	0.018	0.017
	R-21 (21.0)	**0.046**	0.043	0.042	0.040	0.038	0.037	0.035	0.034	0.033	0.032	0.031	0.030	0.029	0.028	0.027	0.027	0.023	0.021	0.019	0.017	0.016
7.25 in.	R-25 (25.0)	**0.039**	0.037	0.036	0.035	0.033	0.032	0.031	0.030	0.029	0.028	0.028	0.027	0.026	0.025	0.025	0.024	0.022	0.019	0.018	0.016	0.015
	R-30C (30.0)	**0.034**	0.033	0.032	0.031	0.030	0.029	0.028	0.027	0.026	0.026	0.025	0.024	0.024	0.023	0.023	0.022	0.020	0.018	0.016	0.015	0.014
9.25 in.	R-30 (30.0)	**0.033**	0.032	0.031	0.030	0.029	0.028	0.027	0.027	0.026	0.025	0.024	0.024	0.023	0.023	0.022	0.022	0.020	0.018	0.016	0.015	0.014
11.25 in.	R-38C (38.0)	**0.027**	0.026	0.025	0.025	0.024	0.024	0.023	0.022	0.022	0.021	0.021	0.020	0.020	0.020	0.019	0.019	0.017	0.016	0.015	0.014	0.013
13.25 in.	R-38 (38.0)	**0.026**	0.026	0.025	0.024	0.024	0.023	0.023	0.022	0.022	0.021	0.021	0.020	0.020	0.019	0.019	0.019	0.017	0.016	0.015	0.014	0.013

TABLE A6.3 Assembly F-Factors for Slab-on-Grade Floors

	Rated R-Value of Insulation												
Insulation Description	**R-0**	**R-5**	**R-7.5**	**R-10**	**R-15**	**R-20**	**R-25**	**R-30**	**R-35**	**R-40**	**R-45**	**R-50**	**R-55**
Unheated Slabs													
None	0.73												
12 in. horizontal		0.72	0.71	0.71	0.71								
24 in. horizontal		0.70	0.70	0.70	0.69								
36 in. horizontal		0.68	0.67	0.66	0.66								
48 in. horizontal		0.67	0.65	0.64	0.63								
12 in. horizontal		0.61	0.60	0.58	0.57	0.567	0.565	0.564					
24 in. horizontal		0.58	0.56	0.54	0.52	0.510	0.505	0.502					
36 in. horizontal		0.56	0.53	0.51	0.48	0.472	0.464	0.460					
48 in. horizontal		0.54	0.51	0.48	0.45	0.434	0.424	0.419					
Fully insulated slab		0.46	0.41	0.36	0.30	0.261	0.233	0.213	0.198	0.186	0.176	0.168	0.161
Heated Slabs													
None	1.35												
12 in. horizontal		1.31	1.31	1.30	1.30								
24 in. horizontal		1.28	1.27	1.26	1.25								
36 in. horizontal		1.24	1.21	1.20	1.18								
48 in. horizontal		1.20	1.17	1.13	1.11								
12 in. horizontal		1.06	1.02	1.00	0.98	0.968	0.964	0.961					
24 in. horizontal		0.99	0.95	0.90	0.86	0.843	0.832	0.827					
36 in. horizontal		0.95	0.89	0.84	0.79	0.762	0.747	0.740					
48 in. horizontal		0.91	0.85	0.78	0.72	0.688	0.671	0.659					
Fully insulated slab		0.74	0.64	0.55	0.44	0.373	0.326	0.296	0.273	0.255	0.239	0.227	0.217

TABLE A8.1A Assembly U-Factors for Unlabeled Skylights

	Product Type	Sloped Installation: Unlabeled Skylight with Curb (Includes glass/plastic, flat/domed, fixed/operable)				Sloped Installation: Unlabeled Skylight without Curb (Includes glass/plastic, flat/domed, fixed/operable)		
	Frame Type	Aluminum without Thermal Break	Aluminum with Thermal Break	Reinforced Vinyl/ Aluminum Clad Wood	Wood/ Vinyl	Aluminum without Thermal Break	Aluminum with Thermal Break	Structural Glazing
ID	Glazing Type							
	Single Glazing							
1	1/8 in. glass	1.98	1.89	1.75	1.47	1.36	1.25	1.25
2	1/4 in. acrylic/polycarb	1.82	1.73	1.60	1.31	1.21	1.10	1.10
3	1/8 in. acrylic/polycarb	1.90	1.81	1.68	1.39	1.29	1.18	1.18
	Double Glazing							
4	1/4 in. airspace	1.31	1.11	1.05	0.84	0.82	0.70	0.66
5	1/2 in. airspace	1.30	1.10	1.04	0.84	0.81	0.69	0.65
6	1/4 in. argon space	1.27	1.07	1.00	0.80	0.77	0.66	0.62
7	1/2 in. argon space	1.27	1.07	1.00	0.80	0.77	0.66	0.62
	Double Glazing, $e = 0.60$ on surface 2 or 3							
8	1/4 in. airspace	1.27	1.08	1.01	0.81	0.78	0.67	0.63
9	1/2 in. airspace	1.27	1.07	1.00	0.80	0.77	0.66	0.62
10	1/4 in. argon space	1.23	1.03	0.97	0.76	0.74	0.63	0.58
11	1/2 in. argon space	1.23	1.03	0.97	0.76	0.74	0.63	0.58
	Double Glazing, $e = 0.40$ on surface 2 or 3							
12	1/4 in. airspace	1.25	1.05	0.99	0.78	0.76	0.64	0.60
13	1/2 in. airspace	1.24	1.04	0.98	0.77	0.75	0.64	0.59
14	1/4 in. argon space	1.18	0.99	0.92	0.72	0.70	0.58	0.54
15	1/2 in. argon space	1.20	1.00	0.94	0.74	0.71	0.60	0.56
	Double Glazing, $e = 0.20$ on surface 2 or 3							
16	1/4 in. airspace	1.20	1.00	0.94	0.74	0.71	0.60	0.56
17	1/2 in. airspace	1.20	1.00	0.94	0.74	0.71	0.60	0.56
18	1/4 in. argon space	1.14	0.94	0.88	0.68	0.65	0.54	0.50
19	1/2 in. argon space	1.15	0.95	0.89	0.68	0.66	0.55	0.51
	Double Glazing, $e = 0.10$ on surface 2 or 3							
20	1/4 in. airspace	1.18	0.99	0.92	0.72	0.70	0.58	0.54
21	1/2 in. airspace	1.18	0.99	0.92	0.72	0.70	0.58	0.54
22	1/4 in. argon space	1.11	0.91	0.85	0.65	0.63	0.52	0.47
23	1/2 in. argon space	1.13	0.93	0.87	0.67	0.65	0.53	0.49
	Double Glazing, $e = 0.05$ on surface 2 or 3							
24	1/4 in. airspace	1.17	0.97	0.91	0.70	0.68	0.57	0.52
25	1/2 in. airspace	1.17	0.98	0.91	0.71	0.69	0.58	0.53
26	1/4 in. argon space	1.09	0.89	0.83	0.63	0.61	0.50	0.45
27	1/2 in. argon space	1.11	0.91	0.85	0.65	0.63	0.52	0.47
	Triple Glazing							
28	1/4 in. airspaces	1.12	0.89	0.84	0.64	0.64	0.53	0.48
29	1/2 in. airspaces	1.10	0.87	0.81	0.61	0.62	0.51	0.45
30	1/4 in. argon spaces	1.09	0.86	0.80	0.60	0.61	0.50	0.44
31	1/2 in. argon spaces	1.07	0.84	0.79	0.59	0.59	0.48	0.42

TABLE A8.1A Assembly U-Factors for Unlabeled Skylights *(continued)*

	Product Type	Sloped Installation						
		Unlabeled Skylight with Curb (Includes glass/plastic, flat/domed, fixed/operable)				Unlabeled Skylight without Curb (Includes glass/plastic, flat/domed, fixed/operable)		
	Frame Type	Aluminum without Thermal Break	Aluminum with Thermal Break	Reinforced Vinyl/ Aluminum Clad Wood	Wood/ Vinyl	Aluminum without Thermal Break	Aluminum with Thermal Break	Structural Glazing
ID	Glazing Type							
	Triple Glazing, $e = 0.20$ on surface 2,3,4, or 5							
32	1/4 in. airspace	1.08	0.85	0.79	0.59	0.60	0.49	0.43
33	1/2 in. airspace	1.05	0.82	0.77	0.57	0.57	0.46	0.41
34	1/4 in. argon space	1.02	0.79	0.74	0.54	0.55	0.44	0.38
35	1/2 in. argon space	1.01	0.78	0.73	0.53	0.54	0.43	0.37
	Triple Glazing, $e = 0.20$ on surfaces 2 or 3 and 4 or 5							
36	1/4 in. airspace	1.03	0.80	0.75	0.55	0.56	0.45	0.39
37	1/2 in. airspace	1.01	0.78	0.73	0.53	0.54	0.43	0.37
38	1/4 in. argon space	0.99	0.75	0.70	0.50	0.51	0.40	0.35
39	1/2 in. argon space	0.97	0.74	0.69	0.49	0.50	0.39	0.33
	Triple Glazing, $e = 0.10$ on surfaces 2 or 3 and 4 or 5							
40	1/4 in. airspace	1.01	0.78	0.73	0.53	0.54	0.43	0.37
41	1/2 in. airspace	0.99	0.76	0.71	0.51	0.52	0.41	0.36
42	1/4 in. argon space	0.96	0.73	0.68	0.48	0.49	0.38	0.32
43	1/2 in. argon space	0.95	0.72	0.67	0.47	0.48	0.37	0.31
	Quadruple Glazing, $e = 0.10$ on surfaces 2 or 3 and 4 or 5							
44	1/4 in. airspace	0.97	0.74	0.69	0.49	0.50	0.39	0.33
45	1/2 in. airspace	0.94	0.71	0.66	0.46	0.47	0.36	0.30
46	1/4 in. argon space	0.93	0.70	0.65	0.45	0.46	0.35	0.30
47	1/2 in. argon space	0.91	0.68	0.63	0.43	0.44	0.33	0.28
48	1/4 in. krypton spaces	0.88	0.65	0.60	0.40	0.42	0.31	0.25

TABLE A8.1B Assembly SHGCs and Assembly Visible Light Transmittances (VLTs) for Unlabeled Skylights

Glass Type	Glazing Type: Number of glazing layers Number and emissivity of coatings (Glazing is glass except where noted)	Unlabeled Skylights (Includes glass/plastic, flat/domed, fixed/operable)					
	Frame:	Metal without Thermal Break		Metal with Thermal Break		Wood/Vinyl/ Fiberglass	
	Characteristic:	SHGC	VLT	SHGC	VLT	SHGC	VLT
Clear	Single glazing, 1/8 in. glass	0.82	0.76	0.78	0.76	0.73	0.73
	Single glazing, 1/4 in. glass	0.78	0.75	0.74	0.75	0.69	0.72
	Single glazing, acrylic/polycarbonate	0.83	0.92	0.83	0.92	0.83	0.92
	Double glazing	0.68	0.66	0.64	0.66	0.59	0.64
	Double glazing, E = 0.40 on surface 2 or 3	0.71	0.65	0.67	0.65	0.62	0.63
	Double glazing, E = 0.20 on surface 2 or 3	0.66	0.61	0.62	0.61	0.57	0.59
	Double glazing, E = 0.10 on surface 2 or 3	0.59	0.63	0.55	0.63	0.51	0.61
	Double glazing, acrylic/polycarbonate	0.77	0.89	0.77	0.89	0.77	0.89
	Triple glazing	0.60	0.59	0.56	0.59	0.52	0.57
	Triple glazing, E = 0.40 on surface 2, 3, 4, or 5	0.64	0.60	0.60	0.60	0.56	0.57
	Triple glazing, E = 0.20 on surface 2, 3, 4, or 5	0.59	0.55	0.55	0.55	0.51	0.53
	Triple glazing, E = 0.10 on surface 2, 3, 4, or 5	0.54	0.56	0.50	0.56	0.46	0.54
	Triple glazing, E = 0.40 on surfaces 3 and 5	0.62	0.57	0.58	0.57	0.53	0.55
	Triple glazing, E = 0.20 on surfaces 3 and 5	0.56	0.51	0.52	0.51	0.48	0.49
	Triple glazing, E = 0.10 on surfaces 3 and 5	0.47	0.54	0.43	0.54	0.40	0.52
	Triple glazing, acrylic/polycarbonate	0.71	0.85	0.71	0.85	0.71	0.85
	Quadruple glazing, E = 0.10 on surfaces 3 and 5	0.41	0.48	0.37	0.48	0.33	0.46
	Quadruple glazing, acrylic/polycarbonate	0.65	0.81	0.65	0.81	0.65	0.81
Tinted	Single glazing, 1/8 in. glass	0.70	0.58	0.66	0.58	0.62	0.56
	Single glazing, 1/4 in. glass	0.61	0.45	0.56	0.45	0.52	0.44
	Single glazing, acrylic/polycarbonate	0.46	0.27	0.46	0.27	0.46	0.27
	Double glazing	0.50	0.40	0.46	0.40	0.42	0.39
	Double glazing, E = 0.40 on surface 2 or 3	0.59	0.50	0.55	0.50	0.50	0.48
	Double glazing, E = 0.20 on surface 2 or 3	0.47	0.37	0.43	0.37	0.39	0.36
	Double glazing, E = 0.10 on surface 2 or 3	0.43	0.38	0.39	0.38	0.35	0.37
	Double glazing, acrylic/polycarbonate	0.37	0.25	0.37	0.25	0.37	0.25
	Triple glazing	0.42	0.22	0.37	0.22	0.34	0.21
	Triple glazing, E = 0.40 on surface 2, 3, 4, or 5	0.53	0.45	0.49	0.45	0.45	0.44
	Triple glazing, E = 0.20 on surface 2, 3, 4, or 5	0.42	0.33	0.38	0.33	0.35	0.32
	Triple glazing, E = 0.10 on surface 2, 3, 4, or 5	0.39	0.34	0.35	0.34	0.31	0.33
	Triple glazing, E = 0.40 on surfaces 3 and 5	0.51	0.43	0.47	0.43	0.43	0.42
	Triple glazing, E = 0.20 on surfaces 3 and 5	0.40	0.31	0.36	0.31	0.32	0.29
	Triple glazing, E = 0.10 on surfaces 3 and 5	0.34	0.32	0.30	0.32	0.27	0.31
	Triple glazing, acrylic/polycarbonate	0.30	0.23	0.30	0.23	0.30	0.23
	Quadruple glazing, E = 0.10 on surfaces 3 and 5	0.30	0.29	0.26	0.29	0.23	0.28
	Quadruple glazing, acrylic/polycarbonate	0.27	0.25	0.27	0.25	0.27	0.25

TABLE A8.2 Assembly U-Factors, Assembly SHGCs, and Assembly Visible Light Transmittances (VLTs) for Unlabeled Vertical Fenestration

Frame Type	Glazing Type	Unlabeled Vertical Fenestration					
		Clear Glass			Tinted Glass		
		U-Factor	SHGC	VLT	U-Factor	SHGC	VLT
All frame types	Single glazing	1.25	0.82	0.76	1.25	0.70	0.58
	Glass block	0.60	0.56	0.56	n.a.	n.a.	n.a.
Wood, vinyl, or fiberglass frames	Double glazing	0.60	0.59	0.64	0.60	0.42	0.39
	Triple glazing	0.45	0.52	0.57	0.45	0.34	0.21
Metal and other frame types	Double glazing	0.90	0.68	0.66	0.90	0.50	0.40
	Triple glazing	0.70	0.60	0.59	0.70	0.42	0.22

A9. DETERMINATION OF ALTERNATE ASSEMBLY U-FACTORS, C-FACTORS, F-FACTORS, OR HEAT CAPACITIES

A9.1 General. Component *U-factors* for other opaque assemblies shall be determined in accordance with Section A9 only if approved by the *building official* in accordance with Section A1.2. The procedures required for each class of construction are specified in Section A9.2. Testing shall be performed in accordance with Section A9.3. Calculations shall be performed in accordance with Section A9.4.

A9.2 Required Procedures. Two- or three-dimensional finite difference and finite volume computer models shall be an acceptable alternative method to calculating the thermal performance values for all assemblies and constructions listed below. The following procedures shall also be permitted to determine all alternative *U-factors*, *F-factors*, and *C-factors*.

a. Roofs

1. *Roofs with insulation entirely above deck*: testing or series calculation method.
2. *Metal building roofs*: testing.
3. *Attic roofs*, wood joists: testing or parallel path calculation method.
4. *Attic roofs*, steel joists: testing or parallel path calculation method using the insulation/framing layer adjustment factors in Table A9.2A or modified zone calculation method.
5. *Attic roofs*, concrete joists: testing or parallel path calculation method if concrete is solid and uniform or isothermal planes calculation method if concrete has hollow sections.
6. Other *attic roofs* and other *roofs*: testing or two-dimensional calculation method.

b. Above-Grade Walls

1. *Mass walls*: testing or isothermal planes calculation method or two-dimensional calculation method. The parallel path calculation method is not acceptable.
2. *Metal building walls*: testing.
3. *Steel-framed walls*: testing or parallel path calculation method using the insulation/framing layer adjustment factors in Table A9.2B or the modified zone method.
4. *Wood-framed walls*: testing or parallel path calculation method.
5. Other *walls*: testing or two-dimensional calculation method.

c. Below-Grade Walls

1. *Mass walls*: testing or isothermal planes calculation method or two-dimensional calculation method. The parallel path calculation method is not acceptable.
2. *Other walls*: testing or two-dimensional calculation method.

d. Floors

1. *Mass floors*: testing or parallel path calculation method if concrete is solid and uniform or isothermal planes calculation method if concrete has hollow sections.
2. *Steel joist floors*: testing or modified zone calculation method.
3. *Wood joist floors*: testing or parallel path calculation method or isothermal planes calculation method.
4. Other *floors*: testing or two-dimensional calculation method.

e. Slab-on-Grade Floors
No testing or calculations allowed.

A9.3 Testing Procedures

A9.3.1 Building Material Thermal Properties. If *building material* R-values or thermal conductivities are determined by testing, one of the following test procedures shall be used:

a. ASTM C177
b. ASTM C518
c. ASTM C1363

For concrete, the oven-dried conductivity shall be multiplied by 1.2 to reflect the moisture content as typically installed.

TABLE A9.2A Effective Insulation/Framing Layer R-Values for Roof and Floor Insulation Installed Between Metal Framing (4 ft on Center)

Rated R-Value of Insulation	Correction Factor	Framing/Cavity R-Value	Rated R-Value of Insulation	Correction Factor	Framing/Cavity R-Value
0.00	1.00	0.00	20.00	0.85	17.00
4.00	0.97	3.88	21.00	0.84	17.64
5.00	0.96	4.80	24.00	0.82	19.68
8.00	0.94	7.52	25.00	0.81	20.25
10.00	0.92	9.20	30.00	0.79	23.70
11.00	0.91	10.01	35.00	0.76	26.60
12.00	0.90	10.80	38.00	0.74	28.12
13.00	0.90	11.70	40.00	0.73	29.20
15.00	0.88	13.20	45.00	0.71	31.95
16.00	0.87	13.92	50.00	0.69	34.50
19.00	0.86	16.34	55.00	0.67	36.85

TABLE A9.2B Effective Insulation/Framing Layer R-Values for Wall Insulation Installed Between Steel Framing

Nominal Depth of Cavity, in.	Actual Depth of Cavity, in.	Rated R-Value of Airspace or Insulation	Effective Framing/Cavity R-Value at 16 in. on Center	Effective Framing/Cavity R-Value at 24 in. on Center
		Empty Cavity, No Insulation		
4	3.5	R-0.91	0.79	0.91
		Insulated Cavity		
4	3.5	R-11	5.5	6.6
4	3.5	R-13	6.0	7.2
4	3.5	R-15	6.4	7.8
6	6.0	R-19	7.1	8.6
6	6.0	R-21	7.4	9.0
8	8.0	R-25	7.8	9.6

A9.3.2 Assembly U-Factors. If assembly *U-factors* are determined by testing, ASTM C1363 test procedures shall be used.

Product samples tested shall be production line material or representative of material as purchased by the consumer or contractor. If the assembly is too large to be tested at one time in its entirety, then either a representative portion shall be tested or different portions shall be tested separately and a weighted average determined. To be representative, the portion tested shall include edges of panels, joints with other panels, typical framing percentages, and thermal bridges.

A9.4 Calculation Procedures and Assumptions. The following procedures and assumptions shall be used for all calculations. R-values for air films, insulation, and *building materials* shall be taken from Sections A9.4.1 through A9.4.3, respectively. In addition, the appropriate assumptions listed in Sections A2 through A8, including framing factors, shall be used.

A9.4.1 Air Films. Prescribed R-values for air films shall be as follows:

R-Value	Condition
0.17	All exterior surfaces
0.46	All semi-exterior surfaces
0.61	Interior horizontal surfaces, heat flow up
0.92	Interior horizontal surfaces, heat flow down
0.68	Interior vertical surfaces

A9.4.1.1 Exterior surfaces are areas exposed to the wind.

A9.4.1.2 Semi-exterior surfaces are protected surfaces that face attics, crawlspaces, and parking garages with natural or mechanical ventilation.

A9.4.1.3 Interior surfaces are surfaces within enclosed spaces.

A9.4.1.4 The R-value for cavity airspaces shall be taken from Table A9.4A based on the emissivity of the cavity from Table A9.4B. No credit shall be given for airspaces in cavities that contain any insulation or are less than 0.5 in. The values for 3.5 in. cavities shall be used for cavities of that width and greater.

TABLE A9.4A Values for Cavity Air Spaces

Component	Airspace Thickness, in.	R-Value Effective Emissivity				
		0.03	0.05	0.20	0.50	0.82
Roof	0.50	2.13	2.04	1.54	1.04	0.77
	0.75	2.33	2.22	1.64	1.09	0.80
	1.50	2.53	2.41	1.75	1.13	0.82
	3.50	2.83	2.66	1.88	1.19	0.85
Wall	0.50	2.54	2.43	1.75	1.13	0.82
	0.75	3.58	3.32	2.18	1.30	0.90
	1.50	3.92	3.62	2.30	1.34	0.93
	3.50	3.67	3.40	2.21	1.31	0.91
Floor	0.50	2.55	1.28	1.00	0.69	0.53
	0.75	1.44	1.38	1.06	0.73	0.54
	1.50	2.49	2.38	1.76	1.15	0.85
	3.50	3.08	2.90	2.01	1.26	0.90

TABLE A9.4B Emittance Values of Various Surfaces and Effective Emittances of Air Spaces

Surface	Average Emittance *e*	Effective Emittance *e eff* of Air Space	
		One Surface *e*; Other, 0.9	Both Surfaces Emittance *e*
Aluminum foil, bright	0.05	0.05	0.03
Aluminum foil, with condensate just visible (>0.7 gr/ft^2)	0.30	0.29	—
Aluminum foil, with condensate clearly visible (>2.9 gr/ft^2)	0.70	0.65	—
Aluminum sheet	0.12	0.12	0.06
Aluminum coated paper, polished	0.20	0.20	0.11
Steel, galv., bright	0.25	0.24	0.15
Aluminum paint	0.50	0.47	0.35
Building materials: wood, paper, masonry, nonmetallic paints	0.90	0.82	0.82
Regular glass	0.84	0.77	0.72

A9.4.2 Insulation R-Values. Insulation R-values shall be determined as follows:

a. For insulation that is not compressed, the *rated R-value of insulation* shall be used.

b. For calculation purposes, the effective R-value for insulation that is uniformly compressed in confined cavities shall be taken from Table A9.4C.

c. For calculation purposes, the effective R-value for insulation installed in cavities in attic roofs with steel joists shall be taken from Table A9.2A.

d. For calculation purposes, the effective R-value for insulation installed in cavities in steel-framed walls shall be taken from Table A9.2B.

A9.4.3 Building Material Thermal Properties. R-values for *building materials* shall be taken from Table A9.4D. Concrete block R-values shall be calculated using the isothermal planes method or a two-dimensional calculation program, thermal conductivities from Table A9.4E, and dimensions from ASTM C90. The parallel path calculation method is not acceptable.

Exception: R-values for *building materials* or thermal conductivities determined from testing in accordance with Section A9.3.

A9.4.4 Building Material Heat Capacities. The *HC* of assemblies shall be calculated using published values for the unit weight and specific heat of all building material components that make up the assembly.

TABLE A9.4C Effective R-Values for Fiberglass

		Insulation R-Value at Standard Thickness							
Rated R-Value		38	30	22	21	19	15	13	11
Standard Thickness, in.		12	9.5	6.5	5.5	6	3.5	3.5	3.5
Nominal Lumber Size, in.	Actual Depth of Cavity, in.	Effective Insulation R-Values when Installed in a Confined Cavity							
2 × 12	11.25	37	—	—	—	—	—	—	—
2 × 10	9.25	32	30	—	—	—	—	—	—
2 × 8	7.25	27	26	22	21	19	—	—	—
2 × 6	5.5	—	21	20	21	18	—	—	—
2 × 4	3.5	—	—	14	—	13	15	13	11
	2.5	—	—	—	—	—	—	9.8	—
	1.5	—	—	—	—	—	—	6.3	6

TABLE A9.4D R-Values for Building Materials

Material	Nominal Size, in.	Actual Size, in.	R-Value
Carpet and rubber pad	—	–	1.23
Concrete at R-0.0625/in.	—	2	0.13
	—	4	0.25
	—	6	0.38
	—	8	0.5
	—	10	0.63
	—	12	0.75
Flooring, wood subfloor	—	0.75	0.94
Gypsum board	—	0.5	0.45
	—	0.625	0.56
Metal deck	—	—	0
Roofing, built-up	—	0.375	0.33
Sheathing, vegetable fiber board, 0.78 in.	—	0.78	2.06
Soil at R-0.104/in.	—	12	1.25
Steel, mild		1	0.0031807
Stucco	—	0.75	0.08
Wood, 2 × 4 at R-1.25/in.	4	3.5	4.38
Wood, 2 × 4 at R-1.25/in.	6	5.5	6.88
Wood, 2 × 4 at R-1.25/in.	8	7.25	9.06
Wood, 2 × 4 at R-1.25/in.	10	9.25	11.56
Wood, 2 × 4 at R-1.25/in.	12	11.25	14.06
Wood, 2 × 4 at R-1.25/in.	14	13.25	16.56

TABLE A9.4E Thermal Conductivity of Concrete Block Material

Concrete Block Density, lb/ft^3	Thermal Conductivity, $Btu \cdot in./h \cdot ft^2 \cdot °F$
80	3.7
85	4.2
90	4.7
95	5.1
100	5.5
105	6.1
110	6.7
115	7.2
120	7.8
125	8.9
130	10.0
135	11.8
140	13.5

(This is a normative appendix and is part of this standard.)

NORMATIVE APPENDIX B—BUILDING ENVELOPE CLIMATE CRITERIA

B1. GENERAL

This normative appendix provides the information to determine both United States and international climate zones. For US locations, use either Figure B-1 or Table B-1 to determine the climate zone number and letter that are required for determining compliance regarding various sections and tables in this standard. Figure B-1 contains the county-by-county climate zone map for the United States. Table B-1 lists each state and major counties within the state and shows the climate number and letter for each county listed.

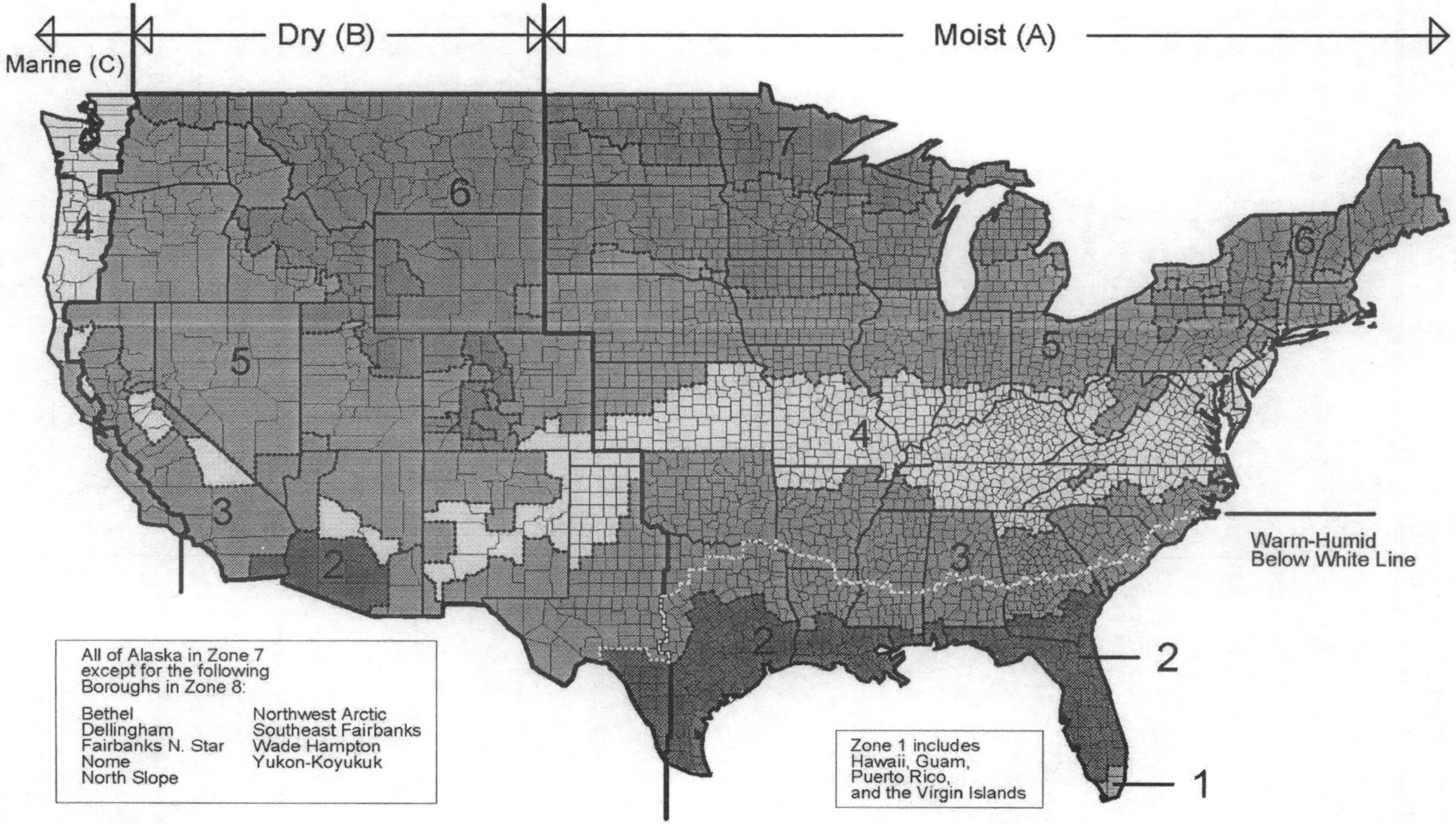

Figure B-1 Climate zones for United States locations.

TABLE B-1 US Climate Zones

State County	Zone
Alabama (AL)	
Zone 3a Except	
Baldwin	2A
Mobile	2A
Alaska (AK)	
Zone 7 Except	
Bethel (CA)	8
Dillingham (CA)	8
Fairbanks North Star	8
Nome (CA)	8
North Slope	8
Northwest Arctic	8
Southeast Fairbanks (CA)	8
Wade Hampton (CA)	8
Yukon-Koyukuk (CA)	8
Arizona (AZ)	
Zone 3B Except	
La Paz	2B
Maricopa	2B
Pima	2B
Pinal	2B
Yuma	2B
Gila	4B
Yavapai	4B
Apache	5B
Coconino	5B
Navajo	5B
Arkansas (AR)	
Zone 3A Except	
Baxter	4A
Benton	4A
Boone	4A
Carroll	4A
Fulton	4A
Izard	4A
Madison	4A
Marion	4A
Newton	4A
Searcy	4A
Stone	4A
(Arkansas cont.)	
Washington	4A
California (CA)	
Zone 3B Except	
Imperial	2B
Alameda	3C
Marin	3C
Mendocino	3C
Monterey	3C
Napa	3C
San Benito	3C
San Francisco	3C
San Luis Obispo	3C
San Mateo	3C
Santa Barbara	3C
Santa Clara	3C
Santa Cruz	3C
Sonoma	3C
Ventura	3C
Amador	4B
Calaveras	4B
Del Norte	4B
El Dorado	4B
Humboldt	4B
Inyo	4B
Lake	4B
Mariposa	4B
Trinity	4B
Tuolumme	4B
Lassen	5B
Modoc	5B
Nevada	5B
Plumas	5B
Sierra	5B
Siskiyou	5B
Alpine	6B
Mono	6B
Colorado (CO)	
Zone 5B Except	
Baca	4B
(Colorado cont.)	
Las Animas	4B
Otero	4B
Alamosa	6B
Archuleta	6B
Chaffee	6B
Conejos	6B
Costilla	6B
Custer	6B
Dolores	6B
Eagle	6B
Moffat	6B
Ouray	6B
Rio Blanco	6B
Saguache	6B
San Miguel	6B
Clear Creek	7
Grand	7
Gunnison	7
Hinsdale	7
Jackson	7
Lake	7
Mineral	7
Park	7
Pitkin	7
Rio Grande	7
Routt	7
San Juan	7
Summitt	7
Connecticut (CT)	
Zone 5A	
Delaware (DE)	
Zone 4A	
District of Columbia (DC)	
Zone 4A	
Florida (FL)	
Zone 2A Except	
Broward	1A
Miami-Dade	1A
Monroe	1A
Georgia (GA)	
Zone 3A Except	
Appling	2A
Atkinson	2A
Bacon	2A
Baker	2A
Berrien	2A
Brantley	2A
Brooks	2A
Bryan	2A
Camden	2A
Charlton	2A
Chatham	2A
Clinch	2A
Colquitt	2A
Cook	2A
Decatur	2A
Echols	2A
Effingham	2A
Evans	2A
Glynn	2A
Grady	2A
Jeff Davis	2A
Lanier	2A
Liberty	2A
Long	2A
Lowndes	2A
McIntosh	2A
Miller	2A
Mitchell	2A
Pierce	2A
Seminole	2A
Tattnall	2A
Thomas	2A
Toombs	2A
Ware	2A
Wayne	2A
Banks	4A
Catoosa	4A
Chattooga	4A

TABLE B-1 US Climate Zones *(continued)*

State County	Zone
(Georgia cont.)	
Dade	4A
Dawson	4A
Fannin	4A
Floyd	4A
Franklin	4A
Gilmer	4A
Gordon	4A
Habersham	4A
Hall	4A
Lumpkin	4A
Murray	4A
Pickens	4A
Rabun	4A
Stephens	4A
Towns	4A
Union	4A
Walker	4A
White	4A
Whitfield	4A
Hawaii (HI)	
Zone 1A	
Idaho (ID)	
Zone 6B Except	
Ada	5B
Benewah	5B
Canyon	5B
Cassia	5B
Clearwater	5B
Elmore	5B
Gem	5B
Gooding	5B
Idaho	5B
Jerome	5B
Kootenai	5B
Latah	5B
Lewis	5B
Lincoln	5B
Minidoka	5B
Nez Perce	5B
Owyhee	5B
(Idaho cont.)	
Payette	5B
Power	5B
Shoshone	5B
Twin Falls	5B
Washington	5B
Illinois (IL)	
Zone 5A Except	
Alexander	4A
Bond	4A
Christian	4A
Clay	4A
Clinton	4A
Crawford	4A
Edwards	4A
Effingham	4A
Fayette	4A
Franklin	4A
Gallatin	4A
Hamilton	4A
Hardin	4A
Jackson	4A
Jasper	4A
Jefferson	4A
Johnson	4A
Lawrence	4A
Macoupin	4A
Madison	4A
Monroe	4A
Montgomery	4A
Perry	4A
Pope	4A
Pulaski	4A
Randolph	4A
Richland	4A
Saline	4A
Shelby	4A
St. Clair	4A
Union	4A
Wabash	4A
Washington	4A
(Illinois cont.)	
Wayne	4A
White	4A
Williamson	4A
Indiana (IN)	
Zone 5A Except	
Brown	4A
Clark	4A
Crawford	4A
Daviess	4A
Dearborn	4A
Dubois	4A
Floyd	4A
Gibson	4A
Greene	4A
Harrison	4A
Jackson	4A
Jefferson	4A
Jennings	4A
Knox	4A
Lawrence	4A
Martin	4A
Monroe	4A
Ohio	4A
Orange	4A
Perry	4A
Pike	4A
Posey	4A
Ripley	4A
Scott	4A
Spencer	4A
Sullivan	4A
Switzerland	4A
Vanderburgh	4A
Warrick	4A
Washington	4A
Iowa (IA)	
Zone 5A Except	
Allamakee	6A
Black Hawk	6A
Bremer	6A
(Iowa cont.)	
Buchanan	6A
Buena Vista	6A
Butler	6A
Calhoun	6A
Cerro Gordo	6A
Cherokee	6A
Chickasaw	6A
Clay	6A
Clayton	6A
Delaware	6A
Dickinson	6A
Emmet	6A
Fayette	6A
Floyd	6A
Franklin	6A
Grundy	6A
Hamilton	6A
Hancock	6A
Hardin	6A
Howard	6A
Humboldt	6A
Ida	6A
Kossuth	6A
Lyon	6A
Mitchell	6A
O'Brien	6A
Osceola	6A
Palo Alto	6A
Plymouth	6A
Pocahontas	6A
Sac	6A
Sioux	6A
Webster	6A
Winnebago	6A
Worth	6A
Wright	6A
Kansas (KS)	
Zone 4A Except	
Cheyenne	5A
Cloud	5A

TABLE B-1 US Climate Zones *(continued)*

State County	Zone
(Kansas cont.)	
Decatur	5A
Ellis	5A
Gove	5A
Graham	5A
Greeley	5A
Hamilton	5A
Jewell	5A
Lane	5A
Logan	5A
Mitchell	5A
Ness	5A
Norton	5A
Osborne	5A
Phillips	5A
Rawlins	5A
Republic	5A
Rooks	5A
Scott	5A
Sheridan	5A
Sherman	5A
Smith	5A
Thomas	5A
Trego	5A
Wallace	5A
Wichita	5A
Kentucky (KY)	
Zone 4A	
Louisiana (LA)	
Zone 2A Except	
Bienville	3A
Bossier	3A
Caddo	3A
Caldwell	3A
Catahoula	3A
Claiborne	3A
Concordia	3A
De Soto	3A
East Carroll	3A
Franklin	3A
Grant	3A
(Louisiana cont.)	
Jackson	3A
La Salle	3A
Lincoln	3A
Madison	3A
Morehouse	3A
Natchitoches	3A
Ouachita	3A
Red River	3A
Richland	3A
Sabine	3A
Tensas	3A
Union	3A
Vernon	3A
Webster	3A
West Carroll	3A
Winn	3A
Maine (ME)	
Zone 6A Except	
Aroostook	7
Maryland (MD)	
Zone 4A Except	
Garrett	5A
Massachusetts (MA)	
Zone 5	
Michigan (MI)	
Zone 5A Except	
Alcona	6A
Alger	6A
Alpena	6A
Antrim	6A
Arenac	6A
Benzie	6A
Charlevoix	6A
Cheboygan	6A
Clare	6A
Crawford	6A
Delta	6A
Dickinson	6A
Emmet	6A
Gladwin	6A
(Michigan cont.)	
Grand Traverse	6A
Huron	6A
Iosco	6A
Isabella	6A
Kalkaska	6A
Lake	6A
Leelanau	6A
Manistee	6A
Marquette	6A
Mason	6A
Mecosta	6A
Menominee	6A
Missaukee	6A
Montmorency	6A
Newaygo	6A
Oceana	6A
Ogemaw	6A
Osceola	6A
Oscoda	6A
Otsego	6A
Presque Isle	6A
Roscommon	6A
Sanilac	6A
Wexford	6A
Baraga	7
Chippewa	7
Gogebic	7
Houghton	7
Iron	7
Keweenaw	7
Luce	7
Mackinac	7
Ontonagon	7
Schoolcraft	7
Minnesota (MN)	
Zone 6A Except	
Aitkin	7
Becker	7
Beltrami	7
Carlton	7
(Minnesota cont.)	
Cass	7
Clay	7
Clearwater	7
Cook	7
Crow Wing	7
Grant	7
Hubbard	7
Itasca	7
Kanabec	7
Kittson	7
Koochiching	7
Lake	7
Lake of the Woods	7
Mahnomen	7
Marshall	7
Mille Lacs	7
Norman	7
Otter Trail	7
Pennington	7
Pine	7
Polk	7
Red Lake	7
Roseau	7
St. Louis	7
Wadena	7
Wilkin	7
Mississippi (MS)	
Zone 3A Except	
Hancock	2A
Harrison	2A
Jackson	2A
Pearl River	2A
Stone	2A
Missouri (MO)	
Zone 4A Except	
Adair	5A
Andrew	5A
Atchison	5A
Buchanan	5A
Caldwell	5A

State County	Zone	State County	Zone	State County	Zone	State County	Zone
(Missouri cont.)		(New Jersey cont.)		(New York cont.)		(North Carolina cont.)	
Chariton	5A	Hunterdon	5A	Cattaraugus	6A	Duplin	3A
Clark	5A	Mercer	5A	Chenango	6A	Edgecombe	3A
Clinton	5A	Morris	5A	Clinton	6A	Gaston	3A
Daviess	5A	Passaic	5A	Delaware	6A	Greene	3A
Gentry	5A	Somerset	5A	Essex	6A	Hoke	3A
Grundy	5A	Sussex	5A	Franklin	6A	Hyde	3A
Harrison	5A	Warren	5A	Fulton	6A	Johnston	3A
Holt	5A	New Mexico (NM)		Hamilton	6A	Jones	3A
Knox	5A	Zone 5B Except		Herkimer	6A	Lenoir	3A
Lewis	5A	Chaves	3B	Jefferson	6A	Martin	3A
Linn	5A	Dona Ana	3B	Lewis	6A	Mecklenberg	3A
Livingston	5A	Eddy	3B	Madison	6A	Montgomery	3A
Macon	5A	Hidalgo	3B	Montgomery	6A	Moore	3A
Marion	5A	Lea	3B	Oneida	6A	New Hanover	3A
Mercer	5A	Luna	3B	Otsego	6A	Onslow	3A
Nodaway	5A	Otero	3B	Schoharie	6A	Pamlico	3A
Pike	5A	Bernalillo	4B	Schuyler	6A	Pasquotank	3A
Putnam	5A	Curry	4B	St. Lawrence	6A	Pender	3A
Ralls	5A	DeBaca	4B	Steuben	6A	Perquimans	3A
Schuyler	5A	Grant	4B	Sullivan	6A	Pitt	3A
Scotland	5A	Guadalupe	4B	Tompkins	6A	Randolph	3A
Shelby	5A	Lincoln	4B	Ulster	6A	Richmond	3A
Sullivan	5A	Quay	4B	Warren	6A	Robeson	3A
Worth	5A	Roosevelt	4B	Wyoming	6A	Rowan	3A
Montana (MT)		Sierra	4B	North Carolina (NC)		Sampson	3A
Zone 6B		Socorro	4B	Zone 4A Except		Scotland	3A
Nebraska (NE)		Union	4B	Anson	3A	Stanly	3A
Zone 5A		Valencia	4B	Beaufort	3A	Tyrrell	3A
Nevada (NV)		New York (NY)		Bladen	3A	Union	3A
Zone 5B Except		Zone 5A Except		Brunswick	3A	Washington	3A
Clark	3B	Bronx	4A	Cabarrus	3A	Wayne	3A
New Hampshire (NH)		Kings	4A	Camden	3A	Wilson	3A
Zone 6A Except		Nassau	4A	Carteret	3A	Alleghany	5A
Cheshire	5A	New York	4A	Chowan	3A	Ashe	5A
Hillsborough	5A	Queens	4A	Columbus	3A	Avery	5A
Rockingham	5A	Richmond	4A	Craven	3A	Mitchell	5A
Strafford	5A	Suffolk	4A	Cumberland	3A	Watauga	5A
New Jersey (NJ)		Westchester	4A	Currituck	3A	Yancey	5A
Zone 4A Except		Allegany	6A	Dare	3A	North Dakota (ND)	
Bergen	5A	Broome	6A	Davidson	3A	Zone 7 Except	

TABLE B-1 US Climate Zones *(continued)*

State / County	Zone
(North Dakota cont.)	
Adams	6A
Billings	6A
Bowman	6A
Burleigh	6A
Dickey	6A
Dunn	6A
Emmons	6A
Golden Valley	6A
Grant	6A
Hettinger	6A
LaMoure	6A
Logan	6A
McIntosh	6A
McKenzie	6A
Mercer	6A
Morton	6A
Oliver	6A
Ransom	6A
Richland	6A
Sargent	6A
Sioux	6A
Slope	6A
Stark	6A
Ohio (OH)	
Zone 5A Except	
Adams	4A
Brown	4A
Clermont	4A
Gallia	4A
Hamilton	4A
Lawrence	4A
Pike	4A
Scioto	4A
Washington	4A
Oklahoma (OK)	
Zone 3A Except	
Beaver	4A
Cimarron	4A
Texas	4A
Oregon (OR)	
Zone 4C Except	
Baker	5B
Crook	5B
Deschutes	5B
Gilliam	5B
Grant	5B
Harney	5B
Hood River	5B
Jefferson	5B
Klamath	5B
Lake	5B
Malheur	5B
Morrow	5B
Sherman	5B
Umatilla	5B
Union	5B
Wallowa	5B
Wasco	5B
Wheeler	5B
Pennsylvania (PA)	
Zone 5A Except	
Bucks	4A
Chester	4A
Delaware	4A
Montgomery	4A
Philadelphia	4A
York	4A
Rhode Island (RI)	
Zone 5A	
South Carolina (SC)	
Zone 3A	
South Dakota (SD)	
Zone 6A Except	
Bennett	5A
Bon Homme	5A
Charles Mix	5A
Clay	5A
Douglas	5A
Gregory	5A
Hutchinson	5A
(South Dakota cont.)	
Jackson	5A
Mellette	5A
Todd	5A
Tripp	5A
Union	5A
Yankton	5A
Tennessee (TN)	
Zone 4A Except	
Chester	3A
Crockett	3A
Dyer	3A
Fayette	3A
Hardeman	3A
Hardin	3A
Haywood	3A
Henderson	3A
Lake	3A
Lauderdale	3A
Madison	3A
McNairy	3A
Shelby	3A
Tipton	3A
Texas (TX)	
Zone 3A Except	
Anderson	2A
Angelina	2A
Aransas	2A
Atascosa	2A
Austin	2A
Bandera	2B
Bastrop	2A
Bee	2A
Bell	2A
Bexar	2A
Bosque	2A
Brazoria	2A
Brazos	2A
Brooks	2A
Burleson	2A
Caldwell	2A
(Texas cont.)	
Calhoun	2A
Cameron	2A
Chambers	2A
Cherokee	2A
Colorado	2A
Comal	2A
Coryell	2A
DeWitt	2A
Dimmit	2B
Duval	2A
Edwards	2B
Falls	2A
Fayette	2A
Fort Bend	2A
Freestone	2A
Frio	2B
Galveston	2A
Goliad	2A
Gonzales	2A
Grimes	2A
Guadalupe	2A
Hardin	2A
Harris	2A
Hays	2A
Hidalgo	2A
Hill	2A
Houston	2A
Jackson	2A
Jasper	2A
Jefferson	2A
Jim Hogg	2A
Jim Wells	2A
Karnes	2A
Kenedy	2A
Kinney	2B
Kleberg	2A
La Salle	2B
Lavaca	2A
Lee	2A
Leon	2A

TABLE B-1 US Climate Zones *(continued)*

State		**State**		**State**		**State**	
County	**Zone**	**County**	**Zone**	**County**	**Zone**	**County**	**Zone**
(Texas cont.)		(Texas cont.)		(Texas cont.)		(Texas cont.)	
Liberty	2A	Brewster	3B	Mason	3B	Hansford	4B
Limestone	2A	Callahan	3B	McCulloch	3B	Hartley	4B
Live Oak	2A	Childress	3B	Menard	3B	Hockley	4B
Madison	2A	Coke	3B	Midland	3B	Hutchinson	4B
Matagorda	2A	Coleman	3B	Mitchell	3B	Lamb	4B
Maverick	2B	Concho	3B	Motley	3B	Lipscomb	4B
McLennan	2A	Cottle	3B	Nolan	3B	Moore	4B
McMullen	2A	Crane	3B	Pecos	3B	Ochiltree	4B
Medina	2B	Crockett	3B	Presidio	3B	Oldham	4B
Milam	2A	Crosby	3B	Reagan	3B	Parmer	4B
Montgomery	2A	Culberson	3B	Reeves	3B	Potter	4B
Newton	2A	Dawson	3B	Runnels	3B	Randall	4B
Nueces	2A	Dickens	3B	Schleicher	3B	Roberts	4B
Orange	2A	Ector	3B	Scurry	3B	Sherman	4B
Polk	2A	El Paso	3B	Shackelford	3B	Swisher	4B
Real	2B	Fisher	3B	Sterling	3B	Yoakum	4B
Refugio	2A	Foard	3B	Stonewall	3B	Utah (UT)	
Robertson	2A	Gaines	3B	Sutton	3B	Zone 5B Except	
San Jacinto	2A	Garza	3B	Taylor	3B	Washington	3B
San Patricio	2A	Glasscock	3B	Terrell	3B	Box Elder	6B
Starr	2A	Hackell	3B	Terry	3B	Cache	6B
Travis	2A	Hall	3B	Throckmorton	3B	Carbon	6B
Trinity	2A	Hardeman	3B	Tom Green	3B	Daggett	6B
Tyler	2A	Haskell	3B	Upton	3B	Duchesne	6B
Uvalde	2B	Hemphill	3B	Ward	3B	Morgan	6B
Val Verde	2B	Howard	3B	Wheeler	3B	Rich	6B
Victoria	2A	Hudspeth	3B	Wilbarger	3B	Summit	6B
Walker	2A	Irion	3B	Winkler	3B	Uintah	6B
Waller	2A	Jeff Davis	3B	Armstrong	4B	Wasatch	6B
Washington	2A	Jones	3B	Bailey	4B	Vermont (VT)	
Webb	2B	Kendall	3B	Briscoe	4B	Zone 6A	
Wharton	2A	Kent	3B	Carson	4B	Virginia (VA)	
Willacy	2A	Kerr	3B	Castro	4B	Zone 4A	
Williamson	2A	King	3B	Cochran	4B	Washington (WA)	
Wilson	2A	Knox	3B	Dallam	4B	Zone 5B Except	
Zapata	2B	Lipscomb	3B	Deaf Smith	4B	Clallam	4C
Zavala	2B	Loving	3B	Donley	4B	Clark	4C
Andrews	3B	Lubbock	3B	Floyd	4B	Cowlitz	4C
Baylor	3B	Lynn	3B	Gray	4B	Grays Harbor	4C
Borden	3B	Martin	3B	Hale	4B	Jefferson	4C

TABLE B-1 US Climate Zones *(continued)*

State County	Zone	State County	Zone
(Washington cont.)		(West Virginia cont.)	
King	4C	Wayne	4A
Kitsap	4C	Wirt	4A
Lewis	4C	Wood	4A
Mason	4C	Wyoming	4A
Pacific	4C	Wisconsin (WI)	
Pierce	4C	Zone 6A Except	
Skagit	4C	Ashland	7A
Snohomish	4C	Bayfield	7A
Thurston	4C	Burnett	7A
Wahkiakum	4C	Douglas	7A
Whatcom	4C	Florence	7A
Ferry	6B	Forest	7A
Okanogan	6B	Iron	7A
Pend Oreille	6B	Langlade	7A
Stevens	6B	Lincoln	7A
West Virginia (WV)		Oneida	7A
Zone 5A Except		Price	7A
Berkeley	4A	Sawyer	7A
Boone	4A	Taylor	7A
Braxton	4A	Vilas	7A
Cabell	4A	Washburn	7A
Calhoun	4A	Wyoming (WY)	
Clay	4A	Zone 6B Except	
Gilmer	4A	Goshen	5B
Jackson	4A	Platte	5B
Jefferson	4A	Lincoln	7B
Kanawha	4A	Sublette	7B
Lincoln	4A	Teton	7B
Logan	4A	Puerto Rico (PR)	
Mason	4A	Zone 1A Except	
McDowell	4A	Barranquitas 2 SSW	2B
Mercer	4A	Cayey 1 E	2B
Mingo	4A	Pacific Islands (PI)	
Monroe	4A	Zone 1A Except	
Morgan	4A	Midway Sand Island	2B
Pleasants	4A	Virgin Islands (VI)	
Putnam	4A	Zone 1A	
Ritchie	4A		
Roane	4A		
Tyler	4A		

Table B-2 shows the climate zone numbers for a wide variety of Canadian locations. When the climate zone letter is required to determine compliance with this standard, refer to Table B-4 and the Major Climate Type Definitions in Section B2 to determine the letter (A, B, or C).

Table B-3 shows the climate zone numbers for a wide variety of other international locations besides Canada. When the climate zone letter is required to determine compliance with this standard, refer to Table B-4 and the Major Climate Type Definitions in Section B2 to determine the letter (A, B, or C).

For all international locations that are not listed either in Table B-2 or B-3, use Table B-4 and the Major Climate Type Definitions in Section B2 to determine both the climate zone letter and number.

Note: CDD50 and HDD65 values may be found in Normative Appendix D.

B2. MAJOR CLIMATE TYPE DEFINITIONS

Use the following information along with Table B-4 to determine climate zone numbers and letters for international climate zones.

Marine (C) definition—Locations meeting all four criteria:

1. Mean temperature of coldest month between 27°F and 65°F.
2. Warmest month mean <72°F.
3. At least four months with mean temperatures over 50°F.
4. Dry season in summer. The month with the heaviest precipitation in the cold season has at least three times as much precipitation as the month with the least precipitation in the rest of the year. The cold season is October through March in the Northern Hemisphere and April through September in the Southern Hemisphere.

Dry (B) definition—Locations meeting the following criteria: not marine and

$$P_{in} < 0.44 \times (TF - 19.5),$$

where

P = annual precipitation, in.; and

T = annual mean temperature, °F.

Moist (A) definition—Locations that are not marine and not dry.

TABLE B-2 Canadian Climatic Zones

Province / City	Zone	Province / City	Zone	Province / City	Zone	Province / City	Zone
Alberta (AB)		(Manitoba cont.)		Ontario (ON)		(Québec cont.)	
Calgary International A	7	Winnipeg International A	7	Belleville	6	Granby	6
Edmonton International A	7	New Brunswick (NB)		Cornwall	6	Montreal Dorval International A	6
Grande Prairie A	7	Chatham A	7	Hamilton RBG	5	Québec City A	7
Jasper	7	Fredericton A	6	Kapuskasing A	7	Rimouski	7
Lethbridge A	6	Moncton A	6	Kenora A	7	Septles A	7
Medicine Hat A	6	Saint John A	6	Kingston A	6	Shawinigan	7
Red Deer A	7	Newfoundland (NF)		London A	6	Sherbrooke A	7
British Columbia (BC)		Corner Brook	6	North Bay A	7	St Jean de Cherbourg	7
Dawson Creek A	7	Gander International A	7	Oshawa WPCP	6	St Jerome	7
Ft Nelson A	8	Goose A	7	Ottawa International A	6	Thetford Mines	7
Kamloops	5	St John's A	6	Owen Sound MOE	6	Trois Rivieres	7
Nanaimo A	5	Stephenville A	6	Peterborough	6	Val d'Or A	7
New Westminster BC Pen	5	Northwest Territories (NW)		St Catharines	5	Valleyfield	6
Penticton A	5	Ft Smith A	8	Sudbury A	7	Saskatchewan (SK)	
Prince George	7	Inuvik A	8	Thunder Bay A	7	Estevan A	7
Prince Rupert A	6	Yellowknife A	8	Timmins A	7	Moose Jaw A	7
Vancouver International A	5	Nova Scotia (NS)		Toronto Downsview A	6	North Battleford A	7
Victoria Gonzales Hts	5	Halifax International A	6	Windsor A	5	Prince Albert A	7
Manitoba (MB)		Kentville CDA	6	Prince Edward Island (PE)		Regina A	7
Brandon CDA	7	Sydney A	6	Charlottetown A	6	Saskatoon A	7
Churchill A	8	Truro	6	Summerside A	6	Swift Current A	7
Dauphin A	7	Yarmouth A	6	Québec (PQ)		Yorkton A	7
Flin Flon	7	Nunavut		Bagotville A	7	Yukon Territory (YT)	
Portage La Prairie A	7	Resolute A	8	Drummondville	6	Whitehorse A	8
The Pas A	7						

TABLE B-3 International Climate Zones

Country City (Province or Region)	Zone
Argentina	
Buenos Aires/Ezeiza	3
Cordoba	3
Tucuman/Pozo	2
Australia	
Adelaide (SA)	4
Alice Springs (NT)	2
Brisbane (AL)	2
Darwin Airport (NT)	1
Perth/Guildford (WA)	3
Sydney/KSmith (NSW)	3
Azores (Terceira)	
Lajes	3
Bahamas	
Nassau	1
Belgium	
Brussels Airport	5
Bermuda	
St. Georges/Kindley	2
Bolivia	
La Paz/El Alto	5
Brazil	
Belem	1
Brasilia	2
Fortaleza	1
Porto Alegre	2
Recife/Curado	1
Rio de Janeiro	1
Salvador/Ondina	1
Sao Paulo	2
Bulgaria	
Sofia	5
Chile	
Concepcion	4
Punta Arenas/Chabunco	6
Santiago/Pedahuel	4
China	
Shanghai/Hongqiao	3
Cuba	
Guantanamo Bay NAS (Ote.)	1
Cyprus	
Akrotiri	3
Larnaca	3
Paphos	3
Czech Republic	
Prague/Libus	5
Dominican Republic	
Santo Domingo	1
Egypt	
Cairo	2
Luxor	1
Finland	
Helsinki/Seutula	7
France	
Lyon/Satolas	4
Marseille	4
Nantes	4
Nice	4
Paris/Le Bourget	4
Strasbourg	5
Germany	
Berlin/Schoenfeld	5
Hamburg	5
Hannover	5
Mannheim	5
Greece	
Souda (Crete)	3
Thessalonika/Mikra	4
Greenland	
Narssarssuaq	7
Hungary	
Budapest/Lorinc	5
Iceland	
Reykjavik	7
India	
Ahmedabad	1
Bangalore	1
Bombay/Santa Cruz	1
Calcutta/Dum Dum	1
Madras	1
Nagpur Sonegaon	1
New Delhi/Safdarjung	1
Indonesia	
Djakarta/Halimperda (Java)	1
Kupang Penfui (Sunda Island)	1
Makassar (Celebes)	1
Medan (Sumatra)	1
Palembang (Sumatra)	1
Surabaja Perak (Java)	1
Ireland	
Dublin Airport	5
Shannon Airport	4
Israel	
Jerusalem	3
Tel Aviv Port	2
Italy	
Milano/Linate	4
Napoli/Capodichino	4
Roma/Fiumicion	4
Jamaica	
Kingston/Manley	1
Montego Bay/Sangster	1
Japan	
Fukaura	5
Sapporo	5
Tokyo	3
Jordan	
Amman	3
Kenya	
Nairobi Airport	3
Korea	
Pyonggang	5
Seoul	4
Malaysia	
Kuala Lumpur	1
Penang/Bayan Lepas	1
Mexico	
Mexico City (Distrito Federal)	3
Guadalajara (Jalisco)	1
Monterrey (Nuevo Laredo)	3
Tampico (Tamaulipas)	1
Veracruz (Veracruz)	4
Merida (Yucatan)	1
Netherlands	
Amsterdam/Schiphol	5
New Zealand	
Auckland Airport	4
Christchurch	4
Wellington	4
Norway	
Bergen/Florida	5
Oslo/Fornebu	6
Pakistan	
Karachi Airport	1
Papua New Guinea	
Port Moresby	1
Paraguay	
Asuncion/Stroessner	1
Peru	
LimaCallao/Chavez	2
San Juan de Marcona	2
Talara	2
Philippines	
Manila Airport (Luzon)	1
Poland	
Krakow/Balice	5
Romania	
Bucuresti/Bancasa	5
Russia	
Kaliningrad (East Prussia)	5
Krasnoiarsk	7
Moscow Observatory	6
Petropavlovsk	7
(Russia cont.)	
RostovNaDonu	5
Vladivostok	6
Volgograd	6
Saudi Arabia	
Dhahran	1
Riyadh	1
Senegal	
Dakar/Yoff	1
Singapore	
Singapore/Changi	1
South Africa	
Cape Town/D F Malan	4
Johannesburg	4
Pretoria	3
Spain	
Barcelona	4
Madrid	4
Valencia/Manises	3
Sweden	
Stockholm/Arlanda	6
Switzerland	
Zurich	5
Syria	
Damascus Airport	3
Taiwan	
Tainan	1
Taipei	2
Tanzania	
Dar es Salaam	1
Thailand	
Bangkok	1
Tunisia	
Tunis/El Aouina	3
Turkey	
Adana	3
Ankara/Etimesgut	4
Istanbul/Yesilkoy	4
United Kingdom	
Birmingham (England)	5
Edinburgh (Scotland)	5
Glasgow Apt (Scotland)	5
London/Heathrow (England)	4
Uruguay	
Montevideo/Carrasco	3
Venezuela	
Caracas/Maiquetia	1
Vietnam	
Hanoi/Gialam	1
Saigon (Ho Chi Minh)	1

TABLE B-4 International Climate Zone Definitions

Zone Number	Name	Thermal Criteria
1	Very Hot–Humid (1A), Dry (1B)	9000 < CDD50°F
2	Hot–Humid (2A), Dry (2B)	6300 < CDD50°F ≤ 9000
3A and 3B	Warm–Humid (3A), Dry (3B)	4500 < CDD50°F ≤ 6300
3C	Warm–Marine	CDD50°F ≤ 4500 and HDD65°F ≤ 3600
4A and 4B	Mixed–Humid (4A), Dry (4B)	CDD50°F ≤ 4500 and 3600 < HDD65°F ≤ 5400
4C	Mixed–Marine	3600 < HDD65°F ≤ 5400
5A, 5B and 5C	Cool–Humid (5A), Dry (5B), Marine (5C)	5400 < HDD65°F ≤ 7200
6A and 6B	Cold–Humid (6A), Dry (6B)	7200 < HDD65°F ≤ 9000
7	Very Cold	9000 < HDD65°F ≤ 12600
8	Subarctic	12600 < HDD65°F

(**This is a normative appendix and is part of this standard.**)

NORMATIVE APPENDIX C
METHODOLOGY FOR BUILDING ENVELOPE TRADE-OFF OPTION IN SUBSECTION 5.6

C1. MINIMUM INFORMATION

The following minimum information shall be specified for the proposed design.

C1.1 At the Building Level. The floor area, broken down by *space-conditioning categories*, shall be specified.

C1.2 At the Exterior Surface Level. The classification, gross area, orientation, *U-factor*, and exterior conditions shall be specified. For *mass walls* only: *HC* and insulation position. Each surface is associated with a *space-conditioning category* as defined in Section C1.1.

C1.3 For Fenestration. The classification, area, *U-factor*, *SHGC*, VLT, overhang *PF* for *vertical fenestration*, and width, depth, and height for *skylight wells* shall be specified. (See Figure C1.3 for definition of width, depth, and height for *skylight wells*.) Each *fenestration* element is associated with a surface (defined in Section C1.2) and has the orientation of that surface.

C1.4 For Opaque Doors. The classification, area, *U-factor*, *HC*, and insulation position shall be specified. Each *opaque door* is associated with a surface (defined in Section C1.2) and has the orientation of that surface.

C1.5 For Below-Grade Walls. The area, average depth to the bottom of the wall, and *C-factor* shall be specified. Each *below-grade wall* is associated with a *space-conditioning category* as defined in C1.1.

C1.6 For Slab-On-Grade Floor. The perimeter length and F-factor shall be specified. Each slab-on-grade floor is associated with a space-conditioning category as defined in Section C1.1.

C2. OUTPUT REQUIREMENTS

Output reports shall contain the following information.

C2.1 Tables summarizing the minimum information described in Section C1.

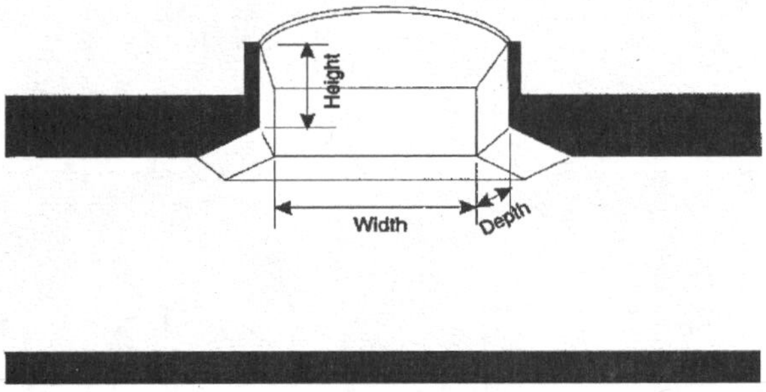

Figure C1.3 Skylight well dimensions.

C2.2 The *envelope performance factor* differential broken down by envelope component. The differential is the difference between the *envelope performance factor* of the proposed building and the *envelope performance factor* of the base envelope design. Envelope components include the *opaque roof*, *skylights*, *opaque above-grade walls* including *vertical fenestration* and *opaque doors*, *below-grade walls*, *floors*, and *slab-on-grade floors*.

C3. BASE ENVELOPE DESIGN SPECIFICATION

C3.1 The base envelope design shall have the same building floor area, *building envelope floor area*, *slab-on-grade floor* perimeter, below-grade floor area, gross *wall area*, *opaque door area*, and *gross roof area* as the proposed design. The distribution of these areas among *space-conditioning categories* shall be the same as the proposed design.

C3.2 The *U-factor* of each *opaque* element of the base envelope design shall be equal to the criteria from Tables 5.5-1 through 5.5-8 for the appropriate climate for each construction classification. *The HC* of *mass wall* elements in the base envelope design shall be identical to the proposed design. *Mass walls* in the base envelope design shall have interior insulation, when required.

C3.3 The *vertical fenestration area* of each *space-conditioning category* in the base envelope design shall be the same as the proposed building or 40% of the *gross wall area*, whichever is less. The distribution of *vertical fenestration* among *space-conditioning categories* and surface orientations shall be the same as the proposed design. If the *vertical fenestration area* of any *space-conditioning category* is greater than 40% of the *gross wall area* of that *space-conditioning category*, then the area of each *fenestration* element shall be reduced in the base envelope design by the same percentage so that the total *vertical fenestration area* is exactly equal to 40% of the *gross wall area*.

C3.4 The *skylight area* of each space category in the base envelope design shall be the same as the proposed building or 5% of the *gross roof area*, whichever is less. This distribution of *skylights* among *space-conditioning categories* shall be the same as the proposed design. If the *skylight area* of any space category is greater than 5% of the *gross roof area* of that *space-conditioning category*, then the area of each *skylight* shall be reduced in the base envelope design by the same percentage so that the total *skylight area* is exactly equal to 5% of the *gross roof area*.

C3.5 The *U-factor* for *fenestration* in the base envelope design shall be equal to the criteria from Tables 5.5-1 through 5.5-8 for the appropriate climate. The *SHGC* for *fenestration* in the base envelope design shall be equal to the criteria from Tables 5.5-1 through 5.5-8. For portions of those tables where there are no requirements, the *SHGC* shall be equal to 0.46 for all *vertical fenestration*, 0.77 for plastic *skylights* on a curb, and 0.72 for all other *skylights* with a curb and without. The VLT for *fenestration* in the base envelope design shall be the VLT factor from Table C3.5 times the *SHGC* criteria as determined in this subsection.

TABLE C3.5 VLT Factor for the Base Envelope Design

Climate Bin	Vertical Fenestration	Glass Skylights	Plastic Skylights
1(A, B)	1.00	1.27	1.20
2(A, B)	1.00	1.27	1.20
3(C)	1.00	1.27	1.20
3(A, B)	1.27	1.27	1.20
4(A, B, C)	1.27	1.27	1.20
5(A, B, C)	1.27	1.27	1.20
6(A, B)	1.27	1.27	1.20
7	1.00	1.00	1.20
8	1.00	1.00	1.20

C4. ZONING AND BUILDING GEOMETRY

No information about thermal zones needs to be entered to perform the calculations, but when the calculations are performed the building shall be divided into thermal zones according to the following procedure.

C4.1 Determine the ratio (*Rc*) of the *gross floor area* to the *gross wall area* for each *space-conditioning category.* The index "c" refers to the *space-conditioning category*, either *nonresidential conditioned*, *residential conditioned*, or *semiheated.*

C4.2 Create a perimeter zone for each unique combination of *space-conditioning category* and *wall* orientation. The *floor area* of each perimeter zone shall be the *gross wall area* of the zone times *Rc* or 1.25, whichever is smaller.

C4.3 For *space-conditioning categories* where *Rc* is greater than 1.25, interior zones shall be created and used in the trade-off procedure. The *floor area* of the interior zone shall be the total floor area for the *space-conditioning category* less the floor area of the perimeter zones created in Section C4.2 for that *space-conditioning category.*

C4.4 *Roof area*, *floor area*, *below-grade wall area*, and *slab-on-grade floor* perimeter associated with each *space-conditioning category* shall be prorated among the zones according to *floor area.*

C4.5 *Skylights* shall be assigned to the interior zone of the *space-conditioning category.* If the *skylight area* is larger than the *roof area* of the interior zone, then the *skylight area* in the interior zone shall be equal to the *roof area* in the interior zone and the remaining *skylight area* shall be prorated among the perimeter zones based on *floor area.*

C5. MODELING ASSUMPTIONS

The following are modeling assumptions for the purposes of this appendix only and are not requirements for building operation.

C5.1 The *residential conditioned* and *nonresidential conditioned space-conditioning categories* shall be modeled with both heating and cooling systems for both the base envelope design and the proposed design. The thermostat setpoints for *residential* and *nonresidential spaces* shall be 70°F for heating and 75°F for cooling, with night setback temperatures of 55°F for heating and 99°F for cooling.

C5.2 The *semiheated* space categories shall be modeled with heating-only systems for both the base envelope design and the proposed design. The thermostat setpoint shall be 50°F for all hours.

C5.3 Both the base envelope design and the proposed design shall be modeled with the same heating, ventilating, and air-conditioning (HVAC) systems. The system shall consist of a packaged rooftop system serving each thermal zone. Cooling shall be provided by a direct expansion air conditioner (EER = 9.5, $COP_{cooling}$ = 2.78). Heating shall be provided by a gas furnace (AFUE = 0.78).

C5.4 The electrical systems shall be the same for both the base envelope design and the proposed design. The *LPD* shall be 1.20 W/ft^2 for *nonresidential conditioned spaces*, 1.00 W/ft^2 for *residential conditioned spaces,* and 0.50 W/ft^2 for *semiheated spaces*. The equipment power density shall be 0.75 W/ft^2 for *nonresidential conditioned spaces*, 0.25 W/ft^2 for *residential conditioned spaces*, and 0.25 W/ft^2 for *semi-heated spaces.* Continuous daylight dimming shall be assumed in all spaces and be activated at 50 fc for *nonresidential conditioned spaces* and *residential conditioned spaces* and 30 fc for *semiheated spaces.*

C5.5 Surface reflectances for daylighting calculations shall be 80% for ceilings, 50% for walls, and 20% for floors.

C5.6 *Envelope performance factor (EPF)* is defined in the following equation.

$$\text{Envelope Performance Factor} = \frac{\text{MBtu} \times 6600 + \text{kWh} \times 80}{\text{Total Building Floor Area}} \quad \text{(C-1)}$$

C5.7 The *U-factor* entered for surfaces adjacent to crawlspaces, attics, and parking garages with mechanical or natural ventilation shall be adjusted by adding R-2 to the *thermal resistance* to account for the buffering effect.

C5.8 Heat transfer for *below-grade walls* shall be based on the temperature difference between indoor and outdoor temperature conditions and a heat transfer path at the average wall depth below grade.

C6. EQUATIONS FOR ENVELOPE TRADE-OFF CALCULATIONS

The procedure defined in this subsection shall be used in all building envelope trade-off calculations.

C6.1 Inputs. Building descriptions shall be converted to equation variables using Table C6.1.

C6.2 Envelope Performance Factor. The *EPF* of a building shall be calculated using Equation C-2.

$$EPF = FAF \times [\Sigma HVAC_{surface} + \Sigma Lighting_{zone}] \quad \text{(C-2)}$$

where

FAF = floor area factor for the entire building

TABLE C6.1 Input Variables

Variable	Description	I-P Units
$Area_{surface}$	Area of surface	ft^2
$Area_{zone}$	Gross floor area of zone as defined in Section C.5	ft^2
C-factor	C-factor for below-grade walls	$Btu/h \cdot ft^2 \cdot °F$
CDD50	CDDs	Base 50°F·day
CDD65	CDDs	Base 65°F·day
CDH80	Cooling degree-hours	Base 80°F·hour
CFA	Conditioned floor area	ft^2
Depth	Depth of bottom of below-grade wall	ft
DI	Artificial lighting design illuminance from Section C.5.4	footcandles
DR	Daily range (average outdoor maximum-minimum in hottest month)	°F
EPD	Miscellaneous equipment power density from Section C.5.4	W/ft^2
F-factor	F-factor for slab-on-grade floors	Btu/h·ft·°F
FAF	Building floor area factor	$1000/CFA$, ft^2
HC	Wall heat capacity	$Btu/ft^2 \cdot °F$
HDD50	HDDs	Base 50°F·day
HDD65	HDDs	Base 65°F·day
Length	Length of slab-on-grade floor perimeter	ft
LPD	*LPD* from Section C.5.4	W/ft^2
R-Value	Effective R-value of soil for below-grade walls	$h \cdot ft^2 \cdot °F/Btu$
U-factor	U-factor	$Btu/h \cdot ft^2 \cdot °F$
VS	Annual average daily incident solar radiation on vertical surface	$Btu/ft^2 \cdot day$

$\Sigma HVAC_{surface}$ = sum of HVAC for each surface calculated using Equation C-3

$\Sigma Lighting_{zone}$ = sum of lighting for each zone calculated using Equation C-4

C6.3 HVAC. The HVAC term for each *exterior* or *semi-exterior* surface in the building shall be calculated using Equation C-3.

$$HVAC_{surface} = COOL + HEAT \quad \text{(C-3)}$$

where

COOL = cooling factor for the surface calculated according to the appropriate equation in C-14, C-19, or C-22

HEAT = heating factor for the surface calculated according to the appropriate equation in C-16, C-18, or C-23

C6.4 Lighting. The lighting term for each zone in the building as defined in Section C4 shall be calculated using Equation C-4.

$$Lighting_{zone} = LPDadj_{zone} \times AREA_{zone} \times 216 \quad \text{(C-4)}$$

where

$LPDadj_{zone}$ = lighting power density for the zone adjusted for daylighting potential using Equation C-9

C6.5 Solar and Visible Aperture

C6.5.1 Solar and Visible Aperture of Vertical Fenestration. The visible aperture (VA), solar aperture for cooling (SA_c), and solar aperture for heating (SA_h) of each *vertical fenestration* shall be calculated using Equations C-5, C-6, and C-7.

$$VA = Area_{vf} \times VLT_{vf} \times (1 + PCC1 \times PF + PCC2 \times PF^2) \quad \text{(C-5)}$$

$$SA_c = Area_{vf} \times 1.163 \times SHGC \times (1 - PCC1 \times PF + PCC2 \times PF^2) \quad \text{(C-6)}$$

$$SA_h = Area_{vf} \times 1.163 \times SHGC \times (1 + PCH1 \times PF + PCH2 \times PF^2) \quad \text{(C-7)}$$

where

$Area_{vf}$ = *glazing area* of the *vertical fenestration*

SHGC = the *solar heat gain coefficient* of the *vertical fenestration* assembly

VLT_{vf} = the visible light transmittance of the *vertical fenestration* assembly

PF = the *projection factor* for the overhang shade on the *vertical fenestration*

PCH1, PCH2, PCC1, and PCC2 = overhang projection coefficients for the *vertical fenestration orientation* from Table C6.5.1

TABLE C6.5.1 Overhang Projection Coefficients

Orientation	PCC1	PCC2	PCH1	PCH2
North	–0.5	0.22	0	0
East, South, West	–0.97	0.38	0	0

C6.5.2 Visible Aperture of Skylights. The VA of a *skylight* shall be calculated using Equation C-8.

$$VA = Area_{sky} \times VLT_{sky} \times 10^{(-0.250 \times (5 \times D \times (W + L) / (W \times L))} \quad \text{(C-8)}$$

where

$Area_{sky}$ = *fenestration area* of the *skylight* assembly

VLT_{sky} = the visible light transmittance of the *skylight* assembly

D = average depth of skylight well from *fenestration* to ceiling

W = width of skylight well

L = length of skylight well

C6.6 Adjusted Lighting Power (*LPD*adj). The adjusted lighting power for each zone shall be calculated using Equation C-9.

$$LPDadj_{zone} = LPD \times (1 - Kd_{zone}) \quad \text{(C-9)}$$

where Kd_{zone} = daylight potential fraction calculated using Equation C-10.

If a zone has both *skylights* and *vertical fenestration*, the larger of the Kd calculated independently for each shall be used to calculate *LPD*adj.

$$Kd_{zone} = \left(\Phi1 + \left(\frac{\Phi2 \times DI \times VA}{Area_{fen}}\right)\right) \times (1 - e^{((\Phi3 + \Phi4 \times DI) \times VA)/Area_{surface}}) \quad \text{(C10)}$$

where

$Area_{fen}$ = total *fenestration area* of the *vertical fenestration* or *skylight* assemblies in the zone

VA = total visible aperture of the *vertical fenestration* or *skylights* in the zone, as calculated in Equation C-5

$Area_{surface}$ = *gross wall area* of the zone for *vertical fenestration* or *gross roof area* of the zone for *skylights*

and the coefficients 1 through 4 are defined in Table C6.6.

TABLE C6.6 Coefficients for Calculating Kd

Coefficient	Skylight	Vertical Fenestration
Φ1	0.589	0.737
Φ2	5.18E–07	–3.17E-04
Φ3	–220	–24.71
Φ4	2.29	0.234

C6.7 Delta Load Factors for Mass Walls in the Exterior Building Envelope. Adjustments to cooling and heating loads for use in Equations C-14 and C-16 due to the mass properties of each *mass wall* component shall be calculated using Equations C-11 and C-12.

$$CMC = 1.43 \times Area_{mw} \times [1 - e^{-CP_1(HC-1)}] \times \left[CP_2 + CP_3U - \left(\frac{CP_4}{1 + (CP_5 + CP_6U)e^{-(CP_7 + CP_8U^2)(HC-1)}}\right)\right] \quad \text{(C-11)}$$

where

CMC = cooling delta load factor

$Area_{mw}$ = net *opaque area* of this *mass* wall

A_c = CDH80/10000 + 2

B = DR/10 + 1

HC = wall heat capacity

DR = average daily temperature range for warmest month

CP_1 = C_5

CP_2 = $C_{15}/B^3 + C_{16}/(A_C{}^2B^2) + C_{17}$

CP_3 = $C_1/A_C{}^3 + C_2B^3 + C_2B^3 + C^3/(A_C{}^2\sqrt{B}) + C_4$

CP_4 = $C_{12}(A_C{}^2B^2) + C_{13}/B^3 + C_{14}$

U = area average of *U-factors* of *mass walls* in the zone

CP_5 = C_{18}

CP_6 = $C_6\sqrt{B}\,LN(A_C) + C_7$

LN = natural logarithm

CP_7 = $C_{19}/(A_C{}^2B^2) + C_{20}/(A_CB) + C_{21}A_C{}^2/\sqrt{B} + C_{22}$

CP_8 = $C_8/(A_C{}^2B^2) + C_9/(A_CB) + C_{10}A_C{}^2/\sqrt{B} + C_{11}$

The coefficients C_1 through C_{22} depend on insulation position in the wall and are taken from Table C6.7A.

$$HMC = 1.43 \times Area_{mw} \times [1 - e^{-HP_1(HC-1)}] \times \left[HP_2 + HP_3U - \left(\frac{HP_4}{1 + (HP_5 + HP_6U)e^{-(HP_7 + HP_8U^2)(HC-1)}}\right)\right] \quad \text{(C-12)}$$

where

HMC = heating delta load factor

HC = wall heat capacity

$Area_{mw}$ = net *opaque area* of this *mass* wall

HP_1 = H_6

A_H = HDD65/100 + 2

HP_2 = $H_{14}LN(A_H) + H_{15}$

LN = natural logarithm

HP_3 = $H_1A_H{}^3 + H_2A_H{}^2 + H_3/\sqrt{A} + H_4\sqrt{A} + H_5$

U = area average of *U-factors* of *mass walls* in the zone

HP_4 = $H_{11}A_H{}^2 + H_{12}/A_H{}^2 + H_{13}$

HP_5 = H_{16}

HP_6 = $H_7A_H + H_8$

HP_7 = $H_{17}/A_H{}^3 + H_{18}$

HP_8 = $H_9/A_H{}^3 + H_{10}$

The coefficients H_1 through H_{18} depend on the position of the insulation in the wall and are taken from Table C6.7B. If the

TABLE C6.7A Cooling Delta Load Coefficients

	Insulation Position		
Variable	Exterior	Integral	Interior
c_1	220.7245	139.1057	181.6168
c_2	–0.0566	–0.0340	–0.0552
c_3	–118.8354	–10.3267	–34.1590
c_4	–13.6744	–20.8674	–25.5919
c_5	0.2364	0.2839	0.0810
c_6	0.9596	0.3059	1.4190
c_7	–0.2550	0.0226	0.4324
c_8	–905.6780	–307.9438	–1882.9268
c_9	425.1919	80.2096	443.1958
c_{10}	–2.5106	0.0500	0.4302
c_{11}	–43.3880	–5.9895	–28.2851
c_{12}	–259.7234	–11.3961	–63.5623
c_{13}	–33.9755	0.3669	20.8447
c_{14}	20.4882	30.2535	9.8175
c_{15}	–26.2092	8.8337	24.4598
c_{16}	–241.1734	–22.2546	–70.3375
c_{17}	18.8978	29.3297	9.8843
c_{18}	–0.3538	–0.0239	–0.1146
c_{19}	156.3056	63.3228	326.3447
c_{20}	–74.0990	–16.3347	–77.6355
c_{21}	0.4454	–0.0111	–0.0748
c_{22}	7.4967	1.2956	5.2041

U-factor of *mass wall* is greater than 0.4 Btu/(h·ft²·°F), then the *U-factor* shall be set to 0.4 Btu/(h·ft²·°F). If the *U-factor* of the *mass wall* is less than 0.05 Btu/(h·ft²·°F), then the *U-Factor* shall be set to 0.05 Btu/(h·ft²·°F). If the wall *HC* of the *mass wall* is greater than 20 Btu/(ft²·°F), then *HC* = 20 Btu/(ft²·°F) shall be used.

C6.8 Walls and Vertical Fenestration in the Exterior Building Envelope. Equations C-14 and C-16 shall be used to calculate COOL and HEAT for *exterior walls* and *vertical fenestration* in the *exterior building envelope* except walls next to crawlspaces, attics, and parking garages with natural or mechanical ventilation. *Walls* next to crawlspaces, attics, and parking garages with natural or mechanical ventilation shall use the equations in Section C6.10 and they shall not be included in calculations in Section C6.8. Zones shall be constructed according to Section C4 and the HEAT and COOL for the combination of all *exterior walls* and *vertical fenestration* in the zone shall be calculated using Equations C-14 and C-16, which include interactive effects. For a zone having a cardinal *orientation* (north, east, south, or west), Equations C-14 and C-15 shall be applied directly. For zones with northeast, northwest, southwest, and southeast *orientations*, EC shall be determined by finding the average of the values for the two closest cardinal *orientations*; for instance, COOL for a *wall* facing northeast is calculated by taking the average of COOL for a north-facing wall and COOL for an east-facing *wall*.

TABLE C6.7B Heating Delta Load Coefficients

	Insulation Position		
Variable	Exterior	Integral	Interior
H_1	0.0000	0.0000	0.0000
H_2	–0.0015	–0.0018	–0.0015
H_3	13.3886	15.1161	19.8314
H_4	1.9332	2.1056	1.4579
H_5	–11.8967	–13.3053	–15.5620
H_6	0.4643	0.1840	0.0719
H_7	0.0094	0.0255	0.0264
H_8	–0.1000	0.0459	0.7754
H_9	–1223.3962	–622.0801	0.2008
H_{10}	–0.9454	–0.5192	-0.6379
H_{11}	–0.0001	–0.0001	0.0000
H_{12}	3.8585	4.1379	2.4243
H_{13}	7.5829	6.2380	7.9804
H_{14}	–0.7774	–0.7711	–0.1699
H_{15}	9.0147	7.7229	8.5854
H_{16}	0.2007	0.2083	–0.0386
H_{17}	206.6382	105.9849	3.1397
H_{18}	0.2573	0.1983	0.1863

C6.8.1 Effective Internal Gain. The effective internal gain to zone G shall be calculated using Equation C-13.

$$G = EPD + LPDadj_{zone} \quad \text{(C-13)}$$

where

$LPDadj_{zone}$ = lighting power density adjusted for daylighting, from Equation C-9

C6.8.2 Cooling Factor. The cooling factor for the surfaces in the zone shall be calculated using Equation C-14.

$$COOL = 0.005447 \times [CLU + CLUO + CLXUO + CLM + CLG + CLS + CLC] \quad \text{(C-14)}$$

where

$$CLU = Area_{opaque} \times U_{ow} \times [CU1 \times CDH80 + CU2 \times CDH80^2 + CU3 \times (VS \times CDH80)^2 + CU4 \text{ x } DR]$$

$$CLUO = Area_{grosswall} \times UO \times [CUO1 \times EA_C \times VS \times CDD50 + CUO2 \times G + CUO3 \times G^2 \times EA_C^2 \times VS \times CDD50 + CUO4 \times G^2 \times EA_C^2 \times VS \times CDD65]$$

$$CLXUO = Area_{grosswall} / UO \times [CXUO1 \times EA_C \times VS \times CDD50 + CXUO2 \times EA_C \times (VS \times CDD50)^2 + CXUO3 \times G \times CDD50 + CXUO4 \times G^2 \times EA_C^2 \times VS \times CDD50 + CXUO5 \times G^2 \times CDD65]$$

$$CLM = Area_{opaque} \times SCMC \times [CM1 + CM2 \times EA_C \times VS \times CDD50 + CM3 \times EA_C \times VS \times CDD65 + CM4 \times EA_C^2 \times VS \times CDD50 + CM5 \times G^2 \times CDD65 + CM6 \times G \times CDD50 + CM7 \times G \times CDD65 + CM8 \times G \times EA_C \times VS \times CDD50]$$

$$CLG = Area_{grosswall} \times \{G \times [CG1 + CG2 \times CDD50 + CG3 \times EA_C \times (VS \times CDD50)^2 + CG4 \times EA_C^2 \times VS \times CDD50 + CG5 \times CDD65 + CG6 \times CDD50^3 + CG7 \times CDD65^3] + G^2 \times [CG8 \times EA_C \times VS \times CDD50 + CG9 \times EA_C^2 \times VS \times CDD50]\}$$

$$CLS = Area_{grosswall} \times \{EA_C \times [CS1 + CS2 \times VS \times CDD50 + CS3 \times (VS \times CDD50)^2 + CS4 \times VS \times CDD65 + CS5 \times (VS \times CDD65)^2] + EA_C^2 \times [CS6 + CS7 \times (VS \times CDD65)^2]\}$$

$$CLC = Area_{grosswall} \times [CC1 \times CDD50 + CC2 \times CDD50^2 + CC3 \times CDH80 + CC4 \times CDH80^2 + CC5 \times CDD65 + CC6 \times (VS \times CDD65)^2 + CC7 \times VS \times CDD50 + CC8 \times (VS \times CDD50)^2 + CC9 \times (VS \times CDH80)^2 + CC10 \times VS + CC11 \times DR + CC12 \times DR^2 + CC13]$$

where

$Area_{grosswall}$	=	total gross area of all *walls* and *vertical fenestration* in the zone, including *opaque* and *fenestration areas*
$Area_{opaque}$	=	total *opaque area* of all *walls* in the zone
U_{ow}	=	area average of *U-factors* of *opaque walls* (including those of mass construction) in the zone
VS	=	annual average daily incident solar energy on surface
DR	=	average daily temperature range for the warmest month
UO	=	area average of *U-factor* of *opaque walls* and *vertical fenestration* in the zone
SCMC	=	sum of the CMC from Equation C-11 for each *mass wall* in the zone
G	=	effective internal gain to space, from Equation C-13
EA_C	=	effective solar aperture fraction for zone calculated using Equation C-15

$$EA_C = \frac{\sum SA_C}{Area_{grosswall}} \qquad \text{(C-15)}$$

where

ΣSA_c = the sum of SA_c from Equation C-6.6 for all *vertical fenestration* in the zone.

The coefficients used in the above equations depend on the *orientation* of the surface and shall be found in Table C6.8.2.

C6.8.3 Heating Factor. The heating factor for the surfaces in the zone shall be calculated using Equation C-16.

$$HEAT = 0.007669 \times [HLU + HLUO + HLXUO + HLM + HLG + HLS + HLC] \qquad \text{(C-16)}$$

where

$$HLU = Area_{opaque} \times U_{ow} \times [HU1 \times HDD50 + HU2 \times (VS \times HDD65)^2]$$

$$HLUO = Area_{grosswall} \times UO \times [HUO1 \times HDD50 + HUO2 \times HDD65 + HUO3 \times EA_H \times VS \times HDD65]$$

$$HLXUO = Area_{grosswall} \times \{(1/UO) \times [HXUO1 \times EA_H \times (VS \times HDD50)^2 + HXUO2 \times EA_H \times (VS \times HDD65)^2] + (1/UO^2) \times [HXUO3 \times EA_H^2 \times VS \times HDD65]\}$$

$$HLM = Area_{opaque} \times SHMC \times [HM1 + HM2 \times G \times UO \times HDD65 + HM3 \times G^2 \times EA_H^2 \times VS \times HDD50 + HM4 \times UO \times EA_H \times VS \times HDD65 + HM5 \times UO \times HDD50 + HM6 \times EA_H \times (VS \times HDD65)^2 + HM7 \times EA_H^2 \times VS \times HDD65/UO]$$

$$HLG = Area_{grosswall} \times \{G \times [HG1 \times HDD65 + HG2 \times UO \times HDD65 + HG3 \times EA_H \times VS \times HDD65 + HG4 \times EA_H^2 \times VS \times HDD50] \times G^2 \times [HG5 \times HDD65 + HG6 \times EA_H^2 \times VS \times HDD65]\}$$

$$HLS = Area_{grosswall} \times \{EA_H \times [HS1 \times VS \times HDD65 + HS2 \times (VS \times HDD50)^2] + EA_H^2 \times [HS3 \times VS \times HDD50 + HS4 \times VS \times HDD65]\}$$

$$HLC = Area_{grosswall} \times [HC1 + HC2 \times HDD65 + HC3 \times HDD65^2 + HC4 \times VS^2 + HC5 \times VS \times HDD50 + HC6 \times VS \times HDD65 + HC7 \times (VS \times HDD50)^2]$$

where

VS	=	annual average daily incident solar energy on surface
SHMC	=	sum of the HMC from Equation C-12 for each *mass wall* in the zone
EA_H	=	effective solar aperture fraction for zone calculated using Equation C-17.

$$EA_H = \frac{\sum SA_H}{Area_{grosswall}} \qquad \text{(C-17)}$$

ΣSA_h = the sum of SA_h from Equation C-7 for all *vertical fenestration* in the zone.

The coefficients used in the above equations depend on the *orientation* of the surface and shall be found in Table C6.8.3. Terms not defined for Equation C-16 are found under Equation C-14.

C6.9 Skylights in the Exterior Building Envelope. HEAT and COOL shall be calculated for *skylights* in *nonresidential conditioned* and *residential conditioned* zones using Equations C-18 and C-19.

$$HEAT = Area_{sky} \times HDD65 \times 0.66 \times (H_2 \times U_{sky} + H_3 \times 1.163 \times SHGC) \qquad \text{(C-18)}$$

$$COOL = Area_{sky} \times C_2 \times CDD50 \times 0.093 \times SHGC \qquad \text{(C-19)}$$

where

$Area_{sky}$	=	*fenestration* area of the *skylight* assembly
SHGC	=	the *solar heat gain coefficient* of the *skylight* assembly
U_{sky}	=	*U-factor* of *skylight* assembly

The coefficients used in the equations depend on the space type and shall be taken from Table C6.9.

TABLE C6.8.2 Cooling Coefficients for the Exterior Wall Equation

Variable	Orientation of Surface			
	North	East	South	West
CU1	0.001539	0.003315	0.003153	0.00321
CU2	–3.0855E-08	–8.9662E-08	–7.1299E-08	–8.1053E-08
CU3	7.99493E-14	3.7928E-14	1.83083E-14	3.3981E-14
CU4	–0.079647	0.163114	0.286458	0.11178
CM1	0.32314	0.515262	0.71477	0.752643
CM2	1.5306E-06	1.38197E-06	1.6163E-06	1.42228E-06
CM3	–2.0432E-06	–1.6024E-06	–2.1106E-06	–1.9794E-06
CM4	–7.5367E-07	–7.6785E-07	–6.6443E-07	–7.4007E-07
CM5	–1.0047E-06	0	8.01057E-06	3.15193E-06
CM6	3.66708E-05	3.56503E-05	4.48106E-05	2.96012E-05
CM7	–6.7305E-05	–6.4094E-05	–0.000119	–7.6672E-05
CM8	–2.3834E-08	–4.7253E-08	–4.9747E-08	0
CUO1	–6.5109E-06	–8.3867E-06	–8.89E-06	–7.5647E-06
CUO2	–1.040207	–1.507235	–1.512625	-1.238545
CUO3	–4.3825E-06	–2.7883E-06	–2.3135E-06	–4.1257E-06
CUO4	0.000012658	8.09874E-06	7.36219E-06	1.06712E-05
CXUO1	1.03744E-06	1.19338E-06	1.18588E-06	1.23251E-06
CXUO2	–1.3218E-13	–1.3466E-13	–1.1625E-13	–1.3E-13
CXUO3	2.75554E-05	2.02621E-05	2.02365E-05	2.36964E-05
CXUO4	9.7409E-08	1.175E-07	9.39207E-08	1.36276E-07
CXUO5	–1.1825E-05	–9.0969E-06	–9.0919E-06	–1.1108E-05
CG1	0.891286	0.583388	0.393756	0.948654
CG2	0.001479	0.001931	0.002081	0.001662
CG3	–5.5204E-13	–2.8214E-13	–2.8477E-13	–4.5572E-13
CG4	2.52311E-06	3.70821E-06	4.30536E-06	5.91511E-06
CG5	–0.001151	–0.001745	–0.001864	–0.00153
CG6	1.95243E-12	0	–2.9606E-12	3.16358E-12
CG7	–8.3581E-12	1.01089E-11	3.30027E-11	0
CG8	1.41022E-06	7.53875E-07	7.133E-07	9.70752E-07
CG9	–2.3889E-06	–1.6496E-06	–1.6393E-06	–1.9736E-06
CS1	46.9871	33.9683	18.32016	29.3089
CS2	3.48091E-05	3.74118E-05	0.000034049	5.02498E-05
CS3	0	0	2.71313E-12	0
CS4	–1.6641E-05	6.94779E-06	–2.8218E-05	–2.7716E-05
CS5	8.42765E-12	0	–3.0468E-12	2.91137E-12
CS6	–56.5446	0	26.9954	14.9771
CS7	–1.3476E-11	–5.881E-12	–6.5009E-12	–7.8922E-12
CC1	0.002747	0	0.010349	0.001865
CC2	0	3.18928E-07	–3.0441E-07	0
CC3	–0.000348	0.000319	0.00024	0.000565
CC4	1.22123E-08	–7.7532E-08	–2.7144E-08	–5.4438E-08
CC5	0.012112	0.011894	0.013248	0.009236
CC6	1.04027E-12	–6.2266E-13	–2.0518E-12	0
CC7	–1.2401E-05	–7.0628E-06	–1.6538E-05	–6.0269E-06
CC8	0	0	8.20869E-13	0
CC9	–3.758E-14	6.06235E-14	1.97598E-14	3.89425E-14
CC10	0.030056	0.023121	0.0265	0.01704
CC11	0	0	–0.271026	–0.244274
CC12	0.002138	0.001103	0.006368	0.007323
CC13	–12.8674	–13.16522	–18.271	–10.1285

TABLE C6.8.3 Heating Coefficients for the Exterior Wall Equation

Variable	Orientation of Surface			
	North	East	South	West
HU1	0.006203	0.007691	0.006044	0.006672
HU2	–1.3587E-12	–5.7162E-13	–2.69E-13	–4.3566E-13
HM1	0.531005	0.545732	0.837901	0.616936
HM2	0.000152	0.000107	0.000208	0.00015
HM3	–5.3183E-07	–1.0619E-07	–6.8253E-07	–2.6457E-07
HM4	–7.7381E-07	–1.4787E-06	2.11938E-06	–4.5783E-07
HM5	–0.000712	–0.000484	–0.001042	–0.000625
HM6	3.34859E-13	4.95762E-14	7.7019E-14	7.37105E-14
HM7	2.39071E-07	2.75045E-07	–3.8989E-07	0
HUO1	0.004943	0.008683	0.009028	0.008566
HUO2	0.013686	0.011055	0.010156	0.01146
HUO3	–1.1018E-05	–8.6896E-06	–7.3232E-06	–8.9867E-06
HXUO1	1.2694E-12	7.85644E-14	–2.8202E-13	3.04904E-14
HXUO2	–7.3058E-13	–8.109E-14	7.45599E-14	–7.4718E-14
HXUO3	1.9709E-07	1.94026E-07	9.87587E-08	1.95776E-07
HG1	–0.001051	–0.000983	–0.000981	–0.000948
HG2	–0.001063	–0.00093	–0.000815	–0.000975
HG3	2.99013E-06	2.62269E-06	2.4188E-06	2.49976E-06
HG4	7.49049E-07	–1.1106E-06	–2.1669E-06	–8.5605E-07
HG5	0.000109	0.000093431	9.75523E-05	8.62389E-05
HG6	–5.5591E-07	–3.158E-07	–2.61E-07	–2.9133E-07
HS1	–2.1825E-05	–2.0922E-05	–2.1089E-05	–2.0205E-05
HS2	3.39179E-12	1.905E-12	1.48388E-12	2.18215E-12
HS3	–6.5325E-06	–2.2341E-05	–1.8473E-05	–2.4049E-05
HS4	2.23087E-05	2.41331E-05	2.45412E-05	2.30538E-05
HC1	–0.106468	–5.19297	–3.66743	–5.29681
HC2	0.00729	0.007684	0.007175	0.007672
HC3	–2.976E-07	–3.0784E-07	–2.6419E-07	–3.0713E-07
HC4	2.01569E-06	6.3035E-06	3.32112E-06	6.43491E-06
HC5	1.29061E-05	4.77552E-06	3.25089E-06	4.83233E-06
HC6	–1.2859E-05	–6.1854E-06	–4.6309E-06	–6.251E-06
HC7	2.75861E-12	8.20051E-13	4.38148E-13	8.09106E-13

C6.10 Calculations for Other Exterior and Semi-Exterior Surfaces. For all *exterior* and *semi-exterior* surfaces not covered in Sections C6.8 and C6.9, the cooling factor, COOL, and heating factor, HEAT, shall be calculated using the procedure in this section.

C6.10.1 U-Factor for Below-Grade Walls. The effective *U-factor* of *below-grade walls* shall be calculated using Equation C-20. R_{soil} shall be selected from Table C6.10.1 based on the average depth of the bottom of the wall below the surface of the ground.

$$\text{U-factor} = 1 / ((1/\text{C-factor}) + 0.85 + R_{soil}) \quad \text{(C-20)}$$

where

R_{soil} = effective R-value of the soil from Table C6.10.1

C6.10.2 Adjustment for Other Protected Elements of the Exterior Envelope. The adjusted *U-factor* for *exterior envelope* surfaces, which are protected from outdoor conditions by crawlspaces, attics, or parking garages with natural or mechanical ventilation, shall be adjusted using Equation C-21 before calculating HEAT and COOL.

$$U_{adj} = 1 / ((1 / \text{U-factor}) + 2) \quad \text{(C-21)}$$

C6.10.3 Calculation of COOL and HEAT. COOL and HEAT shall be calculated for each surface using Equations C-22 and C-23 and coefficients from Table C6.10.2, which depend on surface classification and *space-conditioning category*.

$$\text{COOL} = \text{Size} \times \text{Factor} \times 0.08 \times (\text{Ccoef1} \times \text{CDD50} + \text{Ccoef2}) \quad \text{(C-22)}$$

$$\text{HEAT} = \text{Size} \times \text{Hcoef} \times \text{Factor} \times \text{HDD65} \times 0.66 \quad \text{(C-23)}$$

where

Size = area of surface or length of exposed *slab-on-grade floor* perimeter in the building

Ccoef1, Ccoef2 = coefficients, from Table C6.10.2

Hcoef = coefficient from Table C6.10.2

Factor = *U-factor* except U_{adj} calculated using Equation C-21 for protected surfaces and for *slab-on-grade floors*, perimeter *F-factor*

TABLE C6.9 Heating and Cooling Coefficients for Skylights

Coefficient	Nonresidential	Residential
C_2	1.09E-02	1.64E-02
H_2	2.12E-04	2.91E-04
H_3	–1.68E-04	–2.96E-04

TABLE C6.10.1 Effective R-Value of Soil for Below-Grade Walls

Depth, ft	R_{soil} (h·ft²·°F/Btu)
1	0.86
2	1.6
3	2.2
4	2.9
5	3.4
6	4.0
7	4.5
8	5.1
9	5.6
10	6.1

TABLE C6.10.2 Heating and Cooling Coefficients for Other Exterior and Semi-Exterior Surfaces

Building Envelope Classification	Exterior						Semi-Exterior		
Space-Conditioning Type	Nonresidential			Residential			All		
Surface Type	Ccoef1	Ccoef2	Hcoef	Ccoef1	Ccoef2	Hcoef	Ccoef1	Ccoef2	Hcoef
Roof	0.001153	5.56	2.28E-04	0.001656	9.44	3.37E-04	0	0	8.08E-05
Wall, above-grade, and opaque doors	6.04E-04	0	2.28E-04	1.18E-03	0	3.37E-04	0	0	7.56E-05
Wall, below-grade	2.58E-04	0	2.29E-04	6.80E-04	0	3.35E-04	N/A	0	7.85E-05
Mass floor	6.91E-04	0	2.39E-04	1.01E-03	0	3.60E-04	0	0	7.14E-05
Other floor	7.09E-04	0	2.43E-04	9.54E-04	0	3.66E-04	0	0	7.14E-05
Slab-on-grade floor	0	0	2.28E-04	0	0	3.37E-04	0	0	6.80E-05
Vertical fenestration	N/A	0	N/A	N/A	0	N/A	0	0	7.56E-05
Skylights	N/A	0	N/A	N/A	0	N/A	0	0	8.08E-05

(This is a normative appendix and is part of this standard.)

NORMATIVE APPENDIX D
CLIMATIC DATA

This normative appendix contains the climatic data necessary to determine building envelope and mechanical requirements for various US, Canadian, and international locations. (See Section 5.1.4 for additional information regarding the selection of climatic data.) The following definition applies: N.A. = Not Available.

TABLE D-1 US and US Territory Climatic Data

State City	Latitude	Longitude	Elev., ft	HDD65	CDD50	Heating Design Temperature	Cooling Design Temperature Dry-Bulb	Cooling Design Temperature Wet-Bulb	No. Hrs. 8 a.m.–4 p.m.
						99.6%	1.0%	1.0%	55 < Tdb < 69
Alabama (AL)									
Alexander City	32.95 N	85.93 W	640	2910	5102	N.A.	N.A.	N.A.	N.A.
Anniston FAA AP	33.58 N	85.85 W	611	2854	5217	19	93	76	N.A.
Auburn Agronomy Farm	32.60 N	85.50 W	652	2612	5428	N.A.	N.A.	N.A.	N.A.
Birmingham FAA AP	33.57 N	86.75 W	625	2918	5206	18	92	75	760
Dothan	31.32 N	85.45 W	400	1703	6659	28	93	76	N.A.
Gadsden Steam Plant	34.03 N	86.00 W	565	3317	4805	N.A.	N.A.	N.A.	N.A.
Huntsville WSO AP	34.65 N	86.77 W	624	3323	4855	15	92	74	N.A.
Mobile WSO AP	30.68 N	88.25 W	211	1702	6761	26	92	76	774
Montgomery WSO AP	32.30 N	86.40 W	221	2224	5990	24	93	76	734
Selma	32.42 N	87.00 W	147	2249	6080	N.A.	N.A.	N.A.	N.A.
Talladega	33.43 N	86.08 W	555	2790	5097	N.A.	N.A.	N.A.	N.A.
Tuscaloosa FAA AP	33.23 N	87.62 W	169	2661	5624	20	94	77	N.A.
Alaska (AK)									
Anchorage WSCMO AP	61.17 N	150.02 W	114	10,570	688	–14	68	57	521
Barrow WSO AP	71.30 N	156.78 W	31	20,226	0	–41	52	49	N.A.
Fairbanks WSFO AP	64.82 N	147.87 W	436	13,940	1040	–47	77	59	682
Juneau AP	58.37 N	134.58 W	12	8897	559	4	69	58	540
Kodiak WSO AP	57.75 N	152.50 W	111	8817	451	7	65	56	384
Nome WSO AP	64.50 N	165.43 W	13	14,129	274	-31	65	55	210
Arizona (AZ)									
Douglas FAA AP	31.47 N	109.60 W	4098	2767	4786	N.A.	N.A.	N.A.	N.A.
Flagstaff WSO AP	35.13 N	111.67 W	7006	7131	1661	1	83	55	N.A.
Kingman	35.20 N	114.02 W	3539	3212	5040	22	97	63	N.A.
Nogales	31.42 N	110.95 W	3560	2928	4554	N.A.	N.A.	N.A.	N.A.
Phoenix WSFO AP	33.43 N	112.02 W	1110	1350	8425	34	108	70	746

TABLE D-1 US and US Territory Climatic Data *(continued)*

State City	Latitude	Longitude	Elev., ft	HDD65	CDD50	Heating Design Temperature	Cooling Design Temperature Dry-Bulb	Cooling Design Temperature Wet-Bulb	No. Hrs. 8 a.m.–4 p.m.
						99.6%	1.0%	1.0%	55 < Tdb < 69
(Arizona cont.)									
Prescott	34.57 N	112.43 W	5205	4995	2875	15	91	60	725
Tucson WSO AP	32.13 N	110.93 W	2584	1678	6921	31	102	65	716
Winslow WSO AP	35.02 N	110.73 W	4890	4776	3681	10	93	60	634
Yuma WSO AP	32.67 N	114.60 W	206	927	8897	40	109	72	697
Arkansas (AR)									
Blytheville AFB	35.97 N	89.95 W	256	3656	5133	12	95	77	N.A.
Camden	33.60 N	92.82 W	116	2953	5309	N.A.	N.A.	N.A.	N.A.
Fayetteville	36.00 N	94.17 W	1250	4040	4452	6	93	75	N.A.
Ft Smith WSO AP	35.33 N	94.37 W	449	3478	5078	13	96	76	547
Hot Springs	34.52 N	93.05 W	680	3181	5243	N.A.	N.A.	N.A.	N.A.
Jonesboro	35.88 N	90.70 W	390	3504	5118	N.A.	N.A.	N.A.	N.A.
Little Rock FAA AP	34.73 N	92.23 W	257	3155	5299	16	95	77	626
Pine Bluff	34.22 N	92.02 W	215	3016	5467	N.A.	N.A.	N.A.	N.A.
Texarkana FAA AP	33.45 N	94.00 W	361	2295	6152	20	95	77	N.A.
California (CA)									
Bakersfield WSO AP	35.42 N	119.05 W	495	2182	6049	32	101	69	848
Blythe FAA Airport	33.62 N	114.72 W	390	1144	8789	N.A.	N.A.	N.A.	N.A.
Burbank Hollywood	34.20 N	118.37 W	774	1204	5849	39	95	69	N.A.
Chico University Farm	39.70 N	121.82 W	185	2953	4454	N.A.	N.A.	N.A.	N.A.
Crescent City	41.77 N	124.20 W	40	4397	1628	N.A.	N.A.	N.A.	N.A.
El Centro	32.77 N	115.57 W	–30	1156	8132	N.A.	N.A.	N.A.	N.A.
Eureka WSO City	40.80 N	124.17 W	60	4496	1529	N.A.	N.A.	N.A.	N.A.
Fairfield/Travis AFB	38.27 N	121.93 W	62	2556	4223	31	94	67	N.A.
Fresno WSO AP	36.77 N	119.72 W	328	2556	5350	30	101	70	785
Laguna Beach	33.55 N	117.78 W	35	2157	3881	N.A.	N.A.	N.A.	N.A.
Livermore	37.67 N	121.77 W	480	2909	3810	N.A.	N.A.	N.A.	N.A.
Lompoc	34.65 N	120.45 W	95	2651	3240	N.A.	N.A.	N.A.	N.A.
Long Beach WSO AP	33.82 N	118.15 W	34	1430	5281	40	88	67	1502
Los Angeles WSO AP	33.93 N	118.38 W	100	1458	4777	43	81	64	1849
Merced/Castle AFB	37.37 N	120.57 W	187	2687	4694	30	97	69	N.A.
Monterey	36.60 N	121.90 W	385	3125	2574	N.A.	N.A.	N.A.	N.A.

TABLE D-1 US and US Territory Climatic Data *(continued)*

State City	Latitude	Longitude	Elev., ft	HDD65	CDD50	Heating Design Temperature	Cooling Design Temperature Dry-Bulb	Cooling Design Temperature Wet-Bulb	No. Hrs. 8 a.m.–4 p.m.
						99.6%	1.0%	1.0%	55 < Tdb < 69
(California cont.)									
Napa State Hospital	38.28 N	122.27 W	60	2844	3463	N.A.	N.A.	N.A.	N.A.
Needles FAA Airport	34.77 N	114.62 W	914	1309	8645	N.A.	N.A.	N.A.	N.A.
Oakland/Intl	37.73 N	122.20 W	7	2644	3126	N.A.	N.A.	N.A.	1905
Oceanside Marina	33.22 N	117.40 W	10	2010	4069	N.A.	N.A.	N.A.	N.A.
Ontario/Intl	34.05 N	117.62 W	961	1488	5823	35	98	70	N.A.
Oxnard	34.20 N	119.18 W	49	1992	3980	39	79	64	N.A.
Palm Springs	33.83 N	116.50 W	425	985	8555	N.A.	N.A.	N.A.	N.A.
Palmdale	34.58 N	118.10 W	2596	2948	4863	N.A.	N.A.	N.A.	N.A.
Pasadena	34.15 N	118.15 W	864	1453	5476	N.A.	N.A.	N.A.	N.A.
Petaluma Fire Stn 3	38.23 N	122.63 W	27	3050	3188	N.A.	N.A.	N.A.	N.A.
Pomona Cal Poly	34.07 N	117.82 W	740	1713	5145	N.A.	N.A.	N.A.	N.A.
Redding WSO	40.50 N	122.30 W	502	2855	4964	N.A.	N.A.	N.A.	N.A.
Redlands	34.05 N	117.18 W	1318	1875	5435	N.A.	N.A.	N.A.	N.A.
Richmond	37.93 N	122.35 W	55	2574	3285	N.A.	N.A.	N.A.	N.A.
Riverside/March AFB	33.90 N	117.25 W	1535	1861	5295	34	98	68	N.A.
Sacramento FAA AP	38.52 N	121.50 W	18	2749	4474	30	97	68	990
Salinas FAA AP	36.67 N	121.60 W	69	2964	2951	33	78	62	N.A.
San Bernardino/Norton	34.10 N	117.23 W	1155	1821	5450	34	101	70	N.A.
San Diego WSO AP	32.73 N	117.17 W	13	1256	5223	44	81	67	1911
San Francisco WSO AP	37.62 N	122.38 W	8	3016	2883	37	78	62	1796
San Jose	37.35 N	121.90 W	67	2387	3935	35	89	66	N.A.
San Luis Obispo Poly	35.30 N	120.67 W	315	2498	3492	N.A.	N.A.	N.A.	N.A.
Santa Ana Fire Station	33.75 N	117.87 W	135	1238	5430	N.A.	N.A.	N.A.	N.A.
Santa Barbara FAA AP	34.43 N	119.83 W	9	2438	3449	34	80	64	N.A.
Santa Cruz	36.98 N	122.02 W	130	2969	2913	N.A.	N.A.	N.A.	N.A.
Santa Maria WSO AP	34.90 N	120.45 W	254	2984	2918	32	82	62	2016
Santa Monica Pier	34.00 N	118.50 W	14	1819	4145	N.A.	N.A.	N.A.	N.A.
Santa Paula	34.32 N	119.15 W	237	2039	4114	N.A.	N.A.	N.A.	N.A.
Santa Rosa	38.45 N	122.70 W	167	2883	3432	N.A.	N.A.	N.A.	N.A.
Stockton WSO AP	37.90 N	121.25 W	22	2707	4755	30	97	68	N.A.
Ukiah	39.15 N	123.20 W	623	2954	3868	N.A.	N.A.	N.A.	N.A.

TABLE D-1 US and US Territory Climatic Data *(continued)*

State City	Latitude	Longitude	Elev., ft	HDD65	CDD50	Heating Design Temperature	Cooling Design Temperature		No. Hrs. 8 a.m.–4 p.m.
							Dry-Bulb	Wet-Bulb	
						99.6%	1.0%	1.0%	55 < Tdb < 69
(California cont.)									
Visalia	36.33 N	119.30 W	325	2511	5186	N.A.	N.A.	N.A.	N.A.
Yreka	41.72 N	122.63 W	2625	5386	2611	N.A.	N.A.	N.A.	N.A.
Colorado (CO)									
Alamosa WSO AP	37.45 N	105.87 W	7536	8749	1374	–17	82	55	N.A.
Boulder	40.03 N	105.28 W	5420	5554	2820	N.A.	N.A.	N.A.	N.A.
Colorado Sprgs WSO AP	38.82 N	104.72 W	6090	6415	2312	–2	87	58	725
Denver WSFO AP	39.77 N	104.87 W	5286	6020	2732	–3	90	59	739
Durango	37.28 N	107.88 W	6600	6911	1942	N.A.	N.A.	N.A.	N.A.
Ft Collins	40.58 N	105.08 W	5004	6368	2411	N.A.	N.A.	N.A.	N.A.
Grand Junction WSO AP	39.10 N	108.55 W	4849	5548	3632	2	94	60	518
Greeley UNC	40.42 N	104.70 W	4715	6306	2698	N.A.	N.A.	N.A.	N.A.
La Junta FAA AP	38.05 N	103.52 W	4190	5265	3795	N.A.	N.A.	N.A.	N.A.
Pueblo WSO AP	38.28 N	104.52 W	4640	5413	3358	–1	94	62	720
Sterling	40.62 N	103.22 W	3938	6541	2809	N.A.	N.A.	N.A.	N.A.
Trinidad FAA AP	37.25 N	104.33 W	5746	5483	2976	–2	90	60	N.A.
Connecticut (CT)									
Bridgeport WSO AP	41.17 N	73.13 W	10	5537	2997	8	84	72	N.A.
Hartford-Brainard Fld	41.73 N	72.65 W	15	6155	2768	2	88	72	598
Norwalk Gas Plant	41.12 N	73.42 W	37	5865	2768	N.A.	N.A.	N.A.	N.A.
Norwich Pub Util Plt	41.53 N	72.07 W	20	5869	2687	N.A.	N.A.	N.A.	N.A.
Delaware (DE)									
Dover	39.15 N	75.52 W	30	4337	3894	14	89	75	N.A.
Wilmington WSO AP	39.67 N	75.60 W	79	4937	3557	10	89	74	617
Florida (FL)									
Belle Glade Exp Stn	26.67 N	80.63 W	16	451	8285	N.A.	N.A.	N.A.	N.A.
Daytona Beach WSO AP	29.18 N	81.05 W	29	909	7567	34	90	77	641
Ft Lauderdale	26.07 N	80.15 W	10	171	9735	46	90	78	N.A.
Ft Myers FAA AP	26.58 N	81.87 W	15	418	8924	42	93	77	N.A.
Ft Pierce	27.47 N	80.35 W	25	490	8448	N.A.	N.A.	N.A.	N.A.
Gainesville Mun AP	29.68 N	82.27 W	138	1267	7009	30	92	77	N.A.
Jacksonville WSO AP	30.50 N	81.70 W	26	1434	6847	29	93	77	674

TABLE D-1 US and US Territory Climatic Data *(continued)*

State City	Latitude	Longitude	Elev., ft	HDD65	CDD50	Heating Design Temperature	Cooling Design Temperature		No. Hrs. 8 a.m.–4 p.m.
							Dry-Bulb	Wet-Bulb	
						99.6%	1.0%	1.0%	55 < Tdb < 69
(Florida cont.)									
Key West WSO AP	24.55 N	81.75 W	4	100	10,174	55	89	79	N.A.
Lakeland	28.02 N	81.92 W	145	588	8472	N.A.	N.A.	N.A.	N.A.
Miami WSCMO AP	25.80 N	80.30 W	12	200	9474	46	90	77	259
Ocala	29.20 N	82.08 W	75	930	7696	N.A.	N.A.	N.A.	N.A.
Orlando WSO Mc Coy	28.43 N	81.33 W	91	686	8227	37	93	76	571
Panama City/Tyndall	30.07 N	85.58 W	16	1216	7023	33	89	79	N.A.
Pensacola FAA AP	30.47 N	87.20 W	112	1617	6816	28	92	78	N.A.
St Augustine WFOY	29.90 N	81.32 W	8	1040	7261	N.A.	N.A.	N.A.	N.A.
St Petersburg	27.77 N	82.63 W	8	603	8537	43	93	79	N.A.
Tallahassee WSO AP	30.38 N	84.37 W	55	1705	6639	25	93	76	747
Tampa WSCMO AP	27.97 N	82.53 W	19	725	8239	36	91	77	592
West Palm Beach WSO AP	26.68 N	80.12 W	18	323	9049	43	90	78	308
Georgia (GA)									
Albany	31.53 N	84.13 W	180	2205	6020	27	95	76	N.A.
Americus	32.05 N	84.25 W	490	2430	5634	N.A.	N.A.	N.A.	N.A.
Athens WSO AP	33.95 N	83.32 W	802	2893	5079	20	92	75	N.A.
Atlanta WSO AP	33.65 N	84.43 W	1010	2991	5038	18	91	74	749
Augusta WSO AP	33.37 N	81.97 W	148	2565	5519	21	94	76	774
Brunswick	31.17 N	81.50 W	13	1578	6729	30	91	79	N.A.
Columbus WSO AP	32.52 N	84.95 W	449	2261	6052	23	93	75	N.A.
Dalton	34.75 N	84.95 W	700	3552	4546	N.A.	N.A.	N.A.	N.A.
Dublin	32.50 N	82.90 W	215	2476	5664	N.A.	N.A.	N.A.	N.A.
Gainesville	34.30 N	83.85 W	1170	3500	4310	N.A.	N.A.	N.A.	N.A.
La Grange	33.05 N	85.02 W	715	2667	5216	N.A.	N.A.	N.A.	N.A.
Macon WSO AP	32.70 N	83.65 W	354	2334	5826	23	94	75	787
Savannah WSO AP	32.13 N	81.20 W	46	1847	6389	26	93	76	N.A.
Valdosta/Moody AFB	30.97 N	83.20 W	233	1552	7216	30	94	77	N.A.
Waycross	31.25 N	82.32 W	145	2025	6172	29	94	76	N.A.
Hawaii (HI)									
Hilo (Hawaii)	19.72 N	155.07 W	36	0	8759	61	84	74	153
Honolulu WSFO AP (Oahu)	21.33 N	157.92 W	7	0	9949	61	88	73	69
Kaneohe Mauka (Oahu)	21.42 N	157.82 W	190	0	8955	67	85	74	N.A.

TABLE D-1 US and US Territory Climatic Data *(continued)*

State City	Latitude	Longitude	Elev., ft	HDD65	CDD50	Heating Design Temperature	Cooling Design Temperature Dry-Bulb	Cooling Design Temperature Wet-Bulb	No. Hrs. 8 a.m.–4 p.m.
						99.6%	1.0%	1.0%	55 < Tdb < 69
Idaho (ID)									
Boise WSFO AP	43.57 N	116.22 W	2838	5861	2807	2	94	63	647
Burley FAA AP	42.53 N	113.77 W	4157	6745	2174	–5	90	62	N.A.
Coeur D'Alene R S	47.68 N	116.75 W	2158	6239	2216	N.A.	N.A.	N.A.	N.A.
Idaho Falls FAA AP	43.52 N	112.07 W	4730	8063	1853	–12	89	60	N.A.
Lewiston WSO AP	46.38 N	117.02 W	1436	5270	2964	6	93	64	748
Moscow-Univ of Idaho	46.73 N	116.97 W	2660	6782	1789	N.A.	N.A.	N.A.	N.A.
Mountain Home	43.13 N	115.70 W	3190	6176	2725	0	96	62	N.A.
Pocatello WSO AP	42.92 N	112.60 W	4454	7180	2142	–7	90	60	546
Twin Falls WSO	42.55 N	114.35 W	3960	6769	1995	N.A.	N.A.	N.A.	N.A.
Illinois (IL)									
Aurora	41.75 N	88.35 W	644	6699	2880	N.A.	N.A.	N.A.	N.A.
Belleville/Scott AFB	38.55 N	89.85 W	453	4878	4146	3	93	77	N.A.
Carbondale Sewage Plt	37.73 N	89.17 W	390	4865	3934	N.A.	N.A.	N.A.	N.A.
Champaign	40.03 N	88.28 W	755	5689	3697	N.A.	N.A.	N.A.	N.A.
Chicago Midway AP	41.73 N	87.77 W	620	6176	3251	N.A.	N.A.	N.A.	N.A.
Chicago O'Hare WSO AP	41.98 N	87.90 W	674	6536	2941	–6	88	73	613
Chicago University	41.78 N	87.60 W	594	5753	3391	N.A.	N.A.	N.A.	N.A.
Danville	40.13 N	87.65 W	558	5610	3471	–4	90	77	N.A.
Decatur	39.83 N	89.02 W	620	5522	3652	–2	91	75	N.A.
Dixon	41.83 N	89.52 W	700	6873	2965	N.A.	N.A.	N.A.	N.A.
Freeport Waste Wtr Plt	42.30 N	89.60 W	750	7169	2739	N.A.	N.A.	N.A.	N.A.
Galesburg	40.95 N	90.38 W	771	6314	3249	N.A.	N.A.	N.A.	N.A.
Joliet Brandon Rd Dam	41.50 N	88.10 W	543	6463	3025	N.A.	N.A.	N.A.	N.A.
Moline WSO AP	41.45 N	90.50 W	582	6474	3207	–8	90	74	640
Mt Vernon	38.35 N	88.87 W	490	5189	3818	N.A.	N.A.	N.A.	N.A.
Peoria WSO AP	40.67 N	89.68 W	650	6148	3339	–6	89	74	N.A.
Quincy FAA AP	39.93 N	91.20 W	763	5763	3574	–4	91	75	N.A.
Rantoul	40.32 N	88.17 W	740	6183	3288	N.A.	N.A.	N.A.	N.A.
Rockford WSO AP	42.20 N	89.10 W	724	6969	2852	–10	88	73	N.A.
Springfield WSO AP	39.85 N	89.68 W	594	5688	3635	–4	91	75	600
Waukegan	42.35 N	87.88 W	700	7136	2515	N.A.	N.A.	N.A.	N.A.

TABLE D-1 US and US Territory Climatic Data *(continued)*

State City	Latitude	Longitude	Elev., ft	HDD65	CDD50	Heating Design Temperature	Cooling Design Temperature Dry-Bulb	Cooling Design Temperature Wet-Bulb	No. Hrs. 8 a.m.–4 p.m.
						99.6%	1.0%	1.0%	55 < Tdb < 69
Indiana (IN)									
Anderson Sewage Plant	40.10 N	85.72 W	847	5916	3091	N.A.	N.A.	N.A.	N.A.
Bloomington Indiana U	39.17 N	86.52 W	825	5309	3585	N.A.	N.A.	N.A.	N.A.
Columbus	39.20 N	85.92 W	621	5536	3353	N.A.	N.A.	N.A.	N.A.
Evansville WSO AP	38.05 N	87.53 W	380	4708	4074	3	92	76	611
Ft Wayne WSO AP	41.00 N	85.20 W	797	6273	3077	–4	88	73	601
Goshen College	41.57 N	85.83 W	805	6282	2941	N.A.	N.A.	N.A.	N.A.
Hobart	41.53 N	87.25 W	600	6043	3168	N.A.	N.A.	N.A.	N.A.
Indianapolis WSFO	39.73 N	86.27 W	792	5615	3453	–3	88	74	N.A.
Kokomo	40.42 N	86.05 W	855	6429	2978	N.A.	N.A.	N.A.	N.A.
Lafayette	40.35 N	86.87 W	600	6228	3069	–5	90	75	N.A.
Marion	40.57 N	85.67 W	790	6260	2996	N.A.	N.A.	N.A.	N.A.
Muncie Ball State Univ	40.22 N	85.42 W	940	6027	3196	N.A.	N.A.	N.A.	N.A.
Peru/Grissom AFB	40.65 N	86.15 W	814	5908	3439	–3	89	75	N.A.
Richmond Wtr Wks	39.88 N	84.88 W	1015	5963	3004	N.A.	N.A.	N.A.	N.A.
Shelbyville Sewage Plt	39.52 N	85.78 W	750	5784	3291	N.A.	N.A.	N.A.	N.A.
South Bend WSO AP	41.70 N	86.32 W	773	6331	2920	–2	87	72	635
Terre Haute	39.35 N	87.42 W	555	5581	3490	–3	90	76	N.A.
Valparaiso Waterworks	41.52 N	87.03 W	800	6267	2942	N.A.	N.A.	N.A.	N.A.
Iowa (IA)									
Ames	42.03 N	93.80 W	1099	6776	3079	N.A.	N.A.	N.A.	N.A.
Burlington	40.78 N	91.12 W	597	5943	3601	–4	91	76	649
Cedar Rapids FAA AP	41.88 N	91.70 W	863	6924	3003	–11	89	74	N.A.
Clinton	41.80 N	90.27 W	585	6324	3291	N.A.	N.A.	N.A.	N.A.
Des Moines WSFO AP	41.53 N	93.65 W	938	6497	3371	–9	90	74	667
Dubuque WSO AP	42.40 N	90.70 W	1065	7327	2672	N.A.	N.A.	N.A.	N.A.
Ft Dodge	42.50 N	94.20 W	1115	7261	2902	–13	88	73	N.A.
Iowa City	41.65 N	91.53 W	640	6227	3434	N.A.	N.A.	N.A.	N.A.
Keokuk Lock and Dam	40.40 N	91.37 W	527	5969	3467	N.A.	N.A.	N.A.	N.A.
Marshalltown	42.07 N	92.93 W	870	7170	2813	N.A.	N.A.	N.A.	N.A.
Mason City FAA AP	43.17 N	93.33 W	1194	7837	2653	–15	88	73	610

TABLE D-1 US and US Territory Climatic Data *(continued)*

State City	Latitude	Longitude	Elev., ft	HDD65	CDD50	Heating Design Temperature	Cooling Design Temperature Dry-Bulb	Cooling Design Temperature Wet-Bulb	No. Hrs. 8 a.m.–4 p.m.
						99.6%	1.0%	1.0%	55 < Tdb < 69
(Iowa cont.)									
Newton	41.70 N	93.05 W	938	6783	3131	N.A.	N.A.	N.A.	N.A.
Ottumwa Airport	41.10 N	92.45 W	842	6269	3414	–5	92	75	N.A.
Sioux City WSO AP	42.40 N	96.38 W	1103	6893	3149	–11	90	74	602
Waterloo WSO AP	42.55 N	92.40 W	868	7406	2813	–14	88	73	N.A.
Kansas (KS)									
Atchison	39.57 N	95.12 W	945	5184	3940	N.A.	N.A.	N.A.	N.A.
Chanute FAA Airport	37.67 N	95.48 W	978	4650	4226	N.A.	N.A.	N.A.	N.A.
Dodge City WSO AP	37.77 N	99.97 W	2582	5001	4090	0	97	70	637
El Dorado	37.82 N	96.83 W	1340	4587	4317	N.A.	N.A.	N.A.	N.A.
Garden City FAA AP	37.93 N	100.72 W	2882	5216	3936	–3	97	69	N.A.
Goodland WSO AP	39.37 N	101.70 W	3650	5974	3018	–3	94	66	625
Great Bend	38.35 N	98.77 W	1850	4679	4425	N.A.	N.A.	N.A.	N.A.
Hutchinson	37.93 N	98.03 W	1570	5103	4106	N.A.	N.A.	N.A.	N.A.
Liberal	37.05 N	100.92 W	2834	4706	4185	N.A.	N.A.	N.A.	N.A.
Manhattan	39.20 N	96.58 W	1065	5043	4155	N.A.	N.A.	N.A.	N.A.
Parsons	37.37 N	95.28 W	910	4606	4339	N.A.	N.A.	N.A.	N.A.
Russell FAA AP	38.87 N	98.82 W	1864	5338	3939	–4	96	72	N.A.
Salina FAA AP	38.80 N	97.63 W	1257	5101	4167	–3	97	73	N.A.
Topeka WSFO AP	39.07 N	95.63 W	877	5265	3880	–2	93	75	608
Wichita WSO AP	37.65 N	97.43 W	1321	4791	4351	2	97	73	N.A.
Kentucky (KY)									
Ashland	38.45 N	82.62 W	555	5225	3280	N.A.	N.A.	N.A.	N.A.
Bowling Green FAA AP	36.97 N	86.42 W	547	4328	4132	7	91	75	N.A.
Covington WSO AP	39.07 N	84.67 W	869	5248	3488	1	89	73	661
Hopkinsville/Campbell	36.67 N	87.50 W	571	3928	4654	N.A.	N.A.	N.A.	N.A.
Lexington WSO AP	38.03 N	84.60 W	966	4783	3754	4	89	73	618
Louisville WSFO AP	38.18 N	85.73 W	477	4514	4000	6	90	75	636
Madisonville	37.35 N	87.52 W	440	4167	4290	N.A.	N.A.	N.A.	N.A.
Owensboro	37.77 N	87.15 W	405	4334	4222	N.A.	N.A.	N.A.	N.A.
Paducah WSO	37.07 N	88.77 W	410	4279	4317	7	93	76	N.A.

TABLE D-1 US and US Territory Climatic Data *(continued)*

State City	Latitude	Longitude	Elev., ft	HDD65	CDD50	Heating Design Temperature	Cooling Design Temperature Dry-Bulb	Cooling Design Temperature Wet-Bulb	No. Hrs. 8 a.m.–4 p.m.
						99.6%	1.0%	1.0%	55 < Tdb < 69
Louisiana (LA)									
Alexandria	31.32 N	92.47 W	87	2003	6407	27	94	78	N.A.
Baton Rouge WSO AP	30.53 N	91.13 W	64	1669	6845	27	92	77	677
Bogalusa	30.78 N	89.87 W	100	1911	6457	N.A.	N.A.	N.A.	N.A.
Houma	29.58 N	90.73 W	15	1429	6974	N.A.	N.A.	N.A.	N.A.
Lafayette FAA AP	30.20 N	91.98 W	38	1587	6877	28	93	78	N.A.
Lake Charles WSO AP	30.12 N	93.22 W	9	1616	6813	29	91	78	668
Minden	32.58 N	93.28 W	185	2533	5823	N.A.	N.A.	N.A.	N.A.
Monroe FAA AP	32.52 N	92.05 W	78	2407	6039	22	94	78	N.A.
Natchitoches	31.77 N	93.08 W	130	2152	6273	N.A.	N.A.	N.A.	N.A.
New Orleans WSCMO AP	29.98 N	90.25 W	4	1513	6910	30	92	78	789
Shreveport WSO AP	32.47 N	93.82 W	254	2,264	6166	22	95	77	697
Maine (ME)									
Augusta FAA AP	44.32 N	69.80 W	350	7550	2093	–3	84	69	N.A.
Bangor FAA AP	44.80 N	68.82 W	163	7930	1916	–7	84	69	669
Caribou WSO AP	46.87 N	68.02 W	624	9651	1470	–14	82	67	692
Lewiston	44.10 N	70.22 W	180	7244	2261	N.A.	N.A.	N.A.	N.A.
Millinocket	45.65 N	68.70 W	360	8902	1708	N.A.	N.A.	N.A.	N.A.
Portland WSMO AP	43.65 N	70.32 W	57	7378	1943	–3	83	70	665
Waterville Pmp Stn	44.55 N	69.65 W	90	7382	2180	N.A.	N.A.	N.A.	N.A.
Maryland (MD)									
Baltimore WSO AP	39.18 N	76.67 W	196	4707	3709	11	91	74	N.A.
Cumberland	39.63 N	78.75 W	730	5036	3432	N.A.	N.A.	N.A.	N.A.
Hagerstown	39.65 N	77.73 W	660	5293	3341	N.A.	N.A.	N.A.	N.A.
Salisbury	38.37 N	75.58 W	10	4027	4002	13	90	76	N.A.
Massachusetts (MA)									
Boston WSO AP	42.37 N	71.03 W	20	5641	2897	7	87	71	713
Clinton	42.40 N	71.68 W	398	6698	2457	N.A.	N.A.	N.A.	N.A.
Framingham	42.28 N	71.42 W	170	6262	2695	N.A.	N.A.	N.A.	N.A.
Lawrence	42.70 N	71.17 W	57	6322	2648	N.A.	N.A.	N.A.	N.A.
Lowell	42.65 N	71.37 W	110	6339	2715	N.A.	N.A.	N.A.	N.A.
New Bedford	41.63 N	70.93 W	120	5426	2973	N.A.	N.A.	N.A.	N.A.

TABLE D-1 US and US Territory Climatic Data *(continued)*

State City	Latitude	Longitude	Elev., ft	HDD65	CDD50	Heating Design Temperature	Cooling Design Temperature Dry-Bulb	Cooling Design Temperature Wet-Bulb	No. Hrs. 8 a.m.–4 p.m.
						99.6%	1.0%	1.0%	55 < Tdb < 69
(Massachusetts cont.)									
Springfield	42.10 N	72.58 W	190	5754	3037	N.A.	N.A.	N.A.	N.A.
Taunton	41.90 N	71.07 W	20	6346	2461	N.A.	N.A.	N.A.	N.A.
Worcester WSO AP	42.27 N	71.87 W	986	6979	2203	0	83	69	N.A.
Michigan (MI)									
Adrian	41.92 N	84.02 W	760	6737	2586	N.A.	N.A.	N.A.	N.A.
Alpena WSO AP	45.07 N	83.57 W	689	8284	1779	–7	84	69	695
Battle Creek/Kellogg	42.30 N	85.23 W	942	6416	3399	N.A.	N.A.	N.A.	N.A.
Benton Harbor AP	42.13 N	86.43 W	649	6303	2829	N.A.	N.A.	N.A.	N.A.
Detroit City Airport	42.42 N	83.02 W	625	6167	3046	0	87	72	N.A.
Escanaba	45.75 N	87.03 W	600	8593	1664	N.A.	N.A.	N.A.	N.A.
Flint WSO AP	42.97 N	83.75 W	766	6979	2451	–2	86	71	634
Grand Rapids WSO AP	42.88 N	85.52 W	707	6973	2537	0	86	71	622
Holland	42.80 N	86.12 W	610	6747	2536	N.A.	N.A.	N.A.	N.A.
Jackson FAA AP	42.27 N	84.45 W	1005	6791	2707	–3	86	73	N.A.
Kalamazoo State Hosp	42.28 N	85.60 W	945	6230	3015	N.A.	N.A.	N.A.	N.A.
Lansing WSO AP	42.77 N	84.60 W	841	7101	2449	–3	86	72	N.A.
Marquette	46.55 N	87.38 W	665	8356	1730	–13	82	67	N.A.
Mt Pleasant University	43.58 N	84.77 W	796	7436	2319	N.A.	N.A.	N.A.	N.A.
Muskegon WSO AP	43.17 N	86.23 W	628	6924	2361	3	83	70	N.A.
Pontiac State Hospital	42.65 N	83.30 W	982	6653	2770	N.A.	N.A.	N.A.	N.A.
Port Huron	42.98 N	82.42 W	590	6898	2541	N.A.	N.A.	N.A.	N.A.
Saginaw FAA AP	43.53 N	84.08 W	660	7139	2476	0	87	72	N.A.
Sault Ste Marie WSO	46.47 N	84.37 W	724	9316	1421	–12	80	68	733
Traverse City FAA AP	44.73 N	85.58 W	623	7749	2127	–3	86	70	679
Ypsilanti East Mich U	42.25 N	83.62 W	779	6466	2878	N.A.	N.A.	N.A.	N.A.
Minnesota (MN)									
Albert Lea	43.62 N	93.42 W	1230	8146	2608	N.A.	N.A.	N.A.	N.A.
Alexandria FAA AP	45.87 N	95.38 W	1416	8999	2316	–20	86	70	N.A.
Bemidji Airport	47.50 N	94.93 W	1377	10200	1781	N.A.	N.A.	N.A.	N.A.
Brainerd	46.37 N	94.20 W	1180	9437	1958	–24	85	68	N.A.
Duluth WSO AP	46.83 N	92.18 W	1428	9818	1536	–21	81	67	650

TABLE D-1 US and US Territory Climatic Data *(continued)*

State City	Latitude	Longitude	Elev., ft	HDD65	CDD50	Heating Design Temperature	Cooling Design Temperature Dry-Bulb	Cooling Design Temperature Wet-Bulb	No. Hrs. 8 a.m.–4 p.m.
						99.6%	1.0%	1.0%	55 < Tdb < 69
(Minnesota cont.)									
Faribault	44.30 N	93.27 W	940	8279	2498	N.A.	N.A.	N.A.	N.A.
International Falls WSO AP	48.57 N	93.38 W	1179	10,487	1630	–29	83	67	656
Mankato	44.15 N	94.02 W	836	8005	2691	N.A.	N.A.	N.A.	N.A.
Minneapolis-St Paul WSO AP	44.88 N	93.22 W	834	7981	2680	–16	88	71	566
Rochester WSO AP	43.92 N	92.50 W	1297	8250	2376	–17	85	71	652
St Cloud WSO AP	45.55 N	94.07 W	1037	8928	2149	–20	88	71	N.A.
Virginia	47.50 N	92.55 W	1435	10,024	1583	N.A.	N.A.	N.A.	N.A.
Willmar State Hospital	45.13 N	95.02 W	1128	8637	2465	N.A.	N.A.	N.A.	N.A.
Winona	44.05 N	91.63 W	652	7694	2695	N.A.	N.A.	N.A.	N.A.
Mississippi (MS)									
Biloxi/Keesler AFB	30.42 N	88.92 W	26	1486	6946	31	91	78	N.A.
Clarksdale	34.20 N	90.57 W	173	3188	5357	N.A.	N.A.	N.A.	N.A.
Columbus AFB	33.65 N	88.45 W	220	2769	5565	20	94	78	N.A.
Greenville	33.38 N	91.02 W	132	2778	5661	N.A.	N.A.	N.A.	N.A.
Greenwood FAA AP	33.50 N	90.08 W	155	2698	5760	20	94	78	N.A.
Hattiesburg	31.32 N	89.30 W	161	2180	6085	N.A.	N.A.	N.A.	N.A.
Jackson WSFO AP	32.32 N	90.08 W	330	2467	5900	21	93	76	640
Laurel	31.68 N	89.12 W	225	2327	5893	N.A.	N.A.	N.A.	N.A.
McComb FAA AP	31.23 N	90.47 W	413	2115	6025	23	92	76	N.A.
Meridian WSO AP	32.33 N	88.75 W	294	2444	5804	21	94	76	719
Natchez	31.55 N	91.38 W	195	1903	6378	N.A.	N.A.	N.A.	N.A.
Tupelo WSO AP	34.27 N	88.73 W	361	3079	5224	18	94	76	N.A.
Vicksburg Military Pk	32.35 N	90.85 W	255	2196	6059	N.A.	N.A.	N.A.	N.A.
Missouri (MO)									
Cape Girardeau FAA AP	37.23 N	89.57 W	337	4386	4359	6	94	77	N.A.
Columbia WSO AP	38.82 N	92.22 W	887	5212	3752	–1	92	75	633
Farmington	37.70 N	90.38 W	935	5041	3653	N.A.	N.A.	N.A.	N.A.
Hannibal	39.72 N	91.37 W	712	5628	3685	N.A.	N.A.	N.A.	N.A.
Jefferson City Wtr Plt	38.58 N	92.15 W	670	5302	3705	N.A.	N.A.	N.A.	N.A.
Joplin FAA AP	37.17 N	94.50 W	980	4303	4417	3	94	75	N.A.
Kansas City WSO AP	39.32 N	94.72 W	973	5393	3852	–1	93	75	N.A.

TABLE D-1 US and US Territory Climatic Data *(continued)*

State City	Latitude	Longitude	Elev., ft	HDD65	CDD50	Heating Design Temperature	Cooling Design Temperature		No. Hrs. 8 a.m.–4 p.m.
							Dry-Bulb	Wet-Bulb	
						99.6%	1.0%	1.0%	55 < Tdb < 69
(Missouri cont.)									
Kirksville Radio KIRX	40.22 N	92.58 W	970	5867	3494	N.A.	N.A.	N.A.	N.A.
Mexico	39.18 N	91.88 W	775	5590	3664	N.A.	N.A.	N.A.	N.A.
Moberly Radio KWIX	39.40 N	92.43 W	840	5204	3948	N.A.	N.A.	N.A.	N.A.
Poplar Bluff R S	36.77 N	90.42 W	380	4328	4368	8	92	76	N.A.
Rolla	38.13 N	91.77 W	1148	4748	4186	N.A.	N.A.	N.A.	N.A.
Rolla Univ of MO	37.95 N	91.77 W	1180	4959	3986	N.A.	N.A.	N.A.	N.A.
St Joseph	39.77 N	94.92 W	811	5590	3783	N.A.	N.A.	N.A.	N.A.
St Louis WSCMO AP	38.75 N	90.37 W	535	4758	4283	2	93	75	N.A.
Montana (MT)									
Billings WSO AP	45.80 N	108.53 W	3567	7164	2466	–13	90	62	617
Bozeman	45.82 N	110.88 W	5950	9908	672	–20	87	60	N.A.
Butte FAA AP	45.95 N	112.50 W	5540	9517	1152	–22	84	56	N.A.
Cut Bank FAA AP	48.60 N	112.37 W	3838	8904	1475	–21	84	59	672
Glasgow WSO AP	48.22 N	106.62 W	2284	8745	2244	–22	90	63	570
Glendive	47.10 N	104.72 W	2076	8178	2619	N.A.	N.A.	N.A.	N.A.
Great Falls WSCMO AP	47.48 N	111.37 W	3663	7741	1993	–19	88	60	641
Havre WSO AP	48.55 N	109.77 W	2584	8447	2132	–25	90	62	N.A.
Helena WSO AP	46.60 N	112.00 W	3893	8031	1922	–18	87	59	651
Kalispell WSO AP	48.30 N	114.27 W	2965	8378	1345	–12	86	61	N.A.
Lewistown FAA AP	47.07 N	109.45 W	4132	8479	1580	–18	86	60	673
Livingston FAA AP	45.70 N	110.45 W	4653	7220	1900	N.A.	N.A.	N.A.	N.A.
Miles City FAA AP	46.43 N	105.87 W	2628	7796	2680	–19	93	65	565
Missoula WSO AP	46.92 N	114.08 W	3190	7792	1679	–9	88	61	658
Nebraska (NE)									
Chadron FAA AP	42.83 N	03.08 W	3312	7020	2692	N.A.	N.A.	N.A.	N.A.
Columbus	41.47 N	97.33 W	1450	6543	3345	N.A.	N.A.	N.A.	N.A.
Fremont	41.43 N	96.48 W	1180	6140	3421	N.A.	N.A.	N.A.	N.A.
Grand Island WSO AP	40.97 N	98.32 W	1841	6421	3243	–8	93	72	611
Hastings	40.58 N	98.35 W	1925	6506	3217	N.A.	N.A.	N.A.	N.A.
Kearney	40.73 N	99.02 W	2130	6548	3090	N.A.	N.A.	N.A.	N.A.
Lincoln WSO AP	40.85 N	96.75 W	1190	6278	3455	–7	94	74	N.A.
Mc Cook	40.22 N	100.58 W	2580	6115	3236	N.A.	N.A.	N.A.	N.A.

TABLE D-1 US and US Territory Climatic Data *(continued)*

State City	Latitude	Longitude	Elev., ft	HDD65	CDD50	Heating Design Temperature	Cooling Design Temperature Dry-Bulb	Cooling Design Temperature Wet-Bulb	No. Hrs. 8 a.m.–4 p.m.
						99.6%	1.0%	1.0%	55 < Tdb < 69
(Nebraska cont.)									
Norfolk WSO AP	41.98 N	97.43 W	1551	6873	3072	–11	92	72	N.A.
North Platte WSO AP	41.13 N	100.68 W	2775	6859	2737	–10	92	69	592
Omaha (Eppley Field)	41.30 N	95.90 W	980	6300	3398	–7	92	75	N.A.
Scottsbluff WSO AP	41.87 N	103.60 W	3945	6729	2680	–11	92	64	620
Sidney	41.23 N	103.00 W	4,320	6966	2409	–8	92	63	N.A.
Nevada (NV)									
Carson City	39.15 N	119.77 W	4651	5691	2312	N.A.	N.A.	N.A.	N.A.
Elko FAA AP	40.83 N	115.78 W	5075	7077	2144	–5	92	59	569
Ely WSO AP	39.28 N	114.85 W	6262	7621	1717	–6	87	56	683
Las Vegas WSO AP	36.08 N	115.17 W	2162	2407	6745	27	106	66	719
Lovelock FAA AP	40.07 N	118.55 W	3900	5869	2886	N.A.	N.A.	N.A.	606
Reno WSFO AP	39.50 N	119.78 W	4404	5674	2504	8	92	60	752
Tonopah AP	38.07 N	117.08 W	5426	5733	2840	7	92	57	660
Winnemucca WSO AP	40.90 N	117.80 W	4297	6315	2379	1	94	60	608
New Hampshire (NH)									
Berlin	44.45 N	71.18 W	930	8645	1718	N.A.	N.A.	N.A.	N.A.
Concord WSO AP	43.20 N	71.50 W	346	7554	2087	–8	87	70	683
Keene	42.92 N	72.27 W	480	6948	2398	N.A.	N.A.	N.A.	N.A.
Portsmouth/Pease AFB	43.08 N	70.82 W	102	6572	2418	4	85	70	N.A.
New Jersey (NJ)									
Atlantic City WSO AP	39.45 N	74.57 W	138	5169	3198	8	88	73	N.A.
Long Branch Oakhurst	40.27 N	74.00 W	30	5253	3057	N.A.	N.A.	N.A.	N.A.
Newark WSO AP	40.70 N	74.17 W	30	4888	3748	10	90	73	644
New Mexico (NM)									
Alamogordo/Holloman	32.85 N	106.10 W	4094	3232	4726	20	96	63	N.A.
Albuquerque WSFO AP	35.05 N	106.62 W	5326	4425	3908	13	93	60	703
Artesia	32.77 N	104.38 W	3320	3527	4583	N.A.	N.A.	N.A.	N.A.
Carlsbad FAA AP	32.33 N	104.27 W	3232	2812	5512	19	98	66	N.A.
Clovis/Cannon AFB	34.38 N	103.32 W	4295	3983	4178	10	93	64	N.A.
Farmington	36.73 N	108.23 W	5502	5464	3307	8	92	60	N.A.
Gallup FAA AP	35.52 N	108.78 W	6468	6244	2355	–1	87	56	N.A.
Grants Airport	35.17 N	107.90 W	6520	5907	2481	N.A.	N.A.	N.A.	N.A.

TABLE D-1 US and US Territory Climatic Data *(continued)*

State City	Latitude	Longitude	Elev., ft	HDD65	CDD50	Heating Design Temperature	Cooling Design Temperature Dry-Bulb	Cooling Design Temperature Wet-Bulb	No. Hrs. 8 a.m.–4 p.m.
						99.6%	1.0%	1.0%	55 < Tdb < 69
(New Mexico cont.)									
Hobbs	32.70 N	103.13 W	3615	2851	5160	N.A.	N.A.	N.A.	N.A.
Raton Filter Plant	36.92 N	104.43 W	6932	6103	2187	N.A.	N.A.	N.A.	N.A.
Roswell FAA AP	33.30 N	104.53 W	3669	3267	4962	14	96	65	677
Socorro	34.08 N	106.88 W	4585	4074	3845	N.A.	N.A.	N.A.	N.A.
Tucumcari	35.20 N	103.68 W	4086	3912	4196	9	95	65	710
New York (NY)									
Albany WSFO AP	42.75 N	73.80 W	275	6894	2525	–7	86	70	605
Auburn	42.92 N	76.53 W	770	6782	2531	N.A.	N.A.	N.A.	N.A.
Batavia	42.98 N	78.18 W	890	6657	2536	N.A.	N.A.	N.A.	N.A.
Binghamton WSO AP	42.22 N	75.98 W	1600	7273	2193	–2	82	69	662
Buffalo WSCMO AP	42.93 N	78.73 W	705	6747	2468	2	84	69	697
Cortland	42.60 N	76.18 W	1129	7168	2225	N.A.	N.A.	N.A.	N.A.
Elmira/Chemung Co	42.17 N	76.90 W	951	6845	2420	–2	87	71	N.A.
Geneva Research Farm	42.88 N	77.03 W	718	6939	2364	N.A.	N.A.	N.A.	N.A.
Glens Falls FAA AP	43.35 N	73.62 W	321	7635	2182	–10	85	71	N.A.
Gloversville	43.05 N	74.35 W	812	7664	2118	N.A.	N.A.	N.A.	N.A.
Ithaca Cornell Univ	42.45 N	76.45 W	960	7207	2117	N.A.	N.A.	N.A.	N.A.
Lockport	43.18 N	78.65 W	520	6703	2482	N.A.	N.A.	N.A.	N.A.
Massena FAA AP	44.93 N	74.85 W	214	8255	2046	–15	84	71	627
NY Central Pk WSO City	40.78 N	73.97 W	132	4805	3634	N.A.	N.A.	N.A.	790
NY Kennedy WSO AP	40.65 N	73.78 W	16	5027	3342	11	88	72	N.A.
NY La Guardia WSO AP	40.77 N	73.90 W	11	4910	3547	13	89	73	790
Oswego East	43.47 N	76.50 W	350	6733	2431	N.A.	N.A.	N.A.	N.A.
Plattsburgh AFB	44.65 N	73.47 W	165	7837	2175	–9	83	69	N.A.
Poughkeepsie FAA AP	41.63 N	73.88 W	155	6391	2663	2	88	72	N.A.
Rochester WSO AP	43.12 N	77.67 W	547	6734	2406	1	86	71	608
Rome/Griffiss AFB	43.23 N	75.40 W	505	7244	2344	–5	86	70	N.A.
Schenectady	42.83 N	73.92 W	220	6881	2500	N.A.	N.A.	N.A.	N.A.
Syracuse WSO AP	43.12 N	76.12 W	421	6834	2399	–3	85	71	730
Utica	43.10 N	75.28 W	500	7066	2354	N.A.	N.A.	N.A.	N.A.
Watertown	43.97 N	75.87 W	497	7540	2294	–12	83	70	N.A.

TABLE D-1 US and US Territory Climatic Data *(continued)*

State City	Latitude	Longitude	Elev., ft	HDD65	CDD50	Heating Design Temperature	Cooling Design Temperature		No. Hrs. 8 a.m.–4 p.m.
							Dry-Bulb	Wet-Bulb	
						99.6%	1.0%	1.0%	55 < Tdb < 69
North Carolina (NC)									
Asheville WSO AP	35.43 N	82.55 W	2140	4308	3365	11	85	71	915
Charlotte WSO AP	35.22 N	80.93 W	700	3341	4704	18	91	74	777
Durham	36.03 N	78.97 W	406	3867	4159	N.A.	N.A.	N.A.	N.A.
Elizabeth City FAA AP	36.27 N	76.18 W	10	3139	4765	N.A.	N.A.	N.A.	N.A.
Fayetteville/Pope AFB	35.17 N	79.02 W	217	2917	5308	22	94	76	N.A.
Goldsboro	35.33 N	77.97 W	109	3040	5018	22	94	76	N.A.
Greensboro WSO AP	36.08 N	79.95 W	886	3865	4144	15	90	74	718
Greenville	35.62 N	77.38 W	30	3129	4824	N.A.	N.A.	N.A.	N.A.
Henderson	36.37 N	78.42 W	480	4038	4002	N.A.	N.A.	N.A.	N.A.
Hickory FAA AP	35.73 N	81.38 W	1143	3728	4199	18	91	72	N.A.
Jacksonville/New River	34.70 N	77.43 W	26	2456	5678	23	92	78	N.A.
Lumberton	34.70 N	79.07 W	130	3212	4723	N.A.	N.A.	N.A.	N.A.
New Bern FAA AP	35.07 N	77.05 W	18	2742	5262	22	92	78	N.A.
Raleigh-Durham WSFO AP	35.87 N	78.78 W	376	3457	4499	16	90	75	740
Rocky Mount	35.90 N	77.72 W	110	3321	4586	N.A.	N.A.	N.A.	N.A.
Wilmington WSO AP	34.27 N	77.90 W	72	2470	5557	23	91	78	N.A.
North Dakota (ND)									
Bismarck WSFO AP	46.77 N	100.77 W	1647	8968	2144	–21	90	67	556
Devils Lake KDLR	48.12 N	98.87 W	1464	9950	1973	–23	87	67	N.A.
Dickinson FAA AP	46.78 N	102.80 W	2581	8657	2152	N.A.	N.A.	N.A.	N.A.
Fargo WSO AP	46.90 N	96.80 W	900	9254	2289	–22	88	70	546
Grand Forks FAA AP	47.95 N	97.17 W	847	9733	2084	–20	88	69	N.A.
Jamestown FAA AP	46.92 N	98.68 W	1492	9168	2262	N.A.	N.A.	N.A.	N.A.
Minot FAA AP	48.27 N	101.28 W	1715	9193	2135	–20	88	66	581
Ohio (OH)									
Akron-Canton WSO AP	40.92 N	81.43 W	1208	6160	2779	0	85	71	680
Ashtabula	41.85 N	80.80 W	690	6429	2604	N.A.	N.A.	N.A.	N.A.
Bowling Green	41.38 N	83.62 W	675	6482	2876	N.A.	N.A.	N.A.	N.A.
Cambridge	40.02 N	81.58 W	800	5488	3118	N.A.	N.A.	N.A.	N.A.
Cincinnati-Abbe WSO	39.15 N	84.52 W	760	4988	3733	5	90	75	N.A.

TABLE D-1 US and US Territory Climatic Data *(continued)*

State City	Latitude	Longitude	Elev., ft	HDD65	CDD50	Heating Design Temperature	Cooling Design Temperature		No. Hrs. 8 a.m.–4 p.m.
							Dry-Bulb	Wet-Bulb	
						99.6%	1.0%	1.0%	55 < Tdb < 69
(Ohio cont.)									
Cleveland WSFO AP	41.42 N	81.87 W	770	6201	2755	1	86	72	N.A.
Columbus WSO AP	40.00 N	82.88 W	812	5708	3119	1	88	73	708
Dayton WSCMO AP	39.90 N	84.20 W	995	5708	3249	–1	88	73	611
Defiance	41.28 N	84.38 W	700	6628	2810	N.A.	N.A.	N.A.	N.A.
Findlay FAA AP	41.02 N	83.67 W	797	6302	2907	–2	87	72	N.A.
Fremont	41.33 N	83.12 W	600	6439	2823	N.A.	N.A.	N.A.	N.A.
Lancaster	39.73 N	82.63 W	860	5988	2935	N.A.	N.A.	N.A.	N.A.
Lima Sewage Plant	40.72 N	84.13 W	850	6253	3050	N.A.	N.A.	N.A.	N.A.
Mansfield WSO AP	40.82 N	82.52 W	1295	6258	2818	–1	85	72	N.A.
Marion	40.62 N	83.13 W	965	6407	2836	N.A.	N.A.	N.A.	N.A.
Newark Water Works	40.08 N	82.42 W	835	5657	3107	N.A.	N.A.	N.A.	N.A.
Norwalk	41.27 N	82.62 W	670	6434	2715	N.A.	N.A.	N.A.	N.A.
Portsmouth	38.75 N	82.88 W	540	4913	3581	N.A.	N.A.	N.A.	N.A.
Sandusky	41.45 N	82.72 W	584	6131	2986	N.A.	N.A.	N.A.	N.A.
Springfield New Wtr Wk	39.97 N	83.82 W	930	6254	2790	N.A.	N.A.	N.A.	N.A.
Steubenville	40.38 N	80.63 W	992	5700	3054	N.A.	N.A.	N.A.	N.A.
Toledo Express WSO AP	41.58 N	83.80 W	669	6579	2720	–2	87	72	652
Warren	41.20 N	80.82 W	900	6402	2546	N.A.	N.A.	N.A.	N.A.
Wooster Exp Station	40.78 N	81.92 W	1020	6379	2570	N.A.	N.A.	N.A.	N.A.
Youngstown WSO AP	41.25 N	80.67 W	1178	6544	2536	–1	85	70	679
Zanesville FAA AP	39.95 N	81.90 W	881	5714	3013	2	88	73	N.A.
Oklahoma (OK)									
Ada	34.78 N	96.68 W	1015	3182	5317	N.A.	N.A.	N.A.	N.A.
Altus AFB	34.65 N	99.27 W	1378	3151	5708	13	100	73	N.A.
Ardmore	34.20 N	97.15 W	860	2702	5978	N.A.	N.A.	N.A.	N.A.
Bartlesville	36.75 N	96.00 W	715	3777	4976	N.A.	N.A.	N.A.	N.A.
Chickasha Exp Station	35.05 N	97.92 W	1085	3366	5298	N.A.	N.A.	N.A.	N.A.
Enid	36.42 N	97.87 W	1245	3788	5119	5	98	74	N.A.
Lawton	34.62 N	98.45 W	1150	3457	5268	12	97	73	N.A.
McAlester FAA AP	34.88 N	95.78 W	760	3354	5233	10	96	76	N.A.
Muskogee	35.77 N	95.33 W	583	3413	5185	N.A.	N.A.	N.A.	N.A.

TABLE D-1 US and US Territory Climatic Data *(continued)*

State City	Latitude	Longitude	Elev., ft	HDD65	CDD50	Heating Design Temperature	Cooling Design Temperature Dry-Bulb	Cooling Design Temperature Wet-Bulb	No. Hrs. 8 a.m.–4 p.m.
						99.6%	1.0%	1.0%	55 < Tdb < 69
(Oklahoma cont.)									
Norman	35.18 N	97.45 W	1109	3,295	5272	N.A.	N.A.	N.A.	N.A.
Oklahoma City WSFO AP	35.40 N	97.60 W	1280	3659	4972	9	96	74	733
Ponca City FAA AP	36.73 N	97.10 W	999	4226	4791	N.A.	N.A.	N.A.	N.A.
Seminole	35.23 N	96.67 W	865	3097	5552	N.A.	N.A.	N.A.	N.A.
Stillwater	36.12 N	97.10 W	895	4028	4718	N.A.	N.A.	N.A.	N.A.
Tulsa WSO AP	36.18 N	95.90 W	668	3691	5150	9	97	76	591
Woodward	36.45 N	99.38 W	1900	3900	4884	N.A.	N.A.	N.A.	N.A.
Oregon (OR)									
Astoria WSO AP	46.15 N	123.88 W	8	5158	1437	25	72	62	1236
Baker FAA AP	44.83 N	117.82 W	3368	7155	1741	N.A.	N.A.	N.A.	N.A.
Bend	44.07 N	121.28 W	3660	6926	1405	N.A.	N.A.	N.A.	N.A.
Corvallis State Univ	44.63 N	123.20 W	225	4923	2051	N.A.	N.A.	N.A.	N.A.
Eugene WSO AP	44.12 N	123.22 W	364	4546	2354	21	87	65	N.A.
Grants Pass	42.42 N	123.33 W	960	4219	2986	N.A.	N.A.	N.A.	N.A.
Klamath Falls	42.20 N	121.78 W	4098	6634	1954	4	87	62	N.A.
Medford WSO AP	42.38 N	122.88 W	1300	4611	2989	21	95	66	749
Pendleton WSO AP	45.68 N	118.85 W	1492	5294	2787	3	93	63	N.A.
Portland WSFO AP	45.60 N	122.60 W	21	4522	2517	22	86	66	1060
Roseburg KQEN	43.20 N	123.35 W	465	4312	2607	N.A.	N.A.	N.A.	N.A.
Salem WSO AP	44.92 N	123.02 W	195	4927	2100	20	87	66	916
Pennsylvania (PA)									
Allentown WSO AP	40.65 N	75.43 W	388	5785	3028	5	88	72	710
Altoona FAA AP	40.30 N	78.32 W	1476	6140	2719	5	86	70	N.A.
Chambersburg	39.93 N	77.63 W	640	5574	3060	N.A.	N.A.	N.A.	N.A.
Erie WSO AP	42.08 N	80.18 W	732	6279	2652	2	83	70	716
Harrisburg FAA AP	40.22 N	76.85 W	338	5347	3358	9	89	73	648
Johnstown	40.33 N	78.92 W	1214	5649	3028	N.A.	N.A.	N.A.	N.A.
Lancaster	40.05 N	76.28 W	270	5584	3079	N.A.	N.A.	N.A.	N.A.
Meadville	41.63 N	80.17 W	1065	6934	2209	N.A.	N.A.	N.A.	N.A.
New Castle	41.02 N	80.37 W	825	6542	2502	N.A.	N.A.	N.A.	N.A.
Philadelphia WSCMO AP	39.88 N	75.23 W	10	4954	3623	11	89	74	646

TABLE D-1 US and US Territory Climatic Data *(continued)*

State City	Latitude	Longitude	Elev., ft	HDD65	CDD50	Heating Design Temperature	Cooling Design Temperature Dry-Bulb	Cooling Design Temperature Wet-Bulb	No. Hrs. 8 a.m.–4 p.m.
						99.6%	1.0%	1.0%	55 < Tdb < 69
(Pennsylvania cont.)									
Pittsburgh WSCMO2 AP	40.50 N	80.22 W	1150	5968	2836	2	86	70	700
Reading	40.37 N	75.93 W	270	5796	3021	N.A.	N.A.	N.A.	N.A.
State College	40.80 N	77.87 W	1170	6364	2629	N.A.	N.A.	N.A.	N.A.
Uniontown	39.92 N	79.72 W	956	5684	2913	N.A.	N.A.	N.A.	N.A.
Warren	41.85 N	79.15 W	1210	6890	2334	N.A.	N.A.	N.A.	N.A.
West Chester	39.97 N	75.63 W	450	5283	3288	N.A.	N.A.	N.A.	N.A.
Williamsport WSO AP	41.25 N	76.92 W	524	6087	2796	2	87	71	N.A.
York Pump Station 22	39.92 N	76.75 W	390	5256	3274	N.A.	N.A.	N.A.	N.A.
Rhode Island (RI)									
Newport	41.52 N	71.32 W	20	5659	2548	N.A.	N.A.	N.A.	N.A.
Providence WSO AP	41.73 N	71.43 W	51	5884	2743	5	86	71	684
South Carolina (SC)									
Anderson	34.53 N	82.67 W	800	2965	4900	N.A.	N.A.	N.A.	N.A.
Charleston WSO AP	32.90 N	80.03 W	41	2013	6188	N.A.	N.A.	N.A.	N.A.
Charleston WSO City	32.78 N	79.93 W	10	1866	6303	25	92	77	N.A.
Columbia WSFO AP	33.95 N	81.12 W	213	2649	5508	21	94	75	705
Florence FAA AP	34.18 N	79.72 W	146	2585	5597	23	94	76	N.A.
Georgetown	33.35 N	79.25 W	10	2081	5947	N.A.	N.A.	N.A.	N.A.
Greenville-Spartanburg WSO AP	34.90 N	82.22 W	973	3272	4625	19	91	74	851
Greenwood	34.17 N	82.20 W	615	3288	4673	N.A.	N.A.	N.A.	N.A.
Orangeburg	33.50 N	80.87 W	160	2534	5477	N.A.	N.A.	N.A.	N.A.
Spartanburg	34.98 N	81.88 W	840	2887	5046	N.A.	N.A.	N.A.	N.A.
Sumter/Shaw AFB	33.97 N	80.48 W	240	2506	5453	24	93	75	N.A.
South Dakota (SD)									
Aberdeen WSO AP	45.45 N	98.43 W	1296	8446	2497	N.A.	N.A.	N.A.	N.A.
Brookings	44.32 N	96.77 W	1642	8653	2228	N.A.	N.A.	N.A.	N.A.
Huron WSO AP	44.38 N	98.22 W	1282	7923	2709	–17	91	71	545
Mitchell	43.72 N	98.00 W	1274	7558	2925	N.A.	N.A.	N.A.	N.A.
Pierre FAA AP	44.38 N	100.28 W	1726	7411	2938	–14	95	69	557
Rapid City WSO AP	44.05 N	103.07 W	3162	7301	2412	–11	91	65	572
Sioux Falls WSFO AP	43.57 N	96.73 W	1418	7809	2735	–16	90	72	599

TABLE D-1 US and US Territory Climatic Data *(continued)*

State City	Latitude	Longitude	Elev., ft	HDD65	CDD50	Heating Design Temperature	Cooling Design Temperature Dry-Bulb	Cooling Design Temperature Wet-Bulb	No. Hrs. 8 a.m.–4 p.m.
						99.6%	1.0%	1.0%	55 < Tdb < 69
(South Dakota, cont.)									
Watertown FAA AP	44.92 N	97.15 W	1746	8375	2499	N.A.	N.A.	N.A.	N.A.
Yankton	42.88 N	97.35 W	1180	7304	2935	N.A.	N.A.	N.A.	N.A.
Tennessee (TN)									
Athens	35.43 N	84.58 W	940	4054	4040	N.A.	N.A.	N.A.	N.A.
Bristol WSO AP	36.48 N	82.40 W	1525	4406	3621	9	87	72	N.A.
Chattanooga WSO AP	35.03 N	85.20 W	692	3587	4609	15	92	75	684
Clarksville Sew Plt	36.55 N	87.37 W	382	4159	4241	N.A.	N.A.	N.A.	N.A.
Columbia	35.63 N	87.08 W	650	4206	4047	N.A.	N.A.	N.A.	N.A.
Dyersburg FAA AP	36.02 N	89.40 W	337	3536	5010	N.A.	N.A.	N.A.	N.A.
Greeneville Exp Stn	36.10 N	82.85 W	1320	4392	3710	N.A.	N.A.	N.A.	N.A.
Jackson FAA AP	35.60 N	88.92 W	433	3540	4915	12	93	76	N.A.
Knoxville WSO AP	35.80 N	84.00 W	949	3937	4164	13	90	74	703
Memphis FAA-AP	35.05 N	90.00 W	265	3082	5467	16	94	77	851
Murfreesboro	35.92 N	86.37 W	550	3992	4270	N.A.	N.A.	N.A.	N.A.
Nashville WSO AP	36.12 N	86.68 W	580	3729	4689	10	92	75	749
Tullahoma	35.35 N	86.20 W	1048	3630	4422	N.A.	N.A.	N.A.	N.A.
Texas (TX)									
Abilene WSO AP	32.42 N	99.68 W	1784	2584	6050	16	97	71	648
Alice	27.73 N	98.07 W	201	1062	8121	N.A.	N.A.	N.A.	N.A.
Amarillo WSO AP	35.23 N	101.70 W	3590	4258	4128	6	94	66	680
Austin WSO AP	30.30 N	97.70 W	597	1688	7171	25	96	74	664
Bay City Waterworks	28.98 N	95.98 W	52	1370	7211	N.A.	N.A.	N.A.	N.A.
Beaumont Research Ctr	30.07 N	94.28 W	27	1677	6703	29	92	79	N.A.
Beeville	28.45 N	97.70 W	255	1372	7393	28	98	77	N.A.
Big Spring	32.25 N	101.45 W	2500	2772	5621	N.A.	N.A.	N.A.	N.A.
Brownsville WSO AP	25.90 N	97.43 W	19	635	8777	36	94	77	422
Brownwood	31.72 N	99.00 W	1385	2199	6479	N.A.	N.A.	N.A.	N.A.
Corpus Christi WSO AP	27.77 N	97.50 W	44	1016	8023	32	94	78	543
Corsicana	32.08 N	96.47 W	425	2396	6133	N.A.	N.A.	N.A.	N.A.
Dallas FAA AP	32.85 N	96.85 W	440	2259	6587	17	98	74	N.A.
Del Rio/Laughlin AFB	29.37 N	100.78 W	1079	1565	7207	28	98	73	732

TABLE D-1 US and US Territory Climatic Data *(continued)*

State City	Latitude	Longitude	Elev., ft	HDD65	CDD50	Heating Design Temperature	Cooling Design Temperature		No. Hrs. 8 a.m.–4 p.m.
							Dry-Bulb	Wet-Bulb	
						99.6%	1.0%	1.0%	55 < Tdb < 69
(Texas cont.)									
Denton	33.20 N	97.10 W	630	2665	5816	N.A.	N.A.	N.A.	N.A.
Eagle Pass	28.70 N	100.48 W	805	1441	7682	N.A.	N.A.	N.A.	N.A.
El Paso WSO AP	31.80 N	106.40 W	3918	2708	5488	21	98	64	735
Ft Worth/Meacham	32.82 N	97.35 W	692	2304	6557	19	98	74	N.A.
Galveston WSO City	29.30 N	94.80 W	7	1263	7378	N.A.	N.A.	N.A.	N.A.
Greenville	33.20 N	96.22 W	610	2953	5527	N.A.	N.A.	N.A.	N.A.
Harlingen	26.20 N	97.67 W	38	813	8405	N.A.	N.A.	N.A.	N.A.
Houston /Hobby	29.65 N	95.28 W	50	1371	7357	29	93	77	N.A.
Houston-Bush Intercontinental Airport	29.97 N	95.35 W	96	1599	6876	27	94	77	N.A.
Huntsville	30.72 N	95.55 W	494	1862	6697	N.A.	N.A.	N.A.	N.A.
Killeen/Robert-Gray	31.07 N	97.83 W	1014	2127	6477	20	96	73	N.A.
Lamesa	32.70 N	101.93 W	2965	3159	5107	N.A.	N.A.	N.A.	N.A.
Laredo	27.57 N	99.50 W	430	1025	8495	32	101	74	598
Longview	32.47 N	94.73 W	330	2433	5920	N.A.	N.A.	N.A.	N.A.
Lubbock WSFO AP	33.65 N	101.82 W	3254	3431	4833	11	95	67	743
Lufkin FAA AP	31.23 N	94.75 W	281	1951	6527	23	95	77	681
McAllen	26.20 N	98.22 W	122	778	8597	34	98	76	N.A.
Midland/Odessa WSO AP	31.95 N	102.18 W	2857	2751	5588	17	97	67	729
Mineral Wells FAA AP	32.78 N	98.07 W	934	2625	6015	N.A.	N.A.	N.A.	N.A.
Palestine	31.78 N	95.60 W	465	2005	6454	N.A.	N.A.	N.A.	N.A.
Pampa No 2	35.53 N	100.98 W	3250	4358	4131	N.A.	N.A.	N.A.	N.A.
Pecos	31.42 N	103.50 W	2610	2505	5992	N.A.	N.A.	N.A.	N.A.
Plainview	34.18 N	101.70 W	3370	3717	4462	N.A.	N.A.	N.A.	N.A.
Port Arthur WSO AP	29.95 N	94.02 W	16	1499	6994	N.A.	N.A.	N.A.	697
San Angelo WSO AP	31.37 N	100.50 W	1903	2414	6070	20	97	70	619
San Antonio WSFO	29.53 N	98.47 W	794	1644	7142	26	96	73	N.A.
Sherman	33.63 N	96.62 W	720	2890	5682	N.A.	N.A.	N.A.	721
Snyder	32.72 N	100.92 W	2335	3185	5178	N.A.	N.A.	N.A.	N.A.
Temple	31.08 N	97.37 W	700	2153	6487	N.A.	N.A.	N.A.	N.A.
Tyler	32.35 N	95.40 W	545	2194	6562	N.A.	N.A.	N.A.	N.A.
Vernon	34.08 N	99.30 W	1202	3186	5605	N.A.	N.A.	N.A.	N.A.

TABLE D-1 US and US Territory Climatic Data *(continued)*

State City	Latitude	Longitude	Elev., ft	HDD65	CDD50	Heating Design Temperature	Cooling Design Temperature Dry-Bulb	Cooling Design Temperature Wet-Bulb	No. Hrs. 8 a.m.–4 p.m.
						99.6%	1.0%	1.0%	55 < Tdb < 69
(Texas cont.)									
Victoria WSO AP	28.85 N	96.92 W	104	1296	7507	29	94	76	N.A.
Waco WSO AP	31.62 N	97.22 W	500	2179	6668	22	99	75	622
Wichita Falls WSO AP	33.97 N	98.48 W	994	3042	5717	N.A.	N.A.	N.A.	723
Utah (UT)									
Cedar City FAA AP	37.70 N	113.10 W	5610	5962	2770	2	91	59	629
Logan Utah State Univ	41.75 N	111.80 W	4790	6854	2541	N.A.	N.A.	N.A.	N.A.
Moab	38.60 N	109.60 W	3965	4494	4356	N.A.	N.A.	N.A.	N.A.
Ogden Sugar Factory	41.23 N	112.03 W	4280	5950	3053	N.A.	N.A.	N.A.	N.A.
Richfield Radio KSVC	38.77 N	112.08 W	5270	6367	2300	N.A.	N.A.	N.A.	N.A.
Saint George	37.10 N	113.57 W	2760	3215	5424	N.A.	N.A.	N.A.	N.A.
Salt Lake City NWSFO	40.78 N	111.95 W	4222	5765	3276	6	94	62	586
Vernal Airport	40.45 N	109.52 W	5260	7562	2334	N.A.	N.A.	N.A.	N.A.
Vermont (VT)									
Burlington WSO AP	44.47 N	73.15 W	332	7771	2228	–11	84	69	637
Rutland	43.60 N	72.97 W	620	7066	2345	N.A.	N.A.	N.A.	N.A.
Virginia (VA)									
Charlottesville	38.03 N	78.52 W	870	4224	3902	N.A.	N.A.	N.A.	N.A.
Danville-Bridge St	36.58 N	79.38 W	410	3944	4236	N.A.	N.A.	N.A.	N.A.
Fredericksburg Natl Pk	38.32 N	77.45 W	90	4554	3754	N.A.	N.A.	N.A.	N.A.
Lynchburg WSO AP	37.33 N	79.20 W	916	4340	3728	12	90	74	N.A.
Norfolk WSO AP	36.90 N	76.20 W	22	3495	4478	20	91	76	685
Richmond WSO AP	37.50 N	77.33 W	164	3963	4223	14	92	75	716
Roanoke WSO AP	37.32 N	79.97 W	1149	4360	3715	12	89	72	713
Staunton Sewage Plant	38.15 N	79.03 W	1385	5273	3004	N.A.	N.A.	N.A.	N.A.
Winchester	39.18 N	78.12 W	680	5269	3215	N.A.	N.A.	N.A.	N.A.
Washington (WA)									
Aberdeen	46.97 N	123.82 W	10	5285	1488	N.A.	N.A.	N.A.	N.A.
Bellingham FAA AP	48.80 N	122.53 W	149	5609	1508	15	76	64	N.A.
Bremerton	47.57 N	122.67 W	162	5119	1839	N.A.	N.A.	N.A.	N.A.
Ellensburg	46.97 N	120.55 W	1480	6770	1999	N.A.	N.A.	N.A.	N.A.
Everett	47.98 N	122.18 W	60	5311	1660	N.A.	N.A.	N.A.	N.A.

TABLE D-1 US and US Territory Climatic Data *(continued)*

State City	Latitude	Longitude	Elev., ft	HDD65	CDD50	Heating Design Temperature	Cooling Design Temperature Dry-Bulb	Cooling Design Temperature Wet-Bulb	No. Hrs. 8 a.m.–4 p.m.
						99.6%	1.0%	1.0%	55 < Tdb < 69
(Washington cont.)									
Kennewick	46.22 N	119.10 W	390	4895	3195	N.A.	N.A.	N.A.	N.A.
Longview	46.15 N	122.92 W	12	5094	1858	N.A.	N.A.	N.A.	N.A.
Olympia WSO AP	46.97 N	122.90 W	192	5655	1558	18	83	65	985
Port Angeles	48.12 N	123.40 W	40	5695	1257	N.A.	N.A.	N.A.	N.A.
Seattle EMSU WSO	47.65 N	122.30 W	20	4611	2120	N.A.	N.A.	N.A.	N.A.
Seattle-Tacoma WSCMO AP	47.45 N	122.30 W	450	4908	2021	23	81	64	982
Spokane WSO AP	47.63 N	117.53 W	2356	6842	2032	N.A.	N.A.	N.A.	640
Tacoma/McChord AFB	47.15 N	122.48 W	322	5155	1820	18	82	63	N.A.
Walla Walla FAA AP	46.10 N	118.28 W	1166	4958	3161	4	95	65	N.A.
Wenatchee	47.42 N	120.32 W	640	5579	2956	3	92	65	N.A.
Yakima WSO AP	46.57 N	120.53 W	1064	5967	2348	4	92	64	703
West Virginia (WV)									
Beckley WSO AP	37.78 N	81.12 W	2504	5558	2690	N.A.	N.A.	N.A.	N.A.
Bluefield FAA AP	37.30 N	81.22 W	2870	5230	2907	5	83	69	N.A.
Charleston WSFO AP	38.37 N	81.60 W	1015	4646	3655	6	88	73	704
Clarksburg	39.27 N	80.35 W	945	5512	3014	N.A.	N.A.	N.A.	N.A.
Elkins WSO AP	38.88 N	79.85 W	1992	6120	2360	–2	83	70	N.A.
Huntington WSO AP	38.37 N	82.55 W	827	4665	3615	6	89	73	N.A.
Martinsburg FAA AP	39.40 N	77.98 W	531	5192	3368	8	91	73	N.A.
Morgantown FAA AP	39.65 N	79.92 W	1240	5363	3155	4	87	71	N.A.
Parkersburg	39.27 N	81.57 W	615	5094	3507	4	88	72	N.A.
Wisconsin (WI)									
Appleton	44.25 N	88.37 W	750	7693	2513	N.A.	N.A.	N.A.	N.A.
Ashland Exp Farm	46.57 N	90.97 W	650	8960	1811	N.A.	N.A.	N.A.	N.A.
Beloit	42.50 N	89.03 W	780	7161	2737	N.A.	N.A.	N.A.	N.A.
Eau Claire FAA AP	44.87 N	91.48 W	888	8330	2407	–18	87	71	661
Fond du Lac	43.80 N	88.45 W	760	7541	2573	N.A.	N.A.	N.A.	N.A.
Green Bay WSO AP	44.48 N	88.13 W	682	8089	2177	–13	85	72	651
La Crosse FAA AP	43.87 N	91.25 W	651	7491	2790	–14	88	73	644
Madison WSO AP	43.13 N	89.33 W	858	767	2389	–11	87	72	658
Manitowoc	44.10 N	87.68 W	660	7597	2193	N.A.	N.A.	N.A.	N.A.

TABLE D-1 US and US Territory Climatic Data *(continued)*

State City	Latitude	Longitude	Elev., ft	HDD65	CDD50	Heating Design Temperature	Cooling Design Temperature Dry-Bulb	Cooling Design Temperature Wet-Bulb	No. Hrs. 8 a.m.–4 p.m.
						99.6%	1.0%	1.0%	55 < Tdb < 69
(Wisconsin cont.)									
Marinette	45.10 N	87.63 W	605	8059	2272	N.A.	N.A.	N.A.	N.A.
Milwaukee WSO AP	42.95 N	87.90 W	672	7324	2388	–7	86	72	618
Racine	42.70 N	87.77 W	595	7167	2459	N.A.	N.A.	N.A.	N.A.
Sheboygan	43.75 N	87.72 W	648	7087	2390	N.A.	N.A.	N.A.	N.A.
Stevens Point	44.50 N	89.57 W	1079	8009	2325	N.A.	N.A.	N.A.	N.A.
Waukesha	43.02 N	88.23 W	860	7117	2658	N.A.	N.A.	N.A.	N.A.
Wausau FAA AP	44.92 N	89.62 W	1196	8427	2182	–15	85	70	N.A.
Wyoming (WY)									
Casper WSO AP	42.92 N	106.47 W	5338	7682	2082	–13	89	58	535
Cheyenne WSFO AP	41.15 N	104.82 W	6120	7326	1886	–7	85	57	608
Cody	44.52 N	109.07 W	5050	7431	2057	–14	87	58	N.A.
Evanston	41.27 N	110.95 W	6810	8846	1285	N.A.	N.A.	N.A.	N.A.
Lander WSO AP	42.82 N	108.73 W	5370	7889	2184	–14	87	58	N.A.
Laramie FAA AP	41.32 N	105.68 W	7266	9008	1237	N.A.	N.A.	N.A.	N.A.
Newcastle	43.85 N	104.22 W	4410	7267	2518	N.A.	N.A.	N.A.	N.A.
Rawlins FAA AP	41.80 N	107.20 W	6736	8475	1605	N.A.	N.A.	N.A.	N.A.
Rock Springs FAA AP	41.60 N	109.07 W	6741	8365	1734	–9	84	54	552
Sheridan WSO AP	44.77 N	106.97 W	3964	7804	2023	–14	90	61	574
Torrington Exp Farm	42.08 N	104.22 W	4098	6879	2429	N.A.	N.A.	N.A.	N.A.
District of Columbia (DC)									
R. Reagan Nat'l. Airport	38.85 N	77.03 W	66	4047	4391	15	92	76	657
Puerto Rico (PR)									
San Juan/Isla Verde WSFO	18.43 N	66.00 W	10	0	11,406	69	90	78	N.A.
Pacific Islands (PI)									
Guam (GU) - Andersen AFB	13.58 N	144.93 E	361	0	10,690	74	87	79	N.A.
Marshall Island (MH) - Kwajalein Atoll	8.73 N	167.73 E	26	0	11,670	76	88	79	N.A.
Midway Island (MH) - Midway Island NAF	28.22 N	177.37 W	13	134	8323	59	86	75	N.A.
Samoa (WS) - Pago Pago WSO Airport	14.33 S	170.72 W	9	0	11,018	72	88	80	N.A.
Wake Island - Wake Island WSO Airport	19.28 N	166.65 E	12	0	11,097	71	89	79	N.A.
Philippines									
Philippines (PH) - Angeles, Clark AFB	15.18 N	120.55 E	475	0	11,280	68	95	77	N.A.

TABLE D-2 Canadian Climatic Data

Province City	Latitude	Longitude	Elev., ft	HDD65	CDD50	Heating Design Temperature	Cooling Design Temperature Dry-Bulb	Cooling Design Temperature Wet-Bulb
						99.6%	1.0%	1.0%
Alberta (AB)								
Calgary International A	51.12 N	114.02 W	3533	9885	1167	–22	80	59
Edmonton International A	53.30 N	113.58 W	2345	11,023	1069	–28.1	78	62
Grande Prairie A	55.18 N	118.88 W	2185	11,240	1031	–32	78	60
Jasper	52.88 N	118.07 W	3480	10,244	848	N.A.	N.A.	N.A.
Lethbridge A	49.63 N	112.80 W	3047	8783	1730	–22	84	61
Medicine Hat A	50.02 N	110.72 W	2352	8988	1981	–24	87	62
Red Deer A	52.18 N	113.90 W	2969	10,765	1095	–27	79	61
British Columbia (BC)								
Dawson Creek A	55.73 N	120.18 W	2148	11,435	890	N.A.	N.A.	N.A.
Ft Nelson A	58.83 N	122.58 W	1253	12,941	1013	–33	78	60
Kamloops	50.67 N	120.33 W	1243	6779	2335	–8	88	63
Nanaimo A	49.05 N	123.87 W	98	6054	1469	N.A.	N.A.	N.A.
New Westminster BC Pen	49.22 N	122.90 W	59	5520	1691	N.A.	N.A.	N.A.
Penticton A	49.47 N	119.60 W	1128	6500	2002	5	87	64
Prince George	53.88 N	122.67 W	2267	9495	906	–25	78	59
Prince Rupert A	54.30 N	130.43 W	111	7650	572	7	63	57
Vancouver International A	49.18 N	123.17 W	9	5682	1536	18	74	64
Victoria Gonzales Hts	48.42 N	123.32 W	229	5494	1286	23	75	62
Manitoba (MB)								
Brandon CDA	49.87 N	99.98 W	1190	10,969	1661	–29	84	66
Churchill A	58.73 N	94.07 W	91	16,719	275	–36	72	60
Dauphin A	51.10 N	100.05 W	1000	11,242	1520	–28	84	66
Flin Flon	54.77 N	101.85 W	1099	12,307	1352	N.A.	N.A.	N.A.
Portage La Prairie A	49.90 N	98.27 W	885	10,594	1807	–25	85	67
The Pas A	53.97 N	101.10 W	889	12,490	1231	–32	79	64
Winnipeg International A	49.90 N	97.23 W	784	10,858	1784	–27	84	67
New Brunswick (NB)								
Chatham A	47.02 N	65.45 W	111	9028	1531	–12	83	67
Fredericton A	45.87 N	66.53 W	55	8666	1631	–12	83	68
Moncton A	46.12 N	64.68 W	232	8731	1427	–10	80	67
Saint John A	45.33 N	65.88 W	337	8776	1179	–9	75	64
Newfoundland (NF)								
Corner Brook	48.95 N	57.95 W	16	8756	1075	N.A.	N.A.	N.A.

TABLE D-2 Canadian Climatic Data *(continued)*

Province City	Latitude	Longitude	Elev., ft	HDD65	CDD50	Heating Design Temperature	Cooling Design Temperature Dry-Bulb	Cooling Design Temperature Wet-Bulb
						99.6%	1.0%	1.0%
(Newfoundland cont.)								
Gander International A	48.95 N	54.57 W	495	9354	956	–4	76	63
Goose A	53.32 N	60.42 W	150	12,017	758	–23	77	61
St John's A	47.62 N	52.73 W	439	8888	848	3	73	64
Stephenville A	48.53 N	58.55 W	26	8869	952	–2	71	64
Northwest Territories (NW)								
Ft Smith A	60.02 N	111.95 W	666	14,192	932	–34	78	61
Inuvik A	68.30 N	133.48 W	193	18,409	489	–43	75	59
Yellowknife A	62.47 N	114.45 W	672	15,555	851	–39	74	59
Nova Scotia (NS)								
Halifax International A	44.88 N	63.52 W	416	8133	1464	–2	78	66
Kentville CDA	45.07 N	64.48 W	160	7683	1665	N.A.	N.A.	N.A.
Sydney A	46.17 N	60.05 W	183	8364	1287	–1	78	67
Truro	45.37 N	63.27 W	131	8596	1295	–9	77	67
Yarmouth A	43.83 N	66.08 W	141	7515	1180	7	71	64
Nunavut								
Resolute A	74.72 N	94.98 W	219	22,864	0	–42	48	43
Ontario (ON)								
Belleville	44.15 N	77.40 W	249	7556	2252	N.A.	N.A.	N.A.
Cornwall	45.02 N	74.75 W	209	8062	2187	N.A.	N.A.	N.A.
Hamilton RBG	43.28 N	79.88 W	334	6872	2450	N.A.	N.A.	N.A.
Kapuskasing A	49.42 N	82.47 W	744	11,742	1108	–30	80	65
Kenora A	49.78 N	94.37 W	1335	10,884	1626	–27	81	65
Kingston A	44.22 N	76.60 W	305	7826	1960	N.A.	N.A.	N.A.
London A	43.03 N	81.15 W	912	7565	2126	–3	83	70
North Bay A	46.35 N	79.43 W	1174	9794	1509	–18	78	66
Oshawa WPCP	43.87 N	78.83 W	275	7253	2106	N.A.	N.A.	N.A.
Ottawa International A	45.32 N	75.67 W	380	8571	2045	–13	83	69
Owen Sound MOE	44.58 N	80.93 W	587	7730	1896	N.A.	N.A.	N.A.
Peterborough	44.28 N	78.32 W	636	8037	1975	N.A.	N.A.	N.A.
St Catharines	43.20 N	79.25 W	298	6700	2564	N.A.	N.A.	N.A.
Sudbury A	46.62 N	80.80 W	1141	9990	1557	–19	81	66
Thunder Bay A	48.37 N	89.32 W	652	10,562	1198	–22	80	66
Timmins A	48.57 N	81.37 W	967	11,374	1225	–28	81	65

TABLE D-2 Canadian Climatic Data *(continued)*

Province City	Latitude	Longitude	Elev., ft	HDD65	CDD50	Heating Design Temperature	Cooling Design Temperature Dry-Bulb	Cooling Design Temperature Wet-Bulb
						99.6%	1.0%	1.0%
(Ontario cont.)								
Toronto Downsview A	43.75 N	79.48 W	649	7306	2370	–4	84	70
Windsor A	42.27 N	82.97 W	623	6619	2679	2	86	71
Prince Edward Island (PE)								
Charlottetown A	46.28 N	63.13 W	157	8598	1400	–6	77	67
Summerside A	46.43 N	63.83 W	78	8411	1536	–5	77	66
Québec (PQ)								
Bagotville A	48.33 N	71.00 W	521	10,603	1300	–23	80	65
Drummondville	45.88 N	72.48 W	269	8601	2024	N.A.	N.A.	N.A.
Granby	45.38 N	72.70 W	551	8367	1984	N.A.	N.A.	N.A.
Montreal Dorval International A	45.47 N	73.75 W	101	8285	2146	–12	83	70
Québec City A	46.80 N	71.38 W	229	9449	1571	–16	80	68
Rimouski	48.45 N	68.52 W	118	9665	1215	N.A.	N.A.	N.A.
Sept-Iles A	50.22 N	66.27 W	180	11,287	690	–20	69	59
Shawinigan	46.57 N	72.75 W	400	9246	1720	N.A.	N.A.	N.A.
Sherbrooke A	45.43 N	71.68 W	780	9464	1372	–20	80	68
St Jean de Cherbourg	48.88 N	67.12 W	1151	11,277	801	N.A.	N.A.	N.A.
St Jerome	45.80 N	74.05 W	557	9171	1771	N.A.	N.A.	N.A.
Thetford Mines	46.10 N	71.35 W	1250	9687	1425	N.A.	N.A.	N.A.
Trois Rivieres	46.37 N	72.60 W	173	9124	1766	N.A.	N.A.	N.A.
Val d'Or A	48.07 N	77.78 W	1105	11,256	1193	–27	80	65
Valleyfield	45.28 N	74.10 W	150	8083	2268	N.A.	N.A.	N.A.
Saskatchewan (SK)								
Estevan A	49.22 N	102.97 W	1876	10,092	1793	–25	86	65
Moose Jaw A	50.33 N	105.55 W	1893	9989	1812	–27	87	64
North Battleford A	52.77 N	108.25 W	1797	11,127	1473	–31	82	63
Prince Albert A	53.22 N	105.68 W	1404	12,009	1252	–34	81	64
Regina A	50.43 N	104.67 W	1893	10,773	1620	–29	85	64
Saskatoon A	52.17 N	106.68 W	1643	11,118	1537	–31	84	63
Swift Current A	50.28 N	107.68 W	2683	10,128	1541	–25	84	62
Yorkton A	51.27 N	102.47 W	1633	11,431	1476	–30	82	64
Yukon Territory (YT)								
Whitehorse A	60.72 N	135.07 W	2306	12,797	611	–34	73	55

TABLE D-3 International Climatic Data

Country City	Province or Region	Latitude		Longitude		Elev., ft	HDD65	CDD50	Heating Design Temperature	Cooling Design Temperature Dry-Bulb	Cooling Design Temperature Wet-Bulb
									99.6%	1.0%	1.0%
Argentina											
Buenos Aires/Ezeiza		34.82	S	58.53	W	66	2211	4693	31	90	72
Cordoba		31.32	S	64.22	W	1555	1816	5182	31	91	72
Tucuman/Pozo		26.85	S	65.10	W	1444	1416	6622	N.A.	N.A.	N.A.
Australia											
Adelaide	SA	34.95	S	138.53	E	20	2082	4381	39	92	64
Alice Springs	NT	23.80	S	133.90	E	1782	1142	7777	34	102	64
Brisbane	QL	27.43	S	153.08	E	7	545	7009	44	86	72
Darwin Airport	NT	12.43	S	130.87	E	95	0	11,736	64	92	76
Perth/Guildford	WA	31.92	S	115.97	E	56	1507	5353	41	95	66
Sydney/K Smith	NSW	33.95	S	151.18	E	20	1351	5259	42	85	67
Azores											
Lajes	Terceira	38.75	N	27.08	W	180	1279	4892	46	78	71
Bahamas											
Nassau		25.05	N	77.47	W	10	29	9775	57	90	78
Belgium											
Brussels Airport		50.90	N	4.47	E	128	5460	1862	15	79	66
Bermuda											
St Georges/Kindley		32.37	N	64.68	W	20	170	8365	N.A.	N.A.	N.A.
Bolivia											
La Paz/El Alto		16.50	S	68.18	W	13,287	7189	237	25	62	44
Brazil											
Belem		1.43	S	48.48	W	79	0	11,552	72	90	78
Brasilia		15.77	S	47.93	W	3809	58	7943	48	88	65
Fortaleza		3.72	S	38.55	W	62	1	11,748	72	90	78
Porto Alegre		30.08	S	51.18	W	23	902	7076	40	92	75
Recife/Curado		8.13	S	34.92	W	36	2	10,951	70	91	78
Rio de Janeiro		22.90	S	43.17	W	16	14	9688	59	99	77
Salvador/Ondina		13.00	S	38.52	W	167	0	10,785	68	88	78
Sao Paulo		23.50	S	46.62	W	2608	447	7219	48	88	69
Bulgaria											
Sofia		42.82	N	23.38	E	1952	5629	2508	10	85	65
Chile											
Concepcion		36.77	S	73.05	W	39	3559	2283	35	74	62
Punta Arenas/Chabunco		53.03	S	70.85	W	108	7807	395	23	61	53

TABLE D-3 International Climatic Data *(continued)*

Country City	Province or Region	Latitude		Longitude		Elev., ft	HDD65	CDD50	Heating Design Temperature	Cooling Design Temperature Dry-Bulb	Cooling Design Temperature Wet-Bulb
									99.6%	1.0%	1.0%
(Chile cont.)											
Santiago/Pedahuel		33.38	S	70.88	W	1575	2820	3471	29	88	65
China											
Beijing/Peking	Municipalities	39.93	N	116.28	E	180	5252	4,15	12	92	72
Cangzhou	Municipalities	38.33	N	116.83	E	36	4888	4504	14	92	74
Hong Kong Intl Arpt	Special Admin. Region	22.33	N	114.18	E	79	543	7894	48	91	79
Shanghai	Municipalities	31.40	N	121.47	E	13	3182	5124	29	92	80
Shanghai/Hongqiao	Municipalities	31.17	N	121.43	E	23	3184	5127	26	92	82
Tianjin/Tientsin	Municipalities	39.10	N	117.17	E	16	4948	4450	14	91	74
Anqing	Anhui	30.53	N	117.05	E	66	3093	5476	28	94	80
Bengbu	Anhui	32.95	N	117.37	E	72	3,44	5053	23	93	79
Fuyang	Anhui	32.93	N	115.83	E	128	3639	5004	23	93	79
Hefei/Luogang	Anhui	31.87	N	117.23	E	118	3468	5110	25	93	80
Huang Shan (Mtns)	Anhui	30.13	N	118.15	E	6024	6723	1647	9	70	65
Huoshan	Anhui	31.40	N	116.33	E	223	3516	4907	24	94	80
Changting	Fujian	25.85	N	116.37	E	1020	1902	6289	30	91	77
Fuding	Fujian	27.33	N	120.20	E	125	1868	6277	34	92	80
Fuzhou	Fujian	26.08	N	119.28	E	279	1396	7047	40	94	80
Jiuxian Shan	Fujian	25.72	N	118.10	E	5417	3923	2763	23	74	67
Longyan	Fujian	25.10	N	117.02	E	1119	1120	7248	37	93	75
Nanping	Fujian	26.65	N	118.17	E	420	1551	6986	35	95	78
Pingtan	Fujian	25.52	N	119.78	E	102	1478	6550	43	87	79
Pucheng	Fujian	27.92	N	118.53	E	902	2325	5940	29	93	78
Shaowu	Fujian	27.33	N	117.43	E	630	2075	6232	29	94	78
Xiamen	Fujian	24.48	N	118.08	E	456	1014	7,26	43	91	79
Yong'An	Fujian	25.97	N	117.35	E	669	1570	6917	33	95	77
Dunhuang	Gansu	40.15	N	94.68	E	3740	6531	3272	1	93	64
Hezuo	Gansu	35.00	N	102.90	E	9547	9760	491	–5	70	54
Huajialing	Gansu	35.38	N	105.00	E	8038	9275	871	4	70	56
Jiuquan/Suzhou	Gansu	39.77	N	98.48	E	4849	7316	2473	–2	86	62
Lanzhou	Gansu	36.05	N	103.88	E	4980	5849	2954	11	87	63
Mazong Shan (Mount)	Gansu	41.80	N	97.03	E	5807	9187	1748	–9	84	55
Minqin	Gansu	38.63	N	103.08	E	4485	7045	2830	0	89	61

TABLE D-3 International Climatic Data *(continued)*

Country City	Province or Region	Latitude		Longitude		Elev., ft	HDD65	CDD50	Heating Design Temperature 99.6%	Cooling Design Temperature Dry-Bulb 1.0%	Cooling Design Temperature Wet-Bulb 1.0%
(China cont.)											
Pingliang	Gansu	35.55	N	106.67	E	4423	6248	2407	9	84	64
Ruo'ergai	Gansu	33.58	N	102.97	E	11,289	10,826	232	–8	65	52
Tianshui	Gansu	34.58	N	105.75	E	3750	5192	3073	17	87	67
Wudu	Gansu	33.40	N	104.92	E	3540	3419	4250	28	90	68
Wushaoling (Pass)	Gansu	37.20	N	102.87	E	9987	11,697	263	–5	64	50
Xifengzhen	Gansu	35.73	N	107.63	E	4669	6471	2388	10	82	63
Yumenzhen	Gansu	40.27	N	97.03	E	5010	7614	2367	–3	86	60
Zhangye	Gansu	38.93	N	100.43	E	4865	7288	2439	–2	88	62
Fogang	Guangdong	23.87	N	113.53	E	223	1063	7709	39	92	79
Gaoyao	Guangdong	23.05	N	112.47	E	39	720	8493	44	93	80
Guangzhou/Baiyun	Guangdong	23.13	N	113.32	E	26	737	8352	42	93	80
Heyuan	Guangdong	23.73	N	114.68	E	135	902	8079	40	93	79
Lian Xian	Guangdong	24.78	N	112.38	E	322	1660	7018	35	94	79
Lianping	Guangdong	24.37	N	114.48	E	702	1301	7189	36	92	78
Meixian	Guangdong	24.30	N	116.12	E	276	937	8016	39	94	79
Shangchuan Island	Guangdong	21.73	N	112.77	E	59	514	8621	46	89	81
Shantou	Guangdong	23.40	N	116.68	E	10	779	7743	45	90	80
Shanwei	Guangdong	22.78	N	115.37	E	16	528	8272	46	89	79
Shaoguan	Guangdong	24.80	N	113.58	E	223	1370	7565	37	94	79
Shenzhen	Guangdong	22.55	N	114.10	E	59	531	8597	44	92	80
Xinyi	Guangdong	22.35	N	110.93	E	276	570	8763	43	93	79
Yangjiang	Guangdong	21.87	N	111.97	E	72	547	8470	45	90	80
Zhangjiang	Guangdong	21.22	N	110.40	E	92	423	9002	46	92	80
Beihai	Guangxi	21.48	N	109.10	E	52	621	8826	44	91	80
Bose	Guangxi	23.90	N	106.60	E	794	716	8488	43	96	79
Guilin	Guangxi	25.33	N	110.30	E	545	1971	6549	35	92	78
Guiping	Guangxi	23.40	N	110.08	E	144	957	8084	42	93	80
Hechi/Jnchengjiang	Guangxi	24.70	N	108.05	E	702	1229	7489	40	93	78
Lingling	Guangxi	26.23	N	111.62	E	571	2608	5993	31	94	78
Liuzhou	Guangxi	24.35	N	109.40	E	318	1370	7604	38	94	78
Longzhou	Guangxi	22.37	N	106.75	E	423	681	8596	43	94	80
Mengshan	Guangxi	24.20	N	110.52	E	476	1485	7125	36	92	79

TABLE D-3 International Climatic Data *(continued)*

Country City	Province or Region	Latitude		Longitude		Elev., ft	HDD65	CDD50	Heating Design Temperature	Cooling Design Temperature Dry-Bulb	Cooling Design Temperature Wet-Bulb
									99.6%	1.0%	1.0%
(China cont.)											
Nanning/Wuxu	Guangxi	22.82	N	108.35	E	240	857	8315	42	93	79
Napo	Guangxi	23.30	N	105.95	E	2605	1283	6469	37	87	74
Qinzhou	Guangxi	21.95	N	108.62	E	20	769	8415	43	91	80
Wuzhou	Guangxi	23.48	N	111.30	E	394	1074	7934	39	94	80
Bijie	Guizhou	27.30	N	105.23	E	4957	3837	3496	27	83	68
Dushan	Guizhou	25.83	N	107.55	E	3340	3021	4530	27	83	71
Guiyang	Guizhou	26.58	N	106.72	E	3524	2879	4689	28	85	70
Luodian	Guizhou	25.43	N	106.77	E	1447	1351	7066	38	93	77
Rongjiang/Guzhou	Guizhou	25.97	N	108.53	E	942	1967	6362	34	93	78
Sansui	Guizhou	26.97	N	108.67	E	2005	3322	4659	28	88	75
Sinan	Guizhou	27.95	N	108.25	E	1371	2494	5719	34	93	76
Weining	Guizhou	26.87	N	104.28	E	7336	4632	2342	21	75	60
Xingren	Guizhou	25.43	N	105.18	E	4524	2595	4527	30	83	68
Zunyi	Guizhou	27.70	N	106.88	E	2772	3091	4673	30	88	73
Danxian/Nada	Hainan	19.52	N	109.58	E	554	245	9606	48	94	78
Dongfang/Basuo	Hainan	19.10	N	108.62	E	26	107	10,168	53	91	81
Haikou	Hainan	20.03	N	110.35	E	49	211	9659	51	93	81
Qionghai/Jiaji	Hainan	19.23	N	110.47	E	82	133	9882	52	93	81
Sanhu Island	Hainan	16.53	N	111.62	E	16	0	11,282	69	90	83
Xisha Island	Hainan	16.83	N	112.33	E	16	0	11,221	69	89	82
Yaxian/Sanya	Hainan	18.23	N	109.52	E	23	7	10,735	60	90	80
Baoding	Hebei	38.85	N	115.57	E	62	4949	4411	14	93	73
Chengde	Hebei	40.97	N	117.93	E	1227	6778	3356	0	89	69
Fengning/Dagezhen	Hebei	41.22	N	116.63	E	2169	7891	2574	–5	86	66
Huailai/Shacheng	Hebei	40.40	N	115.50	E	1765	6490	3403	5	89	67
Leting	Hebei	39.43	N	118.90	E	39	5918	3562	8	87	74
Qinglong	Hebei	40.40	N	118.95	E	748	6611	3261	0	88	71
Shijiazhuang	Hebei	38.03	N	114.42	E	266	4695	4469	15	93	73
Tangshan	Hebei	39.67	N	118.15	E	95	5675	3867	8	89	74
Weichang/Zhuizishan	Hebei	41.93	N	117.75	E	2769	8600	2201	–6	83	65
Xingtai	Hebei	37.07	N	114.50	E	256	4506	4626	18	93	73
Yu Xian	Hebei	39.83	N	114.57	E	2986	7948	2545	–9	86	65

TABLE D-3 International Climatic Data *(continued)*

Country City	Province or Region	Latitude		Longitude		Elev., ft	HDD65	CDD50	Heating Design Temperature	Cooling Design Temperature Dry-Bulb	Cooling Design Temperature Wet-Bulb
									99.6%	1.0%	1.0%
(China cont.)											
Zhangjiakou	Hebei	40.78	N	114.88	E	2382	6823	3202	2	88	65
Aihui	Heilongjiang	50.25	N	127.45	E	545	11,840	1840	–28	83	68
Anda	Heilongjiang	46.38	N	125.32	E	492	10,066	2482	–20	86	69
Baoqing	Heilongjiang	46.32	N	132.18	E	272	9731	2379	–17	85	69
Fujin	Heilongjiang	47.23	N	131.98	E	213	10,265	2356	–18	85	70
Hailun	Heilongjiang	47.43	N	126.97	E	787	11,017	2137	–24	83	68
Harbin	Heilongjiang	45.75	N	126.77	E	469	9830	2482	–20	85	69
Hulin	Heilongjiang	45.77	N	132.97	E	338	9977	2228	–17	82	70
Huma	Heilongjiang	51.72	N	126.65	E	587	12,658	1760	–36	84	67
Jixi	Heilongjiang	45.28	N	130.95	E	768	9518	2318	–14	84	69
Keshan	Heilongjiang	48.05	N	125.88	E	778	11,108	2123	–25	84	68
Mudanjiang	Heilongjiang	44.57	N	129.60	E	794	9464	2449	–16	85	69
Qiqihar	Heilongjiang	47.38	N	123.92	E	486	9924	2514	–18	86	69
Shangzhi	Heilongjiang	45.22	N	127.97	E	627	10,340	2189	–26	84	70
Suifenhe	Heilongjiang	44.38	N	131.15	E	1634	10,219	1714	–16	82	68
Sunwu	Heilongjiang	49.43	N	127.35	E	771	12,334	1585	–32	83	68
Tailai	Heilongjiang	46.40	N	123.42	E	492	9431	2663	–16	87	69
Tonghe	Heilongjiang	45.97	N	128.73	E	361	10,618	2210	–24	84	71
Yichun	Heilongjiang	47.72	N	128.90	E	761	11,239	1965	–28	83	68
Anyang/Zhangde	Henan	36.12	N	114.37	E	249	4318	4648	18	93	75
Boxian	Henan	33.88	N	115.77	E	138	4006	4755	20	93	77
Gushi	Henan	32.17	N	115.67	E	190	3567	4964	24	92	80
Lushi	Henan	34.05	N	111.03	E	1870	4572	3865	17	90	73
Nanyang	Henan	33.03	N	112.58	E	430	3779	4750	23	92	77
Xihua	Henan	33.78	N	114.52	E	174	4032	4623	21	93	78
Xinyang	Henan	32.13	N	114.05	E	377	3576	4922	24	92	78
Zhengzhou	Henan	34.72	N	113.65	E	364	4146	4614	19	93	75
Zhumadian	Henan	33.00	N	114.02	E	272	3885	4718	22	93	77
Fangxian	Hubei	32.03	N	110.77	E	1427	3688	4483	24	91	75
Guanghua	Hubei	32.38	N	111.67	E	299	3445	4989	26	93	79
Jiangling/Jingzhou	Hubei	30.33	N	112.18	E	108	3064	5325	29	93	81
Macheng	Hubei	31.18	N	114.97	E	194	3166	5363	27	94	80

TABLE D-3 International Climatic Data *(continued)*

Country City	Province or Region	Latitude		Longitude		Elev., ft	HDD65	CDD50	Heating Design Temperature 99.6%	Cooling Design Temperature Dry-Bulb 1.0%	Cooling Design Temperature Wet-Bulb 1.0%
(China cont.)											
Wuhan/Nanhu	Hubei	30.62	N	114.13	E	75	3140	5433	28	94	81
Yichang	Hubei	30.70	N	111.30	E	440	2812	5476	30	93	79
Zaoyang	Hubei	32.15	N	112.67	E	417	3463	5034	25	93	78
Zhongxiang	Hubei	31.17	N	112.57	E	217	3192	5240	28	92	80
Changde	Hunan	29.05	N	111.68	E	115	2896	5520	30	95	81
Chenzhou	Hunan	25.80	N	113.03	E	607	2496	6255	31	95	78
Nanyue	Hunan	27.30	N	112.70	E	4196	4866	3090	17	77	71
Sangzhi	Hunan	29.40	N	110.17	E	1056	2896	5229	30	93	77
Shaoyang	Hunan	27.23	N	111.47	E	814	2794	5651	30	93	78
Tongdao/Shuangjiang	Hunan	26.17	N	109.78	E	1302	2706	5440	30	90	76
Wugang	Hunan	26.73	N	110.63	E	1115	2854	5424	30	92	77
Yuanling	Hunan	28.47	N	110.40	E	469	2817	5442	30	93	78
Yueyang	Hunan	29.38	N	113.08	E	171	2870	5681	30	92	81
Zhijiang	Hunan	27.45	N	109.68	E	896	2857	5385	30	92	78
Abag Qi/Xin Hot	Inner Mongolia	44.02	N	114.95	E	3701	11,253	1853	–25	84	60
Arxan	Inner Mongolia	47.17	N	119.95	E	3373	13,802	964	–35	77	61
Bailing-Miao	Inner Mongolia	41.70	N	110.43	E	4518	9399	2005	–15	85	59
Bayan Mod	Inner Mongolia	40.75	N	104.50	E	4360	7762	2911	–6	89	59
Bugt	Inner Mongolia	48.77	N	121.92	E	2425	12,243	1187	–22	79	62
Bugt	Inner Mongolia	42.33	N	120.70	E	1316	7853	2855	–4	87	68
Chifeng/Ulanhad	Inner Mongolia	42.27	N	118.97	E	1877	7571	3015	–5	88	67
Dongsheng	Inner Mongolia	39.83	N	109.98	E	4787	8149	2202	–3	83	59
Duolun/Dolonnur	Inner Mongolia	42.18	N	116.47	E	4091	10,403	1547	–18	80	61
Ejin Qi	Inner Mongolia	41.95	N	101.07	E	3087	7313	3592	–5	95	62
Erenhot	Inner Mongolia	43.65	N	112.00	E	3169	9870	2442	–19	89	61
Guaizihu	Inner Mongolia	41.37	N	102.37	E	3150	7189	3769	–4	97	61
Hailar	Inner Mongolia	49.22	N	119.75	E	2005	12,730	1604	–32	82	64
Hails	Inner Mongolia	41.45	N	106.38	E	4954	8903	2317	–11	85	57
Haliut	Inner Mongolia	41.57	N	108.52	E	4232	8927	2305	–9	85	61
Hohhot	Inner Mongolia	40.82	N	111.68	E	3494	8022	2509	–4	86	63
Huade	Inner Mongolia	41.90	N	114.00	E	4869	10,129	1600	–13	80	59
Jartai	Inner Mongolia	39.78	N	105.75	E	3389	6960	3456	–3	93	62

TABLE D-3 International Climatic Data *(continued)*

Country City	Province or Region	Latitude		Longitude		Elev., ft	HDD65	CDD50	Heating Design Temperature	Cooling Design Temperature Dry-Bulb	Cooling Design Temperature Wet-Bulb
									99.6%	1.0%	1.0%
(China cont.)											
Jarud Qi/Lubei	Inner Mongolia	44.57	N	120.90	E	873	8245	2856	–7	89	68
Jining	Inner Mongolia	41.03	N	113.07	E	4646	9276	1709	–9	81	60
Jurh	Inner Mongolia	42.40	N	112.90	E	3780	9067	2401	–13	87	60
Lindong/Bairin Zuoq	Inner Mongolia	43.98	N	119.40	E	1591	8954	2352	–10	87	67
Linhe	Inner Mongolia	40.77	N	107.40	E	3415	7302	2995	–1	89	64
Linxi	Inner Mongolia	43.60	N	118.07	E	2625	9154	2171	–10	84	64
Mandal	Inner Mongolia	42.53	N	110.13	E	4012	8967	2413	–10	87	59
Naran Bulag	Inner Mongolia	44.62	N	114.15	E	3881	11,695	1655	–23	84	60
Nenjiang	Inner Mongolia	49.17	N	125.23	E	797	11,980	1880	–32	83	67
Otog Qi/Ulan	Inner Mongolia	39.10	N	107.98	E	4531	7722	2505	–5	87	60
Tongliao	Inner Mongolia	43.60	N	122.27	E	591	8319	2951	–9	88	70
Tulihe	Inner Mongolia	50.45	N	121.70	E	2405	14,791	902	–42	78	62
Uliastai	Inner Mongolia	45.52	N	116.97	E	2756	11,342	1892	–24	85	62
Xi Ujimqin Qi	Inner Mongolia	44.58	N	117.60	E	3271	11,137	1656	–21	83	62
Xilin Hot/Abagnar	Inner Mongolia	43.95	N	116.07	E	3251	10,480	2051	–20	85	62
Xin Barag Youqi	Inner Mongolia	48.67	N	116.82	E	1824	11,562	1945	–23	85	63
Dongtai	Jiangsu	32.87	N	120.32	E	16	3813	4612	24	91	81
Ganyu/Dayishan	Jiangsu	34.83	N	119.13	E	33	4412	4255	19	89	78
Liyang	Jiangsu	31.43	N	119.48	E	26	3517	4909	25	93	81
Lusi	Jiangsu	32.07	N	121.60	E	33	4613	4572	27	90	81
Qingjiang	Jiangsu	33.60	N	119.03	E	62	4018	4561	21	90	80
Shenyang/Hede	Jiangsu	33.77	N	120.25	E	23	4099	4370	22	90	80
Xuzhou	Jiangsu	34.28	N	117.15	E	138	4081	4695	20	92	77
Ganzhou	Jiangxi	25.85	N	114.95	E	410	1924	6919	34	94	78
Guangchang	Jiangxi	26.85	N	116.33	E	466	2289	6373	30	95	78
Ji'An	Jiangxi	27.12	N	114.97	E	256	2378	6378	32	95	79
Jingdezhen	Jiangxi	29.30	N	117.20	E	197	2620	5889	29	95	80
Lu Shan (Mountain)	Jiangxi	29.58	N	115.98	E	3822	4773	3240	17	80	72
Nanchang	Jiangxi	28.60	N	115.92	E	164	2685	5976	31	94	80
Nancheng	Jiangxi	27.58	N	116.65	E	269	2509	6120	31	94	79
Xiushui	Jiangxi	29.03	N	114.58	E	482	2853	5582	27	95	79
Xunwu	Jiangxi	24.95	N	115.65	E	981	1658	6685	33	92	77

TABLE D-3 International Climatic Data *(continued)*

Country City	Province or Region	Latitude		Longitude		Elev., ft	HDD65	CDD50	Heating Design Temperature 99.6%	Cooling Design Temperature Dry-Bulb 1.0%	Cooling Design Temperature Wet-Bulb 1.0%
(China cont.)											
Yichun	Jiangxi	27.80	N	114.38	E	423	2717	5726	30	94	79
Changbai	Jilin	41.35	N	128.17	E	3340	10,452	1502	–17	78	66
Changchun	Jilin	43.90	N	125.22	E	781	8844	2708	–13	85	70
Changling	Jilin	44.25	N	123.97	E	623	8939	2725	–14	86	69
Dunhua	Jilin	43.37	N	128.20	E	1726	9923	1891	–17	81	68
Huadian	Jilin	42.98	N	126.75	E	866	9326	2484	–26	84	71
Ji'An	Jilin	41.10	N	126.15	E	587	7612	2944	–9	86	72
Linjiang	Jilin	41.72	N	126.92	E	1093	8645	2573	–15	85	71
Qian Gorlos	Jilin	45.12	N	124.83	E	453	9062	2770	–16	86	71
Yanji	Jilin	42.88	N	129.47	E	584	8680	2396	–10	85	70
Chaoyang	Liaoning	41.55	N	120.45	E	577	7072	3397	–5	90	70
Dalian/Dairen/Luda	Liaoning	38.90	N	121.63	E	318	5648	3441	10	86	73
Dandong	Liaoning	40.05	N	124.33	E	46	6642	3014	2	83	74
Haiyang Island	Liaoning	39.05	N	123.22	E	33	5475	3341	13	82	77
Jinzhou	Liaoning	41.13	N	121.12	E	230	6598	3397	2	87	72
Kuandian	Liaoning	40.72	N	124.78	E	856	7744	2667	–10	84	72
Qingyuan	Liaoning	42.10	N	124.95	E	771	8373	2749	–17	87	71
Shenyang/Dongta	Liaoning	41.77	N	123.43	E	141	7218	3325	–8	87	73
Siping	Liaoning	43.18	N	124.33	E	541	8240	2898	–10	86	71
Yingkou	Liaoning	40.67	N	122.20	E	13	6765	3403	0	85	75
Zhangwu	Liaoning	42.42	N	122.53	E	276	7754	3060	–8	87	71
Yanchi	Ningxia	37.78	N	107.40	E	4426	6914	2774	–2	88	61
Yinchuan	Ningxia	38.48	N	106.22	E	3648	6617	2979	1	87	66
Zhongning	Ningxia	37.48	N	105.67	E	3888	6217	3070	3	88	66
Daqaidam	Qinghai	37.85	N	95.37	E	10,413	10,776	734	–11	74	49
Darlag	Qinghai	33.75	N	99.65	E	13,018	12,136	100	–13	62	48
Delingha	Qinghai	37.37	N	97.37	E	9783	9185	1170	–3	77	53
Dulan/Qagan Us	Qinghai	36.30	N	98.10	E	10,472	9668	770	–1	74	50
Gangca/Shaliuhe	Qinghai	37.33	N	100.13	E	10,830	11,792	174	–7	64	50
Golmud	Qinghai	36.42	N	94.90	E	9216	8414	1442	1	79	52
Henan	Qinghai	34.73	N	101.60	E	11,483	11,607	155	–17	64	50
Lenghu	Qinghai	38.83	N	93.38	E	8970	10,060	1142	–8	78	49

TABLE D-3 International Climatic Data *(continued)*

Country City	Province or Region	Latitude		Longitude		Elev., ft	HDD65	CDD50	Heating Design Temperature	Cooling Design Temperature Dry-Bulb	Cooling Design Temperature Wet-Bulb
									99.6%	1.0%	1.0%
(China cont.)											
Madoi/Huangheyan	Qinghai	34.92	N	98.22	E	14,019	14,135	31	–18	58	43
Qumarleb	Qinghai	34.13	N	95.78	E	13,701	13,175	67	–16	62	46
Tongde	Qinghai	35.27	N	100.65	E	10,794	11,220	288	–14	68	51
Tuotuohe/Tanggulash	Qinghai	34.22	N	92.43	E	14,879	14,505	21	–21	60	42
Wudaoliang	Qinghai	35.22	N	93.08	E	15,135	15,114	8	–16	56	40
Xining	Qinghai	36.62	N	101.77	E	7421	7417	1620	3	78	57
Yushu	Qinghai	33.02	N	97.02	E	12,080	9354	550	–2	70	52
Zadoi	Qinghai	32.90	N	95.30	E	13,346	11,257	218	–9	65	48
Ankang/Xing'an	Shaanxi	32.72	N	109.03	E	955	3242	4920	28	93	76
Baoji	Shaanxi	34.35	N	107.13	E	2001	4345	3985	21	92	71
Hanzhong	Shaanxi	33.07	N	107.03	E	1670	3676	4253	27	89	75
Hua Shan (Mount)	Shaanxi	34.48	N	110.08	E	6768	7893	1516	5	72	60
Tongchuan	Shaanxi	35.17	N	109.05	E	2999	5470	3117	14	87	67
Xi'An	Shaanxi	34.30	N	108.93	E	1306	4332	4276	21	93	74
Yan An	Shaanxi	36.60	N	109.50	E	3146	5872	3132	6	89	66
Yulin	Shaanxi	38.23	N	109.70	E	3471	7039	2834	–5	88	64
Chengshantou (Cape)	Shandong	37.40	N	122.68	E	154	5125	3151	20	79	74
Dezhou	Shandong	37.43	N	116.32	E	72	4643	4591	16	91	75
Haiyang	Shandong	36.77	N	121.17	E	210	4943	3742	16	85	74
Heze/Caozhou	Shandong	35.25	N	115.43	E	167	4280	4627	18	92	77
Huimin	Shandong	37.50	N	117.53	E	39	5009	4270	12	91	75
Jinan/Sinan	Shandong	36.68	N	116.98	E	190	4161	5036	18	93	74
Linyi	Shandong	35.05	N	118.35	E	282	4388	4395	18	90	76
Longkou	Shandong	37.62	N	120.32	E	16	5167	3822	17	88	76
Quingdao/Singtao	Shandong	36.07	N	120.33	E	253	4651	3872	19	86	74
Rizhao	Shandong	35.38	N	119.53	E	49	4595	3926	19	85	78
Tai Shan (Mtns)	Shandong	36.25	N	117.10	E	5039	8288	1537	2	71	63
Weifang	Shandong	36.70	N	119.08	E	167	4816	4315	12	91	75
Xinxian	Shandong	36.03	N	115.58	E	154	4619	4426	16	92	77
Yanzhou	Shandong	35.57	N	116.85	E	174	4526	4412	15	92	76
Yiyuan/Nanma	Shandong	36.18	N	118.15	E	991	5093	3949	12	89	72
Datong	Shanxi	40.10	N	113.33	E	3507	7877	2512	–5	86	63

TABLE D-3 International Climatic Data *(continued)*

Country City	Province or Region	Latitude		Longitude		Elev., ft	HDD65	CDD50	Heating Design Temperature 99.6%	Cooling Design Temperature Dry-Bulb 1.0%	Cooling Design Temperature Wet-Bulb 1.0%
(China cont.)											
Hequ	Shanxi	39.38	N	111.15	E	2825	7336	2879	–7	89	66
Jiexiu	Shanxi	37.05	N	111.93	E	2461	5700	3285	8	89	68
Lishi	Shanxi	37.50	N	111.10	E	3120	6542	2959	1	88	66
Taiyuan/Wusu/Wusu	Shanxi	37.78	N	112.55	E	2556	6066	3132	5	88	69
Wutai Shan (Mtn)	Shanxi	39.03	N	113.53	E	9508	14,214	100	–19	63	53
Yangcheng	Shanxi	35.48	N	112.40	E	2162	5057	3714	14	88	69
Yuanping	Shanxi	38.75	N	112.70	E	2749	6705	2943	2	88	66
Yuncheng	Shanxi	35.03	N	111.02	E	1234	4433	4553	18	94	72
Yushe	Shanxi	37.07	N	112.98	E	3419	6482	2777	3	85	64
Barkam	Sichuan	31.90	N	102.23	E	8747	5419	1882	13	79	59
Batang	Sichuan	30.00	N	99.10	E	8494	3599	3267	22	85	59
Chengdu	Sichuan	30.67	N	104.02	E	1667	2708	4843	33	88	76
Da Xian	Sichuan	31.20	N	107.50	E	1020	2498	5455	34	94	78
Daocheng/Dabba	Sichuan	29.05	N	100.30	E	12,234	8614	624	4	68	49
Dawu	Sichuan	30.98	N	101.12	E	9708	6110	1639	11	77	57
Emei Shan	Sichuan	29.52	N	103.33	E	10,003	9458	381	8	61	54
Fengjie	Sichuan	31.05	N	109.50	E	1991	2889	5043	32	92	75
Garze	Sichuan	31.62	N	100.00	E	11,135	7656	991	5	72	53
Jiulong/Gyaisi	Sichuan	29.00	N	101.50	E	9823	5505	1568	18	75	55
Kangding/Dardo	Sichuan	30.05	N	101.97	E	8586	6870	1224	17	71	58
Langzhong	Sichuan	31.58	N	105.97	E	1263	2553	5192	34	92	77
Liangping	Sichuan	30.68	N	107.80	E	1493	2733	5111	33	92	77
Litang	Sichuan	30.00	N	100.27	E	12,959	9367	370	1	65	48
Luzhou	Sichuan	28.88	N	105.43	E	1102	2150	5690	38	93	78
Mianyang	Sichuan	31.47	N	104.68	E	1549	2771	4943	31	90	75
Nanchong	Sichuan	30.80	N	106.08	E	1017	2446	5422	35	93	78
Neijiang	Sichuan	29.58	N	105.05	E	1171	2235	5591	36	93	78
Pingwu	Sichuan	32.42	N	104.52	E	2877	3115	4327	30	88	71
Songpan/Sungqu	Sichuan	32.65	N	103.57	E	9357	7329	1094	8	74	56
Wanyuan	Sichuan	32.07	N	108.03	E	2211	3354	4305	28	90	73
Xichang	Sichuan	27.90	N	102.27	E	5246	1736	5211	35	87	65
Ya'An	Sichuan	29.98	N	103.00	E	2064	2584	4962	34	88	76

TABLE D-3 International Climatic Data *(continued)*

Country City	Province or Region	Latitude		Longitude		Elev., ft	HDD65	CDD50	Heating Design Temperature	Cooling Design Temperature Dry-Bulb	Cooling Design Temperature Wet-Bulb
									99.6%	1.0%	1.0%
(China cont.)											
Yibin	Sichuan	28.80	N	104.60	E	1122	2043	5715	38	92	78
Youyang	Sichuan	28.83	N	108.77	E	2182	3311	4486	29	88	74
Baingoin	Tibet	31.37	N	90.02	E	15,423	12,487	70	–7	60	42
Dengqen	Tibet	31.42	N	95.60	E	12,710	9327	508	4	68	50
Lhasa	Tibet	29.67	N	91.13	E	11,975	6560	1433	14	75	52
Lhunze	Tibet	28.42	N	92.47	E	12,667	7949	864	8	69	49
Nagqu	Tibet	31.48	N	92.07	E	14,790	12,539	64	–11	62	44
Nyingchi	Tibet	29.57	N	94.47	E	9846	5624	1610	19	73	57
Pagri	Tibet	27.73	N	89.08	E	14,111	11,576	12	–5	55	45
Qamdo	Tibet	31.15	N	97.17	E	10,850	6550	1533	10	78	55
Shiquanhe	Tibet	32.50	N	80.08	E	14,039	12,092	517	–14	70	45
Sog Xian	Tibet	31.88	N	93.78	E	13,202	10,546	316	–6	67	49
Tingri/Xegar	Tibet	28.63	N	87.08	E	14,114	9994	456	0	67	46
Xainza	Tibet	30.95	N	88.63	E	15,325	11,849	98	–5	62	42
Xigaze	Tibet	29.25	N	88.88	E	12,589	7635	1064	6	72	51
Akqi	Xinjiang	40.93	N	78.45	E	6516	7653	2055	0	81	57
Alar	Xinjiang	40.50	N	81.05	E	3323	5921	3882	3	92	67
Altay	Xinjiang	47.73	N	88.08	E	2418	9426	2390	–21	85	63
Andir	Xinjiang	37.93	N	83.65	E	4147	6189	3804	–1	96	62
Bachu	Xinjiang	39.80	N	78.57	E	3665	5431	4284	7	94	65
Balguntay	Xinjiang	42.67	N	86.33	E	5751	7609	1963	1	81	56
Bayanbulak	Xinjiang	43.03	N	84.15	E	8068	15,010	204	–37	67	50
Baytik Shan (Mtns)	Xinjiang	45.37	N	90.53	E	5417	10,272	1357	–11	78	53
Fuyun	Xinjiang	46.98	N	89.52	E	2713	10,149	2386	–27	89	60
Hami	Xinjiang	42.82	N	93.52	E	2425	6518	3926	–1	95	66
Hoboksar	Xinjiang	46.78	N	85.72	E	4245	9445	1739	–9	81	57
Hotan	Xinjiang	37.13	N	79.93	E	4511	5069	4215	12	92	65
Jinghe	Xinjiang	44.62	N	82.90	E	1053	7844	3610	–15	94	69
Kaba He	Xinjiang	48.05	N	86.35	E	1752	9156	2491	–20	87	65
Karamay	Xinjiang	45.60	N	84.85	E	1404	7867	4225	–14	95	63
Kashi	Xinjiang	39.47	N	75.98	E	4236	5421	3784	8	90	65
Korla	Xinjiang	41.75	N	86.13	E	3061	5680	4212	7	93	66

TABLE D-3 International Climatic Data *(continued)*

Country City	Province or Region	Latitude		Longitude		Elev., ft	HDD65	CDD50	Heating Design Temperature	Cooling Design Temperature Dry-Bulb	Cooling Design Temperature Wet-Bulb
									99.6%	1.0%	1.0%
(China cont.)											
Kuqa	Xinjiang	41.72	N	82.95	E	3609	5703	3945	6	91	64
Mangnai	Xinjiang	38.25	N	90.85	E	9662	10,445	727	–3	76	48
Pishan	Xinjiang	37.62	N	78.28	E	4514	5337	4071	8	93	65
Qijiaojing	Xinjiang	43.48	N	91.63	E	2867	7117	3691	–2	95	60
Qitai	Xinjiang	44.02	N	89.57	E	2605	8861	2793	–20	90	63
Ruoqiang	Xinjiang	39.03	N	88.17	E	2917	5751	4280	5	98	66
Shache	Xinjiang	38.43	N	77.27	E	4042	5408	3871	9	91	66
Tacheng	Xinjiang	46.73	N	83.00	E	1755	7772	2834	–11	90	64
Tikanlik	Xinjiang	40.63	N	87.70	E	2779	6093	4132	1	96	67
Turpan	Xinjiang	42.93	N	89.20	E	121	5256	6038	7	104	70
Urumqi	Xinjiang	43.78	N	87.62	E	3015	8214	3015	–7	89	61
Yining	Xinjiang	43.95	N	81.33	E	2175	6617	3085	–8	89	66
Yiwu/Araturuk	Xinjiang	43.27	N	94.70	E	5673	9362	1538	–7	78	56
Baoshan	Yunnan	25.13	N	99.22	E	5430	2150	4324	34	81	66
Chuxiong	Yunnan	25.02	N	101.53	E	5817	2102	4413	33	82	63
Dali	Yunnan	25.70	N	100.18	E	6535	2398	3815	34	79	64
Deqen	Yunnan	28.50	N	98.90	E	11,444	7883	668	18	66	53
Guangnan	Yunnan	24.07	N	105.07	E	4104	1837	5381	33	85	67
Huili	Yunnan	26.65	N	102.25	E	5866	2471	4074	30	82	64
Huize	Yunnan	26.42	N	103.28	E	6923	3522	3015	25	78	62
Jiangcheng	Yunnan	22.62	N	101.82	E	3678	757	6438	42	85	68
Jinghong	Yunnan	22.02	N	100.80	E	1814	92	9106	49	93	72
Kunming/Wujiaba	Yunnan	25.02	N	102.68	E	6207	2461	3766	33	79	63
Lancang/Menglangba	Yunnan	22.57	N	99.93	E	3458	491	7158	41	88	66
Lijing	Yunnan	26.83	N	100.47	E	7854	3389	2818	30	76	60
Lincang	Yunnan	23.95	N	100.22	E	4931	1131	5588	39	83	64
Luxi	Yunnan	24.53	N	103.77	E	5604	2254	4341	31	81	63
Mengding	Yunnan	23.57	N	99.08	E	1680	168	8782	46	93	72
Mengla	Yunnan	21.50	N	101.58	E	2077	133	8686	47	91	72
Mengzi	Yunnan	23.38	N	103.38	E	4272	947	6397	39	86	66
Ruili	Yunnan	24.02	N	97.83	E	2546	478	7544	43	88	70
Simao	Yunnan	22.77	N	100.98	E	4275	796	6251	42	85	64

TABLE D-3 International Climatic Data *(continued)*

Country City	Province or Region	Latitude	Longitude	Elev., ft	HDD65	CDD50	Heating Design Temperature	Cooling Design Temperature Dry-Bulb	Cooling Design Temperature Wet-Bulb
							99.6%	1.0%	1.0%
(China cont.)									
Tengchong	Yunnan	25.12 N	98.48 E	5410	2161	4008	34	78	64
Yuanjiang	Yunnan	23.60 N	101.98 E	1306	166	9856	48	98	75
Yuanmou	Yunnan	25.73 N	101.87 E	3675	503	8165	41	93	67
Zhanyi	Yunnan	25.58 N	103.83 E	6234	2526	3855	30	80	61
Zhaotong	Yunnan	27.33 N	103.75 E	6398	4062	2977	23	80	63
Dachen Island	Zhejiang	28.45 N	121.88 E	276	2708	4966	34	84	80
Dinghai	Zhejiang	30.03 N	122.12 E	121	2799	5158	31	88	80
Hangzhou/Jianqiao	Zhejiang	30.23 N	120.17 E	141	3069	5353	28	95	81
Kuocang Shan	Zhejiang	28.82 N	120.92 E	4498	5430	2585	13	77	70
Lishui	Zhejiang	28.45 N	119.92 E	203	2311	6205	30	96	79
Qixian Shan	Zhejiang	27.95 N	117.83 E	4623	4321	3155	19	77	70
Qu Xian	Zhejiang	28.97 N	118.87 E	233	2724	5740	30	95	80
Shengsi/Caiyuanzhen	Zhejiang	30.73 N	122.45 E	266	2955	4905	31	87	79
Shengxian	Zhejiang	29.60 N	120.82 E	354	2999	5431	27	94	80
Shipu	Zhejiang	29.20 N	121.95 E	417	2785	5166	31	88	80
Taishan	Zhejiang	27.00 N	120.70 E	348	2271	5424	38	85	79
Tianmu Shan (Mtns)	Zhejiang	30.35 N	119.42 E	4902	6115	2225	11	75	69
Wenzhou	Zhejiang	28.02 N	120.67 E	23	2104	5981	34	91	81
Cuba									
Guantanamo Bay NAS	Ote.	19.90 N	75.15 W	75	0	11,719	67	93	78
Cyprus									
Akrotiri		34.58 N	32.98 E	75	1287	6147	40	89	72
Larnaca		34.88 N	33.63 E	7	1452	6028	37	91	72
Paphos		34.75 N	32.40 E	30	1279	5924	39	86	76
Czech Republic (Former Czechoslovakia)									
Prague/Libus		50.00 N	14.45 E	1001	6376	1853	3	80	64
Dominican Republic									
Santo Domingo		18.47 N	69.88 W	43	0	10,862	N.A.	N.A.	N.A.
Egypt									
Cairo		30.13 N	31.40 E	243	834	7993	45	97	69
Luxor		25.67 N	32.70 E	289	581	9849	40	108	71
Finland									
Helsinki/Seutula		60.32 N	24.97 E	167	9051	1138	–11	75	61

TABLE D-3 International Climatic Data *(continued)*

Country / City	Province or Region	Latitude	Longitude	Elev., ft	HDD65	CDD50	Heating Design Temperature 99.6%	Cooling Design Temperature Dry-Bulb 1.0%	Cooling Design Temperature Wet-Bulb 1.0%
France									
Lyon/Satolas		45.73 N	5.08 E	814	4930	2609	17	86	69
Marseille		43.45 N	5.22 E	26	3194	3933	25	87	70
Nantes		47.17 N	1.60 W	89	4286	2480	23	83	68
Nice		43.65 N	7.20 E	33	2641	3983	35	83	73
Paris/Le Bourget		48.97 N	2.45 E	217	5046	2211	18	82	68
Strasbourg		48.55 N	7.63 E	502	5533	2193	12	84	68
Germany									
Berlin/Schoenfeld		52.38 N	13.52 E	154	6331	1820	11	82	65
Hamburg		53.63 N	9.98 E	52	6319	1569	11	79	64
Hannover		52.47 N	9.70 E	180	6093	1730	9	80	65
Mannheim		49.53 N	8.50 E	318	5428	2262	N.A.	N.A.	N.A.
Greece									
Souda	Crete	35.55 N	24.12 E	417	1767	5472	39	90	67
Thessalonika/Mikra		40.52 N	22.97 E	26	3389	4115	25	90	69
Greenland									
Narssarssuaq		61.18 N	45.42 W	79	11,521	292	–18	62	49
Hungary									
Budapest/Lorinc		47.43 N	19.18 E	459	5534	2647	8	86	68
Iceland									
Reykjavik		64.13 N	21.93 W	200	9286	293	14	58	52
India									
Ahmedabad		23.07 N	72.63 E	180	31	11,648	52	106	74
Bangalore		12.97 N	77.58 E	3018	2	9409	59	92	67
Bombay/Santa Cruz		19.12 N	72.85 E	26	2	11,372	62	93	74
Calcutta/Dum Dum		22.65 N	88.45 E	16	26	11,064	54	97	79
Madras		13.00 N	80.18 E	52	0	12,403	68	99	77
Nagpur Sonegaon		21.10 N	79.05 E	1014	18	11,274	53	108	71
New Delhi/Safdarjung		28.58 N	77.20 E	702	480	10,060	44	105	72
Indonesia									
Djakarta/Halimperda	Java	6.25 S	106.90 E	98	0	11,477	N.A.	N.A.	N.A.
Kupang Penfui	Sunda Island	10.17 S	123.67 E	354	2	11,686	N.A.	N.A.	N.A.
Makassar	Celebes	5.07 S	119.55 E	56	3	11,481	N.A.	N.A.	N.A.

TABLE D-3 International Climatic Data *(continued)*

Country City	Province or Region	Latitude		Longitude		Elev., ft	HDD65	CDD50	Heating Design Temperature	Cooling Design Temperature Dry-Bulb	Cooling Design Temperature Wet-Bulb
									99.6%	1.0%	1.0%
(Indonesia cont.)											
Medan	Sumatra	3.57	N	98.68	E	85	0	11,491	N.A.	N.A.	N.A.
Palembang	Sumatra	2.90	S	104.70	E	33	0	11,565	N.A.	N.A.	N.A.
Surabaja Perak	Java	7.22	S	112.72	E	10	0	12,088	N.A.	N.A.	N.A.
Ireland											
Dublin Airport		53.43	N	6.25	W	279	5507	1276	29	69	61
Shannon Airport		52.68	N	8.92	W	66	5106	1455	28	71	63
Israel											
Jerusalem		31.78	N	35.22	E	2654	2423	4609	33	86	64
Tel Aviv Port		32.10	N	34.78	E	33	955	6851	44	86	74
Italy											
Milano/Linate		45.43	N	9.28	E	351	4507	3335	21	87	72
Napoli/Capodichino		40.88	N	14.30	E	236	2658	4301	32	89	73
Roma/Fiumicino		41.80	N	12.23	E	7	2684	4173	30	86	74
Jamaica											
Kingston/Manley		17.93	N	76.78	W	46	0	11,860	71	98	78
Montego Bay/Sangster		18.50	N	77.92	W	3	1	10,915	70	90	79
Japan											
Fukaura		40.65	N	139.93	E	223	5522	2933	30	91	78
Sapporo		43.05	N	141.33	E	56	6753	2518	12	81	71
Tokyo		35.68	N	139.77	E	118	2986	4749	31	88	77
Jordan											
Amman		31.98	N	35.98	E	2516	2337	5427	33	92	65
Kenya											
Nairobi Airport		1.32	S	36.93	E	5328	273	6177	49	83	60
Korea											
Pyonggang		38.40	N	127.30	E	1217	6735	2840	3	85	74
Seoul		37.57	N	126.97	E	282	5007	3956	N.A.	N.A.	N.A.
Malaysia											
Kuala Lumpur		3.13	N	101.55	E	56	0	11,530	71	93	78
Penang/Bayan Lepas		5.30	N	100.27	E	10	0	11,472	N.A.	N.A.	N.A.
Mexico											
Mexico City	Distrito Federal	19.40	N	99.20	W	7572	1203	4762	39	82	57

TABLE D-3 International Climatic Data *(continued)*

Country City	Province or Region	Latitude		Longitude		Elev., ft	HDD65	CDD50	Heating Design Temperature 99.6%	Cooling Design Temperature Dry-Bulb 1.0%	Cooling Design Temperature Wet-Bulb 1.0%
(Mexico cont.)											
Guadalajara	Jalisco	20.67	N	103.38	W	5213	701	6121	N.A.	N.A.	N.A.
Monterrey	Nuevo Laredo	25.87	N	100.20	W	1476	844	8326	N.A.	N.A.	N.A.
Tampico	Tamaulipas	22.22	N	97.85	W	39	216	9870	50	90	80
Veracruz	Veracruz	19.15	N	96.12	W	52	17	10,006	57	92	80
Merida	Yucatan	20.98	N	89.65	W	30	10	11,122	57	98	76
Netherlands											
Amsterdam/Schiphol		52.30	N	4.77	E	–13	5691	1619	17	77	65
New Zealand											
Auckland Airport		37.02	S	174.80	E	23	2242	3650	35	76	66
Christchurch		43.48	S	172.55	E	118	4359	2115	28	79	61
Wellington		41.28	S	174.77	E	420	3597	2258	35	71	63
Norway											
Bergen/Florida		60.38	N	5.33	E	128	6882	1014	16	68	57
Oslo/Fornebu		59.90	N	10.62	E	52	8020	1331	0	77	62
Pakistan											
Karachi Airport		24.90	N	67.13	E	75	1155	11,049	N.A.	N.A.	N.A.
Papua New Guinea											
Port Moresby		9.43	S	147.22	E	92	2	11,272	N.A.	N.A.	N.A.
Paraguay											
Asuncion/Stroessner		25.27	S	57.63	W	331	469	9005	41	95	75
Peru											
Lima-Callao/Chavez		12.00	S	77.12	W	43	260	6745	57	84	74
San Juan de Marcona		15.35	S	75.15	W	197	306	6765	N.A.	N.A.	N.A.
Talara		4.57	S	81.25	W	282	4	8973	60	88	75
Philippines											
Manila Airport	Luzon	14.52	N	121.00	E	75	0	11,449	69	93	80
Poland											
Krakow/Balice		50.08	N	19.80	E	778	6924	2007	–1	81	67
Puerto Rico											
San Juan/Isla Verde WSFO		18.43	N	66.00	W	10	0	11,406	69	90	78
Romania											
Bucuresti/Bancasa		44.50	N	26.13	E	308	5461	2948	8	88	70

TABLE D-3 International Climatic Data *(continued)*

Country City	Province or Region	Latitude		Longitude		Elev., ft	HDD65	CDD50	Heating Design Temperature	Cooling Design Temperature Dry-Bulb	Cooling Design Temperature Wet-Bulb
									99.6%	1.0%	1.0%
Russia (Former Soviet Union)											
Kaliningrad	East Prussia	54.70	N	20.62	E	89	7115	1589	–3	77	64
Krasnoiarsk		56.00	N	92.88	E	636	11,278	1351	–29	80	63
Moscow Observatory		55.75	N	37.57	E	512	8596	1708	–10	79	65
Petropavlovsk		53.02	N	158.72	E	23	10,107	530	5	66	58
Rostov-Na-Donu		47.25	N	39.82	E	259	6360	3015	2	86	68
Vladivostok		43.12	N	131.90	E	453	8915	1728	–8	75	67
Volgograd		48.68	N	44.35	E	476	7558	2840	–6	88	65
Saudi Arabia											
Dhahran		26.27	N	50.17	E	72	381	10,936	N.A.	N.A.	N.A.
Riyadh		24.70	N	46.73	E	2005	536	10,725	41	110	64
Senegal											
Dakar/Yoff		14.73	N	17.50	W	89	6	9750	61	88	77
Singapore											
Singapore/Changi		1.37	N	103.98	E	49	0	11,995	73	90	79
South Africa											
Cape Town/D F Malan		33.97	S	18.60	E	151	1685	4454	38	83	67
Johannesburg		26.13	S	28.23	E	5558	1919	4252	34	82	60
Pretoria		25.73	S	28.18	E	4364	1151	5828	39	88	63
Spain											
Barcelona		41.28	N	2.07	E	13	2638	3965	32	84	74
Madrid		40.47	N	3.57	W	1909	3669	3702	24	94	68
Valencia/Manises		39.50	N	0.47	W	203	1942	5045	34	88	72
Sweden											
Stockholm/Arlanda		59.65	N	17.95	E	200	8123	1297	–2	77	61
Switzerland											
Zurich		47.38	N	8.57	E	1867	6015	1995	13	80	65
Syria											
Damascus Airport		33.42	N	36.52	E	2001	2771	5293	25	98	64
Taiwan											
Alisan Shan		23.52	N	120.80	E	7894	4406	1958	N.A.	N.A.	N.A.
Chiayi (TW-AFB)		23.50	N	120.42	E	92	318	8926	48	91	81
Chiayyi		23.47	N	120.38	E	82	275	9288	47	92	82

TABLE D-3 International Climatic Data *(continued)*

Country City	Province or Region	Latitude	Longitude	Elev., ft	HDD65	CDD50	Heating Design Temperature	Cooling Design Temperature Dry-Bulb	Cooling Design Temperature Wet-Bulb
							99.6%	1.0%	1.0%
(Taiwan cont.)									
Chilung		25.13 N	121.75 E	10	472	8554	50	91	79
Chinmen		24.43 N	118.43 E	39	974	7420	N.A.	N.A.	N.A.
Dawu		22.35 N	120.90 E	30	24	10,355	N.A.	N.A.	N.A.
Hengchun		22.00 N	120.75 E	79	23	10,120	60	90	80
Hengchun/Wu Lu Tien		22.03 N	120.72 E	43	21	10,407	N.A.	N.A.	N.A.
Hsinchu/Singjo		24.82 N	120.93 E	26	482	8567	48	91	82
Hua Lien		23.97 N	121.62 E	62	220	8872	N.A.	N.A.	N.A.
Hwalien		24.02 N	121.62 E	49	221	9043	N.A.	N.A.	N.A.
Joyutang		23.88 N	120.85 E	3330	583	7136	N.A.	N.A.	N.A.
Kao Hsiung Intl. Arpt.		22.57 N	120.35 E	26	111	9702	53	91	80
Kao Hsiung		22.62 N	120.27 E	95	70	9940	54	90	81
Kungkuan		24.27 N	120.62 E	666	541	8306	N.A.	N.A.	N.A.
Kungshan		22.78 N	120.25 E	33	158	9526	N.A.	N.A.	N.A.
Lan Yu		22.03 N	121.55 E	1066	95	8765	57	84	80
Makung		23.57 N	119.62 E	102	283	8957	52	89	82
Matsu Island		26.17 N	119.93 E	302	1948	5898	N.A.	N.A.	N.A.
North Pingtung		22.70 N	120.47 E	95	88	10,049	52	93	81
Peng Hu		23.52 N	119.57 E	69	287	9068	N.A.	N.A.	N.A.
Penkaiyu		25.63 N	122.07 E	335	531	8160	N.A.	N.A.	N.A.
Sing Jo		24.80 N	120.97 E	108	534	8480	N.A.	N.A.	N.A.
Sinkung		23.10 N	121.37 E	121	88	9601	N.A.	N.A.	N.A.
South Pingtung		22.67 N	120.45 E	79	71	10,228	53	93	81
Taichung		24.15 N	120.68 E	256	312	8991	49	91	79
Taichung/Shui Nan		24.18 N	120.65 E	364	381	8915	46	93	82
Tainan (TW-AFB)		22.95 N	120.20 E	52	150	9729	50	91	82
Tainan		23.00 N	120.22 E	46	178	9577	51	91	81
Taipei		25.03 N	121.52 E	26	438	8896	48	93	80
Taipei/Chiang Kai Shek		25.08 N	121.23 E	75	594	8456	48	92	80
Taipei/Sungshan		25.07 N	121.53 E	20	506	8454	48	93	81
Taitung		22.75 N	121.15 E	33	74	9754	N.A.	N.A.	N.A.
Taitung/Fongyentsun		22.80 N	121.18 E	121	72	9767	N.A.	N.A.	N.A.
Taoyuan (AB)		25.07 N	121.23 E	164	626	8315	47	92	82

TABLE D-3 International Climatic Data *(continued)*

Country City	Province or Region	Latitude	Longitude	Elev., ft	HDD65	CDD50	Heating Design Temperature	Cooling Design Temperature Dry-Bulb	Cooling Design Temperature Wet-Bulb
							99.6%	1.0%	1.0%
(Taiwan cont.)									
Tung Shih		23.27 N	119.67 E	148	191	9217	N.A.	N.A.	N.A.
Wu-Chi		24.25 N	120.52 E	16	405	8691	50	90	81
Yilan		24.77 N	121.75 E	23	411	8416	N.A.	N.A.	N.A.
Tanzania									
Dar es Salaam		6.88 S	39.20 E	180	4	10,755	N.A.	N.A.	N.A.
Thailand									
Bangkok		13.73 N	100.57 E	52	0	12,430	65	97	79
Tunisia									
Tunis/El Aouina		36.83 N	10.23 E	16	1657	5769	41		73
Turkey									
Adana		37.00 N	35.42 E	217	1847	6098	32	94	71
Ankara/Etimesgut		39.95 N	32.68 E	2644	5162	3077	2	86	63
Istanbul/Yesilkoy		40.97 N	28.82 E	121	3534	3777	26	84	69
United Kingdom									
Birmingham	England	52.45 N	1.73 W	325	5866	1355	21	75	62
Edinburgh	Scotland	55.95 N	3.35 W	135	6347	1001	21	69	60
Glasgow Apt	Scotland	55.87 N	4.43 W	23	6287	1041	21	71	61
London/Heathrow	England	51.48 N	0.45 W	79	5015	1894	25	78	64
Uruguay									
Montevideo/Carrasco		34.83 S	56.03 W	108	2124	4602	35	86	71
Venezuela									
Caracas/Maiquetia		10.60 N	66.98 W	236	9	11,501	70	91	83
Vietnam									
Hanoi/Gialam		21.02 N	105.80 E	26	330	9868	N.A.	N.A.	N.A.
Saigon (Ho Chi Minh)		10.82 N	106.67 E	62	0	12,057	68	94	77

(This appendix is not part of this standard. It is merely informative and does not contain requirements necessary for conformance to the standard. It has not been processed according to the ANSI requirements for a standard and may contain material that has not been subject to public review or a consensus process. Unresolved objectors on informative material are not offered the right to appeal at ASHRAE or ANSI.)

INFORMATIVE APPENDIX E
INFORMATIVE REFERENCES

This appendix contains informative references for the convenience of users of Standard 90.1-2007 and to acknowledge source documents when appropriate. Some documents are also included in Section 12, "Normative References," because there are other citations of those documents within the standard that are normative.

Address/Contact Information

AABC
Associated Air Balance Council
1518 K Street Northwest, Suite 503
Washington, DC 20005
aabchg@aol.com

BLAST
Building Systems Laboratory
University of Illinois
1206 West Green Street
Urbana, IL 61801
www.bso.uiuc.edu/BLAST/index.html

CRRC
Cool Roof Rating Council
1738 Excelsior Avenue
Oakland, CA 94602
(T) 866-465-2523
(T) 510-482-4420
(F) 510-482-4421
www.coolroofs.org

DOE-2
Building Energy Simulation news
http://simulationresearch.lbl.gov/un.html

MICA
Midwest Insulation Contractors Association
16712 Elm Circle
Omaha, NE 68130
www.micainsulation.org

NEBB
National Environmental Balancing Bureau
8575 Grovemont Circle
Gaithersburg, MD 20877
www.nebb.org

SMACNA
Sheet Metal & Air Conditioning Contractors' National Association
4201 Lafayette Center Drive
Chantilly, VA 20151
info@smacna.org
www.smacna.org

TMY2 Data
National Renewable Energy Laboratory
NREL/RReDC
Attn: Pamela Gray-Hann
1617 Cole Blvd., MS-1612
Golden, Colorado, USA 80401
http://rredc.nrel.gov/solar/old_data/nsrdb/tmy2/

WYEC2 Data
American Society of Heating, Refrigerating and Air-Conditioning Engineers, Inc.
ASHRAE Bookstore
1791 Tullie Circle, NE
Atlanta, GA 30329-2305
(T) 404-636-8400
(F) 404-321-5478
www.ashrae.org/bookstore

IWEC Data
American Society of Heating, Refrigerating and Air-Conditioning Engineers, Inc.
ASHRAE Bookstore
1791 Tullie Circle, NE
Atlanta, GA 30329-2305
(T) 404-636-8400
(F) 404-321-5478
www.ashrae.org/bookstore

Subsection No.	Reference	Title/Source
Exception to 5.5.3.1	CRRC-1-2002	Cool Roof Rating Council Product Rating Program
6.4.2	*2001 ASHRAE Handbook—Fundamentals*	ASHRAE
6.4.4.1.1	MICA Insulation Standards—1999	National Commercial and Industrial Insulation Standards
6.4.4.2.1	SMACNA Duct Construction Standards—1995	HVAC Duct Construction Standards, Metal and Flexible
6.4.4.2.2	SMACNA Duct Leakage Test Procedures—1985	*HVAC Air Duct Leakage Test Manual*
6.7.2.3.1	NEBB Procedural Standards—1999	Procedural Standards for Building Systems Commissioning
6.7.2.3.1	AABC 2002	Associated Air Balance Council Test and Balance Procedures
6.7.2.3.1	ASHRAE Standard 111-1988	*Practices for Measurement, Testing, Adjusting and Balancing of Building Heating, Ventilation, Air-Conditioning and Refrigeration Systems*
6.7.2.2	ASHRAE Guideline 4-1993	*Preparation of Operating and Maintenance Documentation for Building Systems*
6.7.2.4	ASHRAE Guideline 1-1996	*The HVAC Commissioning Process*
7.4.1 and 7.5	*2003 ASHRAE Handbook—HVAC Applications*	Chapter 49, Service Water Heating/ASHRAE
11.2.1	DOE-2	Support provided by Lawrence Berkeley National Laboratory at the referenced Web site
11.2.1	BLAST	University of Illinois
11.2.2	IWEC	International Weather for Energy Calculations
11.2.2	TMY 2 Data	Typical Meteorological Year

(This appendix is not part of this standard. It is merely informative and does not contain requirements necessary for conformance to the standard. It has not been processed according to the ANSI requirements for a standard and may contain material that has not been subject to public review or a consensus process. Unresolved objectors on informative material are not offered the right to appeal at ASHRAE or ANSI.)

INFORMATIVE APPENDIX F ADDENDA DESCRIPTION INFORMATION

ASHRAE/IESNA Standard 90.1-2007 incorporates ANSI/ASHRAE/IESNA Standard 90.1-2004 and Addenda a, b, c, d, e, f, g, h, i, j, k, l, m, n, o, p, q, r, s, t, u, v, x, y, aa, ab, ac, ad, ae, af, ag, ah, ai, aj, ak, al, am, an, ap, aq, ar, and av to ANSI/ASHRAE/IESNA Standard 90.1-2004. Table F-1 lists each addendum and describes the way in which the text is affected by the change and states the ASHRAE and ANSI approval dates.

TABLE F-1 Addenda to ANSI/ASHRAE/IESNA Standard 90.1-2004, Changes Identified

Addenda to 90.1-2004	Section(s) Affected	Description of Changes[a]	ASHRAE Standards Committee Approval Date	ASHRAE Board of Directors Approval Date	IESNA Approval Date	ANSI Approval Date
90.1a	Informative Appendix G	This addendum clarifies how windows should be distributed in the baseline simulation model and how uninsulated assemblies should be treated in the baseline simulation model, increases the size range for the use of packaged VAV systems in the baseline model, and provides more detail on how service hot-water systems should be modeled. Many of these changes may affect the ultimate performance rating of buildings using Appendix G. In addition, a reference was added to ASHRAE Standard 140 for the method of testing simulation programs.	1/21/06	1/26/06	1/18/06	4/10/06
90.1b	6. Heating, Ventilating, and Air Conditioning	Revises Table 6.8.1D and adds a definition for *single-package vertical air-conditioner* and *single-package vertical heat pump*.	6/25/05	6/30/05	8/3/05	8/3/05
90.1c	5. Building Envelope	This addendum revises the definition of *building entrance* to include vestibules and clarifies the requirements and exceptions for vestibules in Section 5.4.3.4.	6/25/05	6/30/05	8/3/05	8/3/05
90.1d	12. Normative References	This addendum updates the references applicable to the building envelope and deletes references that are not cited in the standard.	6/25/05	6/30/05	8/3/05	8/3/05
90.1e	9. Lighting	This addendum recognizes that track and busway type lighting systems can be limited by circuit breakers and permanently installed current limiters in Section 9.1.4.	6/25/05	6/30/05	8/3/05	8/3/05
90.1f	6. Heating, Ventilating, and Air Conditioning	This addendum modifies Tables 6.8.1A and 6.8.1B by raising the minimum efficiency for three-phase air-cooled central conditioners and heat pumps to be consistent with federal minimum standards.	1/21/06	1/26/06	1/18/06	4/10/06
90.1g	6. Heating, Ventilating, and Air Conditioning	This addendum amends the minimum efficiency levels of air-cooled air conditioners and heat pumps in Tables 6.8.1 A and 6.8.1B to be consistent with federal minimum standards.	1/21/06	1/26/06	1/18/06	4/10/06
90.1h	6. Heating, Ventilating, and Air Conditioning	This addendum revises the exceptions to Sections 6.4.3.1.2 and 6.4.3.6 by removing data processing centers from having specific exceptions on temperature and humidification dead bands.	1/21/06	1/26/06	1/18/06	4/10/06
90.1i	9. Lighting	This addendum adds language to Section 9.1.4(b) that allows additional flexibility in assigning wattage to luminaires with multi-level ballasts where other luminaire components would restrict lamp size.	1/21/06	1/26/06	1/18/06	4/10/06
90.1j	9. Lighting	This addendum to Section 9.4.1.3 allows additional flexibility in complying with the controls requirements by allowing additional combinations of commonly available control equipment.	1/21/06	1/26/06	1/18/06	4/10/06
90.1k	Informative Appendix A	This addendum revises Table A2.3 to add U-factors for screw-down roofs with R-19 insulation.	1/21/06	1/26/06	1/18/06	4/10/06
90.1l	12. Normative References	This addendum updates the reference to ASHRAE Standard 140.	1/21/06	1/26/06	1/18/06	4/10/06

TABLE F-1 Addenda to ANSI/ASHRAE/IESNA Standard 90.1-2004, Changes Identified *(continued)*

Addenda to 90.1-2004	Section(s) Affected	Description of Changes*	ASHRAE Standards Committee Approval Date	ASHRAE Board of Directors Approval Date	IESNA Approval Date	ANSI Approval Date
90.1m	9. Lighting	This addendum revises the exception to Section 9.2.2.3 to provide an option for compliance that exempts the commonly used furniture mounted track lighting if it incorporates automatic shutoff.	1/21/06	1/26/06	1/18/06	4/10/06
90.1n	5. Building Envelope	This addendum revises Section 5.5.4.4.1 to provide an exception to allow a user to take credit for overhangs towards compliance with the maximum SHGC requirements.	1/21/06	1/26/06	1/18/06	4/10/06
90.1o	Normative Appendix D	This addendum increases the amount of international climatic data in Appendix D.	1/21/06	1/26/06	1/18/06	4/10/06
90.1p	9. Lighting	This addendum modifies Exception (g) to Section 9.2.2.3 to allow for increased lighting for medical- and age-related issues in addition to visual impairment.	1/21/06	1/26/06	1/18/06	4/10/06
90.1q	6. Heating, Ventilating, and Air Conditioning	This addendum removes Exception (a) to Section 6.4.3.2 for HVAC systems serving hotel/motel rooms and guest rooms.	1/21/06	1/26/06	1/18/06	4/10/06
90.1r	12. Normative References	This addendum updates the reference to ARI 340/260 from the 2000 edition to the 2004 edition.	1/21/06	1/26/06	1/18/06	4/10/06
90.1s	6. Heating, Ventilating, and Air Conditioning and 12. Normative References	This addendum updates language in the standard based on differences between Standard 62-1999 and 62.1-2004. The reference has also been updated.	1/21/06	1/26/06	1/18/06	4/10/06
90.1t	6. Heating, Ventilating, and Air Conditioning and 12. Normative References	This addendum changes Table 6.8.1F to add an additional requirement of combustion efficiency to the current requirement of thermal efficiency for boilers, which will increase minimum efficiency. The reference in Section 12 has also been changed to reflect the change in the table.	1/21/06	1/26/06	1/18/06	4/10/06
90.1u	Informative Appendix G	This addendum provides guidance for complying with the intent of the baseline building design for HVAC systems 5, 6, 7, and 8, which shall be modeled as floor-by-floor HVAC systems.	1/21/06	1/26/06	1/18/06	4/10/06
90.1v	6. Heating, Ventilating, and Air Conditioning	This addendum modifies the provisions of Section 6.4.3.8 to allow for demand control ventilation.	1/21/06	1/26/06	1/18/06	5/10/06
90.1x	12. Normative References and Informative Appendix G	This addendum updates the normative references in Section 12 and Informative Appendix G for ATM-02 to ATM-04.	1/21/06	1/26/06	1/18/06	4/10/06
90.1y	5. Envelope, 12. Normative References, and Informative Appendix G	This addendum adds a reference and method of test for deriving SRI (ASTM Test Method E, 1980) for high albedo roofs. The changes in the standard were in both Section 5 and Informative Appendix G.	6/24/06	6/29/06	6/18/06	3/3/07
90.1aa	9. Lighting	This addendum modifies Section 9.1 to clarify some lighting requirements.	6/24/06	6/29/06	6/18/06	3/3/07
90.1ab	11. Energy Cost Budget Method and Informative Appendix G	This addendum corrects the referenced section in Tables 11.3.1 and G3.1; Heating, Ventilating, and Air Conditioning to Sections 9.1.3, 9.1.4, and 9.2.	6/24/06	6/29/06	6/18/06	3/3/07
90.1ac	3. Definitions and 6. Heating, Ventilating, and Air Conditioning	This addendum modifies the fan power limitation requirements in Section 6.5.3.	1/27/07	3/2/07	1/18/07	3/27/07
90.1ad	5. Building Envelope	This addendum changes the exception to Section 5.3.1.1 to add a requirement that the values for solar reflectance and thermal emittance be determined by a laboratory accredited by a nationally recognized accreditation organization, such as the Cool Roof Rating Council.	2/5/05	2/10/05	2/3/05	3/14/05
90.1ae	9. Lighting	Change to Section 9.2.1.1, "Space Control."	1/27/07	3/2/07	1/18/07	3/27/07

TABLE F-1 Addenda to ANSI/ASHRAE/IESNA Standard 90.1-2004, Changes Identified *(continued)*

Addenda to 90.1-2004	Sections Affected	Description of Changes*	ASHRAE Standards Committee Approval Date	ASHRAE Board of Directors Approval Date	IESNA Approval Date	ANSI Approval Date
90.1ag	Informative Appendix G	This addendum clarifies that only HVAC fans that provide outdoor air for ventilation need to be modeled as running continuously.	6/24/06	6/29/06	6/18/06	3/3/07
90.1ah	11. Energy Cost Budget Method	This addendum modifies the requirements in Table 11.3.1 for condenser heat recovery.	6/24/06	6/29/06	6/18/06	3/3/07
90.1ai	9. Lighting	This addendum modifies the interior lighting power requirements for retail display lighting in Section 9.6.2.	1/27/07	3/2/07	1/18/07	3/27/07
90.1 aj	5. Building Envelope	This addendum modifies the exception to Section 5.5.3.1 by adding the ASTM Test Method E 1980—Standard Practice for Calculating Solar Reflectance Index (SRI) of Horizontal and Low Sloped Opaque Surfaces.	6/24/06	6/29/06	6/18/06	3/3/07
90.1ak	Table 6.2.1G, Performance Requirements for Heat Rejection Equipment, and Section 6.2.1	This addendum changes Table 6.2.1G to add requirements for cooling towers to be tested to CTI test procedures and to update the corresponding references in Section 6.2.1.	6/24/06	6/29/06	6/18/06	3/3/07
90.1al	Normative Appendix A	This addendum corrects the terminology used in Section A2.3 for metal building roofs.	6/24/06	6/29/06	6/18/06	3/3/07
90.1am	11. Energy Cost Budget Method and Informative Appendix G	This addendum modifies the VAV turndown requirements in Section 11 and Informative Appendix G in accordance to the requirements in Section 6.5.2.1.	6/24/06	6/29/06	6/18/06	3/3/07
90.1an	6. Heating, Ventilating, and Air Conditioning	This addendum modifies the equipment efficiency requirements for commercial boilers in Table 6.8.1F.	1/27/07	3/2/07	1/18/07	3/27/07
90.1ao	6. Heating, Ventilating, and Air Conditioning	This addendum adds a footnote for increasing unit heater efficiency requirements (requiring intermittent ignition devices, power venting, or flue dampers) to comply with federal law.	1/27/07	3/2/07	1/18/07	3/27/07
90.1ap	9. Lighting	This addendum clarifies the intent of a "sales area" space in Table 9.6.1.	1/27/07	3/2/07	1/18/07	3/3/07
90.1aq	12. Normative References	This addendum updates the references to CTI documents.	1/27/07	3/2/07	1/18/07	3/3/07
90.1ar	6. Mechanical	This addendum lowers the part-load fan power limitation from 15 HP to 10 HP in Section 6.5.3.2.1.	1/27/07	3/2/07	1/18/07	3/3/07
90.1as	5. Building Envelope	This addendum modifies the opaque assembly requirements in Tables 5.5-1 through 5.5-8.	5/18/07	6/4/07	6/4/07	12/18/07
90.1at	5. Building Envelope	This addendum modifies the fenestration requirements in Tables 5.5-1 through 5.5-8.	5/18/07	6/4/07	6/4/07	12/18/07
90.1av	5. Building Envelope	This addendum adds an exception to Section 5.5.4.4.1 to allow credit for overhangs toward compliance with the maximum SHGC requirements.	1/27/07	3/2/07	1/18/07	3/3/07

[a] These descriptions may not be complete and are provided for information only.

NOTE

When addenda, interpretations, or errata to this standard have been approved, they can be downloaded free of charge from the ASHRAE Web site at http://www.ashrae.org.

(This appendix is not part of this standard. It is merely informative and does not contain requirements necessary for conformance to the standard. It has not been processed according to the ANSI requirements for a standard and may contain material that has not been subject to public review or a consensus process. Unresolved objectors on informative material are not offered the right to appeal at ASHRAE or ANSI.)

INFORMATIVE APPENDIX G PERFORMANCE RATING METHOD

G1. GENERAL

G1.1 Performance Rating Method Scope. This building performance rating method is a modification of the Energy Cost Budget (ECB) Method in Section 11 and is intended for use in rating the energy *efficiency* of building designs that exceed the requirements of this standard. This appendix does NOT offer an alternative compliance path for minimum standard compliance; that is the intent of Section 11, Energy Cost Budget Method. Rather, this appendix is provided for those wishing to use the methodology developed for this standard to quantify performance that substantially exceeds the requirements of Standard 90.1. It may be useful for evaluating the performance of all *proposed designs*, including *alterations* and *additions* to *existing buildings*, except designs with no mechanical systems.

G1.2 Performance Rating. This performance rating method requires conformance with the following provisions:

All requirements of Sections 5.4, 6.4, 7.4, 8.4, 9.4, and 10.4 are met. These sections contain the mandatory provisions of the standard and are prerequisites for this rating method. The improved performance of the proposed building design is calculated in accordance with provisions of this appendix using the following formula:

Percentage improvement
= 100 × (*Baseline building performance*
– *Proposed building performance*) / *Baseline building performance*

Notes:

1. Both the *proposed building performance* and the *baseline building performance* shall include all end-use load components, such as receptacle and process loads.
2. Neither the *proposed building performance* nor the *baseline building performance* are predictions of actual energy consumption or costs for the *proposed design* after construction. Actual experience will differ from these calculations due to variations such as occupancy, building operation and maintenance, weather, energy use not covered by this procedure, changes in energy rates between design of the building and occupancy, and the precision of the calculation tool.

G1.3 Trade-Off Limits. When the proposed modifications apply to less than the whole building, only parameters related to the systems to be modified shall be allowed to vary. Parameters relating to unmodified existing conditions or to future building components shall be identical for determining both the *baseline building performance* and the *proposed building performance*. Future building components shall meet the prescriptive requirements of Sections 5.5, 6.5, 7.5, 9.5, and 9.6.

G1.4 Documentation Requirements. Simulated performance shall be documented, and documentation shall be submitted to the *rating authority*. The information submitted shall include the following:

a. Calculated values for the *baseline building performance,* the *proposed building performance*, and the percentage improvement.
b. A list of the energy-related features that are included in the design and on which the performance rating is based. This list shall document all energy features that differ between the models used in the *baseline building performance* and *proposed building performance* calculations.
c. Input and output report(s) from the *simulation program* or compliance software including a breakdown of energy usage by at least the following components: lights, internal equipment loads, service water heating equipment, space heating equipment, space cooling and heat rejection equipment, fans, and other HVAC equipment (such as pumps). The output reports shall also show the amount of time any loads are not met by the HVAC system for both the *proposed design* and *baseline building design*.
d. An explanation of any error messages noted in the *simulation program* output.

G2. SIMULATION GENERAL REQUIREMENTS

G2.1 Performance Calculations. The *proposed building performance* and *baseline building performance* shall be calculated using the following:

a. the same *simulation program*
b. the same weather data
c. the same energy rates

G2.2 Simulation Program. The *simulation program* shall be a computer-based program for the analysis of energy consumption in buildings (a program such as, but not limited to, DOE-2, BLAST, or EnergyPlus). The *simulation program* shall include calculation methodologies for the building components being modeled. For components that cannot be modeled by the simulation program, the exceptional calculation methods requirements in Section G2.5 may be used.

G2.2.1 The *simulation program* shall be approved by the *rating authority* and shall, at a minimum, have the ability to explicitly model all of the following:

a. 8760 hours per year
b. hourly variations in occupancy, lighting power, miscellaneous equipment power, thermostat setpoints, and HVAC system operation, defined separately for each day of the week and holidays
c. thermal mass effects
d. ten or more thermal zones
e. part-load performance curves for mechanical equipment

f. capacity and *efficiency* correction curves for mechanical heating and cooling equipment
g. air-side economizers with integrated control
h. *baseline building design* characteristics specified in Section G3

G2.2.2 The *simulation program* shall have the ability to either (1) directly determine the *proposed building performance* and *baseline building performance* or (2) produce hourly reports of energy use by an energy source suitable for determining the *proposed building performance* and *baseline building performance* using a separate calculation engine.

G2.2.3 The *simulation program* shall be capable of performing design load calculations to determine required HVAC equipment capacities and air and water flow rates in accordance with generally accepted engineering standards and handbooks (for example, *ASHRAE Handbook—Fundamentals*) for both the *proposed design* and *baseline building design*.

G2.2.4 The simulation program shall be tested according to ASHRAE Standard 140, and the results shall be furnished by the software provider.

G2.3 Climatic Data. The *simulation program* shall perform the simulation using hourly values of climatic data, such as temperature and humidity from representative climatic data, for the site in which the *proposed design* is to be located. For cities or urban regions with several climatic data entries, and for locations where weather data are not available, the designer shall select available weather data that best represent the climate at the construction site. The selected weather data shall be approved by the *rating authority*.

G2.4 Energy Rates. Annual energy costs shall be determined using either actual rates for purchased energy or state average energy prices published by DOE's Energy Information Administration (EIA) for commercial building customers, but rates from different sources may not be mixed in the same project.

Note: The above provision allows users to gain credit for features that yield load management benefits. Where such features are not present, users can simply use state average unit prices from EIA, which are updated annually and readily available on EIA's Web site (www.eia.doe.gov).

Exception: On-site renewable energy sources or site-recovered energy shall not be considered to be purchased energy and shall not be included in the *proposed building performance*. Where on-site renewable or site-recovered sources are used, the *baseline building performance* shall be based on the energy source used as the backup energy source or on the use of electricity if no backup energy source has been specified.

G2.5 Exceptional Calculation Methods. Where no simulation program is available that adequately models a design, material, or device, the *rating authority* may approve an exceptional calculation method to demonstrate above-standard performance using this method. Applications for approval of an exceptional method shall include documentation of the calculations performed and theoretical and/or empirical information supporting the accuracy of the method.

G3. CALCULATION OF THE PROPOSED AND BASELINE BUILDING PERFORMANCE

G3.1 Building Performance Calculations. The simulation model for calculating the proposed and *baseline building performance* shall be developed in accordance with the requirements in Table G3.1.

G3.1.1 Baseline HVAC System Type and Description. HVAC systems in the *baseline building design* shall be based on usage, number of floors, conditioned floor area, and heating source as specified in Table G3.1.1A and shall conform with the system descriptions in Table G3.1.1B. For systems 1, 2, 3, and 4, each thermal block shall be modeled with its own HVAC system. For systems 5, 6, 7, and 8, each floor shall be modeled with a separate HVAC system. Floors with identical thermal blocks can be grouped for modeling purposes.

Exceptions:

a. Use additional system type(s) for nonpredominant conditions (i.e., residential/nonresidential or heating source) if those conditions apply to more than 20,000 ft^2 of conditioned floor area.
b. If the baseline HVAC system type is 5, 6, 7, or 8, use separate single-zone systems conforming with the requirements of System 3 or System 4 (depending on building heating source) for any spaces that have occupancy or process loads or schedules that differ significantly from the rest of the building. Peak thermal loads that differ by 10 Btu/h·ft^2 or more from the average of other spaces served by the system or schedules that differ by more than 40 equivalent full-load hours per week from other spaces served by the system are considered to differ significantly. Examples where this exception may be applicable include, but are not limited to, computer server rooms, natatoriums, and continually occupied security areas.
c. If the baseline HVAC system type is 5, 6, 7, or 8, use separate single-zone systems conforming with the requirements of System 3 or System 4 (depending on building heat source) for any zones having special pressurization relationships, cross-contamination requirements, or code-required minimum circulation rates.
d. For laboratory spaces with a minimum of 5000 cfm of exhaust, use system type 5 or 7 that reduce the exhaust and makeup air volume to 50% of design values during unoccupied periods. For all-electric buildings, the heating shall be electric resistance.

G3.1.1.1 Purchased Heat. For systems using purchased hot water or steam, hot water or steam costs shall be based on actual utility rates, and on-site boilers shall not be modeled in the *baseline building design*.

G3.1.2 General *Baseline* HVAC System Requirements. HVAC systems in the *baseline building design* shall conform with the general provisions in this section.

TABLE G3.1 Modeling Requirements for Calculating Proposed and Baseline Building Performance

No. Proposed Building Performance	Baseline Building Performance
1. Design Model	
a. The simulation model of the *proposed design* shall be consistent with the design documents, including proper accounting of fenestration and opaque envelope types and areas; interior lighting power and controls; HVAC system types, sizes, and controls; and service water heating systems and controls. All end-use load components within and associated with the building shall be modeled, including, but not limited to, exhaust fans, parking garage ventilation fans, snow-melt and freeze-protection equipment, facade lighting, swimming pool heaters and pumps, elevators and escalators, refrigeration, and cooking. Where the simulation program does not specifically model the functionality of the installed system, spreadsheets or other documentation of the assumptions shall be used to generate the power demand and operating schedule of the systems. b. All conditioned spaces in the *proposed design* shall be simulated as being both heated and cooled even if no heating or cooling system is to be installed, and temperature and humidity control setpoints and schedules shall be the same for *proposed* and *baseline building designs*. c. When the *performance rating method* is applied to buildings in which energy-related features have not yet been designed (e.g., a lighting system), those yet-to-be-designed features shall be described in the *proposed design* exactly as they are defined in the *baseline building design*. Where the space classification for a space is not known, the space shall be categorized as an office space.	The *baseline building design* shall be modeled with the same number of floors and identical conditioned floor area as the *proposed design*.
2. Additions and Alterations	
It is acceptable to predict performance using building models that exclude parts of the *existing building* provided that all of the following conditions are met: a. Work to be performed in excluded parts of the building shall meet the requirements of Sections 5 through 10. b. Excluded parts of the building are served by HVAC systems that are entirely separate from those serving parts of the building that are included in the building model. c. Design space temperature and HVAC system operating setpoints and schedules on either side of the boundary between included and excluded parts of the building are essentially the same. d. If a declining block or similar utility rate is being used in the analysis and the excluded and included parts of the building are on the same utility meter, the rate shall reflect the utility block or rate for the building plus the *addition*.	Same as Proposed Design
3. Space Use Classification	
Usage shall be specified using the building type or space type lighting classifications in accordance with Section 9.5.1 or 9.6.1. The user shall specify the space use classifications using either the building type or space type categories but shall not combine the two types of categories. More than one building type category may be used in a building if it is a mixed-use facility. If space type categories are used, the user may simplify the placement of the various space types within the building model, provided that building-total areas for each space type are accurate.	Same as Proposed Design
4. Schedules	
Schedules capable of modeling hourly variations in occupancy, lighting power, miscellaneous equipment power, thermostat setpoints, and HVAC system operation shall be used. The schedules shall be typical of the proposed building type as determined by the designer and approved by the *rating authority*. **HVAC Fan Schedules**. Schedules for HVAC fans that provide outdoor air for ventilation shall run continuously whenever spaces are occupied and shall be cycled on and off to meet heating and cooling loads during unoccupied hours. **Exceptions:** a. Where no heating and/or cooling system is to be installed and a heating or cooling system is being simulated only to meet the requirements described in this table, heating and/or cooling system fans shall not be simulated as running continuously during occupied hours but shall be cycled on and off to meet heating and cooling loads during all hours. b. HVAC fans shall remain on during occupied and unoccupied hours in spaces that have health and safety mandated minimum ventilation requirements during unoccupied hours.	Same as Proposed Design **Exception:** Schedules may be allowed to differ between *proposed design* and *baseline building design* when necessary to model nonstandard *efficiency* measures, provided that the revised schedules have the approval of the rating authority. Measures that may warrant use of different schedules include, but are not limited to, lighting controls, natural ventilation, demand control ventilation, and measures that reduce service water heating loads.

No. Proposed Building Performance	Baseline Building Performance
5. Building Envelope	
All components of the *building envelope* in the *proposed design* shall be modeled as shown on architectural drawings or as built for existing building envelopes. **Exceptions:** The following building elements are permitted to differ from architectural drawings. a. All uninsulated assemblies (e.g., projecting balconies, perimeter edges of intermediate floor stabs, concrete floor beams over parking garages, roof parapet) shall be separately modeled using either of the following techniques: 1. Separate model of each of these assemblies within the energy simulation model. 2. Separate calculation of the U-factor for each of these assemblies. The U-factors of these assemblies are then averaged with larger adjacent surfaces using an area-weighted average method. This average U-factor is modeled within the energy simulation model. Any other envelope assembly that covers less than 5% of the total area of that assembly type (e.g., exterior walls) need not be separately described provided that it is similar to an assembly being modeled. If not separately described, the area of an envelope assembly shall be added to the area of an assembly of that same type with the same orientation and thermal properties. b. Exterior surfaces whose azimuth orientation and tilt differ by less than 45 degrees and are otherwise the same may be described as either a single surface or by using multipliers. c. For exterior roofs, the roof surface may be modeled with a reflectance of 0.45 if the reflectance of the *proposed design* roof is greater than 0.70 and its emittance is greater than 0.75 or has a minimum SRI of 82. Reflectance values shall be based on testing in accordance with ASTM C1549, ASTM E903, or ASTM E1918, and emittance values shall be based on testing in accordance with ASTM C1371 or ASTM E408, and SRI shall be based on ASTM E1980 calculated at medium wind speed. All other roof surfaces shall be modeled with a reflectance of 0.30. d. Manual fenestration shading devices such as blinds or shades shall not be modeled. Automatically controlled fenestration shades or blinds may be modeled. Permanent shading devices such as fins, overhangs, and light shelves may be modeled.	Equivalent dimensions shall be assumed for each exterior envelope component type as in the *proposed design*; i.e., the total gross area of exterior walls shall be the same in the *proposed* and *baseline building designs*. The same shall be true for the areas of roofs, floors, and doors, and the exposed perimeters of concrete slabs on grade shall also be the same in the *proposed* and *baseline building designs*. The following additional requirements shall apply to the modeling of the *baseline building design*: a. **Orientation.** The *baseline building performance* shall be generated by simulating the building with its actual orientation and again after rotating the entire building 90, 180, and 270 degrees, then averaging the results. The building shall be modeled so that it does not shade itself. b. **Opaque Assemblies.** Opaque assemblies used for new buildings or *additions* shall conform with the following common, lightweight assembly types and shall match the appropriate assembly maximum U-factors in Tables 5.5-1 through 5.5-8: • Roofs—Insulation entirely above deck • *Above-grade walls*—Steel-framed • Floors—Steel-joist • Opaque door types shall match the proposed design and conform to the U-factor requirements from the same tables. • Slab-on-grade floors shall match the F-factor for unheated slabs from the same tables. Opaque assemblies used for *alterations* shall conform with Section 5.1.3. c. **Vertical Fenestration.** Vertical fenestration areas for new buildings and *additions* shall equal that in the *proposed design* or 40% of gross above-grade wall area, whichever is smaller, and shall be distributed on each face of the building in the same proportions in the *proposed design*. Fenestration U-factors shall match the appropriate requirements in Tables 5.5-1 through 5.5-8. Fenestration SHGC shall match the appropriate requirements in Tables 5.5-1 through 5.5-8. All vertical glazing shall be assumed to be flush with the exterior wall, and no shading projections shall be modeled. Manual window shading devices such as blinds or shades shall not be modeled. The fenestration areas for envelope *alterations* shall reflect the limitations on area, U-factor, and SHGC as described in Section 5.1.3. d. **Skylights and Glazed Smoke Vents.** Skylight area shall be equal to that in the proposed building design or 5% of the gross roof area that is part of the *building envelope*, whichever is smaller. If the skylight area of the proposed building design is greater than 5% of the gross roof area, baseline skylight area shall be decreased by an identical percentage in all roof components in which skylights are located to reach the 5% skylight-to-roof ratio. Skylight orientation and tilt shall be the same as in the proposed building design. Skylight U-factor and SHGC properties shall match the appropriate requirements in Tables 5.5-1 through 5.5-8. e. **Roof albedo.** All roof surfaces shall be modeled with a reflectivity of 0.30. f. **Existing Buildings.** For existing *building envelopes*, the *baseline building design* shall reflect existing conditions prior to any revisions that are part of the scope of work being evaluated.

No. Proposed Building Performance	Baseline Building Performance
6. Lighting	
Lighting power in the *proposed design* shall be determined as follows: a. Where a complete lighting system exists, the actual lighting power for each thermal block shall be used in the model. b. Where a lighting system has been designed, lighting power shall be determined in accordance with Sections 9.1.3 and 9.1.4. c. Where lighting neither exists nor is specified, lighting power shall be determined in accordance with the Building Area Method for the appropriate building type. d. Lighting system power shall include all lighting system components shown or provided for on the plans (including lamps and ballasts and task and furniture-mounted fixtures). **Exception:** For multifamily *dwelling units*, hotel/motel guest rooms, and other spaces in which lighting systems are connected via receptacles and are not shown or provided for on building plans, assume identical lighting power for the *proposed* and *baseline building designs* in the simulations. e. Lighting power for parking garages and building facades shall be modeled. f. Credit may be taken for the use of automatic controls for daylight utilization but only if their operation is either modeled directly in the building simulation or modeled in the building simulation through schedule adjustments determined by a separate daylighting analysis approved by the *rating authority*. g. For automatic lighting controls in addition to those required for minimum code compliance under Section 9.4.1, credit may be taken for automatically controlled systems by reducing the connected lighting power by the applicable percentages listed in Table G3.2. Alternatively, credit may be taken for these devices by modifying the lighting schedules used for the *proposed design*, provided that credible technical documentation for the modifications are provided to the *rating authority*.	Lighting power in the *baseline building design* shall be determined using the same categorization procedure (building area or space function) and categories as the *proposed design* with lighting power set equal to the maximum allowed for the corresponding method and category in Section 9.2. No automatic lighting controls (e.g., programmable controls or automatic controls for daylight utilization) shall be modeled in the *baseline building design*, as the lighting schedules used are understood to reflect the mandatory control requirements in this standard.
7. Thermal Blocks—HVAC Zones Designed	
Where HVAC zones are defined on HVAC design drawings, each HVAC zone shall be modeled as a separate *thermal block*. **Exception:** Different HVAC zones may be combined to create a single *thermal block* or identical *thermal blocks* to which multipliers are applied, provided that all of the following conditions are met: a. The space use classification is the same throughout the *thermal block*. b. All HVAC zones in the *thermal block* that are adjacent to glazed exterior walls face the same orientation or their orientations vary by less than 45 degrees. c. All of the zones are served by the same HVAC system or by the same kind of HVAC system.	Same as Proposed Design.
8. Thermal Blocks—HVAC Zones Not Designed	
Where the HVAC zones and systems have not yet been designed, *thermal blocks* shall be defined based on similar internal load densities, occupancy, lighting, thermal and space temperature schedules, and in combination with the following guidelines: a. Separate *thermal blocks* shall be assumed for interior and perimeter spaces. Interior spaces shall be those located greater than 15 ft from an exterior wall. Perimeter spaces shall be those located within 15 ft of an exterior wall. b. Separate *thermal blocks* shall be assumed for spaces adjacent to glazed exterior walls; a separate zone shall be provided for each orientation, except that orientations that differ by less than 45 degrees may be considered to be the same orientation. Each zone shall include all floor area that is 15 ft or less from a glazed perimeter wall, except that floor area within 15 ft of glazed perimeter walls having more than one orientation shall be divided proportionately between zones. c. Separate *thermal blocks* shall be assumed for spaces having floors that are in contact with the ground or exposed to ambient conditions from zones that do not share these features. d. Separate *thermal blocks* shall be assumed for spaces having exterior ceiling or roof assemblies from zones that do not share these features.	Same as Proposed Design.
9. Thermal Blocks—Multifamily Residential Buildings	
Residential spaces shall be modeled using at least one *thermal block* per *dwelling unit*, except that those units facing the same orientations may be combined into one *thermal block*. Corner units and units with roof or floor loads shall only be combined with units sharing these features.	Same as Proposed Design.

TABLE G3.1 Modeling Requirements for Calculating Proposed and Baseline Building Performance *(continued)*

No. Proposed Building Performance	Baseline Building Performance
10. HVAC Systems	
The HVAC system type and all related performance parameters in the *proposed design*, such as equipment capacities and efficiencies, shall be determined as follows: a. Where a complete HVAC system exists, the model shall reflect the actual system type using actual component capacities and efficiencies. b. Where an HVAC system has been designed, the HVAC model shall be consistent with design documents. Mechanical equipment efficiencies shall be adjusted from actual design conditions to the standard rating conditions specified in Section 6.4.1 if required by the simulation model. c. Where no heating system exists or no heating system has been specified, the heating system classification shall be assumed to be electric, and the system characteristics shall be identical to the system modeled in the *baseline building design*. d. Where no cooling system exists or no cooling system has been specified, the cooling system shall be identical to the system modeled in the *baseline building design*.	The HVAC system(s) in the *baseline building design* shall be of the type and description specified in Section G3.1.1, shall meet the general HVAC system requirements specified in Section G3.1.2, and shall meet any system-specific requirements in Section G3.1.3 that are applicable to the baseline HVAC system type(s).
11. Service Hot-Water Systems	
The service hot-water system type and all related performance parameters, such as equipment capacities and efficiencies, in the *proposed design* shall be determined as follows: a. Where a complete service hot-water system exists, the *proposed design* shall reflect the actual system type using actual component capacities and efficiencies. b. Where a service hot-water system has been specified, the service hot-water model shall be consistent with design documents. c. Where no service hot-water system exists or has been specified but the building will have service hot-water loads, a service hot-water system shall be modeled that matches the system in the *baseline building design* and serves the same hot-water loads. d. For buildings that will have no service hot-water loads, no service hot-water system shall be modeled.	The service hot-water system in the *baseline building design* shall use the same energy source as the corresponding system in the *proposed design* and shall conform with the following conditions: a. Where the complete service hot-water system exists, the *baseline building design* shall reflect the actual system type using the actual component capacities and efficiencies. b. Where a new service hot-water system has been specified, the system shall be sized according to the provisions of Section 7.4.1 and the equipment shall match the minimum *efficiency* requirements in Section 7.4.2. Where the energy source is electricity, the heating method shall be electrical resistance. c. Where no service hot-water system exists or has been specified but the building will have service hot-water loads, a service water system(s) using electrical-resistance heat and matching minimum *efficiency* requirements of Section 7.4.2 shall be assumed and modeled identically in the *proposed* and *baseline building designs*. d. For buildings that will have no service hot-water loads, no service hot-water heating shall be modeled. e. Where a combined system has been specified to meet both space heating and service water heating loads, the baseline building system shall use separate systems meeting the minimum *efficiency* requirements applicable to each system individually. f. For large, 24-hour-per-day facilities that meet the prescriptive criteria for use of condenser heat recovery systems described in Section 6.5.6.2, a system meeting the requirements of that section shall be included in the *baseline building design* regardless of the exceptions to Section 6.5.6.2. **Exception:** If a condenser heat recovery system meeting the requirements described in Section 6.5.6.2 cannot be modeled, the requirement for including such a system in the actual building shall be met as a prescriptive requirement in accordance with Section 6.5.6.2, and no heat-recovery system shall be included in the *proposed* or *baseline building designs*. g. Service hot-water energy consumption shall be calculated explicitly based upon the volume of service hot water required and the entering makeup water and the leaving service hot-water temperatures. Entering water temperatures shall be estimated based upon the location. Leaving temperatures shall be based upon the end-use requirements. h. Where recirculation pumps are used to ensure prompt availability of service hot water at the end use, the energy consumption of such pumps shall be calculated explicitly. i. Service water loads and usage shall be the same for both the *baseline building design* and the *proposed design* and shall be documented by the calculation procedures described in Section 7.2.1. **Exceptions:** 1. Service hot-water usage can be demonstrated to be reduced by documented water conservation measures that reduce the physical volume of service water required. Examples include low-flow shower heads. Such reduction shall be demonstrated by calculations. 2. Service hot-water energy consumption can be demonstrated to be reduced by reducing the required temperature of service mixed water, by increasing the temperature, or by increasing the temperature of the entering makeup water. Examples include alternative sanitizing technologies for dishwashing and heat recovery to entering makeup water. Such reduction shall be demonstrated by calculations.

No. Proposed Building Performance	Baseline Building Performance
	3. Service hot-water usage can be demonstrated to be reduced by reducing the hot fraction of mixed water to achieve required operational temperature. Examples include shower or laundry heat recovery to incoming cold-water supply, reducing the hot-water fraction required to meet required mixed-water temperature. Such reduction shall be demonstrated by calculations.
12. Receptacle and Other Loads	
Receptacle and process loads, such as those for office and other equipment, shall be estimated based on the building type or space type category and shall be assumed to be identical in the *proposed* and *baseline building designs*, except as specifically authorized by the *rating authority.* These loads shall be included in simulations of the building and shall be included when calculating the *baseline building performance* and *proposed building performance*.	Other systems, such as motors covered by Section 10, and miscellaneous loads shall be modeled as identical to those in the *proposed design* including schedules of operation and control of the equipment. Where there are specific *efficiency* requirements in Section 10, these systems or components shall be modeled as having the lowest *efficiency* allowed by those requirements.Where no efficiency requirements exist, power and energy rating or capacity of the equipment shall be identical between the *baseline building* and the *proposed design* with the following exception: variations of the power requirements, schedules, or control sequences of the equipment modeled in the *baseline building* from those in the *proposed design* may be allowed by the *rating authority* based upon documentation that the equipment installed in the *proposed design* represents a significant verifiable departure from documented conventional practice. The burden of this documentation is to demonstrate that accepted conventional practice would result in *baseline building* equipment different from that installed in the *proposed design*. Occupancy and occupancy schedules may not be changed.
13. Modeling Limitations to the Simulation Program	
If the simulation program cannot model a component or system included in the *proposed design* explicitly, substitute a thermodynamically similar component model that can approximate the expected performance of the component that cannot be modeled explicitly.	Same as Proposed Design.

TABLE G3.1.1A Baseline HVAC System Types

Building Type	Fossil Fuel, Fossil/Electric Hybrid, and Purchased Heat	Electric and Other
Residential	System 1—PTAC	System 2—PTHP
Nonresidential and 3 Floors or Less and <25,000 ft^2	System 3—PSZ-AC	System 4—PSZ-HP
Nonresidential and 4 or 5 Floors and <25,000 ft^2 or 5 Floors or Less and 25,000 ft^2 to 150,000 ft^2	System 5—Packaged VAV with Reheat	System 6—Packaged VAV with PFP Boxes
Nonresidential and More than 5 Floors or >150,000 ft^2	System 7—VAV with Reheat	System 8—VAV with PFP Boxes

Notes:
Residential building types include dormitory, hotel, motel, and multifamily. Residential space types include guest rooms, living quarters, private living space, and sleeping quarters. Other building and space types are considered nonresidential.
Where no heating system is to be provided or no heating energy source is specified, use the "Electric and Other" heating source classification.
Where attributes make a building eligible for more than one *baseline* system type, use the predominant condition to determine the system type for the entire building.
For laboratory spaces with a minimum of 5000 cfm of exhaust, use system type 5 or 7 and reduce the exhaust and makeup air volume to 50% of design values during unoccupied periods. For all-electric buildings, the heating shall be electric resistance.

TABLE G3.1.1B Baseline System Descriptions

System No.	System Type	Fan Control	Cooling Type	Heating Type
1. PTAC	Packaged terminal air conditioner	Constant volume	Direct expansion	Hot-water fossil fuel boiler
2. PTHP	Packaged terminal heat pump	Constant volume	Direct expansion	Electric heat pump
3. PSZ-AC	Packaged rooftop air conditioner	Constant volume	Direct expansion	Fossil fuel furnace
4. PSZ-HP	Packaged rooftop heat pump	Constant volume	Direct expansion	Electric heat pump
5. Packaged VAV with Reheat	Packaged rooftop VAV with reheat	VAV	Direct expansion	Hot-water fossil fuel boiler
6. Packaged VAV with PFP Boxes	Packaged rooftop VAV with reheat	VAV	Direct expansion	Electric resistance
7. VAV with Reheat	Packaged rooftop VAV with reheat	VAV	Chilled water	Hot-water fossil fuel boiler
8. VAV with PFP Boxes	VAV with reheat	VAV	Chilled water	Electric resistance

G3.1.2.1 Equipment Efficiencies. All HVAC equipment in the *baseline building design* shall be modeled at the minimum *efficiency* levels, both part load and full load, in accordance with Section 6.4. Where *efficiency* ratings, such as EER and COP, include fan energy, the descriptor shall be broken down into its components so that supply fan energy can be modeled separately.

G3.1.2.2 Equipment Capacities. The equipment capacities for the *baseline building design* shall be based on sizing runs for each orientation (per Table G3.1, No. 5a) and shall be oversized by 15% for cooling and 25% for heating, i.e., the ratio between the capacities used in the annual simulations and the capacities determined by the sizing runs shall be 1.15 for cooling and 1.25 for heating. Unmet load hours for the *proposed design* or *baseline building designs* shall not exceed 300 (of the 8760 hours simulated), and unmet load hours for the *proposed design* shall not exceed the number of unmet load hours for the *baseline building design* by more than 50. If unmet load hours in the *proposed design* exceed the unmet load hours in the *baseline building* by more than 50, simulated capacities in the *baseline building* shall be decreased incrementally and the building resimulated until the unmet load hours are within 50 of the unmet load hours of the *proposed design*. If unmet load hours for the *proposed design* or *baseline building design* exceed 300, simulated capacities shall be increased incrementally, and the building with unmet loads resimulated until unmet load hours are reduced to 300 or less. Alternatively, unmet load hours exceeding these limits may be accepted at the discretion of the *rating authority* provided that sufficient justification is given indicating that the accuracy of the simulation is not significantly compromised by these unmet loads.

G3.1.2.2.1 Sizing Runs. Weather conditions used in sizing runs to determine *baseline* equipment capacities may be based either on hourly historical weather files containing typical peak conditions or on design days developed using 99.6% heating design temperatures and 1% dry-bulb and 1% wet-bulb cooling design temperatures.

G3.1.2.3 Preheat Coils. If the HVAC system in the *proposed design* has a preheat coil and a preheat coil can be modeled in the *baseline* system, the *baseline* system shall be modeled with a preheat coil controlled in the same manner as the *proposed design*.

G3.1.2.4 Fan System Operation. Supply and return fans shall operate continuously whenever spaces are occupied and shall be cycled to meet heating and cooling loads during unoccupied hours. If the supply fan is modeled as cycling and fan energy is included in the energy-efficiency rating of the equipment, fan energy shall not be modeled explicitly. Supply, return, and/or exhaust fans will remain on during occupied and unoccupied hours in spaces that have health and safety mandated minimum ventilation requirements during unoccupied hours.

G3.1.2.5 Ventilation. Minimum *outdoor air* ventilation rates shall be the same for the *proposed* and *baseline building designs*.

Exception: When modeling demand-control ventilation in the *proposed design* when its use is not required by Section 6.4.3.8.

G3.1.2.6 Economizers. Outdoor air economizers shall not be included in *baseline* HVAC Systems 1 and 2. *Outdoor air* economizers shall be included in *baseline* HVAC Systems 3 through 8 based on climate as specified in Table G3.1.2.6A.

Exceptions: Economizers shall not be included for systems meeting one or more of the exceptions listed below.

a. Systems that include gas-phase air cleaning to meet the requirements of Section 6.1.2 in Standard 62.1. This exception shall be used only if the system in the *proposed design* does not match the *building design*.

b. Where the use of *outdoor air* for cooling will affect supermarket open refrigerated casework systems. This exception shall only be used if the system in the *proposed design* does not use an economizer. If the exception is used, an economizer shall not be included in the *baseline building design*.

G3.1.2.7 Economizer High-Limit Shutoff. The high-limit shutoff shall be a dry-bulb switch with setpoint temperatures in accordance with the values in Table G3.1.2.6B.

G3.1.2.8 Design Airflow Rates. System design supply airflow rates for the *baseline building design* shall be based on a supply-air-to-room-air temperature difference of 20°F or the required ventilation air or makeup air, whichever is greater. If return or relief fans are specified in the *proposed design*, the *baseline building design* shall also be modeled with fans serving the same functions and sized for the *baseline* system supply fan air quantity less the minimum *outdoor air*, or 90% of the supply fan air quantity, whichever is larger.

G3.1.2.9 System Fan Power. System fan electrical power for supply, return, exhaust, and relief (excluding power to fan-

TABLE G3.1.2.6A Climate Conditions under which Economizers are Included for Baseline Systems 3 through 8

Climate Zone	Conditions
1a, 1b, 2a, 3a, 4a	N.R.
Others	Economizer Included

N.R. means that there is no conditioned building floor area for which economizers are included for the type of zone and climate.

TABLE G3.1.2.6B Economizer High-Limit Shutoff

Climate Zone	High-Limit Shutoff
1b, 2b, 3b, 3c, 4b, 4c, 5b, 5c, 6b, 7, 8	75°F
5a, 6a, 7a	70°F
Others	65°F

powered VAV boxes) shall be calculated using the following formulas:

For Systems 1 and 2,

$$P_{fan} = CFM_S \cdot 0.3 .$$

For systems 3 through 8,

$$P_{fan} = \text{bhp} \times 746 \,/\, \text{Fan Motor Efficiency} .$$

where

P_{fan}	=	electric power to fan motor (watts) and
bhp	=	brake horsepower of *baseline* fan motor from Table G3.1.2.9
Fan Motor Efficiency	=	the efficiency from Table 10.8 for the next motor size greater than the bhp using the enclosed motor at 1800 rpm.
CFM_S	=	the baseline system maximum design supply fan airflow rate in cfm

G3.1.2.10 Exhaust Air Energy Recovery. Individual fan systems that have both a design supply air capacity of 5000 cfm or greater and have a minimum outdoor air supply of 70% or greater of the design supply air quantity shall have an energy recovery system with at least 50% recovery effectiveness. Fifty percent energy recovery effectiveness shall mean a change in the enthalpy of the *outdoor air* supply equal to 50% of the difference between the *outdoor air* and return air at design conditions. Provision shall be made to bypass or control the heat-recovery system to permit air economizer operation, where applicable.

Exceptions: If any of these exceptions apply, exhaust air energy recovery shall not be included in the *baseline building design*:

a. Systems serving spaces that are not cooled and that are heated to less than 60°F.
b. Systems exhausting toxic, flammable, or corrosive fumes or paint or dust. This exception shall only be used if exhaust air energy recovery is not used in the *proposed design*.
c. Commercial kitchen hoods (grease) classified as Type 1 by NFPA 96. This exception shall only be used if exhaust air energy recovery is not used in the *proposed design*.
d. Heating systems in climate zones 1 through 3.
e. Cooling systems in climate zones 3c, 4c, 5b, 5c, 6b, 7, and 8.
f. Where the largest exhaust source is less than 75% of the design outdoor airflow. This exception shall only be used if exhaust air energy recovery is not used in the *proposed design*.
g. Systems requiring dehumidification that employ energy recovery in series with the cooling coil. This exception shall only be used if exhaust air energy recovery and series-style energy recovery coils are not used in the *proposed design*.
h. Systems serving laboratories with exhaust rates of 5000 cfm or greater.

G3.1.3 System-Specific Baseline HVAC System Requirements. *Baseline* HVAC systems shall conform with provisions in this section, where applicable, to the specified *baseline* system types as indicated in section headings.

G3.1.3.1 Heat Pumps (Systems 2 and 4). Electric air-source heat pumps shall be modeled with electric auxiliary heat. The systems shall be controlled with multistage space thermostats and an *outdoor air* thermostat wired to energize auxiliary heat only on the last thermostat stage and when outdoor air temperature is less than 40°F.

G3.1.3.2 Type and Number of Boilers (Systems 1, 5, and 7). The boiler plant shall use the same fuel as the *proposed design* and shall be natural draft, except as noted in Section G3.1.1.1. The *baseline building design* boiler plant shall be modeled as having a single boiler if the *baseline building design* plant serves a conditioned floor area of 15,000 ft^2 or less and as having two equally sized boilers for plants serving more than 15,000 ft^2. Boilers shall be staged as required by the load.

G3.1.3.3 Hot-Water Supply Temperature (Systems 1, 5, and 7). Hot-water design supply temperature shall be modeled as 180°F and design return temperature as 130°F.

G3.1.3.4 Hot-Water Supply Temperature Reset (Systems 1, 5, and 7). Hot-water supply temperature shall be reset based on outdoor dry-bulb temperature using the following schedule: 180°F at 20°F and below, 150°F at 50°F and above, and ramped linearly between 180°F and 150°F at temperatures between 20°F and 50°F.

G3.1.3.5 Hot-Water Pumps (Systems 1, 5, and 7). The *baseline building design* hot-water pump power shall be 19 W/gpm. The pumping system shall be modeled as primary-only with continuous variable flow. Hot-water systems serving 120,000 ft^2 or more shall be modeled with variable-speed drives, and systems serving less than 120,000 ft^2 shall be modeled as riding the pump curve.

TABLE G3.1.2.9 Baseline Fan Brake Horsepower

Baseline Fan Motor Brake Horsepower	
Constant Volume Systems 3–4	**Variable Volume Systems 5–8**
$CFM_s \cdot 0.00094 + A$	$CFM_s \cdot 0.0013 + A$

Where *A* is calculated according to Section 6.5.3.1.1 using the pressure drop adjustment from the proposed building design and the design flow rate of the baseline building system. Do not include pressure drop adjustments for evaporative coolers or heat recovery devices that are not required in the baseline building system by Section G3.1.2.10.

G3.1.3.6 Piping Losses (Systems 1, 5, 7, and 8). Piping losses shall not be modeled in either the *proposed* or *baseline building designs* for hot water, chilled water, or steam piping.

G3.1.3.7 Type and Number of Chillers (Systems 7 and 8). Electric chillers shall be used in the *baseline building design* regardless of the cooling energy source, e.g., direct-fired absorption, absorption from purchased steam, or purchased chilled water. The *baseline building design's* chiller plant shall be modeled with chillers having the number and type as indicated in Table G3.1.3.7 as a function of building peak cooling load.

G3.1.3.8 Chilled-Water Design Supply Temperature (Systems 7 and 8). Chilled-water design supply temperature shall be modeled at 44°F and return water temperature at 56°F.

G3.1.3.9 Chilled-Water Supply Temperature Reset (Systems 7 and 8). Chilled-water supply temperature shall be reset based on outdoor dry-bulb temperature using the following schedule: 44°F at 80°F and above, 54°F at 60°F and below, and ramped linearly between 44°F and 54°F at temperatures between 80°F and 60°F.

G3.1.3.10 Chilled-Water Pumps (Systems 7 and 8). The *baseline building design* pump power shall be 22 W/gpm. Chilled-water systems with a cooling capacity of 300 tons or more shall be modeled as primary/secondary systems with variable-speed drives on the secondary pumping loop. Chilled-water pumps in systems serving less than 300 tons cooling capacity shall be modeled as a primary/secondary systems with secondary pump riding the pump curve.

G3.1.3.11 Heat Rejection (Systems 7 and 8). The heat rejection device shall be an axial fan cooling tower with two-speed fans. Condenser water design supply temperature shall be 85°F or 10°F approaching design wet-bulb temperature, whichever is lower, with a design temperature rise of 10°F. The tower shall be controlled to maintain a 70°F leaving water temperature where weather permits, floating up to leaving water temperature at design conditions. The *baseline building design* condenser-water pump power shall be 19 W/gpm. Each chiller shall be modeled with separate condenser water and chilled-water pumps interlocked to operate with the associated chiller.

TABLE G3.1.3.7 Type and Number of Chillers

Building Peak Cooling Load	Number and Type of Chiller(s)
≤300 tons	1 water-cooled screw chiller
>300 tons, <600 tons	2 water-cooled screw chillers sized equally
≥600 tons	2 water-cooled centrifugal chillers minimum with chillers added so that no chiller is larger than 800 tons, all sized equally

G3.1.3.12 Supply Air Temperature Reset (Systems 5 through 8). The air temperature for cooling shall be reset higher by 5°F under the minimum cooling load conditions.

G3.1.3.13 VAV Minimum Flow Setpoints (Systems 5 and 7). Minimum volume setpoints for VAV reheat boxes shall be 0.4 cfm/ft^2 of floor area served or the minimum ventilation rate, whichever is larger.

G3.1.3.14 Fan Power (Systems 6 and 8). Fans in parallel VAV fan-powered boxes shall be sized for 50% of the peak design flow rate and shall be modeled with 0.35 W/cfm fan power. Minimum volume setpoints for fan-powered boxes shall be equal to 30% of peak design flow rate or the rate required to meet the minimum outdoor air ventilation requirement, whichever is larger. The supply air temperature setpoint shall be constant at the design condition.

G3.1.3.15 VAV Fan Part-Load Performance (Systems 5 through 8). VAV system supply fans shall have variable-speed drives, and their part-load performance characteristics shall be modeled using either Method 1 or Method 2 specified in Table G3.1.3.15.

TABLE G3.1.3.15 Part-Load Performance for VAV Fan Systems

Method 1—Part-Load Fan Power Data

Fan Part-Load Ratio	Fraction of Full-Load Power
0.00	0.00
0.10	0.03
0.20	0.07
0.30	0.13
0.40	0.21
0.50	0.30
0.60	0.41
0.70	0.54
0.80	0.68
0.90	0.83
1.00	1.00

Method 2—Part-Load Fan Power Equation

$$P_{fan} = 0.0013 + 0.1470 \times PLR_{fan} + 0.9506 \times (PLR_{fan})^2 - 0.0998 \times (PLR_{fan})^3$$

where

P_{fan} = fraction of full-load fan power and

PLR_{fan} = fan part-load ratio (current cfm/design cfm).

TABLE G3.2 Power Adjustment Percentages for Automatic Lighting Controls

Automatic Control Device(s)	Non-24-h and ≤ 5000 ft^2	All Other
1. Programmable timing control	10%	0%
2. Occupancy sensor	15%	10%
3. Occupancy sensor and programmable timing control	15%	10%

Note: The 5000 ft^2 condition pertains to the total conditioned floor area of the building.

NOTICE

INSTRUCTIONS FOR SUBMITTING A PROPOSED CHANGE TO THIS STANDARD UNDER CONTINUOUS MAINTENANCE

This standard is maintained under continuous maintenance procedures by a Standing Standard Project Committee (SSPC) for which the Standards Committee has established a documented program for regular publication of addenda or revisions, including procedures for timely, documented, consensus action on requests for change to any part of the standard. SSPC consideration will be given to proposed changes within 13 months of receipt by the manager of standards (MOS).

Proposed changes must be submitted to the MOS in the latest published format available from the MOS. However, the MOS may accept proposed changes in an earlier published format if the MOS concludes that the differences are immaterial to the proposed change submittal. If the MOS concludes that a current form must be utilized, the proposer may be given up to 20 additional days to resubmit the proposed changes in the current format.

ELECTRONIC PREPARATION/SUBMISSION OF FORM FOR PROPOSING CHANGES

An electronic version of each change, which must comply with the instructions in the Notice and the Form, is the preferred form of submittal to ASHRAE Headquarters at the address shown below. The electronic format facilitates both paper-based and computer-based processing. Submittal in paper form is acceptable. The following instructions apply to change proposals submitted in electronic form.

Use the appropriate file format for your word processor and save the file in either a recent version of Microsoft Word (preferred) or another commonly used word-processing program. Please save each change proposal file with a different name (for example, "prop01.doc," "prop02.doc," etc.). If supplemental background documents to support changes submitted are included, it is preferred that they also be in electronic form as word-processed or scanned documents.

ASHRAE will accept the following as equivalent to the signature required on the change submittal form to convey non-exclusive copyright:

Files attached to an e-mail:	Electronic signature on change submittal form (as a picture; *.tif, or *.wpg).
Files on a CD:	Electronic signature on change submittal form (as a picture; *.tif, or *.wpg) or a letter with submitter's signature accompanying the CD or sent by facsimile (single letter may cover all of proponent's proposed changes).

Submit an e-mail or a CD containing the change proposal files to:
Manager of Standards
ASHRAE
1791 Tullie Circle, NE
Atlanta, GA 30329-2305
E-mail: change.proposal@ashrae.org
(Alternatively, mail paper versions to ASHRAE address or fax to 404-321-5478.)

The form and instructions for electronic submittal may be obtained from the Standards section of ASHRAE's Home Page, www.ashrae.org, or by contacting a Standards Secretary, 1791 Tullie Circle, NE, Atlanta, GA 30329-2305.
Phone: 404-636-8400. Fax: 404-321-5478. E-mail: standards.section@ashrae.org.

FORM FOR SUBMITTAL OF PROPOSED CHANGE TO AN ASHRAE STANDARD UNDER CONTINUOUS MAINTENANCE

NOTE: Use a separate form for each comment. Submittals (Microsoft Word preferred) may be attached to e-mail (preferred), submitted on a CD, or submitted in paper by mail or fax to ASHRAE, Manager of Standards, 1791 Tullie Circle, NE, Atlanta, GA 30329-2305. E-mail: change.proposal@ashrae.org. Fax: +1-404/321-5478.

1. Submitter:

Affiliation:

Address: City: State: Zip: Country:

Telephone: Fax: E-Mail:

I hereby grant the American Society of Heating, Refrigerating and Air-Conditioning Engineers, Inc. (ASHRAE) the non-exclusive royalty rights, including non-exclusive rights in copyright, in my proposals. I understand that I acquire no rights in publication of the standard in which my proposals in this or other analogous form is used. I hereby attest that I have the authority and am empowered to grant this copyright release.

Submitter's signature: ______________________________ Date: ____________________

All electronic submittals must have the following statement completed:

I *(insert name)* ______________________________, through this electronic signature, hereby grant the American Society of Heating, Refrigerating and Air-Conditioning Engineers (ASHRAE) the non-exclusive royalty rights, including non-exclusive rights in copyright, in my proposals. I understand that I acquire no rights in publication of the standard in which my proposals in this or other analogous form is used. I hereby attest that I have the authority and am empowered to grant this copyright release.

2. Number and year of standard:

3. Page number and clause (section), subclause, or paragraph number:

4. I propose to: *(check one)*

[ ] Change to read as follows
[ ] Add new text as follows
[ ] Delete and substitute as follows
[ ] Delete without substitution

Use underscores to show material to be added (added) and strike through material to be deleted (~~deleted~~). Use additional pages if needed.

5. Proposed change:

6. Reason and substantiation:

7. Will the proposed change increase the cost of engineering or construction? If yes, provide a brief explanation as to why the increase is justified.

[ ] Check if additional pages are attached. Number of additional pages: _______

[ ] Check if attachments or referenced materials cited in this proposal accompany this proposed change. Please verify that all attachments and references are relevant, current, and clearly labeled to avoid processing and review delays. *Please list your attachments here:*

POLICY STATEMENT DEFINING ASHRAE'S CONCERN FOR THE ENVIRONMENTAL IMPACT OF ITS ACTIVITIES

ASHRAE is concerned with the impact of its members' activities on both the indoor and outdoor environment. ASHRAE's members will strive to minimize any possible deleterious effect on the indoor and outdoor environment of the systems and components in their responsibility while maximizing the beneficial effects these systems provide, consistent with accepted standards and the practical state of the art.

ASHRAE's short-range goal is to ensure that the systems and components within its scope do not impact the indoor and outdoor environment to a greater extent than specified by the standards and guidelines as established by itself and other responsible bodies.

As an ongoing goal, ASHRAE will, through its Standards Committee and extensive technical committee structure, continue to generate up-to-date standards and guidelines where appropriate and adopt, recommend, and promote those new and revised standards developed by other responsible organizations.

Through its *Handbook*, appropriate chapters will contain up-to-date standards and design considerations as the material is systematically revised.

ASHRAE will take the lead with respect to dissemination of environmental information of its primary interest and will seek out and disseminate information from other responsible organizations that is pertinent, as guides to updating standards and guidelines.

The effects of the design and selection of equipment and systems will be considered within the scope of the system's intended use and expected misuse. The disposal of hazardous materials, if any, will also be considered.

ASHRAE's primary concern for environmental impact will be at the site where equipment within ASHRAE's scope operates. However, energy source selection and the possible environmental impact due to the energy source and energy transportation will be considered where possible. Recommendations concerning energy source selection should be made by its members.

Product Code: 86147 4/09
Errata noted in the list dated 10/10/2008 have been corrected.

ANSI/ASHRAE/IESNA Addenda a, b, c, g, h, i, j, k, l, m, n, p, q, s, t, u, w, y, ad, and aw to ANSI/ASHRAE/IESNA Standard 90.1-2007

ASHRAE ADDENDA

2008 SUPPLEMENT

Energy Standard for Buildings Except Low-Rise Residential Buildings

See Appendix for approval dates.

This standard is under continuous maintenance by a Standing Standard Project Committee (SSPC) for which the Standards Committee has established a documented program for regular publication of addenda or revisions, including procedures for timely, documented, consensus action on requests for change to any part of the standard. The change submittal form, instructions, and deadlines may be obtained in electronic form from the ASHRAE Web site, http://www.ashrae.org, or in paper form from the Manager of Standards. The latest edition of an ASHRAE Standard may be purchased from ASHRAE Customer Service, 1791 Tullie Circle, NE, Atlanta, GA 30329-2305. E-mail: orders@ashrae.org. Fax: 404-321-5478. Telephone: 404-636-8400 (worldwide), or toll free 1-800-527-4723 (for orders in US and Canada).

ISSN 1041-2336

American Society of Heating, Refrigerating and Air-Conditioning Engineers, Inc.
1791 Tullie Circle NE, Atlanta, GA 30329
www.ashrae.org

SPECIAL NOTE

This American National Standard (ANS) is a national voluntary consensus standard developed under the auspices of the American Society of Heating, Refrigerating and Air-Conditioning Engineers (ASHRAE). *Consensus* is defined by the American National Standards Institute (ANSI), of which ASHRAE is a member and which has approved this standard as an ANS, as "substantial agreement reached by directly and materially affected interest categories. This signifies the concurrence of more than a simple majority, but not necessarily unanimity. Consensus requires that all views and objections be considered, and that an effort be made toward their resolution." Compliance with this standard is voluntary until and unless a legal jurisdiction makes compliance mandatory through legislation.

ASHRAE obtains consensus through participation of its national and international members, associated societies, and public review.

ASHRAE Standards are prepared by a Project Committee appointed specifically for the purpose of writing the Standard. The Project Committee Chair and Vice-Chair must be members of ASHRAE; while other committee members may or may not be ASHRAE members, all must be technically qualified in the subject area of the Standard. Every effort is made to balance the concerned interests on all Project Committees.

The Assistant Director of Technology for Standards and Special Projects of ASHRAE should be contacted for:

a. interpretation of the contents of this Standard,
b. participation in the next review of the Standard,
c. offering constructive criticism for improving the Standard, or
d. permission to reprint portions of the Standard.

ASHRAE INDUSTRIAL ADVERTISING POLICY ON STANDARDS

ASHRAE Standards and Guidelines are established to assist industry and the public by offering a uniform method of testing for rating purposes, by suggesting safe practices in designing and installing equipment, by providing proper definitions of this equipment, and by providing other information that may serve to guide the industry. The creation of ASHRAE Standards and Guidelines is determined by the need for them, and conformance to them is completely voluntary.

In referring to this Standard or Guideline and in marking of equipment and in advertising, no claim shall be made, either stated or implied, that the product has been approved by ASHRAE.

ASHRAE Standing Standard Project Committee 90.1
Cognizant TC: TC 7.6, Systems Energy Utilization
SPLS Liaisons: Hugh F. Crowther and Ross Montgomery
ASHRAE Staff Liaison: Steven C. Ferguson
IESNA Liaison: Rita M. Harrold

Jerry W. White, Jr., *Chair**
James M. Calm, *Vice-Chair**
Michael Schwedler, *Vice-Chair**
Karim Amrane*
Wagdy Anis*
Anthony Arbore
William P. Bahnfleth*
Peter A. Baselici*
Denise M. Beach
Donald L. Beaty*
Donald M. Brundage*
Ken Brendan*
Ernest A. Conrad*
Charles C. Cottrell*
Roy Crane*
Keith I. Emerson*
Drake H. Erbe*
Charles R. Foster, III
Allan Fraser*
James A. Garrigus*
Jason J. Glazer*
S. Pekka Hakkarainen*
Susanna S. Hanson
Richard V. Heinisch*
Ned B. Heminger
John F. Hogan*
William G. Holy*
Hyman M. Kaplan*
Larry Kouma*
Ronald D. Kurtz*
Samantha H. LaFleur*
Michael D. Lane*
Dean E. Lewis*
Richard Lord*
Kenneth Luther*
Ronald Majette*
Itzhak H. Maor*
Merle F. McBride*
Michael W. Mehl*
Harry P. Misuriello
John Montgomery*
Frank T. Morrison
Frank Myers*
Eric E. Richman*
Michael I. Rosenberg*
Steven Rosenstock*
David A. Schaaf, Jr.*
Leonard C. Sciarra*
Dennis Sczomak*
Bipin Vadilal Shah*
Stephen V. Skalko*
Maria Spinu*
Frank A. Stanonik*
Cedric S. Trueman*
Martha G. VanGeem
McHenry Wallace, Jr.*
Richard D. Watson*
David Weitz*
Robin Wilson*

**Denotes members of voting status when this standard was approved for publication.*

ASHRAE STANDARDS COMMITTEE 2006–2007

David E. Knebel, *Chair*
Stephen D. Kennedy, *Vice-Chair*
Michael F. Beda
Donald L. Brandt
Steven T. Bushby
Paul W. Cabot
Hugh F. Crowther
Samuel D. Cummings, Jr.
Robert G. Doerr
Roger L. Hedrick
John F. Hogan
Eli P. Howard, III
Frank E. Jakob
Jay A. Kohler
James D. Lutz
Carol E. Marriott
Merle F. McBride
Mark P. Modera
Ross D. Montgomery
H. Michael Newman
Stephen V. Santoro
Lawrence J. Schoen
Stephen V. Skalko
Bodh R. Subherwal
Jerry W. White, Jr.
James E. Woods
Richard D. Hermans, *BOD ExO*
Hugh D. McMillan, III, *CO*

Claire B. Ramspeck, *Assistant Director of Technology for Standards and Special Projects*

Committee Listings for Addenda g, h, i, j, k, l, n, p, q, y, ad, and aw

ASHRAE Standing Standard Project Committee 90.1
Cognizant TC: TC 7.6, Systems Energy Utilization
SPLS Liaison: Frank Jakob • ASHRAE Staff Liaison: Steven C. Ferguson • IESNA Liaison: Rita M. Harrold

Schwedler, Mick, Chair*
Hydeman, Mark, Vice Chair*
Skalko, Stephen V., Vice Chair*
Jerine Ahmed
Karim Amrane *
Wagdy Anis*
Susan Anderson
Anthony Arbore
Peter Baselici*
Randy Blanchette
Jeff Boldt
Larry Brown *
Dave Branson*
Donald Brundage*
Ken Brendan *
Jim Calm*
Ernest Conrad *
Charles Cottrell *
Craig Drumheller *
Keith Emerson *
Drake Erbe*
Charles Foster *
Allan Fraser *
Jim Garrigus*
Jason Glazer*
Pekka Hakkarainen *
Susanna Hanson
Richard Heinisch *
Ned Heminger
John Hogan*
Larry Kouma*
Ronald Kurtz *
Michael Lane*
John Lewis *
Richard Lord*
Ken Luther
Ronald Majette*
Itzhak Maor*
Merle McBride *
Ray McGowan
Michael Mehl*
Harry Misuriello
John Montgomery*
Frank Morrison*
Eric Richman*
Michael Rosenberg*
Steven Rosenstock *
David Schaaf*
Leonard Sciarra*
Dennis Sczomak*
Maria Spinu*
Frank Stanonik*
Jeff Stein
Christian Taber
Mike Tillou
Martha VanGeem *
McHenry Wallace*
Richard Watson*
David Weitz*
Jerry White*
Robin Wilson*

**Denotes members of voting status when the document was approved for publication.*

ASHRAE STANDARDS COMMITTEE 2007–2008

Stephen D. Kennedy, *Chair*
Hugh F. Crowther, *Vice-Chair*
Robert G. Baker
Michael F. Beda
Donald L. Brandt
Steven T. Bushby
Paul W. Cabot
Kenneth W. Cooper
Samuel D. Cummings, Jr.
K. William Dean
Robert G. Doerr
Roger L. Hedrick
Eli P. Howard, III
Frank E. Jakob
Nadar R. Jayaraman
Byron W. Jones
Jay A. Kohler
James D. Lutz
Carol E. Marriott
R. Michael Martin
Merle F. McBride
Frank Myers
H. Michael Newman
Lawrence J. Schoen
Bodh R. Subherwal
Jerry W. White, Jr.
Bjarne W. Olesen, *BOD ExO*
Lynn G. Bellenger, *CO*

Claire B. Ramspeck, *Assistant Director of Technology for Standards and Special Projects*

ASHRAE Standard Project Committee 90.1
Cognizant TC: TC 7.6, Systems Energy Utilization
SPLS Liaison: Frank Jakob · IESNA Liaison: Rita M. Harrold · ASHRAE Staff Liaison: Steven C. Ferguson

Schwedler, Mick, Chair*
Hydeman, Mark, Vice Chair*
Skalko, Stephen V., Vice Chair*
Jerine Ahmed
Karim Amrane *
Wagdy Anis*
Susan Anderson
Anthony Arbore
Peter Baselici*
Randy Blanchette
Jeff Boldt
Larry Brown *
Dave Branson*
Donald Brundage*
Ken Brendan *
Jim Calm*
Ernest Conrad *
Charles Cottrell *
Craig Drumheller *
Keith Emerson *
Drake Erbe*
Charles Foster *
Allan Fraser *
Jim Garrigus*
Jason Glazer*
Pekka Hakkarainen *
Susanna Hanson
Richard Heinisch *
Ned Heminger
John Hogan*
Larry Kouma*
Ronald Kurtz *
Michael Lane*
John Lewis *
Richard Lord*
Ken Luther
Ronald Majette*
Itzhak Maor*
Merle McBride *
Ray McGowan
Michael Mehl*
Harry Misuriello
John Montgomery*
Frank Morrison*
Eric Richman*
Michael Rosenberg*
Steven Rosenstock *
David Schaaf*
Leonard Sciarra*
Dennis Sczomak
Maria Spinu*
Frank Stanonik*
Jeff Stein
Christian Taber
Mike Tillou
Martha VanGeem *
McHenry Wallace*
Richard Watson*
David Weitz*
Jerry White*
Robin Wilson*

**Denotes members of voting status when the document was approved for publication.*

ASHRAE STANDARDS COMMITTEE 2008–2009

Hugh F. Crowther, *Chair*
Steven T. Bushby, *Vice-Chair*
Robert G. Baker
Michael F. Beda
Donald L. Brandt
Paul W. Cabot
Kenneth W. Cooper
Samuel D. Cummings, Jr.
K. William Dean
Robert G. Doerr
Allan B. Fraser
Nadar R. Jayaraman
Byron W. Jones
Jay A. Kohler
Carol E. Marriott
R. Michael Martin
Merle F. McBride
Frank Myers
H. Michael Newman
Janice C. Peterson
Douglas T. Reindl
Lawrence J. Schoen
Boggarm S. Setty
Bodh R. Subherwal
William F. Walter
Michael W. Woodford
David E. Knebel, *BOD ExO*
Andrew K. Persily, *CO*

Claire B. Ramspeck, *Director of Technology*

CONTENTS

ANSI/ASHRAE/IESNA Addenda to ANSI/ASHRAE/IESNA Standard 90.1-2007, Energy Standard for Buildings Except Low-Rise Residential Buildings

NOTE

When addenda, interpretations, or errata to this standard have been approved, they can be downloaded free of charge from the ASHRAE Web site at www.ashrae.org.

1791 Tullie Circle NE
Atlanta, GA 30329
www.ashrae.org

(This foreword is not part of this standard. It is merely informative and does not contain requirements necessary for conformance to the standard. It has not been processed according to the ANSI requirements for a standard and may contain material that has not been subject to public review or a consensus process. Unresolved objectors on informative material are not offered the right to appeal at ASHRAE or ANSI.)

FOREWORD

Efficiency and certification requirements for open cooling towers were first incorporated into the 2001 edition of Standard 90.1. At the time, closed circuit cooling towers were known as "fluid coolers" with no established certification program and were not covered by these requirements. Since then, however, fluid coolers have become known as "closed circuit cooling towers" and the Cooling Technology Institute adopted a certification standard that covers this equipment. This has led to confusion in the industry with consulting engineers and inspectors on occasion trying to apply the current open circuit cooling tower requirements in the standard to closed circuit cooling towers. This addendum seeks to clarify that the current cooling tower requirements in the standard apply to open circuit cooling towers only, until such time that separate requirements for closed circuit cooling towers are established in the standard.

Note: In this addendum, changes to the current standard are indicated in the text by *underlining* (for additions) and ~~strike-through~~ (for deletions) unless the instructions specifically mention some other means of indicating the changes.

Addendum a to 90.1-2007

Revise Table 6.8.1G as follows (I-P units):

TABLE 6.8.1G Performance Requirements for Heat Rejection Equipment

Equipment Type[d]	Total System Heat Rejection Capacity at Rated Conditions	Subcategory or Rating Condition	Performance Required[a,b]	Test Procedure[c]
Propeller or Axial Fan Open Cooling Towers	All	95°F Entering Water 85°F Leaving Water 75°F wb *Outdoor air*	38.2 gpm/hp	CTI ATC-105 and CTI STD-201
Centrifugal Fan Open Cooling Towers	All	95°F Entering Water 85°F Leaving Water 75°F wb *Outdoor air*	20.0 gpm/hp	CTI ATC-105 and CTI STD-201
Air-Cooled Condensers	All	125°F Condensing Temperature R-22 Test Fluid 190°F Entering Gas Temperature 15°F Subcooling 95°F Entering db	176,000 Btu/h·hp	ARI 460

[a] For purposes of this table, *open cooling tower performance* is defined as the ~~maximum~~ flow rating of the tower at the thermal rating condition listed in Table 6.8.1G divided by the fan nameplate rated motor power.

[b] For purposes of this table, *air-cooled condenser performance* is defined as the heat rejected from the refrigerant divided by the fan nameplate rated motor power.

[c] Section 12 contains a complete specification of the referenced test procedure, including the referenced year version of the test procedure.

[d] The efficiencies for open cooling towers listed in Table 6.8.1G are not applicable for closed-circuit cooling towers.

Revise Table 6.8.1G as follows (S-I units):

TABLE 6.8.1G Performance Requirements for Heat Rejection Equipment

Equipment Type [d]	Total System Heat Rejection Capacity at Rated Conditions	Subcategory or Rating Condition	Performance Required[a,b]	Test Procedure[c]
Propeller or Axial Fan Open Cooling Towers	All	35°C Entering Water 29°C Leaving Water 24°C wb *Outdoor air*	3.23 L/s·kW	CTI ATC-105 and CTI STD-201
Centrifugal Fan Open Cooling Towers	All	35°C Entering Water 29°C Leaving Water 24°C wb *Outdoor air*	1.7 L/s·kW	CTI ATC-105 and CTI STD-201
Air-Cooled Condensers	All	52°C Condensing Temperature R-22 Test Fluid 88°C Entering Gas Temperature 8°C Subcooling 35°C Entering db	69 COP	ARI 460

[a] For purposes of this table, *open cooling tower performance* is defined as the ~~maximum~~ flow rating of the tower at the thermal rating condition listed in Table 6.8.1G divided by the fan nameplate rated motor power.

[b] For purposes of this table, *air-cooled condenser performance* is defined as the heat rejected from the refrigerant divided by the fan nameplate rated motor power.

[c] Section 12 contains a complete specification of the referenced test procedure, including the referenced year version of the test procedure.

[d] The efficiencies for open cooling towers listed in Table 6.8.1G are not applicable for closed-circuit cooling towers.

(This foreword is not part of this standard. It is merely informative and does not contain requirements necessary for conformance to the standard. It has not been processed according to the ANSI requirements for a standard and may contain material that has not been subject to public review or a consensus process. Unresolved objectors on informative material are not offered the right to appeal at ASHRAE or ANSI.)

FOREWORD

Some facilities covered by Standard 90.1 are challenged to demonstrate compliance with fan power limitations requirements of Standard 90.1 while including design features protecting the safety of inhabitants and compliance of other applicable standards, codes, laws, or regulations. These facilities often require compliance with NIH, NFPA, and other standards with air control and conditioning more stringent than Standard 90.1 and 62.1 requirements. An example of these facilities is vivariums. In exception section 6.5.2.3 (a) of Standard 90.1-2004, the reference to the requirements of Standard 62.1 as the minimum ventilation required is an example of this conflict. This addendum corrects the reference by eliminating the specific section and denoting only Standard 62.1 and allows for another, higher outdoor ventilation rate to be set by the regulating body for these specific applications.

Note: In this addendum, changes to the current standard are indicated in the text by underlining (for additions) and ~~strikethrough~~ (for deletions) unless the instructions specifically mention some other means of indicating the changes.

Addendum b to 90.1-2007

Revise the exceptions in Section 6.5.2.3 as follows (I-P and S-I units):

6.5.2.3

Exceptions:

a. The system is capable of reducing supply air volume to 50% or less of the design airflow rate or the minimum outdoor air ventilation rate specified in ~~6.2 of~~ ASHRAE Standard 62.1 or other applicable federal, state or local code or recognized standard, whichever is larger, before simultaneous heating and cooling takes place.

The remainder of Section 6.5.2.3 remains unchanged.

(This foreword is not part of this standard. It is merely informative and does not contain requirements necessary for conformance to the standard. It has not been processed according to the ANSI requirements for a standard and may contain material that has not been subject to public review or a consensus process. Unresolved objectors on informative material are not offered the right to appeal at ASHRAE or ANSI.)

FOREWORD

Some facilities covered by Standard 90.1 are challenged to demonstrate compliance with fan power limitations requirements of Standard 90.1 while including design features protecting the safety of inhabitants and compliance of other applicable standards, codes, laws, or regulations. These facilities often require compliance with NIH, NFPA, and other standards with air control and conditioning more stringent than Standard 90.1 and 62.1 requirements. An example of these facilities is vivariums. In ASHRAE Standard 90.1-2004 Section 6.5.2.3, Exception (d), this application was not included. This addendum adds vivariums to the list of spaces that require specific humidity levels to satisfy process needs.

Note: In this addendum, changes to the current standard are indicated in the text by underlining (for additions) and ~~strikethrough~~ (for deletions) unless the instructions specifically mention some other means of indicating the changes.

Addendum c to 90.1-2007

Revise the exceptions in Section 6.5.2.3 as follows (I-P and S-I units):

6.5.2.3

Exceptions:

d. Systems serving spaces where specific humidity levels are required to satisfy process needs such as vivariums, museums, surgical suites and buildings with refrigerating systems such as supermarkets, refrigerated warehouses and ice arenas. This exception also applies to other applications for which fan volume controls in accordance with Exception (a) are proven to be impractical to the enforcement agency.

The remainder of Section 6.5.2.3 remains unchanged.

(This foreword is not part of this standard. It is merely informative and does not contain requirements necessary for conformance to the standard. It has not been processed according to the ANSI requirements for a standard and may contain material that has not been subject to public review or a consensus process. Unresolved objectors on informative material are not offered the right to appeal at ASHRAE or ANSI.)

FOREWORD

This addendum updates the building envelope criteria for metal buildings for the first time since Standard 90.1-1999. Other envelope criteria were updated through addenda as and at to Standard 90.1-2004.

Note: In this addendum, changes to the current standard are indicated in the text by underlining (for additions) and ~~strikethrough~~ (for deletions) unless the instructions specifically mention some other means of indicating the changes.

Addendum g to Standard 90.1-2007

Revise the standard as follows (I-P and SI units):

Add the following definition to Section 3.2:

liner system (Ls): a continuous vapor barrier liner installed below the purlins and uninterrupted by framing members.

Add the following abbreviation to Section 3.3:

Ls liner system

Modify Section 5 as follows (I-P units):

TABLE 5.5-1 Building Envelope Requirements For Climate Zone 1 (A, B)*

Opaque Elements	Nonresidential		Residential		Semiheated	
	Assembly Maximum	Insulation Min. R-Value	Assembly Maximum	Insulation Min. R-Value	Assembly Maximum	Insulation Min. R-Value
Roofs						
Insulation Entirely above Deck	U-0.063	R-15.0 c.i.	U-0.048	R-20.0 c.i.	U-0.218	R-3.8 ci
Metal Building[a]	U-0.065	R-19.0	U-0.065	R-19.0	~~U-1.280~~ U-0.167	~~NR~~ R-6.0
Attic and Other	U-0.034	R-30.0	U-0.027	R-38.0	U-0.081	R-13.0
Walls, Above-Grade						
Mass	U-0.580	NR	U-0.151[ab]	R-5.7 c.i.[ab]	U-0.580	NR
Metal Building	~~U-0.113~~ U-0.093	~~R-13.0~~ R-16.0	~~U-0.113~~ U-0.093	~~R-13.0~~ R-16.0	~~U-1.180~~ U-0.113	~~NR~~ R-13.0
Steel-Framed	U-0.124	R-13.0	U-0.124	R-13.0	U-0.352	NR
Wood-Framed and Other	U-0.089	R-13.0	U-0.089	R-13.0	U-0.292	NR
Walls, Below-Grade						
Below-Grade Wall	C-1.140	NR	C-1.140	NR	C-1.140	NR
Floors						
Mass	U-0.322	NR	U-0.322	NR	U-0.322	NR
Steel-Joist	U-0.350	NR	U-0.350	NR	U-0.350	NR
Wood-Framed and Other	U-0.282	NR	U-0.282	NR	U-0.282	NR
Slab-On-Grade Floors						
Unheated	F-0.730	NR	F-0.730	NR	F-0.730	NR
Heated	F-1.020	R-7.5 for 12 in.	F-1.020	R-7.5 for 12 in.	F-1.020	R-7.5 for 12 in.
Opaque Doors						
Swinging	U-0.700		U-0.700		U-0.700	
Nonswinging	U-1.450		U-1.450		U-1.450	

Fenestration	Assembly Max. U	Assembly Max. SHGC	Assembly Max. U	Assembly Max. SHGC	Assembly Max. U	Assembly Max. SHGC
Vertical Glazing, 0%–40% of Wall						
Nonmetal framing (all)[bc]	U-1.20	SHGC-0.25 all	U-1.20	SHGC-0.25 all	U-1.20	SHGC-NR all
Metal framing (curtainwall/storefront)[cd]	U-1.20		U-1.20		U-1.20	
Metal framing (entrance door)[cd]	U-1.20		U-1.20		U-1.20	
Metal framing (all other)[cd]	U-1.20		U-1.20		U-1.20	
Skylight with Curb, Glass, % of Roof						
0%–2.0%	U_{all}-1.98	$SHGC_{all}$-0.36	U_{all}-1.98	$SHGC_{all}$-0.19	U_{all}-1.98	$SHGC_{all}$-NR
2.1%–5.0%	U_{all}-1.98	$SHGC_{all}$-0.19	U_{all}-1.98	$SHGC_{all}$-0.16	U_{all}-1.98	$SHGC_{all}$-NR
Skylight with Curb, Plastic, % of Roof						
0%–2.0%	U_{all}-1.90	$SHGC_{all}$-0.34	U_{all}-1.90	$SHGC_{all}$-0.27	U_{all}-1.90	$SHGC_{all}$-NR
2.1%–5.0%	U_{all}-1.90	$SHGC_{all}$-0.27	U_{all}-1.90	$SHGC_{all}$-0.27	U_{all}-1.90	$SHGC_{all}$-NR
Skylight without Curb, All, % of Roof						
0%–2.0%	U_{all}-1.36	$SHGC_{all}$-0.36	U_{all}-1.36	$SHGC_{all}$-0.19	U_{all}-1.36	$SHGC_{all}$-NR
2.1%–5.0%	U_{all}-1.36	$SHGC_{all}$-0.19	U_{all}-1.36	$SHGC_{all}$-0.19	U_{all}-1.36	$SHGC_{all}$-NR

*The following definitions apply: c.i. = continuous insulation (see Section 3.2), NR = no (insulation) requirement.

[a] When using R-value compliance method, a thermal spacer block is required; otherwise use the U-factor compliance method. See Table A2.3.

[ab] Exception to Section A3.1.3.1 applies.

[bc] Nonmetal framing includes framing materials other than metal with or without metal reinforcing or cladding.

[cd] Metal framing includes metal framing with or without thermal break. The "all other" subcategory includes operable windows, fixed windows, and non-entrance doors.

TABLE 5.5-2 Building Envelope Requirements For Climate Zone 2 (A, B)*

Opaque Elements	Nonresidential Assembly Maximum	Nonresidential Insulation Min. R-Value	Residential Assembly Maximum	Residential Insulation Min. R-Value	Semiheated Assembly Maximum	Semiheated Insulation Min. R-Value
Roofs						
Insulation Entirely above Deck	U-0.048	R-20.0 c.i.	U-0.048	R-20.0 c.i.	U-0.218	R-3.8 c.i.
Metal Building[a]	~~U-0.065~~ U-0.055	~~R-19.0~~ R-13.0 + R-13.0	~~U-0.065~~ U-0.055	~~R-19.0~~ R-13.0 + R-13.0	~~U-0.167~~ U-0.097	~~R-6.0~~ R-10.0
Attic and Other	U-0.027	R-38.0	U-0.027	R-38.0	U-0.081	R-13.0
Walls, Above-Grade						
Mass	U-0.151[ab]	R5.7 c.i.[ab]	U-0.123	R-7.6 c.i.	U-0.580	NR
Metal Building	~~U-0.113~~ U-0.093	~~R-13.0~~ R-16.0	~~U-0.113~~ U-0.093	~~R-13.0~~ R-16.0	~~U-0.184~~ U-0.113	~~R-6.0~~ R-13.0
Steel-Framed	U-0.124	R-13.0	U-0.064	R-13.0 + R-7.5 c.i.	U-0.124	R-13.0
Wood-Framed and Other	U-0.089	R-13.0	U-0.089	R-13.0	U-0.089	R-13.0
Walls, Below-Grade						
Below-Grade Wall	C-1.140	NR	C-1.140	NR	C-1.140	NR
Floors						
Mass	U-0.107	R-6.3 c.i.	U-0.087	R-8.3 c.i.	U-0.322	NR
Steel-Joist	U-0.052	R-19.0	U-0.052	R-19.0	U-0.069	R-13.0
Wood-Framed and Other	U-0.051	R-19.0	U-0.033	R-30.0	U-0.066	R-13.0
Slab-On-Grade Floors						
Unheated	F-0.730	NR	F-0.730	NR	F-0.730	NR
Heated	F-1.020	R-7.5 for 12 in.	F-1.020	R-7.5 for 12 in.	F-1.020	R-7.5 for 12 in.
Opaque Doors						
Swinging	U-0.700		U-0.700		U-0.700	
Nonswinging	U-1.450		U-0.500		U-1.450	
Fenestration	**Assembly Max. U**	**Assembly Max. SHGC**	**Assembly Max. U**	**Assembly Max. SHGC**	**Assembly Max. U**	**Assembly Max. SHGC**
Vertical Glazing, 0%–40% of Wall						
Nonmetal framing (all)[bc]	U-0.75	SHGC-0.25 all	U-0.75	SHGC-0.25 all	U-1.20	SHGC-NR all
Metal framing (curtainwall/storefront)[cd]	U-0.70		U-0.70		U-1.20	
Metal framing (entrance door)[cd]	U-1.10		U-1.10		U-1.20	
Metal framing (all other)[cd]	U-0.75		U-0.75		U-1.20	
Skylight with Curb, Glass, % of Roof						
0%–2.0%	U_{all}-1.98	$SHGC_{all}$-0.36	U_{all}-1.98	$SHGC_{all}$-0.19	U_{all}-1.98	$SHGC_{all}$-NR
2.1%–5.0%	U_{all}-1.98	$SHGC_{all}$-0.19	U_{all}-1.98	$SHGC_{all}$-0.19	U_{all}-1.98	$SHGC_{all}$-NR
Skylight with Curb, Plastic, % of Roof						
0%–2.0%	U_{all}-1.90	$SHGC_{all}$-0.39	U_{all}-1.90	$SHGC_{all}$-0.27	U_{all}-1.90	$SHGC_{all}$-NR
2.1%–5.0%	U_{all}-1.90	$SHGC_{all}$-0.34	U_{all}-1.90	$SHGC_{all}$-0.27	U_{all}-1.90	$SHGC_{all}$-NR
Skylight without Curb, All, % of Roof						
0%–2.0%	U_{all}-1.36	$SHGC_{all}$-0.36	U_{all}-1.36	$SHGC_{all}$-0.19	U_{all}-1.36	$SHGC_{all}$-NR
2.1%–5.0%	U_{all}-1.36	$SHGC_{all}$-0.19	U_{all}-1.36	$SHGC_{all}$-0.19	U_{all}-1.36	$SHGC_{all}$-NR

*The following definitions apply: c.i. = continuous insulation (see Section 3.2), NR = no (insulation) requirement.

[a] When using R-value compliance method, a thermal spacer block is required; otherwise use the U-factor compliance method. See Table A2.3.

[ab] Exception to Section A3.1.3.1 applies.

[bc] Nonmetal framing includes framing materials other than metal with or without metal reinforcing or cladding.

[cd] Metal framing includes metal framing with or without thermal break. The "all other" subcategory includes operable windows, fixed windows, and non-entrance doors.

TABLE 5.5-3 Building Envelope Requirements For Climate Zone 3 (A, B, C)*

Opaque Elements	Nonresidential Assembly Maximum	Nonresidential Insulation Min. R-Value	Residential Assembly Maximum	Residential Insulation Min. R-Value	Semiheated Assembly Maximum	Semiheated Insulation Min. R-Value
Roofs						
Insulation Entirely above Deck	U-0.048	R-20.0 c.i.	U-0.048	R-20.0 c.i.	U-0.173	R-5.0 c.i.
Metal Building[a]	~~U-0.065~~ U-0.055	~~R-19.0~~ R-13.0 + R-13.0	~~U-0.065~~ U-0.055	~~R-19.0~~ R-13.0 + R-13.0	U-0.097	R-10.0
Attic and Other	U-0.027	R-38.0	U-0.027	R-38.0	U-0.053	R-19.0
Walls, Above-Grade						
Mass	U-0.123	R-7.6 c.i.	U-0.104	R-9.5 c.i.	U-0.580	NR
Metal Building	~~U-0.113~~ U-0.084	~~R-13.0~~ R-19.0	~~U-0.113~~ U-0.084	~~R-13.0~~ R-19.0	~~U-0.184~~ U-0.113	~~R-6.0~~ R-13.0
Steel-Framed	U-0.084	R-13.0 + R-3.8 c.i.	U-0.064	R-13.0 + R-7.5 c.i.	U-0.124	R-13.0
Wood-Framed and Other	U-0.089	R-13.0	U-0.089	R-13.0	U-0.089	R-13.0
Walls, Below-Grade						
Below-Grade Wall	C-1.140	NR	C-1.140	NR	C-1.140	NR
Floors						
Mass	U-0.107	R-6.3 c.i.	U-0.087	R-8.3 c.i.	U-0.322	NR
Steel-Joist	U-0.052	R-19.0	U-0.052	R-19.0	U-0.069	R-13.0
Wood-Framed and Other	U-0.051	R-19.0	U-0.033	R-30.0	U-0.066	R-13.0
Slab-On-Grade Floors						
Unheated	F-0.730	NR	F-0.730	NR	F-0.730	NR
Heated	F-0.900	R-10 for 24 in.	F-0.900	R-10 for 24 in.	F-1.020	R-7.5 for 12 in.
Opaque Doors						
Swinging	U-0.700		U-0.700		U-0.700	
Nonswinging	U-1.450		U-0.500		U-1.450	

Fenestration	Nonresidential Assembly Max. U	Nonresidential Assembly Max. SHGC	Residential Assembly Max. U	Residential Assembly Max. SHGC	Semiheated Assembly Max. U	Semiheated Assembly Max. SHGC
Vertical Glazing, 0%–40% of Wall						
Nonmetal framing (all)[~~b~~c]	U-0.65	SHGC-0.25 all	U-0.65	SHGC-0.25 all	U-1.20	SHGC-NR all
Metal framing (curtainwall/storefront)[~~c~~d]	U-0.60		U-0.60		U-1.20	
Metal framing (entrance door)[~~c~~d]	U-0.90		U-0.90		U-1.20	
Metal framing (all other)[~~c~~d]	U-0.65		U-0.65		U-1.20	
Skylight with Curb, Glass, % of Roof						
0%–2.0%	U_{all}-1.17	$SHGC_{all}$-0.39	U_{all}-1.17	$SHGC_{all}$-0.36	U_{all}-1.98	$SHGC_{all}$-NR
2.1%–5.0%	U_{all}-1.17	$SHGC_{all}$-0.19	U_{all}-1.17	$SHGC_{all}$-0.19	U_{all}-1.98	$SHGC_{all}$-NR
Skylight with Curb, Plastic, % of Roof						
0%–2.0%	U_{all}-1.30	$SHGC_{all}$-0.65	U_{all}-1.30	$SHGC_{all}$-0.27	U_{all}-1.90	$SHGC_{all}$-NR
2.1%–5.0%	U_{all}-1.30	$SHGC_{all}$-0.34	U_{all}-1.30	$SHGC_{all}$-0.27	U_{all}-1.90	$SHGC_{all}$-NR
Skylight without Curb, All, % of Roof						
0%–2.0%	U_{all}-0.69	$SHGC_{all}$-0.39	U_{all}-0.69	$SHGC_{all}$-0.36	U_{all}-1.36	$SHGC_{all}$-NR
2.1%–5.0%	U_{all}-0.69	$SHGC_{all}$-0.19	U_{all}-0.69	$SHGC_{all}$-0.19	U_{all}-1.36	$SHGC_{all}$-NR

*The following definitions apply: c.i. = continuous insulation (see Section 3.2), NR = no (insulation) requirement.

[a]When using R-value compliance method, a thermal spacer block is required; otherwise use the U-factor compliance method. See Table A2.3.

[~~a~~b]Exception to Section A3.1.3.1 applies.

[~~b~~c]Nonmetal framing includes framing materials other than metal with or without metal reinforcing or cladding.

[~~c~~d]Metal framing includes metal framing with or without thermal break. The "all other" subcategory includes operable windows, fixed windows, and non-entrance doors.

TABLE 5.5-4 Building Envelope Requirements For Climate Zone 4 (A, B, C)*

Opaque Elements	Nonresidential Assembly Maximum	Nonresidential Insulation Min. R-Value	Residential Assembly Maximum	Residential Insulation Min. R-Value	Semiheated Assembly Maximum	Semiheated Insulation Min. R-Value
Roofs						
Insulation Entirely above Deck	U-0.048	R-20.0 c.i.	U-0.048	R-20.0 c.i.	U-0.173	R-5.0 c.i.
Metal Building[a]	~~U-0.065~~ U-0.055	~~R-19.0~~ R-13.0 + R-13.0	~~U-0.065~~ U-0.055	~~R-19.0~~ R-13.0 + R-13.0	U-0.097	R-10.0
Attic and Other	U-0.027	R-38.0	U-0.027	R-38.0	U-0.053	R-19.0
Walls, Above-Grade						
Mass	U-0.104	R-9.5 c.i.	U-0.090	R-11.4 c.i.	U-0.580	NR
Metal Building	~~U-0.113~~ U-0.084	~~R-13.0~~ R-19.0	~~U-0.113~~ U-0.084	~~R-13.0~~ R-19.0	~~U-0.134~~ U-0.113	~~R-10.0~~ R-13.0
Steel-Framed	U-0.064	R-13.0 + R-7.5 c.i.	U-0.064	R-13.0 + R-7.5 c.i.	U-0.124	R-13.0
Wood-Framed and Other	U-0.089	R-13.0	U-0.064	R-13.0 + R-3.8 c.i.	U-0.089	R-13.0
Walls, Below-Grade						
Below-Grade Wall	C-1.140	NR	C-0.119	R-7.5 c.i.	C-1.140	NR
Floors						
Mass	U-0.087	R-8.3 c.i.	U-0.074	R-10.4 c.i.	U-0.137	R-4.2 c.i.
Steel-Joist	U-0.038	R-30.0	U-0.038	R-30.0	U-0.069	R-13.0
Wood-Framed and Other	U-0.033	R-30.0	U-0.033	R-30.0	U-0.066	R-13.0
Slab-On-Grade Floors						
Unheated	F-0.730	NR	F-0.540	R-10 for 24 in.	F-0.730	NR
Heated	F-0.860	R-15 for 24 in.	F-0.860	R-15 for 24in.	F-1.020	R-7.5 for 12 in.
Opaque Doors						
Swinging	U-0.700		U-0.700		U-0.700	
Nonswinging	U-1.500		U-0.500		U-1.450	

Fenestration	Nonresidential Assembly Max. U	Nonresidential Assembly Max. SHGC	Residential Assembly Max. U	Residential Assembly Max. SHGC	Semiheated Assembly Max. U	Semiheated Assembly Max. SHGC
Vertical Glazing, 0%–40% of Wall						
Nonmetal framing (all)[~~b~~c]	U-0.40	SHGC-0.40 all	U-0.40	SHGC-0.40 all	U-1.20	SHGC-NR all
Metal framing (curtainwall/storefront)[~~c~~d]	U-0.50		U-0.50		U-1.20	
Metal framing (entrance door)[~~c~~d]	U-0.85		U-0.85		U-1.20	
Metal framing (all other)[~~c~~d]	U-0.55		U-0.55		U-1.20	
Skylight with Curb, Glass, % of Roof						
0%–2.0%	U_{all}-1.17	$SHGC_{all}$-0.49	U_{all}-0.98	$SHGC_{all}$-0.36	U_{all}-1.98	$SHGC_{all}$-NR
2.1%–5.0%	U_{all}-1.17	$SHGC_{all}$-0.39	U_{all}-0.98	$SHGC_{all}$-0.19	U_{all}-1.98	$SHGC_{all}$-NR
Skylight with Curb, Plastic, % of Roof						
0%–2.0%	U_{all}-1.30	$SHGC_{all}$-0.65	U_{all}-1.30	$SHGC_{all}$-0.62	U_{all}-1.90	$SHGC_{all}$-NR
2.1%–5.0%	U_{all}-1.30	$SHGC_{all}$-0.34	U_{all}-1.30	$SHGC_{all}$-0.27	U_{all}-1.90	$SHGC_{all}$-NR
Skylight without Curb, All, % of Roof						
0%–2.0%	U_{all}-0.69	$SHGC_{all}$-0.49	U_{all}-0.58	$SHGC_{all}$-0.36	U_{all}-1.36	$SHGC_{all}$-NR
2.1%–5.0%	U_{all}-0.69	$SHGC_{all}$-0.39	U_{all}-0.58	$SHGC_{all}$-0.19	U_{all}-1.36	$SHGC_{all}$-NR

*The following definitions apply: c.i. = continuous insulation (see Section 3.2), NR = no (insulation) requirement.

[a] When using R-value compliance method, a thermal spacer block is required; otherwise use the U-factor compliance method. See Table A2.3.

[~~a~~b] Exception to Section A3.1.3.1 applies.

[~~b~~c] Nonmetal framing includes framing materials other than metal with or without metal reinforcing or cladding.

[~~c~~d] Metal framing includes metal framing with or without thermal break. The "all other" subcategory includes operable windows, fixed windows, and non-entrance doors.

TABLE 5.5-5 Building Envelope Requirements For Climate Zone 5 (A, B, C)*

Opaque Elements	Nonresidential Assembly Maximum	Nonresidential Insulation Min. R-Value	Residential Assembly Maximum	Residential Insulation Min. R-Value	Semiheated Assembly Maximum	Semiheated Insulation Min. R-Value
Roofs						
Insulation Entirely above Deck	U-0.048	R-20.0 c.i.	U-0.048	R-20.0 c.i.	U-0.119	R-7.6 c.i.
Metal Building[a]	~~U-0.065~~ U-0.055	~~R-19.0~~ R-13.0 + R-13.0	~~U-0.065~~ U-0.055	~~R-19.0~~ R-13.0 + R-13.0	~~U-0.097~~ U-0.083	~~R-10.0~~ R-13.0
Attic and Other	U-0.027	R-38.0	U-0.027	R-38.0	U-0.053	R-19.0
Walls, Above-Grade						
Mass	U-0.090	R-11.4 c.i.	U-0.080	R-13.3 c.i.	U-0.151[ab]	R-5.7 c.i.[ab]
Metal Building	~~U-0.113~~ U-0.069	~~R-13.0~~ R-13.0 + R- 5.6 c.i.	~~U-0.057~~ U-0.069	~~R-13.0 + R-13.0~~ R-13.0 + R- 5.6 c.i.	~~U-0.123~~ U-0.113	~~R-11.0~~ R-13.0
Steel-Framed	U-0.064	R-13.0 + R-7.5 c.i.	U-0.064	R-13.0 + R-7.5 c.i.	U-0.124	R-13.0
Wood-Framed and Other	U-0.064	R-13.0 + R-3.8 c.i.	U-0.051	R-13.0 + R-7.5 c.i.	U-0.089	R-13.0
Walls, Below-Grade						
Below-Grade Wall	C-0.119	R-7.5 c.i.	C-0.119	R-7.5 c.i.	C-1.140	NR
Floors						
Mass	U-0.074	R-10.4 c.i.	U-0.064	R-12.5 c.i.	U-0.137	R-4.2 c.i.
Steel-Joist	U-0.038	R-30.0	U-0.038	R-30.0	U-0.052	R-19.0
Wood-Framed and Other	U-0.033	R-30.0	U-0.033	R-30.0	U-0.051	R-19.0
Slab-On-Grade Floors						
Unheated	F-0.730	NR	F-0.540	R-10 for 24 in.	F-0.730	NR
Heated	F-0.860	R-15 for 24 in.	F-0.860	R-15 for 24 in.	F-1.020	R-7.5 for 12 in.
Opaque Doors						
Swinging	U-0.700		U-0.500		U-0.700	
Nonswinging	U-0.500		U-0.500		U-1.450	

Fenestration	Nonresidential Assembly Max. U	Nonresidential Assembly Max. SHGC	Residential Assembly Max. U	Residential Assembly Max. SHGC	Semiheated Assembly Max. U	Semiheated Assembly Max. SHGC
Vertical Glazing, % of Wall						
Nonmetal framing (all)[bc]	U-0.35	SHGC-0.40 all	U-0.35	SHGC-0.40 all	U-1.20	SHGC-NR all
Metal framing (curtainwall/storefront)[ed]	U-0.45		U-0.45		U-1.20	
Metal framing (entrance door)[ed]	U-0.80		U-0.80		U-1.20	
Metal framing (all other)[ed]	U-0.55		U-0.55		U-1.20	
Skylight with Curb, Glass, % of Roof						
0%–2.0%	U_{all}–1.17	$SHGC_{all}$–0.49	U_{all}–1.17	$SHGC_{all}$–0.49	U_{all}–1.98	$SHGC_{all}$–NR
2.1%–5.0%	U_{all}–1.17	$SHGC_{all}$–0.39	U_{all}–1.17	$SHGC_{all}$–0.39	U_{all}–1.98	$SHGC_{all}$–NR
Skylight with Curb, Plastic, % of Roof						
0%–2.0%	U_{all}–1.10	$SHGC_{all}$–0.77	U_{all}–1.10	$SHGC_{all}$–0.77	U_{all}–1.90	$SHGC_{all}$–NR
2.1%–5.0%	U_{all}–1.10	$SHGC_{all}$–0.62	U_{all}–1.10	$SHGC_{all}$–0.62	U_{all}–1.90	$SHGC_{all}$–NR
Skylight without Curb, All, % of Roof						
0%–2.0%	U_{all}–0.69	$SHGC_{all}$–0.49	U_{all}–0.69	$SHGC_{all}$–0.49	U_{all}–1.36	$SHGC_{all}$–NR
2.1%–5.0%	U_{all}–0.69	$SHGC_{all}$–0.39	U_{all}–0.69	$SHGC_{all}$–0.39	U_{all}–1.36	$SHGC_{all}$–NR

*The following definitions apply: c.i. = continuous insulation (see Section 3.2), NR = no (insulation) requirement.
[a] When using R-value compliance method, a thermal spacer block is required; otherwise use the U-factor compliance method. See Table A2.3.
[ab] Exception to Section A3.1.3.1 applies.
[bc] Nonmetal framing includes framing materials other than metal with or without metal reinforcing or cladding.
[ed] Metal framing includes metal framing with or without thermal break. The "all other" subcategory includes operable windows, fixed windows, and non-entrance doors.

TABLE 5.5-6 Building Envelope Requirements For Climate Zone 6 (A, B)*

Opaque Elements	Nonresidential Assembly Maximum	Nonresidential Insulation Min. R-Value	Residential Assembly Maximum	Residential Insulation Min. R-Value	Semiheated Assembly Maximum	Semiheated Insulation Min. R-Value
Roofs						
Insulation Entirely above Deck	U-0.048	R-20.0 c.i.	U-0.048	R-20.0 c.i.	U-0.093	R-10.0 c.i.
Metal Building[a]	~~U-0.065~~ U-0.049	~~R-19.0~~ R-13.0 + R-19.0	~~U-0.065~~ U-0.049	~~R-19.0~~ R-13.0 + R-19.0	~~U-0.097~~ U-0.072	~~R-10.0~~ R-16.0
Attic and Other	U-0.027	R-38.0	U-0.027	R-38.0	U-0.034	R-30.0
Walls, Above-Grade						
Mass	U-0.080	R-13.3 c.i.	U-0.071	R-15.2 c.i.	U-0.151[a]	R-5.7 c.i.[a]
Metal Building	~~U-0.113~~ U-0.069	~~R-13.0~~ R-13.0 + R-5.6 c.i.	~~U-0.057~~ U-0.069	~~R-13.0 + R-13.0~~ R-13.0 + R-5.6 c.i.	U-0.113	R-13.0
Steel-Framed	U-0.064	R-13.0 + R-7.5 c.i.	U-0.064	R-13.0 + R-7.5 c.i.	U-0.124	R-13.0
Wood-Framed and Other	U-0.051	R-13.0 + R-7.5 c.i.	U-0.051	R-13.0 + R-7.5 c.i.	U-0.089	R-13.0
Walls, Below-Grade						
Below-Grade Wall	C-0.119	R-7.5 c.i.	C-0.119	R-7.5 c.i.	C-1.140	NR
Floors						
Mass	U-0.064	R-12.5 c.i.	U-0.057	R-14.6 c.i.	U-0.137	R-4.2 c.i.
Steel-Joist	U-0.038	R-30.0	U-0.032	R-38.0	U-0.052	R-19.0
Wood-Framed and Other	U-0.033	R-30.0	U-0.033	R-30.0	U-.0051	R-19.0
Slab-On-Grade Floors						
Unheated	F-0.540	R-10 for 24 in.	F-0.520	R-15 for 24 in.	F-0.730	NR
Heated	F-0.860	R-15 for 24 in.	F-0.688	R-20 for 48 in.	F-1.020	R-7.5 for 12 in.
Opaque Doors						
Swinging	U-0.700		U-0.500		U-0.700	
Nonswinging	U-0.500		U-0.500		U-1.450	
Fenestration	**Assembly Max. U**	**Assembly Max. SHGC**	**Assembly Max. U**	**Assembly Max. SHGC**	**Assembly Max. U**	**Assembly Max. SHGC**
Vertical Glazing, 0%–40% of Wall						
Nonmetal framing (all)[~~b~~c]	U-0.35		U-0.35		U-0.65	
Metal framing (curtainwall/storefront)[~~e~~d]	U-0.45	SHGC-0.40 all	U-0.45	SHGC-0.40 all	U-0.60	SHGC-NR all
Metal framing (entrance door)[~~e~~d]	U-0.80		U-0.80		U-0.90	
Metal framing (all other)[~~e~~d]	U-0.55		U-0.55		U-0.65	
Skylight with Curb, Glass, % of Roof						
0%–2.0%	U_{all}-1.17	$SHGC_{all}$-0.49	U_{all}-0.98	$SHGC_{all}$-0.46	U_{all}-1.98	$SHGC_{all}$-NR
2.1%–5.0%	U_{all}-1.17	$SHGC_{all}$-0.49	U_{all}-0.98	$SHGC_{all}$-0.36	U_{all}-1.98	$SHGC_{all}$-NR
Skylight with Curb, Plastic, % of Roof						
0%–2.0%	U_{all}-0.87	$SHGC_{all}$-0.71	U_{all}-0.74	$SHGC_{all}$-0.65	U_{all}-1.90	$SHGC_{all}$-NR
2.1%–5.0%	U_{all}-0.87	$SHGC_{all}$-0.58	U_{all}-0.74	$SHGC_{all}$-0.55	U_{all}-1.90	$SHGC_{all}$-NR
Skylight without Curb, All, % of Roof						
0%–2.0%	U_{all}-0.69	$SHGC_{all}$-0.49	U_{all}-0.58	$SHGC_{all}$-0.49	U_{all}-1.36	$SHGC_{all}$-NR
2.1%–5.0%	U_{all}-0.69	$SHGC_{all}$-0.49	U_{all}-0.58	$SHGC_{all}$-0.39	U_{all}-1.36	$SHGC_{all}$-NR

*The following definitions apply: c.i. = continuous insulation (see Section 3.2), NR = no (insulation) requirement.

[a] When using R-value compliance method, a thermal spacer block is required; otherwise use the U-factor compliance method. See Table A2.3.

[~~a~~b] Exception to Section A3.1.3.1 applies.

[~~b~~c] Nonmetal framing includes framing materials other than metal with or without metal reinforcing or cladding.

[~~e~~d] Metal framing includes metal framing with or without thermal break. The "all other" subcategory includes operable windows, fixed windows, and non-entrance doors.

TABLE 5.5-7 Building Envelope Requirements For Climate Zone 7*

Opaque Elements	Nonresidential		Residential		Semiheated	
	Assembly Maximum	Insulation Min. R-Value	Assembly Maximum	Insulation Min. R-Value	Assembly Maximum	Insulation Min. R-Value
Roofs						
Insulation Entirely above Deck	U-0.048	R-20.0 c.i.	U-0.048	R-20.0 c.i.	U-0.093	R-10.0 c.i.
Metal Building[a]	U-0.065 U-0.049	R-19.0 R-13.0 + R-19.0	U-0.065 U-0.049	R-19.0 R-13.0 + R-19.0	U-0.097 U-0.072	R-10.0 R-16.0
Attic and Other	U-0.027	R-38.0	U-0.027	R-38.0	U-0.034	R-30.0
Walls, Above-Grade						
Mass	U-0.071	R-15.2 c.i.	U-0.071	R-15.2 c.i.	U-0.123	R-7.6 c.i.
Metal Building	U-0.057	~~R-13.0 + R-13.0~~ R-19.0 + R-5.6 c.i.	U-0.057	~~R-13.0 + R-13.0~~ R-19.0 + R-5.6 c.i.	U-0.113	R-13.0
Steel-Framed	U-0.064	R-13.0 + R-7.5 c.i.	U-0.042	R-13.0 + R-15.6 c.i.	U-0.124	R-13.0
Wood-Framed and Other	U-0.051	R-13.0 + R-7.5 c.i.	U-0.051	R-13.0 + R-7.5 c.i.	U-0.089	R-13.0
Walls, Below-Grade						
Below-Grade Wall	C-0.119	R-7.5 c.i.	C-0.092	R-10.0 c.i.	C-1.140	NR
Floors						
Mass	U-0.064	R-12.5 c.i.	U-0.051	R-16.7 c.i.	U-0.107	R-6.3 c.i.
Steel-Joist	U-0.038	R-30.0	U-0.032	R-38.0	U-0.052	R-19.0
Wood-Framed and Other	U-0.033	R-30.0	U-0.033	R-30.0	U-0.051	R-19.0
Slab-On-Grade Floors						
Unheated	F-0.520	R-15 for 24 in.	F-0.520	R-15 for 24 in.	F-0.730	NR
Heated	F-0.843	R-20 for 24in.	F-0.688	R-20 for 48 in.	F-0.900	R-10 for 24 in.
Opaque Doors						
Swinging	U-0.500		U-0.500		U-0.700	
Nonswinging	U-0.500		U-0.500		U-1.450	
Fenestration	**Assembly Max. U**	**Assembly Max. SHGC**	**Assembly Max. U**	**Assembly Max. SHGC**	**Assembly Max. U**	**Assembly Max. SHGC**
Vertical Glazing, 0%–40% of Wall						
Nonmetal framing (all)[bc]	U-0.35		U-0.35		U-0.65	
Metal framing (curtainwall/storefront)[ed]	U-0.40	SHGC-0.45 all	U-0.40	SHGC-NR all	U-0.60	SHGC-NR all
Metal framing (entrance door)[ed]	U-0.80		U-0.80		U-0.90	
Metal framing (all other)[ed]	U-0.45		U-0.45		U-0.65	
Skylight with Curb, Glass, % of Roof						
0%–2.0%	U_{all}-1.17	$SHGC_{all}$-0.68	U_{all}-1.17	$SHGC_{all}$-0.64	U_{all}-1.98	$SHGC_{all}$-NR
2.1%–5.0%	U_{all}-1.17	$SHGC_{all}$-0.64	U_{all}-1.17	$SHGC_{all}$-0.64	U_{all}-1.98	$SHGC_{all}$-NR
Skylight with Curb, Plastic, % of Roof						
0%–2.0%	U_{all}-0.87	$SHGC_{all}$-0.77	U_{all}-0.61	$SHGC_{all}$-0.77	U_{all}-1.90	$SHGC_{all}$-NR
2.1%–5.0%	U_{all}-0.87	$SHGC_{all}$-0.71	U_{all}-0.61	$SHGC_{all}$-0.77	U_{all}-1.90	$SHGC_{all}$-NR
Skylight without Curb, All, % of Roof						
0%–2.0%	U_{all}-0.69	$SHGC_{all}$-0.68	U_{all}-0.69	$SHGC_{all}$-0.64	U_{all}-1.36	$SHGC_{all}$-NR
2.1%–5.0%	U_{all}-0.69	$SHGC_{all}$-0.64	U_{all}-0.69	$SHGC_{all}$-0.64	U_{all}-1.36	$SHGC_{all}$-NR

*The following definitions apply: c.i. = continuous insulation (see Section 3.2), NR = no (insulation) requirement.
[a]When using R-value compliance method, a thermal spacer block is required; otherwise use the U-factor compliance method. See Table A2.3.
[ab]Exception to Section A3.1.3.1 applies.
[bc]Nonmetal framing includes framing materials other than metal with or without metal reinforcing or cladding.
[ed]Metal framing includes metal framing with or without thermal break. The "all other" subcategory includes operable windows, fixed windows, and non-entrance doors.

TABLE 5.5-8 Building Envelope Requirements For Climate Zone 8*

Opaque Elements	Nonresidential		Residential		Semiheated	
	Assembly Maximum	**Insulation Min. R-Value**	**Assembly Maximum**	**Insulation Min. R-Value**	**Assembly Maximum**	**Insulation Min. R-Value**
Roofs						
Insulation Entirely above Deck	U-0.048	R-20.0 c.i.	U-0.048	R-20.0 c.i.	U-0.063	R-15.0 c.i.
Metal Building[a]	~~U-0.049~~ U-0.035	~~R-13.0 + R-19.0~~ R-11.0 + R-19.0 ls	~~U-0.049~~ U-0.035	~~R-13.0 + R-19.0~~ R-11.0 + R-19.0 ls	~~U-0.072~~ U-0.065	~~R-16.0~~ R-19.0
Attic and Other	U-0.021	R-49.0	U-0.021	R-49.0	U-0.034	R-30.0
Walls, Above-Grade						
Mass	U-0.071	R-15.2 c.i.	U-0.052	R-25.0 c.i.	U-0.104	R-9.5 c.i.
Metal Building	U-0.057	~~R-13.0 + R-13.0~~ R-19.0 + R-5.6 c.i.	U-0.057	~~R-13.0 + R-13.0~~ R-19.0 + R-5.6 c.i.	U-0.113	R-13.0
Steel-Framed	U-0.064	R-13.0 + R-7.5 c.i.	U-0.037	R-13.0 + R-18.8 c.i.	U-0.084	R-13.0 + R-3.8 c.i.
Wood-Framed and Other	U-0.036	R-13.0 + R-15.6 c.i.	U-0.036	R-13.0 + R-15.6 c.i.	U-0.089	R-13.0
Walls, Below-Grade						
Below-Grade Wall	C-0.119	R-7.5 c.i.	C-0.075	R-12.5 c.i.	C-1.140	NR
Floors						
Mass	U-0.057	R-14.6 c.i.	U-0.051	R-16.7 c.i.	U-0.087	R-8.3 c.i.
Steel-Joist	U-0.032	R-38.0	U-0.032	R-38.0	U-0.052	R-19.0
Wood-Framed and Other	U-0.033	R-30.0	U-0.033	R-30.0	U-0.033	R-30.0
Slab-On-Grade Floors						
Unheated	F-0.520	R-15 for 24 in.	F-0.510	R-20 for 24 in.	F-0.730	NR
Heated	F-0.688	R-20 for 48 in.	F-0.688	R-20 for 48 in.	F-0.900	R-10.0 for 24 in.
Opaque Doors						
Swinging	U-0.500		U-0.500		U-0.700	
Nonswinging	U-0.500		U-0.500		U-0.500	
Fenestration	**Assembly Max. U**	**Assembly Max. SHGC**	**Assembly Max. U**	**Assembly Max. SHGC**	**Assembly Max. U**	**Assembly Max. SHGC**
Vertical Glazing, 0%–40% of Wall						
Nonmetal framing (all)[~~b~~c]	U-0.35	SHGC-0.45 all	U-0.35	SHGC-NR all	U-0.65	SHGC-NR all
Metal framing (curtainwall/storefront)[~~c~~d]	U-0.40		U-0.40		U-0.60	
Metal framing (entrance door)[~~c~~d]	U-0.80		U-0.80		U-0.90	
Metal framing (all other)[~~c~~d]	U-0.45		U-0.45		U-0.65	
Skylight with Curb, Glass, % of Roof						
0%–2.0%	U_{all}-0.98	$SHGC_{all}$-NR	U_{all}-0.98	$SHGC_{all}$-NR	U_{all}-1.30	$SHGC_{all}$-NR
2.1%–5.0%	U_{all}-0.98	$SHGC_{all}$-NR	U_{all}-0.98	$SHGC_{all}$-NR	U_{all}-1.30	$SHGC_{all}$-NR
Skylight with Curb, Plastic, % of Roof						
0%–2.0%	U_{all}-0.61	$SHGC_{all}$-NR	U_{all}-0.61	$SHGC_{all}$-NR	U_{all}-1.10	$SHGC_{all}$-NR
2.1%–5.0%	U_{all}-0.61	$SHGC_{all}$-NR	U_{all}-0.61	$SHGC_{all}$-NR	U_{all}-1.10	$SHGC_{all}$-NR
Skylight without Curb, All, % of Roof						
0%–2.0%	U_{all}-0.58	$SHGC_{all}$-NR	U_{all}-0.58	$SHGC_{all}$-NR	U_{all}-0.81	$SHGC_{all}$-NR
2.1%–5.0%	U_{all}-0.58	$SHGC_{all}$-NR	U_{all}-0.58	$SHGC_{all}$-NR	U_{all}-0.81	$SHGC_{all}$-NR

*The following definitions apply: c.i. = continuous insulation (see Section 3.2), NR = no (insulation) requirement.

[a]When using R-value compliance method, a thermal spacer block is required; otherwise use the U-factor compliance method. See Table A2.3.

[~~a~~b]Exception to Section A3.1.3.1 applies.

[~~b~~c]Nonmetal framing includes framing materials other than metal with or without metal reinforcing or cladding.

[~~c~~d]Metal framing includes metal framing with or without thermal break. The "all other" subcategory includes operable windows, fixed windows, and non-entrance doors.

Modify Section 5 as follows (SI units):

TABLE 5.5-1 Building Envelope Requirements for Climate Zone 1 (A, B)*

Opaque Elements	Nonresidential Assembly Maximum	Nonresidential Insulation Min. R-Value	Residential Assembly Maximum	Residential Insulation Min. R-Value	Semiheated Assembly Maximum	Semiheated Insulation Min. R-Value
Roofs						
Insulation Entirely above Deck	U-0.360	R-2.6 c.i.	U-0.273	R-3.5 c.i.	U-1.240	R-0.7 ci
Metal Building[a]	U-0.369	R-3.3	U-0.369	R-3.3	~~U-7.268~~ U-0.947	~~NR~~ R-1.1
Attic and Other	U-0.192	R-5.3	U-0.153	R-6.7	U-0.459	R-2.3
Walls, Above-Grade						
Mass	U-3.293	NR	U-0.857[ab]	R-1.0 c.i.[ab]	U-3.293	NR
Metal Building	~~U-0.642~~ U-0.527	~~R-2.3~~ R-2.8	~~U-0.642~~ U-0.527	~~R-2.3~~ R-2.8	~~U-6.700~~ U-0.642	~~NR~~ R-2.3
Steel-Framed	U-0.705	R-2.3	U-0.705	R-2.3	U-1.998	NR
Wood-Framed and Other	U-0.504	R-2.3	U-0.504	R-2.3	U-1.660	NR
Walls, Below-Grade						
Below-Grade Wall	C-6.473	NR	C-6.473	NR	C-6.473	NR
Floors						
Mass	U-1.825	NR	U-1.825	NR	U-1.825	NR
Steel-Joist	U-1.986	NR	U-1.986	NR	U-1.986	NR
Wood-Framed and Other	U-1.599	NR	U-1.599	NR	U-1.599	NR
Slab-On-Grade Floors						
Unheated	F-1.264	NR	F-1.264	NR	F-1.264	NR
Heated	U-1.766	R-1.3 for 300 mm	F-1.766	R-1.3 for 300 mm	F-1.766	R-1.3 for 300 mm
Opaque Doors						
Swinging	U-3.975		U-3.975		U-3.975	
Nonswinging	U-8.233		U-8.233		U-8.233	
Fenestration	**Assembly Max. U**	**Assembly Max. SHGC**	**Assembly Max. U**	**Assembly Max. SHGC**	**Assembly Max. U**	**Assembly Max. SHGC**
Vertical Glazing, 0%–40% of Wall						
Nonmetal framing (all)[b]	U-6.81	SHGC-0.25 all	U-6.81	SHGC-0.25 all	U-6.81	SHGC-NR all
Metal framing (curtainwall/storefront)[c]	U-6.81		U-6.81		U-6.81	
Metal framing (entrance door)[c]	U-6.81		U-6.81		U-6.81	
Metal framing (all other)[c]	U-6.81		U-6.81		U-6.81	
Skylight with Curb, Glass, % of Roof						
0%–2.0%	U_{all}-11.24	$SHGC_{all}$-0.36	U_{all}-11.24	$SHGC_{all}$-0.19	U_{all}-11.24	$SHGC_{all}$-NR
2.1%–5.0%	U_{all}-11.24	$SHGC_{all}$-0.19	U_{all}-11.24	$SHGC_{all}$-0.16	U_{all}-11.24	$SHGC_{all}$-NR
Skylight with Curb, Plastic, % of Roof						
0%–2.0%	U_{all}-10.79	$SHGC_{all}$-0.34	U_{all}-10.79	$SHGC_{all}$-0.27	U_{all}-10.79	$SHGC_{all}$-NR
2.1%–5.0%	U_{all}-10.79	$SHGC_{all}$-0.27	U_{all}-10.79	$SHGC_{all}$-0.27	U_{all}-10.79	$SHGC_{all}$-NR
Skylight without Curb, All, % of Roof						
0%–2.0%	U_{all}-7.72	$SHGC_{all}$-0.36	U_{all}-7.72	$SHGC_{all}$-0.19	U_{all}-7.72	$SHGC_{all}$-NR
2.1%–5.0%	U_{all}-7.72	$SHGC_{all}$-0.19	U_{all}-7.72	$SHGC_{all}$-0.19	U_{all}-7.72	$SHGC_{all}$-NR

*The following definitions apply: c.i. = continuous insulation (see Section 3.2), NR = no (insulation) requirement.

[a]When using R-value compliance method, a thermal spacer block is required, otherwise use the U-factor compliance method. See Table A2.3.

[ab]Exception to Section A3.1.3.1 applies.

[bc]Nonmetal framing includes framing materials other than metal with or without metal reinforcing or cladding.

[cd]Metal framing includes metal framing with or without thermal break. The "all other" subcategory includes operable windows, fixed windows, and non-entrance doors.

TABLE 5.5-2 Building Envelope Requirements for Climate Zone 2 (A, B)*

Opaque Elements	Nonresidential Assembly Maximum	Nonresidential Insulation Min. R-Value	Residential Assembly Maximum	Residential Insulation Min. R-Value	Semiheated Assembly Maximum	Semiheated Insulation Min. R-Value
Roofs						
Insulation Entirely above Deck	U-0.273	R-3.5 c.i.	U-0.273	R-3.5 c.i.	U-1.240	R-0.7 c.i.
Metal Building[a]	~~U-0.369~~ U-0.312	~~R-3.3~~ R-2.3 + R-2.3	~~U-0.369~~ U-0.312	~~R-3.3~~ R-2.3 + R-2.3	~~U-0.948~~ U-0.551	~~R-1.1~~ R-1.8
Attic and Other	U-0.153	R-6.7	U-0.153	R-6.7	U-0.459	R-2.3
Walls, Above-Grade						
Mass	U-0.857[a]	R-1.0 c.i.[a]	U-0.701[b]	R-1.3 c.i.[b]	U-3.293	NR
Metal Building	~~U-0.642~~ U-0.528	~~R-2.3~~ R-2.8	~~U-0.642~~ U-0.528	~~R-2.3~~ R-2.8	U-0.642	R-2.3
Steel-Framed	U-0.705	R-2.3	U-0.365	R-2.3 + R-1.3 c.i.	U-0.705	R-2.3
Wood-Framed and Other	U-0.504	R-2.3	U-0.504	R-2.3	U-0.504	R-2.3
Walls, Below-Grade						
Below-Grade Wall	C-6.473	NR	C-6.473	NR	C-6.473	NR
Floors						
Mass	U-0.606	R-1.1 c.i.	U-0.496	R-1.5 c.i.	U-1.825	NR
Steel-Joist	U-0.296	R-3.3	U-0.296	R-3.3	U-0.390	R-2.3
Wood-Framed and Other	U-0.288	R-3.3	U-0.188	R-5.3	U-0.376	R-2.3
Slab-On-Grade Floors						
Unheated	F-1.264	NR	F-1.264	NR	F-1.264	NR
Heated	F-1.766	R-1.3 for 300 mm	F-1.766	R-1.3 for 300 mm	F-1.766	R-1.3 for 300 mm
Opaque Doors						
Swinging	U-3.975		U-3.975		U-3.975	
Nonswinging	U-8.233		U-2.839		U-8.233	
Fenestration	**Assembly Max. U**	**Assembly Max. SHGC**	**Assembly Max. U**	**Assembly Max. SHGC**	**Assembly Max. U**	**Assembly Max. SHGC**
Vertical Glazing, 0%–40% of Wall						
Nonmetal framing (all)[b]	U-4.26	SHGC-0.25 all	U-4.26	SHGC-0.25 all	U-6.81	SHGC-NR all
Metal framing (curtainwall/storefront)[c]	U-3.97		U-3.97		U-6.81	
Metal framing (entrance door)[c]	U-6.25		U-6.25		U-6.81	
Metal framing (all other)[c]	U-4.26		U-4.26		U-6.81	
Skylight with Curb, Glass, % of Roof						
0%–2.0%	U_{all}-11.24	$SHGC_{all}$-0.36	U_{all}-11.24	$SHGC_{all}$-0.19	U_{all}-11.24	$SHGC_{all}$-NR
2.1%–5.0%	U_{all}-11.24	$SHGC_{all}$-0.19	U_{all}-11.24	$SHGC_{all}$-0.19	U_{all}-11.24	$SHGC_{all}$-NR
Skylight with Curb, Plastic, % of Roof						
0%–2.0%	U_{all}-10.79	$SHGC_{all}$-0.39	U_{all}-10.79	$SHGC_{all}$-0.27	U_{all}-10.79	$SHGC_{all}$-NR
2.1%–5.0%	U_{all}-10.79	$SHGC_{all}$-0.34	U_{all}-10.79	$SHGC_{all}$-0.27	U_{all}-10.79	$SHGC_{all}$-NR
Skylight without Curb, All, % of Roof						
0%–2.0%	U_{all}-7.72	$SHGC_{all}$-0.36	U_{all}-7.72	$SHGC_{all}$-0.19	U_{all}-7.72	$SHGC_{all}$-NR
2.1%–5.0%	U_{all}-7.72	$SHGC_{all}$-0.19	U_{all}-7.72	$SHGC_{all}$-0.19	U_{all}-7.72	$SHGC_{all}$-NR

*The following definitions apply: c.i. = continuous insulation (see Section 3.2), NR = no (insulation) requirement.
[a] When using R-value compliance method, a thermal spacer block is required, otherwise use the U-factor compliance method. See Table A2.3.
[~~a~~b] Exception to Section A3.1.3.1 applies.
[~~b~~c] Nonmetal framing includes framing materials other than metal with or without metal reinforcing or cladding.
[~~c~~d] Metal framing includes metal framing with or without thermal break. The "all other" subcategory includes operable windows, fixed windows, and non-entrance doors.

TABLE 5.5-3 Building Envelope Requirements for Climate Zone 3 (A, B, C)*

Opaque Elements	Nonresidential		Residential		Semiheated	
	Assembly Maximum	Insulation Min. R-Value	Assembly Maximum	Insulation Min. R-Value	Assembly Maximum	Insulation Min. R-Value
Roofs						
Insulation Entirely above Deck	U-0.273	R-3.5 c.i.	U-0.273	R-3.5 c.i.	U-0.982	R-0.9 c.i.
Metal Building[a]	~~U-0.369~~ U-0.312	~~R-3.3~~ R-2.3 + R-2.3	~~U-0.369~~ U-0.312	~~R-3.3~~ R-2.3 + R-2.3	U-0.551	R-1.8
Attic and Other	U-0.153	R-6.7	U-0.153	R-6.7	U-0.300	R-3.3
Walls, Above-Grade						
Mass	U-0.701	R-1.3 c.i.	U-0.592	R-1.7 c.i.	U-3.293	NR
Metal Building	~~U-0.642~~ U-0.477	~~R-2.3~~ R-3.3	~~U-0.642~~ U-0.476	~~R-2.3~~ R-3.3	~~U-1.045~~ U-0.642	~~R-1.1~~ R-2.3
Steel-Framed	U-0.479	R-2.3 + R-0.7 c.i.	U-0.365	R-2.3 + R-1.3 c.i.	U-0.705	R-2.3
Wood-Framed and Other	U-0.504	R-2.3	U-0.504	R-2.3	U-0.504	R-2.3
Walls, Below-Grade						
Below-Grade Wall	C-6.473	NR	C-6.473	NR	C-6.473	NR
Floors						
Mass	U-0.606	R-1.1	U-0.496	R-1.5	U-1.825	NR
Steel-Joist	U-0.296	R-3.3	U-0.296	R-3.3	U-0.390	R-2.3
Wood-Framed and Other	U-0.288	R-3.3	U-0.188	R-5.3	U-0.376	R-2.3
Slab-On-Grade Floors						
Unheated	F-1.264	NR	F-1.264	NR	F-1.264	NR
Heated	F-1.558	R-1.8 for 600 mm	F-1.558	R-1.8 for 600 mm	F-1.766	R-1.3 for 300 mm
Opaque Doors						
Swinging	U-3.975		U-3.975		U-3.975	
Nonswinging	U-8.233		U-2.839		U-8.233	
Fenestration	**Assembly Max. U**	**Assembly Max. SHGC**	**Assembly Max. U**	**Assembly Max. SHGC**	**Assembly Max. U**	**Assembly Max. SHGC**
Vertical Glazing, 0%–40% of Wall						
Nonmetal framing (all)[b]	U-3.69	SHGC-0.25 all	U-3.69	SHGC-0.25 all	U-6.81	SHGC-NR all
Metal framing (curtainwall/storefront)[c]	U-3.41		U-3.41		U-6.81	
Metal framing (entrance door)[c]	U-5.11		U-5.11		U-6.81	
Metal framing (all other)[c]	U-3.69		U-3.69		U-6.81	
Skylight with Curb, Glass, % of Roof						
0%–2.0%	U_{all}-6.64	$SHGC_{all}$-0.39	U_{all}-6.64	$SHGC_{all}$-0.36	U_{all}-11.24	$SHGC_{all}$-NR
2.1%–5.0%	U_{all}-6.64	$SHGC_{all}$-0.19	U_{all}-6.64	$SHGC_{all}$-0.19	U_{all}-11.24	$SHGC_{all}$-NR
Skylight with Curb, Plastic, % of Roof						
0%–2.0%	U_{all}-7.38	$SHGC_{all}$-0.65	U_{all}-7.38	$SHGC_{all}$-0.27	U_{all}-10.79	$SHGC_{all}$-NR
2.1%–5.0%	U_{all}-7.38	$SHGC_{all}$-0.34	U_{all}-7.38	$SHGC_{all}$-0.27	U_{all}-10.79	$SHGC_{all}$-NR
Skylight without Curb, All, % of Roof						
0%–2.0%	U_{all}-3.92	$SHGC_{all}$-0.39	U_{all}-3.92	$SHGC_{all}$-0.36	U_{all}-7.72	$SHGC_{all}$-NR
2.1%–5.0%	U_{all}-3.92	$SHGC_{all}$-0.19	U_{all}-3.92	$SHGC_{all}$-0.19	U_{all}-7.72	$SHGC_{all}$-NR

*The following definitions apply: c.i. = continuous insulation (see Section 3.2), NR = no (insulation) requirement.

[a] When using R-value compliance method, a thermal spacer block is required, otherwise use the U-factor compliance method. See Table A2.3.

[~~a~~b] Exception to A3.1.3.1 applies.

[~~b~~c] Nonmetal framing includes framing materials other than metal with or without metal reinforcing or cladding.

[~~c~~d] Metal framing includes metal framing with or without thermal break. The "all other" subcategory includes operable windows, fixed windows, and non-entrance doors.

TABLE 5.5-4 Building Envelope Requirements for Climate Zone 4 (A, B, C)*

Opaque Elements	Nonresidential Assembly Maximum	Nonresidential Insulation Min. R-Value	Residential Assembly Maximum	Residential Insulation Min. R-Value	Semiheated Assembly Maximum	Semiheated Insulation Min. R-Value
Roofs						
Insulation Entirely above Deck	U-0.273	R-3.5 c.i.	U-0.273	R-3.5 c.i.	U-0.982	R-0.9 c.i.
Metal Building[a]	U-0.369 U-0.312	~~R-3.3~~ R-2.3 + R-2.3	~~U-0.369~~ U-0.312	~~R-3.3~~ R-2.3 + R-2.3	U-0.551	R-1.8
Attic and Other	U-0.0.153	R-6.7	U-0.153	R-6.7	U-0.300	R-3.3
Walls, Above-Grade						
Mass	U-0.592	R-1.7 c.i.	U-0.513	R-2.0 c.i.	U-3.293	NR
Metal Building	~~U-0.642~~ U-0.477	~~R-2.3~~ R-3.3	~~U-0.642~~ U-0.476	~~R-2.3~~ R-3.3	U-0.642	R-2.3
Steel-Framed	U-0.365	R-2.3 + R-1.3	U-0.365	R-2.3 + R-1.3 c.i.	U-0.705	R-2.3
Wood-Framed and Other	U-0.504	R-2.3	U-0.365	R-2.3 + R-0.7 c.i.	U-0.504	R-2.3
Walls, Below-Grade						
Below-Grade Wall	C-6.473	NR	C-0.678	R-1.3 c.i.	C-6.473	NR
Floors						
Mass	U-0.496	R-1.5 c.i.	U-0.420	R-1.8 c.i.	U-0.780	R-0.7 c.i.
Steel-Joist	U-0.214	R-5.3	U-0.214	R-5.3	U-0.390	R-2.3
Wood-Framed and Other	U-0.188	R-5.3	U-0.188	R-5.3	U-0.376	R-2.3
Slab-On-Grade Floors						
Unheated	F-1.264	NR	F-0.935	R-1.8 for 600 mm	F-1.264	NR
Heated	F-1.489	R-2.6 for 600 mm	F-1.489	R-2.6 for 600 mm	F-1.766	R-1.3 for 300 mm
Opaque Doors						
Swinging	U-3.975		U-3.975		U-3.975	
Nonswinging	U-2.839		U-2.839		U-8.233	

Fenestration	Nonresidential Assembly Max. U	Nonresidential Assembly Max. SHGC	Residential Assembly Max. U	Residential Assembly Max. SHGC	Semiheated Assembly Max. U	Semiheated Assembly Max. SHGC
Vertical Glazing, 0%–40% of Wall						
Nonmetal framing (all)[b]	U-2.27	SHGC-0.40 all	U-2.27	SHGC-0.40 all	U-6.81	SHGC-NR all
Metal framing (curtainwall/storefront)[c]	U-2.84		U-2.84		U-6.81	
Metal framing (entrance door)[c]	U-4.83		U-4.83		U-6.81	
Metal framing (all other)[c]	U-3.12		U-3.12		U-6.81	
Skylight with Curb, Glass, % of Roof						
0%–2.0%	U_{all}-6.64	$SHGC_{all}$-0.49	U_{all}-5.56	$SHGC_{all}$-0.36	U_{all}-11.24	$SHGC_{all}$-NR
2.1%–5.0%	U_{all}-6.64	$SHGC_{all}$-0.39	U_{all}-5.56	$SHGC_{all}$-0.19	U_{all}-11.24	$SHGC_{all}$-NR
Skylight with Curb, Plastic, % of Roof						
0%–2.0%	U_{all}-7.38	$SHGC_{all}$-0.65	U_{all}-7.38	$SHGC_{all}$-0.62	U_{all}-10.79	$SHGC_{all}$-NR
2.1%–5.0%	U_{all}-7.38	$SHGC_{all}$-0.34	U_{all}-7.38	$SHGC_{all}$-0.27	U_{all}-10.79	$SHGC_{all}$-NR
Skylight without Curb, All, % of Roof						
0%–2.0%	U_{all}-3.92	$SHGC_{all}$-0.49	U_{all}-3.29	$SHGC_{all}$-0.36	U_{all}-7.72	$SHGC_{all}$-NR
2.1%–5.0%	U_{all}-3.92	$SHGC_{all}$-0.39	U_{all}-3.29	$SHGC_{all}$-0.19	U_{all}-7.72	$SHGC_{all}$-NR

*The following definitions apply: c.i. = continuous insulation (see Section 3.2), NR = no (insulation) requirement.

[a] When using R-value compliance method, a thermal spacer block is required, otherwise use the U-factor compliance method. See Table A2.3.

[~~a~~b] Exception to A3.1.3.1 applies.

[~~b~~c] Nonmetal framing includes framing materials other than metal with or without metal reinforcing or cladding.

[~~c~~d] Metal framing includes metal framing with or without thermal break. The "all other" subcategory includes operable windows, fixed windows, and non-entrance doors.

TABLE 5.5-5 Building Envelope Requirements for Climate Zone 5 (A, B, C)*

Opaque Elements	Nonresidential		Residential		Semiheated	
	Assembly Maximum	**Insulation Min. R-Value**	**Assembly Maximum**	**Insulation Min. R-Value**	**Assembly Maximum**	**Insulation Min. R-Value**
Roofs						
Insulation Entirely above Deck	U-0.273	R-3.5 c.i.	U-0.273	R-3.5 c.i.	U-0.677	R-1.3 c.i.
Metal Building[a]	~~U-0.369~~ U-0.312	~~R-3.3~~ R-2.3 +R-2.3	~~U-0.369~~ U-0.312	~~R-3.3~~ R-2.3 + R-2.3	~~U-0.551~~ U-0.471	~~R-1.8~~ R-2.3
Attic and Other	U-0.153	R-6.7	U-0.153	R-6.7	U-0.300	R-3.3
Walls, Above-Grade						
Mass	U-0.513	R-2.0 c.i.	U-0.453	R-2.3 c.i.	U-0.857[a]	R-1.0 c.i.[a]
Metal Building	~~U-0.642~~ U-0.391	~~R-2.3~~ R-2.3 + R-1.0 c.i.	~~U-0.324~~ U-0.391	~~R-2.3 + R-2.3~~ R-2.3 + R-1.0 c.i.	~~U-0.698~~ U-0.642	~~R-1.9~~ R-2.3
Steel-Framed	U-0.365	R-2.3 + R-1.3 c.i.	U-0.365	R-2.3 +R-1.3 c.i.	U-0.705	R-2.3
Wood-Framed and Other	U-0.365	R-2.3 + R-0.7 c.i.	U-0.291	R-2.3 + R-1.3 c.i.	U-0.504	R-2.3
Walls, Below-Grade						
Below-Grade Wall	C-0.678	R-1.3 c.i.	C-0.678	R-1.3 c.i.	C-6.473	NR
Floors						
Mass	U-0.420	R-1.8 c.i.	U-0.363	R-2.2 c.i.	U-0.780	R-0.7 c.i.
Steel-Joist	U-0.214	R-5.3	U-0.214	R-5.3	U-0.296	R-3.3
Wood-Framed and Other	U-0.188	R-5.3	U-0.188	R-5.3	U-0.288	R-3.3
Slab-On-Grade Floors						
Unheated	F-1.264	NR	F-0.935	R-1.8 for 600 mm	F-1.264	NR
Heated	F-1.489	R-2.6 for 600 mm	F-1.489	R-2.6 for 600 mm	F-1.766	R-1.3 for 300 mm
Opaque Doors						
Swinging	U-3.975		U-2.839		U-3.975	
Nonswinging	U-2.839		U-2.839		U-8.233	
Fenestration	**Assembly Max. U**	**Assembly Max. SHGC**	**Assembly Max. U**	**Assembly Max. SHGC**	**Assembly Max. U**	**Assembly Max. SHGC**
Vertical Glazing, % of Wall						
Nonmetal framing (all)[b]	U-1.99	SHGC-0.40 all	U-1.99	SHGC-0.40 all	U-6.81	SHGC-NR all
Metal framing (curtainwall/storefront)[c]	U-2.56		U-2.56		U-6.81	
Metal framing (entrance door)[c]	U-4.54		U-4.54		U-6.81	
Metal framing (all other)[c]	U-3.12		U-3.12		U-6.81	
Skylight with Curb, Glass, % of Roof						
0%–2.0%	U_{all}-6.64	$SHGC_{all}$-0.49	U_{all}-6.64	$SHGC_{all}$-0.49	U_{all}-11.24	$SHGC_{all}$-NR
2.1%–5.0%	U_{all}-6.64	$SHGC_{all}$-0.39	U_{all}-6.64	$SHGC_{all}$-0.39	U_{all}-11.24	$SHGC_{all}$-NR
Skylight with Curb, Plastic, % of Roof						
0%–2.0%	U_{all}-6.25	$SHGC_{all}$-0.77	U_{all}-6.25	$SHGC_{all}$-0.77	U_{all}-10.79	$SHGC_{all}$-NR
2.1%–5.0%	U_{all}-6.25	$SHGC_{all}$-0.62	U_{all}-6.25	$SHGC_{all}$-0.62	U_{all}-10.79	$SHGC_{all}$-NR
Skylight without Curb, All, % of Roof						
0%–2.0%	U_{all}-3.92	$SHGC_{all}$-0.49	U_{all}-3.92	$SHGC_{all}$-0.49	U_{all}-7.72	$SHGC_{all}$-NR
2.1%–5.0%	U_{all}-3.92	$SHGC_{all}$-0.39	U_{all}-3.92	$SHGC_{all}$-0.39	U_{all}-7.72	$SHGC_{all}$-NR

*The following definitions apply: c.i. = continuous insulation (see Section 3.2), NR = no (insulation) requirement.

[a] When using R-value compliance method, a thermal spacer block is required, otherwise use the U-factor compliance method. See Table A2.3.

[~~a~~b] Exception to Section A3.1.3.1 applies.

[~~b~~c] Nonmetal framing includes framing materials other than metal with or without metal reinforcing or cladding.

[~~c~~d] Metal framing includes metal framing with or without thermal break. The "all other" subcategory includes operable windows, fixed windows, and non-entrance doors.

TABLE 5.5-6 Building Envelope Requirements for Climate Zone 6 (A, B)*

Opaque Elements	Nonresidential		Residential		Semiheated	
	Assembly Maximum	Insulation Min. R-Value	Assembly Maximum	Insulation Min. R-Value	Assembly Maximum	Insulation Min. R-Value
Roofs						
Insulation Entirely above Deck	U-0.273	R-3.5 c.i.	U-0.273	R-3.5 c.i.	U-0.527	R-1.8 c.i.
Metal Building[a]	~~U-0.369~~ U-0.278	~~R-3.3~~ R-2.3 + R-3.3	~~U-0.369~~ U-0.278	~~R-3.3~~ R-2.3 + R-3.3	~~U-0.551~~ U-0.409	~~R-1.8~~ R-2.8
Attic and Other	U-0.153	R-6.7	U-0.153	R-6.7	U-0.192	R-5.3
Walls, Above-Grade						
Mass	U-0.453	R-2.3 c.i.	U-0.404	R-2.7 c.i.	U-0.857[a]	R-1.0 c.i.[a]
Metal Building	~~U-0.642~~ U-0.392	~~R-2.3~~ R-2.3 + R-1.0 c.i.	~~U-0.324~~ U-0.392	~~R-2.3 + R-2.3~~ R-2.3 + R-1.0 c.i.	U-0.642	R-2.3
Steel-Framed	U-0.365	R-2.3 + R-1.3 c.i.	U-0.365	R-2.3 + R-1.3 c.i.	U-0.705	R-2.3
Wood-Framed and Other	U-0.291	R-2.3 + R-1.3 c.i.	U-0.291	R-2.3 + R-1.3 c.i.	U-0.504	R-2.3
Walls, Below-Grade						
Below-Grade Wall	C-0.678	R-1.3 c.i.	C-0.678	R-1.3c.i.	C-6.473	NR
Floors						
Mass	U-0.363	R-2.2 c.i.	U-0.321	R-2.6 c.i.	U-0.780	R-0.7 c.i.
Steel-Joist	U-0.214	R-5.3	U-0.183	R-6.7	U-0.296	R-3.3
Wood-Framed and Other	U-0.188	R-5.3	U-0.188	R-5.3	U-0.288	R-3.3
Slab-On-Grade Floors						
Unheated	F-0.935	R-1.8 for 600 mm	F-0.900	R-2.6 for 600 mm	F-1.264	NR
Heated	F-1.489	R-2.6 for 600 mm	F-1.191	R-3.5 for 1200 mm	F-1.766	R-1.3 for 300 mm
Opaque Doors						
Swinging	U-3.975		U-2.839		U-3.975	
Nonswinging	U-2.839		U-2.839		U-8.233	
Fenestration	**Assembly Max. U**	**Assembly Max. SHGC**	**Assembly Max. U**	**Assembly Max. SHGC**	**Assembly Max. U**	**Assembly Max. SHGC**
Vertical Glazing, 0%–40% of Wall						
Nonmetal framing (all)[b]	U-1.99	SHGC-0.40 all	U-1.99	SHGC-0.40 all	U-3.69	SHGC-NR all
Metal framing (curtainwall/storefront)[c]	U-2.56		U-2.56		U-3.41	
Metal framing (entrance door)[c]	U-4.54		U-4.54		U-5.11	
Metal framing (all other)[c]	U-3.12		U-3.12		U-3.69	
Skylight with Curb, Glass, % of Roof						
0%–2.0%	U_{all}-6.64	$SHGC_{all}$-0.49	U_{all}-5.56	$SHGC_{all}$-0.46	U_{all}-11.24	$SHGC_{all}$-NR
2.1%–5.0%	U_{all}-6.64	$SHGC_{all}$-0.49	U_{all}-5.56	$SHGC_{all}$-0.36	U_{all}-11.24	$SHGC_{all}$-NR
Skylight with Curb, Plastic, % of Roof						
0%–2.0%	U_{all}-4.94	$SHGC_{all}$-0.71	U_{all}-4.20	$SHGC_{all}$-0.65	U_{all}-10.79	$SHGC_{all}$-NR
2.1%–5.0%	U_{all}-4.94	$SHGC_{all}$-0.58	U_{all}-4.20	$SHGC_{all}$-0.55	U_{all}-10.79	$SHGC_{all}$-NR
Skylight without Curb, All, % of Roof						
0%–2.0%	U_{all}-3.92	$SHGC_{all}$-0.49	U_{all}-3.29	$SHGC_{all}$-0.49	U_{all}-7.72	$SHGC_{all}$-NR
2.1%–5.0%	U_{all}-3.92	$SHGC_{all}$-0.49	U_{all}-3.29	$SHGC_{all}$-0.39	U_{all}-7.72	$SHGC_{all}$-NR

*The following definitions apply: c.i. = continuous insulation (see Section 3.2), NR = no (insulation) requirement.

[a] When using R-value compliance method, a thermal spacer block is required, otherwise use the U-factor compliance method. See Table A2.3.

[~~a~~b] Exception to Section A3.1.3.1 applies.

[~~b~~c] Nonmetal framing includes framing materials other than metal with or without metal reinforcing or cladding.

[~~c~~d] Metal framing includes metal framing with or without thermal break. The "all other" subcategory includes operable windows, fixed windows, and non-entrance doors.

TABLE 5.5-7 Building Envelope Requirements for Climate Zone 7*

Opaque Elements	Nonresidential Assembly Maximum	Nonresidential Insulation Min. R-Value	Residential Assembly Maximum	Residential Insulation Min. R-Value	Semiheated Assembly Maximum	Semiheated Insulation Min. R-Value
Roofs						
Insulation Entirely above Deck	U-0.273	R-3.5 c.i.	U-0.273	R-3.5 c.i.	U-0.527	R-1.8 c.i.
Metal Building[a]	~~U-0.369~~ U-0.278	~~R-3.3~~ R-2.3 + R-3.3	~~U-0.369~~ U-0.278	~~R-3.3~~ R-2.3 + R-3.3	~~U-0.551~~ U-0.409	~~R-1.8~~ R-2.8
Attic and Other	U-0.153	R-6.7	U-0.153	R-6.7	U-0.192	R-5.3
Walls, Above-Grade						
Mass	U-0.404	R-2.7 c.i.	U-0.404	R-2.7 c.i.	U-0.701	R-1.3 c.i.
Metal Building	U-0.324	~~R-2.3 + R-2.3~~ R-3.3 + R-1.0 c.i.	U-0.324	~~R-2.3 + R-2.3~~ R-3.3 + R-1.0 c.i.	U-0.642	R-2.3
Steel-Framed	U-0.365	R-2.3 + R-1.3 c.i.	U-0.240	R-2.3 + R-2.7 c.i.	U-0.705	R-2.3
Wood-Framed and Other	U-0.291	R-2.3 + R-1.3 c.i.	U-0.291	R-2.3 +R-1.3 c.i.	U-0.504	R-2.3
Walls, Below-Grade						
Below-Grade Wall	C-0.678	R-1.3 c.i.	C-0.522	R-1.8 c.i.	C-6.473	NR
Floors						
Mass	U-0.363	R-2.2 c.i.	U-0.287	R-2.9 c.i.	U-0.606	R-1.1 c.i.
Steel-Joist	U-0.214	R-5.3	U-0.183	R-6.7	U-0.296	R-3.3
Wood-Framed and Other	U-0.188	R-5.3	U-0.188	R-5.3	U-0.288	R-3.3
Slab-On-Grade Floors						
Unheated	F-0.900	R-2.6 for 600 mm	F-0.900	R-2.6 for 600 mm	F-1.264	NR
Heated	F-1.459	R-3.5 for 600 mm	F-1.191	R-3.5 for 1200	F-1.558	R-1.8 for 600 mm
Opaque Doors						
Swinging	U-2.839		U-2.839		U-3.975	
Nonswinging	U-2.839		U-2.839		U-8.233	

Fenestration	Nonresidential Assembly Max. U	Nonresidential Assembly Max. SHGC	Residential Assembly Max. U	Residential Assembly Max. SHGC	Semiheated Assembly Max. U	Semiheated Assembly Max. SHGC
Vertical Glazing, 0%–40% of Wall						
Nonmetal framing (all)[b]	U-1.99	SHGC-0.45 all	U-1.99	SHGC-NR all	U-3.69	SHGC-NR all
Metal framing (curtainwall/storefront)[c]	U-2.27		U-2.27		U-3.41	
Metal framing (entrance door)[c]	U-4.54		U-4.54		U-5.11	
Metal framing (all other)[c]	U-2.56		U-2.56		U-3.69	
Skylight with Curb, Glass, % of Roof						
0%–2.0%	U_{all}-6.64	$SHGC_{all}$-0.68	U_{all}-6.64	$SHGC_{all}$-0.64	U_{all}-11.24	$SHGC_{all}$-NR
2.1%–5.0%	U_{all}-6.64	$SHGC_{all}$-0.64	U_{all}-6.64	$SHGC_{all}$-0.64	U_{all}-11.24	$SHGC_{all}$-NR
Skylight with Curb, Plastic, % of Roof						
0%–2.0%	U_{all}-4.94	$SHGC_{all}$-0.77	U_{all}-3.46	$SHGC_{all}$-0.77	U_{all}-10.79	$SHGC_{all}$-NR
2.1%–5.0%	U_{all}-4.94	$SHGC_{all}$-0.71	U_{all}-3.46	$SHGC_{all}$-0.77	U_{all}-10.79	$SHGC_{all}$-NR
Skylight without Curb, All, % of Roof						
0%–2.0%	U_{all}-3.92	$SHGC_{all}$-0.68	U_{all}-3.92	$SHGC_{all}$-0.64	U_{all}-7.72	$SHGC_{all}$-NR
2.1%–5.0%	U_{all}-3.92	$SHGC_{all}$-0.64	U_{all}-3.92	$SHGC_{all}$-0.64	U_{all}-7.72	$SHGC_{all}$-NR

*The following definitions apply: c.i. = continuous insulation (see Section 3.2), NR = no (insulation) requirement.

[a] When using R-value compliance method, a thermal spacer block is required, otherwise use the U-factor compliance method. See Table A2.3.

[~~a~~b] Exception to Section A3.1.3.1 applies.

[~~b~~c] Nonmetal framing includes framing materials other than metal with or without metal reinforcing or cladding.

[~~c~~d] Metal framing includes metal framing with or without thermal break. The "all other" subcategory includes operable windows, fixed windows, and non-entrance doors.

TABLE 5.5-8 Building Envelope Requirements for Climate Zone 8*

Opaque Elements	Nonresidential Assembly Maximum	Nonresidential Insulation Min. R-Value	Residential Assembly Maximum	Residential Insulation Min. R-Value	Semiheated Assembly Maximum	Semiheated Insulation Min. R-Value
Roofs						
Insulation Entirely above Deck	U-0.273	R-3.5 c.i.	U-0.273	R-3.5 c.i.	U-0.360	R-2.6 c.i.
Metal Building[a]	~~U-0.278~~ U-0.199	~~R-2.3 + R-3.3~~ R-1.9 + R-3.3 ls	~~U-0.278~~ U-0.199	~~R-2.3 + R-3.3~~ R-1.9 +R-3.3 ls	~~U-0.409~~ U-0.369	~~R-2.8~~ R-3.3
Attic and Other	U-0.119	R-8.6	U-0.119	R-8.6	U-0.192	R-5.3
Walls, Above-Grade						
Mass	U-0.404	R-2.7 c.i.	U-0.295	R-4.4 c.i.	U-0.592	R-1.7 c.i.
Metal Building	U-0.324	~~R-2.3 + R-2.3~~ R-3.3 +R-1.0 c.i.	U-0.324	~~R-2.3 + R-2.3~~ R-3.3 +R-1.0 c.i.	U-0.642	R-2.3
Steel-Framed	U-0.365	R-2.3 + R-1.3 c.i.	U-0.212	R-2.3 + R-3.3 c.i.	U-0.479	R-2.3 + R-0.7 c.i.
Wood-Framed and Other	U-0.203	R-2.3 +R-2.7 c.i.	U-0.203	R-2.3 + R-2.7 c.ic.	U-0.504	R-2.3
Walls, Below-Grade						
Below-Grade Wall	C-0.678	R-1.3 c.i.	C-0.425	R-2.2 c.i.	C-6.473	NR
Floors						
Mass	U-0.321	R-2.6 c.i.	U-0.287	R-2.9 c.i.	U-0.496	R-1.5 c.i.
Steel-Joist	U-0.183	R-6.7	U-0.183	R-6.7	U-0.296	R-3.3
Wood-Framed and Other	U-0.188	R-5.3	U-0.188	R-5.3	U-0.188	R-5.3
Slab-On-Grade Floors						
Unheated	F-0.900	R-2.6 for 600 mm	F-0.883	R-3.5 for 600 mm	F-1.264	NR
Heated	F-1.191	R-3.5 for 1200 mm	F-1.191	R-3.5 for 1200 mm	F-1.558	R-1.8 for 600 mm
Opaque Doors						
Swinging	U-2.839		U-2.839		U-3.975	
Nonswinging	U-2.839		U-2.839		U-2.839	

Fenestration	Nonresidential Assembly Max. U	Nonresidential Assembly Max. SHGC	Residential Assembly Max. U	Residential Assembly Max. SHGC	Semiheated Assembly Max. U	Semiheated Assembly Max. SHGC
Vertical Glazing, 0%–40% of Wall						
Nonmetal framing (all)[b]	U-1.99	SHGC-0.45 all	U-1.99	SHGC-NR all	U-3.69	SHGC-NR all
Metal framing (curtainwall/storefront)[c]	U-2.27		U-2.27		U-3.41	
Metal framing (entrance door)[c]	U-4.54		U-4.54		U-5.11	
Metal framing (all other)[c]	U-2.56		U-2.56		U-3.69	
Skylight with Curb, Glass, % of Roof						
0%–2.0%	U_{all}-5.56	$SHGC_{all}$-NR	U_{all}-5.56	$SHGC_{all}$-NR	U_{all}-7.38	$SHGC_{all}$-NR
2.1%–5.0%	U_{all}-5.56	$SHGC_{all}$-NR	U_{all}-5.56	$SHGC_{all}$-NR	U_{all}-7.38	$SHGC_{all}$-NR
Skylight with Curb, Plastic, % of Roof						
0%–2.0%	U_{all}-3.46	$SHGC_{all}$-NR	U_{all}-3.46	$SHGC_{all}$-NR	U_{all}-6.25	$SHGC_{all}$-NR
2.1%–5.0%	U_{all}-3.46	$SHGC_{all}$-NR	U_{all}-3.46	$SHGC_{all}$-NR	U_{all}-6.25	$SHGC_{all}$-NR
Skylight without Curb, All, % of Roof						
0%–2.0%	U_{all}-3.29	$SHGC_{all}$-NR	U_{all}-3.29	$SHGC_{all}$-NR	U_{all}-4.60	$SHGC_{all}$-NR
2.1%–5.0%	U_{all}-3.29	$SHGC_{all}$-NR	U_{all}-3.29	$SHGC_{all}$-NR	U_{all}-4.60	$SHGC_{all}$-NR

*The following definitions apply: c.i. = continuous insulation (see Section 3.2), NR = no (insulation) requirement.

[a]When using R-value compliance method, a thermal spacer block is required, otherwise use the U-factor compliance method. See Table A2.3.

~~[a]~~[b]Exception to Section A3.1.3.1 applies.

~~[b]~~[c]Nonmetal framing includes framing materials other than metal with or without metal reinforcing or cladding.

~~[c]~~[d]Metal framing includes metal framing with or without thermal break. The "all other" subcategory includes operable windows, fixed windows, and non-entrance doors.

Modify Appendix A as follows (I-P units):

A2.3 Metal Building Roofs

A2.3.1 General. For the purpose of A1.2, the base assembly is a *roof* with *thermal spacer blocks* where the insulation is draped over the steel structure (purlins), spaced nominally 5 ft on center and ~~then~~ compressed when the metal roof panels are attached to the steel structure (purlins). ~~Additional assemblies include *continuous insulation*, uncompressed and uninterrupted by framing.~~

A2.3.2 Rated R-Value of insulation

A2.3.2.1 The first *rated R-value of insulation* is for insulation draped over purlins and then compressed when the metal roof panels are attached, or for insulation hung between the purlins. A minimum R-3.5~~1 in.~~ thermal spacer block between the purlins and the metal roof panels is required when specified in Table A2.3.

A2.3.2.2 For double-layer installations, the second *rated R-value of insulation* is for insulation installed parallel to the purlins.

A2.3.2.3 For continuous insulation (e.g., insulation boards or blankets), it is assumed that the insulation is installed below the purlins and is uninterrupted by framing members. Insulation exposed to the *conditioned space* or *semiheated space* shall have a facing, and all insulation seams shall be continuously sealed to provide a continuous air barrier.

A2.3.2.4 ***Liner System* (Ls).** A continuous vapor barrier liner is installed below the purlins and uninterrupted by framing members. Uncompressed, unfaced insulation rests on top of the liner between the purlins. For multilayer installations, the *first rated R-Value of insulation* is for unfaced insulation draped over purlins and then compressed when the metal roof panels are attached. A minimum R-3.5 thermal spacer block between the purlins and the metal roof panels is required when specified in Table A2.3.

A2.3.3 U-Factor. *U-factors* for *metal building roofs* shall be taken from Table A2.3 It is not acceptable to use these *U-factors* if additional insulated sheathing is not continuous.

...

...

Exception to A3.1.3.1: For *mass walls*, where the requirement in Tables 5.5-1 through 5.5-8 is for a maximum assembly U-0.151 followed by footnote "~~a~~b," ...

The remainder of Appendix A is unchanged.

Modify Tables A2.3 and A3.2 as follows (I-P units):

TABLE A2.3 Assembly U-Factors for Metal Building Roofs

Insulation System	Rated R-Value of Insulation	Total Rated R-Value of Insulation	Overall U-Factor for Entire Base Roof Assembly	Overall U-Factor for Assembly of Base Roof Plus Continuous Insulation (Uninterrupted by Framing) — Rated R-Value of Continuous Insulation					
				R-5.6	R-11.2	R-16.8	R-22.4	R-28.0	R-33.6
Standing Seam Roofs with Thermal Spacer Blocks									
	None	0	**U-1.280**	~~0.162~~ 0.157	~~0.087~~ 0.083	~~0.059~~ 0.057	~~0.045~~ 0.043	~~0.036~~ 0.035	~~0.030~~ 0.029
Single Layer	R-6	6	**U-0.167**	0.086	0.058	0.044	0.035	0.029	0.025
	R-10	10	**U-0.097**	0.063	0.046	0.037	0.031	0.026	0.023
	R-11	11	**U-0.092**	0.061	0.045	0.036	0.030	0.026	0.022
	R-13	13	**U-0.083**	0.057	0.043	0.035	0.029	0.025	0.022
	R-16	16	**U-0.072**	0.051	0.040	0.033	0.028	0.024	0.021
	R-19	19	**U-0.065**	0.048	0.038	0.031	0.026	0.023	0.020
Double Layer	R-10 + R-10	20	**U-0.063**	0.047	0.037	0.031	0.026	0.023	0.020
	R-10 + R-11	21	**U-0.061**	0.045	0.036	0.030	0.026	0.023	0.020
	R-11 + R-11	22	**U-0.060**	0.045	0.036	0.030	0.026	0.022	0.020
	R-10 + R-13	23	**U-0.058**	0.044	0.035	0.029	0.025	0.022	0.020
	R-11 + R-13	24	**U-0.057**	0.043	0.035	0.029	0.025	0.022	0.020
	R-13 + R-13	26	**U-0.055**	0.042	0.034	0.029	0.025	0.022	0.019
	R-10 + R-19	29	**U-0.052**	0.040	0.033	0.028	0.024	0.021	0.019
	R-11 + R-19	30	**U-0.051**	0.040	0.032	0.027	0.024	0.021	0.019
	R-13 + R-19	32	**U-0.049**	0.038	0.032	0.027	0.023	0.021	0.019
	R-16 + R-19	35	**U-0.047**	0.037	0.031	0.026	0.023	0.020	0.018
	R-19 + R-19	38	**U-0.046**	0.037	0.030	0.026	0.023	0.020	0.018
Liner System	R-11 + R-19	30	**U-0.035**						
	R-11 + R-25	36	**U-0.031**						
	R-11 + R-30	41	**U-0.029**						
	R-11 + R-11 + R-25	47	**U-0.026**						
Standing Seam Roofs without Thermal Spacer Blocks									
Liner System	R-11 + R-19	30	**U-0.040**		0.028	0.024	0.021	0.020	0.017
Filled Cavity with Thermal Spacer Blocks									
	R-19 + R-10	29	**U-0.041**	0.033	0.028	0.024	0.021	~~0.020~~0.019	0.017
(Multiple R-values are listed in order from inside to outside.)									
Thru-Fastened without Thermal Spacer Blocks									
	R-10	10	**U-0.153**						
	R-11	11	**U-0.139**						
	R-13	13	**U-0.130**						
	R-16	16	**U-0.106**						
	R-19	19	**U-0.098**						
Liner System	R-11 + R-19	30	**U-0.044**						

TABLE A3.2 Assembly U-Factors for Metal Building Walls

Insulation System	Rated R-Value of Insulation	Total Rated R-Value of Insulation	Overall U-Factor for Entire Base Wall Assembly	Overall U-Factor for Assembly of Base Wall Plus Continuous Insulation (Uninterrupted by Framing) — Rated R-Value of Continuous Insulation: R-5.6	R-11.2	R-16.8	R-22.4	R-28.0	R-33.6
Single Layer of Mineral Fiber									
	None	0	1.180	0.161	0.086	0.059	0.045	0.036	0.030
	R-6	6	0.184	0.091	0.060	0.045	0.036	0.030	0.026
	R-10	10	0.134	0.077	0.054	0.051	0.033	0.028	0.024
	R-11	11	0.123	0.073	0.052	0.040	0.033	0.028	0.024
	R-13	13	0.113	0.069	0.050	0.039	0.032	0.027	0.024
	R-16	16	0.093	0.061	0.046	0.036	0.030	0.026	0.023
	R-19	19	0.084	0.057	0.043	0.035	0.029	0.025	0.022
Double Layer of Mineral Fiber									
(Second layer inside of girts)									
(Multiple layers are listed in order from inside to outside)									
	R-6 + R-13	19	0.070	N/A	N/A	N/A	N/A	N/A	N/A
	R-10 + R-13	23	0.061	N/A	N/A	N/A	N/A	N/A	N/A
	R-13 + R-13	26	0.057	N/A	N/A	N/A	N/A	N/A	N/A
	R-19 + R-13	32	0.048	N/A	N/A	N/A	N/A	N/A	N/A

Modify Appendix A as follows (SI units):

A2.3 Metal Building Roofs

A2.3.1 General. For the purpose of A1.2, the base assembly is a *roof* with *thermal spacer blocks* where the insulation is draped over the steel structure (purlins), spaced nominally 1.52 m on center, and ~~then~~ compressed when the metal roof panels are attached to the steel structure (purlins). ~~Additional assemblies include *continuous insulation*, uncompressed and uninterrupted by framing.~~

A2.3.2 Rated R-Value of insulation

A2.3.2.1 The first *rated R-value of insulation* is for insulation draped over purlins and then compressed when the metal roof panels are attached, or for insulation hung between the purlins. A minimum R-0.6~~1 in.~~ thermal spacer block between the purlins and the metal roof panels is required when specified in Table A2.3.

A2.3.2.2 For double-layer installations, the second *rated R-value of insulation* is for insulation installed parallel to the purlins.

A2.3.2.3 For continuous insulation (e.g., insulation boards or blankets), it is assumed that the insulation is installed below the purlins and is uninterrupted by framing members. Insulation exposed to the *conditioned space* or *semiheated space* shall have a facing, and all insulation seams shall be continuously sealed to provide a continuous air barrier.

A2.3.2.4 *Liner System* (Ls). A continuous vapor barrier liner is installed below the purlins and uninterrupted by framing members. Uncompressed, unfaced insulation rests on top of the liner between the purlins. For multilayer installations, the *first rated R-value of insulation* is for unfaced insulation draped over purlins and then compressed when the metal roof panels are attached. A minimum R-0.6 thermal spacer block between the purlins and the metal roof panels is required when specified in Table A2.3.

A2.3.3 U-Factor. *U-factors* for *metal building roofs* shall be taken from Table A2.3 It is not acceptable to use these *U-factors* if additional insulated sheathing is not continuous.

...

...

Exception to A3.1.3.1: For *mass walls*, where the requirement in Tables 5.5-1 through 5.5-8 is for a maximum assembly U-0.857 followed by footnote "~~a~~b," ...

The remainder of Appendix A is unchanged.

Modify Tables A2.3 and A3.2 as follows (SI units):

TABLE A2.3 Assembly U-Factors for Metal Building Roofs

Insulation System	Rated R-Value of Insulation	Total Rated R-Value of Insulation	Overall U-Factor for Entire Base Roof Assembly	Overall U-Factor for Assembly of Base Roof Plus Continuous Insulation (Uninterrupted by Framing) — Rated R-Value of Continuous Insulation: R-1.0	R-2.0	R-3.0	R-4.0	R-4.9	R-5.9
Standing Seam Roofs with Thermal Spacer Blocks									
Single Layer	None	0	**U-7.258**	~~0.919~~ 0.891	~~0.493~~ 0.471	~~0.335~~ 0.324	~~0.255~~ 0.244	~~0.204~~ 0.199	~~1.070~~ 0.165
	R-1.1	1.1	**U-0.947**	0.489	0.330	0.249	0.200	0.167	0.143
	R-1.8	1.8	**U-0.550**	0.356	0.264	0.209	0.173	0.148	0.129
	R-1.9	1.9	**U-0.522**	0.344	0.257	0.205	0.170	0.146	0.128
	R-2.3	2.3	**U-0.471**	0.321	0.244	0.197	0.165	0.142	0.124
	R-2.8	2.8	**U-0.408**	0.291	0.226	0.185	0.156	0.135	0.119
	R-3.3	3.3	**U-0.369**	0.270	0.213	0.176	0.150	0.131	0.116
Double Layer	R-1.8 + R-1.8	3.5	**U-0.357**	0.264	0.209	0.174	0.148	0.129	0.115
	R-1.8 + R-1.9	3.7	**U-0.346**	0.258	0.205	0.171	0.146	0.128	0.113
	R-1.9 + R-1.9	3.9	**U-0.340**	0.255	0.203	0.169	0.145	0.127	0.113
	R-1.8 + R-2.3	4.1	**U-0.329**	0.248	0.199	0.167	0.143	0.125	0.112
	R-1.9 + R-2.3	4.2	**U-0.323**	0.245	0.197	0.165	0.142	0.124	0.111
	R-2.3 + R-2.3	4.6	**U-0.312**	0.238	0.193	0.162	0.140	0.123	0.109
	R-1.8 + R-3.3	5.1	**U-0.295**	0.228	0.186	0.157	0.136	0.120	0.107
	R-1.9 + R-3.3	5.3	**U-0.289**	0.225	0.184	0.156	0.135	0.119	0.107
	R-2.3 + R-3.3	5.6	**U-0.278**	0.218	0.179	0.152	0.132	0.117	0.105
	R-2.8 + R-3.3	6.2	**U-0.266**	0.211	0.175	0.149	0.130	0.115	0.103
	R-3.3 + R-3.3	6.7	**U-0.261**	0.207	0.172	0.147	0.128	0.114	0.102
Liner System	R-1.9 + R-3.3	5.3	**0.198**						
	R-1.9 + R-4.4	6.3	**0.176**						
	R-1.9 + R-5.3	7.2	**0.165**						
	R-1.9 + R-1.9 + R-4.4	8.3	**0.148**						
Standing Seam Roofs without Thermal Spacer Blocks									
Liner System	R-1.9 + R-3.3	5.3	**0.227**	0.185	0.157	0.136	0.120	0.107	0.097
Filled Cavity with Thermal Spacer Blocks									
	R-3.3 + R-1.8	5.1	**U-0.232**	0.189	0.159	0.138	0.121	0.108	0.098
(Multiple R-values are listed in order from inside to outside.)									
Thru-Fastened Roofs without Thermal Spacer Blocks									
	R-1.8	1.8	**U-0.868**						
	R-1.9	1.9	**U-0.788**						
	R-2.3	2.3	**U-0.737**						
	R-2.8	2.8	**U-0.660**						
	R-3.3	3.3	**U-0.550**						
Liner System	R-1.9 + R-3.3	5.3	**0.249**						

TABLE A3.2 Assembly U-Factors for Metal Building Walls

Insulation System	Rated R-Value of Insulation	Total Rated R-Value of Insulation	Overall U-Factor for Entire Base Wall Assembly	Overall U-Factor for Assembly of Base Wall Plus Continuous Insulation (Uninterrupted by Framing) — Rated R-Value of Continuous Insulation					
				R-1.0	**R-2.0**	**R-3.0**	**R-4.0**	**R-4.9**	**R-5.9**
Single Layer of Mineral Fiber									
	None	0.0	**6.69**	~~0.91~~ 0.88	~~0.49~~ 0.47	~~0.33~~ 0.32	~~0.26~~ 0.24	0.20	~~0.17~~ 0.16
	R-1.1	1.1	**1.04**	0.51	0.34	0.26	0.20	0.17	0.15
	R-1.8	1.8	**0.76**	0.43	0.30	0.23	0.19	0.16	0.14
	R-1.9	1.9	**0.70**	0.41	0.29	0.23	0.19	0.16	0.14
	R-2.3	2.3	**0.64**	0.39	0.28	0.22	0.18	0.15	0.13
	R-2.8	2.8	**0.53**	0.35	0.26	0.20	0.17	0.15	0.13
	R-3.3	3.3	**0.48**	0.32	0.24	0.20	0.16	0.14	0.12
Double Layer of Mineral Fiber (Second layer inside of girts) (Multiple layers are listed in order from inside to outside)									
	R-1.1 + R-2.3	3.4	**0.40**	N/A	N/A	N/A	N/A	N/A	N/A
	R-1.8 + R-2.3	4.1	**0.35**	N/A	N/A	N/A	N/A	N/A	N/A
	R-2.3 + R-2.3	4.6	**0.32**	N/A	N/A	N/A	N/A	N/A	N/A
	R-3.3 + R-2.3	5.6	**0.27**	N/A	N/A	N/A	N/A	N/A	N/A

(This foreword is not part of this standard. It is merely informative and does not contain requirements necessary for conformance to the standard. It has not been processed according to the ANSI requirements for a standard and may contain material that has not been subject to public review or a consensus process. Unresolved objectors on informative material are not offered the right to appeal at ASHRAE or ANSI.)

FOREWORD

This change includes a new exception to Section 6.5.2.1 that is geared toward zones with direct digital controls (DDC). The new exception (exception b) largely addresses the apparent conflict between Standards 55, 62.1, and 90.1, and also takes advantage of the energy-saving potential of DDC controls in order to save about $0.20/ft^2/yr with a simple payback of less than two years. The apparent conflict is that the current 30% reheat maximum typically requires very high supply air temperatures (e.g., >100°F) to meet peak heating load. High supply air temperatures result in poor comfort per Standard 55 and poor ventilation effectiveness per Standard 62.1. The new exception allows reheat to increase from 30% to 50%, which means lower supply air temperatures and better comfort and ventilation effectiveness. The energy savings come from the fact that maximum airflow in deadband is being lowered from 30% to 20%. This saves fan energy and cooling energy in deadband, and also reduces the amount of time when the zone will be overcooled in deadband and forced into reheat.

This new exception will also alleviate a common problem where engineers feel compelled to violate the current 30% exception in order to provide adequate heating. In addition to poor comfort and ventilation effectiveness, high supply air temperatures also lead to short-circuiting. When hot supply air short circuits directly from the supply to the return, the space takes longer to warm up and may not warm up at all. Therefore, it is very common for designers and contractors to disregard the current 30% requirement and use 40% or 50% minimum flow setpoints to ensure adequate heating. No one likes to disregard the code, but if the choice is between code and comfort, comfort wins. The new exception allows users to achieve comfort, meet the code, and save energy at minimal cost.

Because not all zones have DDC controls and because this is a fairly significant shift in zone controls, the existing 30% exception is left in the standard. However, two clauses from the existing exception are deleted. The 0.4 cfm/ft^2 exception is deleted because it implies that a minimum air speed in the occupied space is required for comfort. ASHRAE Standard 55, however, indicates that no minimum air speed is required for comfort. Furthermore. 0.4 cfm/ft^2 does not guarantee any particular air speed because 0.4 cfm/ft^2 can be a small fraction (e.g., 10%) or a large fraction (e.g., 50%) of the design flow rate and, thus, can result in a low or high air speed. The 300 cfm exception is deleted because the situation that it was intended to address has been largely eliminated by the new 50% exception described above. This criterion was intended to address the following applications: the occasional small zone in a VAV reheat system for which 30% is insufficient to handle heating loads, such as spaces with large north-facing glass areas..

Addendum h to Standard 90.1-2007

Revise the standard as follows (I-P and SI units):

Revise exceptions to Section 6.5.2.1 as follows:

~~**Exceptions to 6.5.2.1:**~~

~~a. Zones for which the volume of air that is reheated, recooled, or mixed is no greater than the larger of the following:~~

~~1. The volume of *outdoor air* required to meet the ventilation requirements of Section 6.2 of ASHRAE Standard 62.1 for the *zone*,~~
~~2. 0.4 cfm/ft2 [2L/s/m2] of the *zone* conditioned floor area,~~
~~3. 30% of the zone design peak supply rate,~~
~~4. 300 cfm [140L/s] this exception is for zones whose peak flow rate totals no more than 10% of the total fan system flow rate,~~
~~5. Any higher rate that can be demonstrated, to the satisfaction of the *authority having jurisdiction*, to reduce overall system annual energy usage by offsetting reheat/recool energy losses through a reduction in outdoor air intake for the system.~~

Exceptions to 6.5.2.1:

a. *Zones* for which the volume of air that is reheated, recooled, or mixed is less than the larger of the following:

1. 30% of the *zone* design peak supply rate;
2. The volume of *outdoor air* required to meet the ventilation requirements of Section 6.2 of ASHRAE Standard 62.1 for the *zone*;
3. Any higher rate that can be demonstrated, to the satisfaction of the *authority having jurisdiction*, to reduce overall system annual energy usage by offsetting reheat/recool energy losses through a reduction in *outdoor air* intake.

b. *Zones* that comply with all of the following:

1. The volume of air that is reheated, recooled, or mixed in *dead band* between heating and cooling does not exceed the larger of the following:

    a. 20% of the *zone* design peak supply rate;
    b. the volume of *outdoor air* required to meet the ventilation requirements of Section 6.2 of ASHRAE Standard 62.1 for the *zone*;
    c. any higher rate that can be demonstrated, to the satisfaction of the *authority having jurisdiction*, to reduce overall system

annual energy usage by offsetting reheat/recool energy losses through a reduction in *outdoor air* intake.

2. The volume of air that is reheated, recooled, or mixed in peak heating demand shall be less than 50% of the *zone* design peak supply rate.
3. Airflow between *dead band* and full heating or full cooling shall be modulated.

~~b~~c.) *Zones* where special pressurization relationships, cross-contamination requirements, or code-required minimum circulation rates are such that variable-air-volume systems are impractical.

~~c~~d.) *Zones* where at least 75% of the energy for reheating or for providing warm air in mixing systems is provided from a *site-recovered* (including condenser heat) or *site- solar energy source.*

(This foreword is not part of this standard. It is merely informative and does not contain requirements necessary for conformance to the standard. It has not been processed according to the ANSI requirements for a standard and may contain material that has not been subject to public review or a consensus process. Unresolved objectors on informative material are not offered the right to appeal at ASHRAE or ANSI.)

FOREWORD

This proposal will apply a four-zone lighting power density approach to exterior lighting requirements. This approach recognizes the varying lighting needs and design differences associated with different building locations. It is acceptable and prudent to reduce the light levels as the designer leaves the downtown city center entering into mixed commercial/high-rise residential districts, then enters into residential areas, and then into rural areas. Several organizations, including the IESNA have been working to develop a zonal approach to exterior lighting recommended practice and this change in the standard will follow that guidance.

The specific IESNA documents used in this proposal are RP-20, DG-5, IESNA Handbook, RP-2, G-1 and RP-33. There are some instances where IESNA recommendations in these documents are available for all four zone criteria, but in many cases only three light level recommendations were found and referenced. Other standards use a multizone system to either classify LPD, lumen, or light trespass requirements—California T-24 (4-zone W/sf), the upcoming MOL (5-zone Lumen/sf, and LEED (light trespass). These standards were evaluated and in some cases incorporated into this proposal.

The first change in Section 9.4.5 is the deletion of the 5% additional power allowances, which is replaced by a base wattage allowance per site. The second change to this section is to define the four zones and apply appropriate requirements. The four zones are based on IESNA and other group definitions to match other requirements and guidance expected to be encountered by designers. The majority of building sites will fall into LZ3, LZ2, or LZ1, and the sites that remain in LZ4 will generally be of relatively small sizes. The added "Base Site Allowance" for each zone takes into account that most sites are not rectangular or match the iso-diagram of typical light luminaires.

The associated energy change from this proposal comes from the lower illuminance requirements for primarily zones 1–3, where the majority of buildings are constructed. Numerous point-by-point lighting calculations were performed for parking lots, walkways, stairways, pedestrian tunnels, entries (with and without canopies), sales canopies, service stations, and auto dealerships for the four zones. In the initial calculations for the parking lots, there was a noticeable difficulty in achieving the recommended light level when the space was 20,000 ft^2 and lower, without any additional power allowance (this was especially true in zone 4). Six odd shaped parking lots were modeled for all four lighting zones to verify that the requirement would cover varying design needs. This modeling was used to determine appropriate base site allowances. Because of the base site allowance, the actual LPDs are on a sliding scale, as shown below.

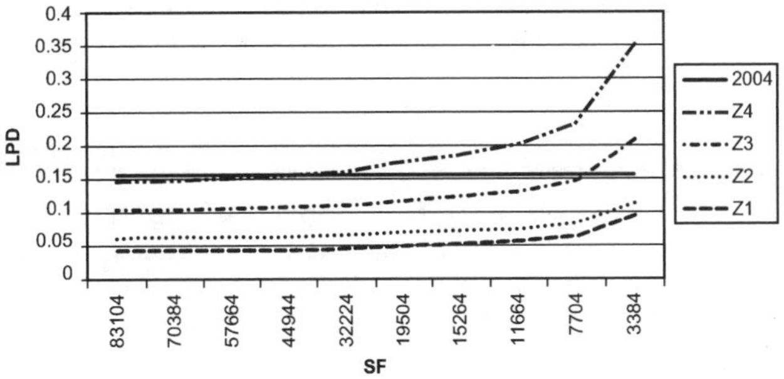

The energy savings from this zone approach is shown in the chart below, with total energy used in each lighting zone for the various square footages. The solid line is the current 2004 standard.

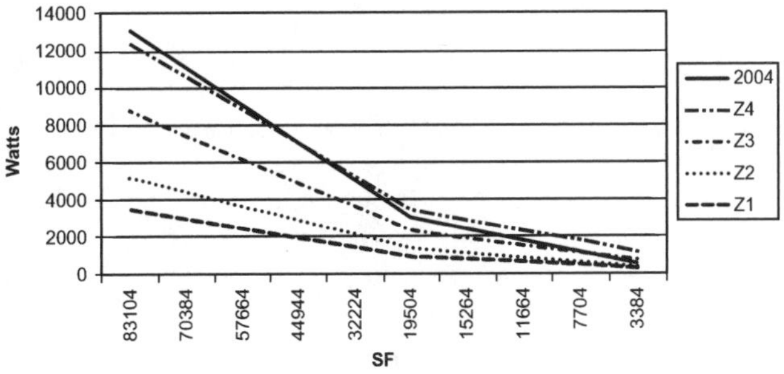

Addendum i to Standard 90.1-2007

Revise the Standard as follows (I-P units):

9.4.5 Exterior Building Lighting Power. The total *exterior lighting power allowance* for all exterior building applications is the sum of the base site allowance plus the individual allowances for areas that are designed to be illuminated and are ~~lighting power densities~~ permitted in Table 9.4.6~~5~~ for the applicable lighting zone ~~for these applications plus an additional unrestricted allowance of 5% of that sum.~~ Trade-offs are allowed only among exterior lighting applications listed in the Table 9.4.6~~5~~ "Tradable Surfaces" section. The lighting zone for the building exterior is determined from Table 9.4.5 unless otherwise specified by the local jurisdiction.

Exceptions to 9.4.5: Lighting used for the following exterior applications is exempt when equipped with a *control device* that complies with the requirements of Section 9.4.1.3 and is independent of the control of the nonexempt lighting:

a. Specialized signal, directional, and marker lighting associated with transportation.
b. Advertising signage or directional signage.
c. Lighting integral to *equipment* or instrumentation and installed by its *manufacturer*.

TABLE 9.4.5 Exterior Lighting Zones

Lighting Zone	Description
1	Developed areas of national parks, state parks, forest land, and rural areas
2	Areas predominantly consisting of residential zoning, neighborhood business districts, light industrial with limited night-time use and residential mixed use areas
3	All other areas
4	High activity commercial districts in major metropolitan areas as designated by the local jurisdiction

TABLE 9.4.6~~.5~~ Individual Lighting Power Allowances~~Densities~~ for Building Exteriors

		Zone 1	Zone 2	Zone 3	Zone 4
Base Site Allowance (base allowance may be used in tradable or non-tradable surfaces)		500 W	600 W	750 W	1300 W
Tradable Surfaces (Lighting power densities for uncovered parking areas, building grounds, building entrances and exits, canopies and overhangs and outdoor sales areas may be traded.)	**Uncovered Parking Areas**				
	Parking ~~Lots~~ areas and drives	0.04 W/ft^2	0.06 W/ft^2	0.10 W/ft^2	~~0.15~~0.13 W/ft^2
	Building Grounds				
	Walkways less than 10 feet wide	0.7 W/linear foot	0.7 W/linear foot	0.8 W/linear foot	1.0 W/linear foot
	Walkways 10 feet wide or greater Plaza areas Special Feature Areas	0.14 W/ft^2	0.14 W/ft^2	0.16 W/ft^2	0.2 W/ft^2
	Stairways	0.75 W/ft^2	1.0 W/ft^2	1.0 W/ft^2	1.0 W/ft^2
	Pedestrian tunnels	0.15 W/ft^2	0.15 W/ft^2	0.2 W/ft^2	0.3 W/ft^2
	Landscaping	0.04 W/ft^2	0.05 W/ft^2	0.05 W/ft^2	0.05 W/ft^2
	Building Entrances and Exits				
	Main entries	20 W/linear foot of door width	20 W/linear foot of door width	30 W/linear foot of door width	30 W/linear foot of door width
	Other doors	20 W/linear foot of door width	20 W/linear foot of door width	20 W/linear foot of door width	20 W/linear foot of door width
	Entry canopies	0.25 W/ft^2	0.25 W/ft^2	0.4 W/ft^2	~~1.25~~0.4 W/ft^2
	Sales Canopies ~~and Overhangs~~				
	~~Canopies (~~free standing and attached ~~and overhangs)~~	0.6 W/ft^2	0.6 W/ft^2	0.8 W/ft^2	~~1.25~~1.0 W/ft^2
	Outdoor Sales				
	Open areas (including vehicle sales lots)	0.25 W/ft^2	0.25 W/ft^2	0.5 W/ft^2	0.~~5~~7 W/ft^2
	Street frontage for vehicle sales lots in addition to "open area" allowance	**No allowance**	10 W/linear foot	10 W/linear foot	~~20~~ 30 W/linear foot

TABLE 9.4.6.~~5~~ Individual Lighting Power Allowances~~Densities~~ for Building Exteriors (Continued)

	Building Facades	**No allowance**	0.1 W/ft^2 for each illuminated wall or surface or 2.5 W/linear foot for each illuminated wall or surface length	0.15 W/ft^2 for each illuminated wall or surface or 3.75 W/linear foot for each illuminated wall or surface length	0.2 W/ft^2 for each illuminated wall or surface or 5.0 W/linear foot for each illuminated wall or surface length
Non-Tradable Surfaces (Lighting power density calculations for the following applications can be used only for the specific application and can-not be traded between surfaces or with other exterior lighting. The following allowances are in addition to any allowance otherwise permitted in the "tradable Surfaces" section of this table.)	**Automated teller machines and night depositories**	270 W per location plus 90 W per additional ATM per location	270 W per location plus 90 W per additional ATM per location	270 W per location plus 90 W per additional ATM per location	270 W per location plus 90 W per additional ATM per location
	Entrances and gatehouse inspection stations at guarded facilities	0.75 W/ft^2 of covered and uncovered area	0.75 W/ft^2 of covered and uncovered area	0.75 W/ft^2 of covered and uncovered area	~~1.25~~0.75 W/ft^2 of covered and uncovered area ~~(covered areas are included in the "Canopies and Overhangs" section of "Tradable Surfaces")~~
	Loading areas for law enforcement, fire, ambulance and other emergency service vehicles	0.5 W/ft^2 of covered and uncovered area	0.5 W/ft^2 of covered and uncovered area	0.5 W/ft^2 of covered and uncovered area	0.5 W/ft^2 of covered and uncovered area ~~(covered areas are included in the "Canopies and Overhangs" section of "Tradable Surfaces")~~
	Drive-up windows/doors ~~at fast food restaurants~~	400 W per drive-through	400 W per drive-through	400 W per drive-through	400 W per drive-through
	Parking near 24-hour retail entrances	800 W per main entry	800 W per main entry	800 W per main entry	800 W per main entry

d. Lighting for theatrical purposes, including performance, stage, film production, and video production.
e. Lighting for athletic playing areas.
f. Temporary lighting.
g. Lighting for industrial production, material handling, transportation sites, and associated storage areas.
h. Theme elements in theme/amusement parks.
i. Lighting used to highlight features of public monuments and registered *historic* landmark structures or *buildings*.

Revise the Standard as follows (SI units):

9.4.5 Exterior Building Lighting Power. The total *exterior lighting power allowance* for all exterior building applications is the sum of the base site allowance plus the individual allowances for areas that are designed to be illuminated and are ~~lighting power densities~~ permitted in Table 9.4.6~~5~~ for the applicable lighting zone ~~for these applications plus an additional unrestricted allowance of 5% of that sum.~~ Trade-offs are allowed only among exterior lighting applications listed in the Table 9.4.6~~5~~ "Tradable Surfaces" section. The lighting zone for the building exterior is determined from Table 9.4.5 unless otherwise specified by the local jurisdiction.

Exceptions to 9.4.5: Lighting used for the following exterior applications is exempt when equipped with a *control device* that complies with the requirements of Section 9.4.1.3 and is independent of the control of the nonexempt lighting:

a. Specialized signal, directional, and marker lighting associated with transportation.
b. Advertising signage or directional signage.
c. Lighting integral to *equipment* or instrumentation and installed by its *manufacturer*.
d. Lighting for theatrical purposes, including performance, stage, film production, and video production.
e. Lighting for athletic playing areas.
f. Temporary lighting.
g. Lighting for industrial production, material handling, transportation sites, and associated storage areas.
h. Theme elements in theme/amusement parks.
i. Lighting used to highlight features of public monuments and registered *historic* landmark structures or *buildings*.

TABLE 9.4.5 Exterior Lighting Zones

Lighting Zone	Description
1	Developed areas of national parks, state parks, forest land, and rural areas
2	Areas predominantly consisting of residential zoning, neighborhood business districts, light industrial with limited nighttime use and residential mixed use areas
3	All other areas
4	High activity commercial districts in major metropolitan areas as designated by the local jurisdiction

TABLE 9.4.6~~.5~~ Individual Lighting Power Allowances~~Densities~~ for Building Exteriors

		Zone 1	Zone 2	Zone 3	Zone 4
Base Site Allowance (base allowance may be used in tradable or non-tradable surfaces)		500 W	600 W	750 W	1300 W
Tradable Surfaces (Lighting power densities for uncovered parking areas, building grounds, building entrances and exits, canopies and overhangs and outdoor sales areas may be traded.)	**Uncovered Parking Areas**				
	Parking ~~Lots~~ areas and drives	0.43 W/m^2	0.65 W/m^2	1.1 W/m^2	~~0.15~~1.4 W/m^2
	Building Grounds				
	Walkways less than 10 feet wide	2.3 W/linear meter	2.3 W/linear meter	2.6 W/linear meter	3.3 W/linear meter
	Walkways 10 feet wide or greater Plaza areas Special Feature Areas	1.5 W/m^2	1.5 W/m^2	1.7 W/m^2	2.2 W/m^2
	Stairways	8.1 W/m^2	10.8 W/m^2	10.8 W/m^2	10.8 W/m^2
	Pedestrian tunnels	1.6 W/m^2	1.6 W/m^2	2.2 W/m^2	3.2 W/m^2
	Landscaping	0.43 W/m^2	0.54 W/m^2	0.54 W/m^2	0.54 W/m^2
	Building Entrances and Exits				
	Main entries	66 W/linear meter of door width	66 W/linear meter of door width	98 W/linear meter of door width	98 W/linear meter of door width
	Other doors	66 W/linear meter of door width	66 W/linear meter of door width	66 W/linear meter of door width	66 W/linear meter of door width
	Entry canopies	2.7 W/m^2	2.7 W/m^2	4.3 W/m^2	~~1.25~~4.3 W/m^2
	Sales Canopies ~~and Overhangs~~				
	~~Canopies (~~free standing and attached ~~and over-hangs)~~	6.5 W/m^2	6.5 W/m^2	8.6 W/m^2	~~1.25~~10.8 W/m^2
	Outdoor Sales				
	Open areas (including vehicle sales lots)	2.7 W/m^2	2.7 W/m^2	5.4 W/m^2	0.~~5-~~7.5 W/m^2
	Street frontage for vehicle sales lots in addition to "open area" allowance	**No allowance**	33 W/linear meter	33 W/linear meter	~~20~~ 98 W/linear meter

TABLE 9.4.6.5 Individual Lighting Power AllowancesDensities for Building Exteriors (Continued)

Non-Tradable Surfaces (Lighting power density calculations for the following applications can be used only for the specific application and can-not be traded between surfaces or with other exterior lighting. The following allowances are in addition to any allowance otherwise permitted in the "tradable Surfaces" section of this table.)	**Building Facades**	**No allowance**	1.1 W/m^2 for each illuminated wall or surface or 8.2 W/linear meter for each illuminated wall or surface length	1.6 W/m^2 for each illuminated wall or surface or 12.3 W/linear meter for each illuminated wall or surface length	2.2 W/m^2 for each illuminated wall or surface or 16.4 W/linear meter for each illuminated wall or surface length
	Automated teller machines and night depositories	270 W per location plus 90 W per additional ATM per location	270 W per location plus 90 W per additional ATM per location	270 W per location plus 90 W per additional ATM per location	270 W per location plus 90 W per additional ATM per location
	Entrances and gatehouse inspection stations at guarded facilities	8.1 W/m^2 of covered and uncovered area	8.1 W/m^2 of covered and uncovered area	8.1 W/m^2 of covered and uncovered area	1.258.1 W/m^2 of covered and uncovered area (covered areas are included in the "Canopies and Overhangs" section of "Tradable Surfaces")
	Loading areas for law enforcement, fire, ambulance and other emergency service vehicles	5.4 W/m^2 of covered and uncovered area	5.4 W/m^2 of covered and uncovered area	5.4 W/m^2 of covered and uncovered area	5.4 W/m^2 of covered and uncovered area (covered areas are included in the "Canopies and Overhangs" section of "Tradable Surfaces")
	Drive-up windows/doors at fast food restaurants	400 W per drive-through	400 W per drive-through	400 W per drive-through	400 W per drive-through
	Parking near 24-hour retail entrances	800 W per main entry	800 W per main entry	800 W per main entry	800 W per main entry

(This foreword is not part of this standard. It is merely informative and does not contain requirements necessary for conformance to the standard. It has not been processed according to the ANSI requirements for a standard and may contain material that has not been subject to public review or a consensus process. Unresolved objectors on informative material are not offered the right to appeal at ASHRAE or ANSI.)

FOREWORD

This addendum is intended to update the mechanical test procedures and references in ANSI/ASHRAE/IESNA Standard 90.1-2007. The changes modify a reference in Table 6.8.1E, the normative references in Chapter 12, and the informative references in Informative Appendix E.

Note: In this addendum, changes to the current standard are indicated in the text by underlining (for additions) and ~~strikethrough~~ (for deletions) unless the instructions specifically mention some other means of indicating the changes.

Addendum j to Standard 90.1-2007

Revise as follows (I-P and SI units):

Modify Section 12, Normative References, as follows.

Reference	Title
~~AMCA 500 D-98~~ AMCE 500-D-07	~~Test Methods for Louvers, Dampers and Shutters~~ Laboratory Methods of Testing Dampers for Rating
ANSI Z21.10.3-~~1998~~ 2004	Gas Water Heater, Volume 3, Storage, with Input Ratings above 75,000 Btu/h, Circulating and Instantaneous Water Heaters
ANSI Z21.47-~~2001~~ 2006	Gas-Fired Central Furnaces (Except Direct Vent and Separated Combustions System Furnaces)
ANSI Z21.83.8-~~2002~~ 2006	Gas Unit Heaters and Duct Furnaces
ANSI/AHAM RAC-1-~~87~~ 2003	Room Air Conditioners
ARI 210/240-~~2003~~2006	Unitary Air Conditioning and Air-Source Heat Pump Equipment
ARI 460-~~2000~~ 2005	Remote Mechanical Draft Air Cooled Refrigerant Condensers
ARI 550/590-~~98~~ 2003	Water-Chilling Packages Using the Vapor Compression Cycle
UL 181A-~~94~~ 2005	Closure Systems for Use with Rigid Air Ducts and Air Connectors
UL 181B-~~95~~ 2006	Closure Systems for Use with Flexible Air Ducts and Air Connectors
ANSI/ASHRAE 146-~~1998~~ 2006	Method of Testing for Rating Pool Heaters

Modify Informative Appendix E as follows.

(This appendix is not part of this standard. It is merely informative and does not contain requirements necessary for conformance to the standard. It has not been processed according to the ANSI requirements for a standard and may contain material that has not been subject to public review or a consensus process. Unresolved objectors on informative material are not offered the right to appeal at ASHRAE or ANSI.)

CRRC
Cool Roof Rating Council
1738 Excelsior Avenue
Oakland, CA 94602
(T) 866-465-2523
~~(T) 510-482-4420~~
(F) 510-482-4421
www.coolroofs.org

CRRC-1-~~2002~~ 2006	Cool Roof Rating Council Product Rating Program
200~~1~~5 ASHRAE Handbook—Fundamentals	ASHRAE
MICA Insulation Standards~~—1999~~ 6th Edition	National Commercial and Industrial Insulation Standards
SMACNA Duct Construction Standards~~—1995~~ 2005	HVAC Duct Construction Standards, Metal and Flexible
200~~3~~7 ASHRAE Handbook—HVAC Applications	Chapter 49, Service Water Heating/ASHRAE
AABC 2002	~~Associated Air Balance Council Test and Balance Procedures~~ Associated Air Balance Council, National Standards for Total System Balance

(This foreword is not part of this standard. It is merely informative and does not contain requirements necessary for conformance to the standard. It has not been processed according to the ANSI requirements for a standard and may contain material that has not been subject to public review or a consensus process. Unresolved objectors on informative material are not offered the right to appeal at ASHRAE or ANSI.)

FOREWORD

This addendum revises Tables 6.8.1E and 7.8 in ANSI/ASHRAE/IESNA Standard 90.1-2007, identifying the specific sections of the referenced standards. Table 7.8 is also updated to reflect the current federal efficiency levels for residential water heaters. Additionally, a requirement in Table 7.8 for electric table top water heaters has been added.

Note: In this addendum, changes to the current standard are indicated in the text by underlining (for additions) and ~~strikethrough~~ (for deletions) unless the instructions specifically mention some other means of indicating the changes.

Addendum k to Standard 90.1-2007

Revise as follows (I-P units):

TABLE 6.8.1E Warm Air Furnaces and Combination Warm Air Furnaces/Air-Conditioning Units, Warm Air Duct Furnaces, and Unit Heaters

Equipment Type	Size Category (Input)	Subcategory or Rating Condition	Minimum Efficiency[a]	Test Procedure[b]
Warm Air Furnace, Gas-Fired	<225,000 Btu/h		78% AFUE or 80% E_{td}	DOE 10 CFR Part 430 or Section 2.39, Thermal Efficiency, of ANSI Z21.47
	≥225,000 Btu/h	Maximum Capacity[d]	80% E_{cc}	Section 2.39, Thermal Efficiency, of ANSI Z21.47
Warm Air Furnace, Oil-Fired	<225,000 Btu/h		78% AFUE or 80%E_{td}	DOE 10 CFR Part 430 or Section 42, Combustion, of UL 727
	≥225,000 Btu/h	Maximum Capacity[e]	81% E_{tf}	Section 42, Combustion, of UL 727
Warm Air Duct Furnaces, Gas- Fired	All Capacities	Maximum Capacity[e]	80% E_{cg}	Section 2.10, Efficiency, of ANSI Z83.~~9~~8
Warm Air Unit Heaters, Gas-Fired	All Capacities	Maximum Capacity[e]	80% E_{cg}	Section 2.10, Efficiency, of ANSI Z83.8
Warm Air Unit Heaters, Oil-Fired	All Capacities	Maximum Capacity[e]	80% E_{cg}	Section 40, Combustion, of UL 731

[a] E_t= thermal *efficiency*. See test procedure for detailed discussion.
[b] Section 12 contains a complete specification of the referenced test procedure, including the referenced year version of the test procedure.
[c] E_c = combustion *efficiency*. Units must also include an interrupted or intermittent ignition device (IID), have jacket losses not exceeding 0.75% of the input rating, and have either power venting or a flue damper. A vent damper is an acceptable alternative to a flue damper for those furnaces where combustion air is drawn from the conditioned space.
[d] Combination units not covered by NAECA (three-phase power or cooling capacity greater than or equal to 65,000 Btu/h) may comply with either rating.
[e] Minimum and maximum ratings as provided for and allowed by the unit's controls.
[f] E_t = thermal *efficiency*. Units must also include an interrupted or intermittent ignition device (IID), have jacket losses not exceeding 0.75% of the input rating, and have either power venting or a flue damper. A vent damper is an acceptable alternative to a flue damper for those furnaces where combustion air is drawn from the conditioned space.
[g] E_c = combustion *efficiency* (100% less flue losses). See test procedure for detailed discussion.

Revise as follows (I-P units):

TABLE 7.8 Performance Requirements for Water Heating Equipment

Equipment Type	Size Category (Input)	Subcategory or Rating Condition	Performance Required[a]	Test Procedure[b,c]
Electric Table Top Water Heaters	≤12 kW	Resistance ≥20 gal	0.93-0.00132V EF	DOE 10 CFR Part 430
Electric Water Heaters	≤12 kW	Resistance ≥20 gal	0.9~~3~~7-0.00132V EF	DOE 10 CFR Part 430
	>12 kW	Resistance≥20 gal	20 + 35 $\sqrt{V}$ SL, Btu/h	Section G.2 of ANSI Z21.10.3
	≤24 Amps and ≤250 Volts	Heat Pump	0.93-0.00132V EF	DOE 10 CFR Part 430
Gas Storage Water Heaters	≤75,000 Btu/h	≥20 gal	0.6~~2~~7-0.0019V EF	DOE 10 CFR Part 430
	>75,000 Btu/h	<4000 (Btu/h)/gal	80% E_t (Q/800 + 110 $\sqrt{V}$) SL, Btu/h	Sections G.1 and G.2 of ANSI Z21.10.3
Gas Instantaneous Water Heaters	>50,000 Btu/h and <200,000 Btu/h	≥4000 (Btu/h)/gal and <2 gal	0.62-0.0019V EF	DOE 10 CFR Part 430
	≥200,000 Btu/h[e,d]	≥4000 (Btu/h)/gal and <10 gal	80% E_t	Sections G.1 and G.2 of ANSI Z21.10.3
	≥200,000 Btu/h	≥4000 (Btu/h)/gal and ≥10 gal	80% E_t (Q/800 + 110 $\sqrt{V}$) SL, Btu/h	
Oil Storage Water Heaters	≤105,000 Btu/h	≥20 gal	0.59-0.0019V EF	DOE 10 CFR Part 430
	>105,000 Btu/h	<4000 (Btu/h)/gal	78% E_t (Q/800 + 110 $\sqrt{V}$) SL, Btu/h	Sections G.1 and G.2 of ANSI Z21.10.3
Oil Instantaneous Water Heaters	≤210,000 Btu/h	≥4000 (Btu/h)/gal and <2 gal	0.59-0.0019V EF	DOE 10 CFR Part 430
	>210,000 Btu/h	≥4000 (Btu/h)/gal and <10 gal	80% E_t	Sections G.1 and G.2 of ANSI Z21.10.3
	>210,000 Btu/h	≥4000 (Btu/h)/gal and ≥10 gal	78% E_t (Q/800 + 110 $\sqrt{V}$) SL, Btu/h	
Hot Water Supply Boilers, Gas and Oil	≥300,000 Btu/h and <12,500,000 Btu/h	≥4000 (Btu/h)/gal and <10 gal	80% E_t	Sections G.1 and G.2 of ANSI Z21.10.3
Hot Water Supply Boilers, Gas		≥4000 (Btu/h)/gal and ≥10 gal	80% E_t (Q/800 + 110 $\sqrt{V}$) SL, Btu/h	
Hot Water Supply Boilers, Oil		≥4000 (Btu/h)/gal and ≥10 gal	78% E_t (Q/800 + 110 $\sqrt{V}$) SL, Btu/h	
Pool Heaters Oil and Gas	All		78% E_t	ASHRAE 146
Heat Pump Pool Heaters	All		4.0 COP	ASHRAE 146
Unfired Storage Tanks	All		R-12.5	(none)

[a] Energy factor (EF) and thermal efficiency (E_t) are minimum requirements, while standby loss (SL) is maximum Btu/h based on a 70°F temperature difference between stored water and ambient requirements. In the EF equation, V is the rated volume in gallons. In the SL equation, V is the rated volume in gallons and Q is the nameplate input rate in Btu/h.

[b] Section 12 contains a complete specification, including the year version, of the referenced test procedure.

[c] Section G1 is titled "Test Method for Measuring Thermal Efficiency" and Section G2 is titled "Test Method for Measuring Standby Loss."

[~~e~~d] Instantaneous water heaters with input rates below 200,000 Btu/h must comply with these requirements if the water heater is designed to heat water to temperatures 180°F or higher.

Revise as follows (SI units):

TABLE 6.8.1E Warm Air Furnaces and Combination Warm Air Furnaces/Air-Conditioning Units, Warm Air Duct Furnaces, and Unit Heaters

Equipment Type	Size Category (Input)	Subcategory or Rating Condition	Minimum Efficiency[a]	Test Procedure[b]
Warm Air Furnace, Gas-Fired	<66 kW		78% AFUE or 80% E_{td}	DOE 10 CFR Part 430 or Section 2.39, Thermal Efficiency, of ANSI Z21.47
	≥66 kW	Maximum Capacity[d]	80% E_{cc}	Section 2.39, Thermal Efficiency, of ANSI Z21.47
Warm Air Furnace, Oil-Fired	<66 kW		78% AFUE or 80%E_{td}	DOE 10 CFR Part 430 or Section 42, Combustion, of UL 727
	≥66 kW	Maximum Capacity[e]	81% E_{tf}	Section 42, Combustion, of UL 727
Warm Air Duct Furnaces, Gas-Fired	All Capacities	Maximum Capacity[e]	80% E_{cg}	Section 2.10, Efficiency, of ANSI Z83.8
Warm Air Unit Heaters, Gas-Fired	All Capacities	Maximum Capacity[e]	80% E_{cg}	Section 2.10, Efficiency, of ANSI Z83.8
Warm Air Unit Heaters, Oil-Fired	All Capacities	Maximum Capacity[e]	80% E_{cg}	Section 40, Combustion, of UL 731

[a] E_t= thermal *efficiency.* See test procedure for detailed discussion.
[b] Section 12 contains a complete specification of the referenced test procedure, including the referenced year version of the test procedure.
[c] E_c = combustion *efficiency.* Units must also include an interrupted or intermittent ignition device (IID), have jacket losses not exceeding 0.75% of the input rating, and have either power venting or a flue damper. A vent damper is an acceptable alternative to a flue damper for those furnaces where combustion air is drawn from the conditioned space.
[d] Combination units not covered by NAECA (3-phase power or cooling capacity greater than or equal to 19 kW) may comply with either rating.
[e] Minimum and maximum ratings as provided for and allowed by the unit's controls.
[f] E_t = thermal *efficiency.* Units must also include an interrupted or intermittent ignition device (IID), have jacket losses not exceeding 0.75% of the input rating, and have either power venting or a flue damper. A vent damper is an acceptable alternative to a flue damper for those furnaces where combustion air is drawn from the conditioned space.
[g] E_c = combustion *efficiency* (100% less flue losses). See test procedure for detailed discussion.

Revise as follows (SI units):

TABLE 7.8 Performance Requirements for Water Heating Equipment

Equipment Type	Size Category (Input)	Subcategory or Rating Condition	(SI Criteria) Performance Required[a]	Test Procedure[b,c]
Electric Table Top Water Heaters	≤12 kW	Resistance ≥75.7 L	0.93-0.00035V EF	DOE 10 CFR Part 430
Electric Water Heaters	≤12 kW	Resistance ≥75.7 L	0.9~~3~~7-0.00035V EF	DOE 10 CFR Part 430
	>12 kW	Resistance≥75.7 L	5.9 + 5.3 √V SL, W	Section G.2 of ANSI Z21.10.3
	≤24 Amps and ≤250 Volts	Heat Pump	0.93-0.00035V EF	DOE 10 CFR Part 430
Gas Storage Water Heaters	≤22.98 kW	≥75.7 L	0.6~~2~~7-0.0005V EF	DOE 10 CFR Part 430
	>22.98 kW	<309.75 W/L	80% E_t (Q/800 + 16.6√V) SL, W	Sections G.1 and G.2 of ANSI Z21.10.3
Gas Instantaneous Water Heaters	>14.66 kW and <58.62 kW	≥309.75 W/L and <7.57 L	0.62-0.0005V EF	DOE 10 CFR Part 430
	≥58.62 kW[~~e~~d]	≥309.75 W/L and <37.85 L	80% E_t	Sections G.1 and G.2 of ANSI Z21.10.3
	≥58.62 kW	≥309.75 W/L and ≥37.85 L	80% E_t (Q/800 + 16.6 √V) SL, W	
Oil Storage Water Heaters	≤30.78 kW	≥75.7 L	0.59-0.0005V EF	DOE 10 CFR Part 430
	>30.78 kW	<309.75 W/L	78% E_t (Q/800 + 16.6√V) SL, W	Sections G.1 and G.2 of ANSI Z21.10.3
Oil Instantaneous Water Heaters	≤61.55 kW	≥309.75 W/L and <7.57 L	0.59-0.0005V EF	DOE 10 CFR Part 430
	>61.55 kW	≥309.75 W/L and <37.85 L	80% E_t	Sections G.1 and G.2 of ANSI Z21.10.3
	>61.55 kW	≥309.75 W/L and ≥37.85 L	78% E_t (Q/800 + 16.6√V) SL, W	
Hot Water Supply Boilers, Gas and Oil	≥87.93 kW and <3663.8 kW	≥309.75 W/L and <37.85 L	80% E_t	Sections G.1 and G.2 of ANSI Z21.10.3
Hot Water Supply Boilers, Gas		≥309.75 W/L and ≥37.85 L	80% E_t (Q/800 + 16.6√V) SL, W	
Hot Water Supply Boilers, Oil		≥309.75 W/L and ≥37.85 L	78% E_t (Q/800 + 16.6√V) SL, W	
Pool Heaters Oil and Gas	All		78% E_t	ASHRAE 146
Heat Pump Pool Heaters	All		4.0 COP	ASHRAE 146
Unfired Storage Tanks	All		R-2.2	(none)

[a] Energy factor (EF) and thermal efficiency (E_t) are minimum requirements, while standby loss (SL) is maximum W based on a 39°C temperature difference between stored water and ambient requirements. In the EF equation, *V* is the rated volume in liters. In the SL equation, *V* is the rated volume in liters and *Q* is the nameplate input rate in Watts.
[b] Section 12 contains a complete specification, including the year version, of the referenced test procedure.
[c] Section G1 is titled "Test Method for Measuring Thermal Efficiency" and Section G2 is titled "Test Method for Measuring Standby Loss."
[~~e~~d] Instantaneous water heaters with input rates below 58.62 kW must comply with these requirements if the water heater is designed to heat water to temperatures 82.2°C or higher.

(This foreword is not part of this standard. It is merely informative and does not contain requirements necessary for conformance to the standard. It has not been processed according to the ANSI requirements for a standard and may contain material that has not been subject to public review or a consensus process. Unresolved objectors on informative material are not offered the right to appeal at ASHRAE or ANSI.)

FOREWORD

The purpose of this addendum is to add minimum efficiency and certification requirements for both axial and centrifugal fan closed-circuit cooling towers (also known as fluid coolers) into Table 6.8.1G. In addition, a reference to ATC-105S, The Cooling Technology Institute (CTI) test standard for closed-circuit cooling towers, has been added to Section 12, Normative References. A subcommittee of ASHRAE TC 8.6, Technical Committee on Cooling Towers and Evaporative Condensers, developed this addendum, which has been unanimously supported by the entire TC.

Closed-circuit cooling towers differ from open-circuit cooling towers in that the process fluid is kept isolated from the open loop spray water and airflow by an intermediate heat exchanger, typically a coil. Closed-circuit devices also have an integral spray pump to recirculate the spray water over the coil. To account for this, the gpm/hp value for closed-circuit cooling towers includes both the unit fan and spray pump motors, where hp equals the sum of the fan motor and integral spray pump motor nameplate horsepower. Lastly, the minimum efficiency values for closed-circuit cooling towers are based on typical water-source heat pump conditions, as the water-source heat pump industry is the largest HVAC market for this type of equipment.

The addition of minimum efficiency and certification requirements will provide consulting engineers, system designers, and contractors guidelines for the selection of independently certified, energy-efficient closed-circuit cooling towers. This change will also complement the existing minimum efficiency and certification requirements for open-circuit cooling towers, helping to prevent confusion between the requirements for these two different types of cooling towers.

Note: In this addendum, changes to the current standard are indicated in the text by underlining (for additions) and ~~strikethrough~~ (for deletions) unless the instructions specifically mention some other means of indicating the changes.

Addendum I to Standard 90.1-2007

Revise as follows (I-P units):

TABLE 6.8.1G Performance Requirements for Heat Rejection Equipment

Equipment Type[~~d~~]	Total System Heat Rejection Capacity at Rated Conditions	Subcategory or Rating Condition	Performance Required[a,b,c]	Test Procedure[~~e~~d,e]
Propeller or Axial Fan Open-Circuit Cooling Towers	All	95°F Entering Water 85°F Leaving Water 75°F ~~wb~~ ~~*Outdoor air*~~ *Entering wb*	≥38.2 gpm/hp	CTI ATC-105 and CTI STD-201
Centrifugal Fan Open-Circuit Cooling Towers	All	95°F Entering Water 85°F Leaving Water 75°F ~~wb~~ ~~*Outdoor air*~~ *Entering wb*	≥20.0 gpm/hp	CTI ATC-105 and CTI STD-201
Propeller or Axial Fan Closed-Circuit Cooling Towers	All	102°F Entering Water 90°F Leaving Water 75°F *Entering wb*	≥14.0 gpm/hp	CTI ATC-105S and CTI STD-201
Centrifugal Closed-Circuit Cooling Towers	All	102°F Entering Water 90°F Leaving Water 75°F *Entering wb*	≥7.0 gpm/hp	CTI ATC-105S and CTI STD-201
Air-Cooled Condensers	All	125°F Condensing Temperature R-22 Test Fluid 190°F Entering Gas Temperature 15°F Subcooling 95°F Entering db	≥176,000 Btu/h·hp	ARI 460

[a] For purposes of this table, open-circuit cooling tower performance is defined as the water flow rating of the tower at the thermal rating condition listed in Table 6.8.1G divided by the fan ~~nameplate rated~~ motor nameplate power.

[b] For purposes of this table, *closed-circuit cooling tower performance* is defined as the process water flow rating of the tower at the thermal rating condition listed in Table 6.8.1G divided by the sum of the fan motor nameplate power and the integral spray pump motor nameplate power.

[~~b~~c] For purposes of this table, *air-cooled condenser performance* is defined as the heat rejected from the refrigerant divided by the fan ~~nameplate rated~~ motor nameplate power.

[~~c~~d] Section 12 contains a complete specification of the referenced test procedure, including the referenced year version of the test procedure.

[~~d~~e] ~~The efficiencies for open cooling towers listed in Table 6.8.1G are not applicable for closed-circuit cooling towers.~~ The efficiencies and test procedures for both open- and closed-circuit cooling towers are not applicable to hybrid cooling towers that contain a combination of separate wet and dry heat exchange sections.

Revise Section 12 as follows:

Reference	Title
CTI ATC-105 (00)	Acceptance Test Code for Water Cooling Towers
CTI ATC-105S (96)	Acceptance Test Code for Closed Circuit Cooling Towers
CTI STD-201 (04)	Standard for the Certification of Water-Cooling Tower Thermal Performance

Revise as follows (SI units):

TABLE 6.8.1G Performance Requirements for Heat Rejection Equipment

Equipment Type[d]	Total System Heat Rejection Capacity at Rated Conditions	Subcategory or Rating Condition	Performance Required[a,b,c]	Test Procedure[~~d~~,e]
Propeller or Axial Fan Open-Circuit Cooling Towers	All	35°C Entering Water 29°C Leaving Water 24°C ~~wb~~ ~~*Outdoor air*~~ *Entering wb*	≥3.23 L/s·kW	CTI ATC-105 and CTI STD-201
Centrifugal Fan Open-Circuit Cooling Towers	All	35°C Entering Water 29°C Leaving Water 24°C ~~wb~~ ~~*Outdoor air*~~ *Entering wb*	≥1.70 L/s·kW	CTI ATC-105 and CTI STD-201
Propeller or Axial Fan Closed-Circuit Cooling Towers	All	39°C Entering Water 32°C Leaving Water 24°C *Entering wb*	≥1.18 L/s·kW	CTI ATC-105S and CTI STD-201
Centrifugal Closed-Circuit Cooling Towers	All	39°C Entering Water 32°C Leaving Water 24°C *Entering wb*	≥0.59 L/s·kW	CTI ATC-105S and CTI STD-201
Air-Cooled Condensers	All	52°C Condensing Temperature R-22 Test Fluid 88°C Entering Gas Temperature 8°C Subcooling 35°C Entering db	≥69 COP	ARI 460

[a] For purposes of this table, open-circuit cooling tower performance is defined as the water flow rating of the tower at the thermal rating condition listed in Table 6.8.1G divided by the fan ~~nameplate rated~~ motor nameplate power.

[b] For purposes of this table, *closed-circuit cooling tower performance* is defined as the process water flow rating of the tower at the thermal rating condition listed in Table 6.8.1G divided by the sum of the fan motor nameplate power and the integral spray pump motor nameplate power.

[~~b~~c] For purposes of this table, *air-cooled condenser performance* is defined as the heat rejected from the refrigerant divided by the fan ~~nameplate rated~~ motor nameplate power.

[~~c~~d] Section 12 contains a complete specification of the referenced test procedure, including the referenced year version of the test procedure.

[~~d~~e] ~~The efficiencies for open cooling towers listed in Table 6.8.1G are not applicable for closed-circuit cooling towers.~~ The efficiencies and test procedures for both open- and closed-circuit cooling towers are not applicable to hybrid cooling towers that contain a combination of separate wet and dry heat exchange sections.

Revise Section 12 as follows:

Reference	Title
CTI ATC-105 (00)	Acceptance Test Code for Water Cooling Towers
CTI ATC-105S (96)	Acceptance Test Code for Closed Circuit Cooling Towers
CTI STD-201 (04)	Standard for the Certification of Water-Cooling Tower Thermal Performance

(This foreword is not part of this standard. It is merely informative and does not contain requirements necessary for conformance to the standard. It has not been processed according to the ANSI requirements for a standard and may contain material that has not been subject to public review or a consensus process. Unresolved objectors on informative material are not offered the right to appeal at ASHRAE or ANSI.)

FOREWORD

Product development for water-cooled chillers in recent years has focused on improving both full-load and part-load performance. Variable-speed drives (VSDs) have gone through significant technology advancements and are now finding application in water-cooled chillers. The use of VSDs shows significant improvement of chiller integrated part-load value (IPLV). Improvements of up to 30% in IPLV are possible. Partially offsetting the part-load performance improvement is a small decrease in full-load efficiency at design conditions, nominally up to 4%. The decrease in full-load efficiency is due to inherent electronic drive losses and power line filters.

This addendum establishes—effective January 1, 2010—an additional path of compliance for water-cooled chillers. Path A is intended for applications where significant operating time is expected at full-load conditions. On the other hand, Path B is an alternative set of efficiency levels for water-cooled chillers intended for applications where significant time is expected at part load. Compliance with the standard can be achieved by meeting the requirements of either Path A or Path B. However, both full-load and IPLV levels must be met to fulfill the requirements of Path A or Path B.

The addendum also combines all water-cooled positive displacement chillers into one category and adds a new size category for centrifugal chillers at or above 600 tons. The air-cooled chiller without condenser equipment category has been eliminated. All air-cooled chillers without condensers must now be rated with matching condensers. The minimum efficiencies of air-cooled chillers have also been updated. The minimum efficiencies for absorption chillers were left unchanged, as efficiencies have not improved over the last few years and the absorption market has been shrinking, with less than 150 units sold in the US in 2006. Efficiencies in the I-P version of the standard are now expressed in EER for air-cooled chillers, kW/ton for water-cooled chillers, and COP for absorption chillers to reflect industry practices. Tables 6.8.1H through 6.8.1J, listing minimum full-load and NPLV efficiencies of water-cooled centrifugal chillers at nonstandard rating conditions, have been eliminated and replaced by an algebraic equation. The tables will now be included in the 90.1 User's Manual.

The effective date of the new efficiency standards is January 1, 2010, to coincide with the phase-out date of HCFC-22 mandated under the Clean Air Act of 1992. This addendum is expected to save 457.6 GWh of energy per year compared to the requirements of ASHRAE/IESNA Standard 90.1-2004. This represents an annual energy saving of 13.3%.

Note: In this addendum, changes to the current standard are indicated in the text by underlining (for additions) and ~~strikethrough~~ (for deletions) unless the instructions specifically mention some other means of indicating the changes.

Addendum m to 90.1-2007

Modify Section 6.4.1.2 as follows (I-P units).

6.4.1.2 Minimum Equipment Efficiencies—Listed Equipment—Nonstandard Conditions. Water-cooled centrifugal water-chilling packages that are not designed for operation at ARI Standard 550/590 test conditions (and thus cannot be tested to meet the requirements of Table 6.8.1C) of 44°F leaving chilled-water temperature and 85°F entering condenser water temperature with 3 gpm/ton condenser water flow shall have ~~minimum~~ maximum full-load kW/ton ~~COP~~ and ~~minimum~~ *NPLV* ratings ~~as shown in the tables referenced below.~~ adjusted using the following equation:

~~a. Centrifugal chillers <150 tons shall meet the minimum full-load COP and IPLV/NPLV in Table 6.8.1H.~~

~~b. Centrifugal chillers ≥150 tons and <300 tons shall meet the minimum full-load COP and IPLV/NPLV in Table 6.8.1I.~~

~~c. Centrifugal chillers ≥300 tons shall meet the minimum full-load COP and IPLV/NPLV in Table 6.8.1J.~~

Adjusted maximum full-load kW/ton rating = (full-load kW/ton from Table 6.8.1C)/K_{adj}

Adjusted maximum NPLV rating = (IPLV from Table 6.8.1C)/K_{adj}

where

K_{adj}	=	$6.174722 - 0.303668(X) + 0.00629466(X)^2 - 0.000045780(X)^3$
X	=	DT_{std} + LIFT
DT_{std}	=	(24 + (full-load kW/ton from Table 6.8.1C) × 6.83)/Flow
Flow	=	Condenser water flow (gpm)/Cooling full-load capacity (tons)
LIFT	=	CEWT – CLWT
CEWT	=	Full-load condenser entering water temperature (°F)
CLWT	=	Full-load leaving chilled-water temperature (°F)

The ~~table~~ adjusted full-load and *NPLV* values are only applicable over the following full-load design ranges:

- Minimum Leaving Chiller-Water Temperature: ~~40°F to 48°F~~ 38°F
- Maximum Condenser Entering ~~Condenser~~ Water Temperature: ~~75°F to 85°F~~ 102°F
- Condenser Water ~~Temperature Rise: 5°F to 15°F~~ Flow: 1 to 6 gpm/ton
- $X \geq 39°F$ and $\leq 60°F$

Chillers designed to operate outside of these ranges or applications utilizing fluids or solutions with secondary coolants (e.g., glycol solutions or brines) with a freeze point of 27°F or lower for freeze protection are not covered by this standard.

Example: Path A 600 ton centrifugal chiller
Table 6.8.1C efficiencies as of 1/1/2010
Full Load = 0.570 kW/ton
IPLV = 0.539 kW/ton
CEWT = 80°F
Flow = 2.5 gpm/ton
CLWT = 42°F

LIFT = 80 – 42 = 38°F

$DT = (24 + 0.570 \times 6.83)/2.5 = 11.16$°F

$X = 38 + 11.16 = 49.16$°F

$K_{adj} = 6.174772 - 0.303668(49.16) + 0.00629466(49.16)^2 - 0.00004578(49.16)^3 = 1.020$

Adjusted full load = 0.570/1.020 = 0.559 kW/ton

NPLV = 0.539/1.020 = 0.528 kW/ton

Delete Tables 6.8.1C, 6.8.1H, 6.8.1I, and 6.8.1J in their entirety.

Insert new Table 6.8.1C as follows and renumber the remaining tables appropriately.

TABLE 6.8.1C Water Chilling Packages—Efficiency Requirements[a]

Equipment Type	Size Category	Units	Before 1/1/2010		As of 1/1/2010[c]				Test Procedure[b]
					Path A		Path B		
			Full Load	IPLV	Full Load	IPLV	Full Load	IPLV	
Air-Cooled Chillers	<150 tons	EER	≥9.562	≥10.416	≥9.562	≥12.500	NA[d]	NA[d]	ARI 550/590
	≥150 tons	EER			≥9.562	≥12.750	NA[d]	NA[d]	
Air-Cooled without Condenser, Electrical Operated	All Capacities	EER	≥10.586	≥11.782	Air-cooled chillers without condensers must be rated with matching condensers and comply with the air-cooled chiller efficiency requirements				
Water Cooled, Electrically Operated, Reciprocating	All Capacities	kW/ton	≤0.837	≤0.696	Reciprocating units must comply with water cooled positive displacement efficiency requirements				
Water Cooled, Electrically Operated, Positive Displacement	<75 tons	kW/ton	≤0.790	≤0.676	≤0.780	≤0.630	≤0.800	≤0.600	
	≥75 tons and < 150 tons	kW/ton	≤0.790	≤0.676	≤0.775	≤0.615	≤0.790	≤0.586	
	≥150 tons and < 300 tons	kW/ton	≤0.717	≤0.627	≤0.680	≤0.580	≤0.718	≤0.540	
	≥300 tons	kW/ton	≤0.639	≤0.571	≤0.620	≤0.540	≤0.639	≤0.490	
Water Cooled, Electrically Operated, Centrifugal	<150 tons	kW/ton	≤0.703	≤0.669	≤0.634	≤0.596	≤0.639	≤0.450	
	≥150 tons and < 300 tons	kW/ton	≤0.634	≤0.596	≤0.634	≤0.596	≤0.639	≤0.450	
	≥300 tons and < 600 tons	kW/ton	≤0.576	≤0.549	≤0.576	≤0.549	≤0.600	≤0.400	
	≥600 tons	kW/ton	≤0.576	≤0.549	≤0.570	≤0.539	≤0.590	≤0.400	
Air Cooled Absorption Single Effect	All Capacities	COP	≥0.600	NR[e]	≥0.600	NR[e]	NA[d]	NA[d]	ARI 560
Water-Cooled Absorption Single Effect	All Capacities	COP	≥0.700	NR[e]	≥0.700	NR[e]	NA[d]	NA[d]	
Absorption Double Effect Indirect-Fired	All Capacities	COP	≥1.000	≥1.050	≥1.000	≥1.050	NA[d]	NA[d]	
Absorption Double Effect Direct-Fired	All Capacities	COP	≥1.000	≥1.000	≥1.000	≥1.000	NA[d]	NA[d]	

a. The chiller equipment requirements do not apply for chillers used in low-temperature applications where the design leaving fluid temperature is < 40°F.
b. Section 12 contains a complete specification of the referenced test procedure, including the referenced year version of the test procedure.
c. Compliance with this standard can be obtained by meeting the minimum requirements of Path A or Path B. However, both the full load and IPLV must be met to fulfill the requirements of Path A or Path B.
d. *NA* means that this requirement is not applicable and cannot be used for compliance.
e. *NR* means that there are no minimum requirements for this category.

Modify Section 6.4.1.2 as follows (SI units).

6.4.1.2 Minimum Equipment Efficiencies—Listed Equipment—Nonstandard Conditions. Water-cooled centrifugal water-chilling packages that are not designed for operation at ARI Standard 550/590 test conditions (and thus cannot be tested to meet the requirements of Table 6.8.1C) of 6.7°C leaving chilled-water temperature and 29.4°C entering condenser water temperature with ~~15.3 l/min·kW~~ 0.054 L/s·kW condenser water flow shall have ~~a~~ minimum full-load COP and ~~a minimum~~ *NPLV* ratings ~~as shown in tables referenced below.~~ adjusted using the following equation:

~~a. (a) Centrifugal chillers <528 kW shall meet the minimum full-load COP and IPLV/NPLV in Table 6.8.1H.~~

~~b. (b) Centrifugal chillers ≥528 kW and <1055 kW shall meet the minimum full-load COP and IPLV/NPLV in Table 6.8.1I~~

~~c. (c) Centrifugal chillers ≥1055 kW shall meet the minimum full-load COP and IPLV/NPLV in Table 6.8.1J.~~

Adjusted minimum full-load COP rating
= (full-load COP from Table 6.8.1C) × K_{adj}

Adjusted maximum NPLV rating
= (IPLV from Table 6.8.1C) × K_{adj}

where

K_{adj} = $6.174722 - 0.5466024(X) + 0.020394698(X)^2 - 0.000266989(X)^3$

X = DT_{std} + LIFT

DT_{std} = (0.267114 + 0.267088/(Full-load COP from Table 6.8.1C))/Flow

Flow = Condenser water flow (L/s)/Cooling full load capacity (kW)

LIFT = CEWT – CLWT (°C)

CEWT = Full load condenser entering water temperature (°C)

CLWT = Full load leaving chilled-water temperature (°C)

The ~~table~~ adjusted full-load and *NPLV* values are only applicable over the following full-load design ranges:

- Minimum Leaving Chiller-Water Temperature: ~~4.4°C to 8.9°C~~ 3.3°C
- Maximum Condenser Entering ~~Condenser~~ Water Temperature: ~~23.9°C to 29.4°C~~ 39°C
- Condenser Water ~~Temperature Rise: 2.8°C to 8.3°F~~ Flow: .036 to .0721 L/s·kW
- $X \geq 21.7$°C and ≤ 33.3°C

Chillers designed to operate outside of these ranges or applications utilizing fluids or solutions with secondary coolants (e.g., glycol solutions or brines) with a freeze point of –2.8°C or lower for freeze protection are not covered by this standard.

Example: Path A 2110 kW centrifugal chiller
Table 6.8.1C efficiencies as of 1/1/2010
Full Load = 6.170 COP
IPLV = 6.525 COP
CEWT = 26°C
Flow = 0.05 L/s·kW
CLWT = 5.5°C

LIFT = 26 – 5.5 = 20.50°C
DT = (0.267114 + 0.267088/6.170)/0.05 = 6.208°C
X = 21.11 + 6.208 = 27.319°F
$K_{adj} = 6.174722 - 0.5466024(27.319) + 0.020394698(27.319)^2 - 0.000266989(27.319)^3 = 1.031$
Adjusted full load = 6.170 × 1.031 = 6.359 COP
NPLV =6.525 × 1.031 = 6.725 COP

Delete Tables 6.8.1C, 6.8.1H, 6.8.1I, and 6.8.1J in their entirety.

Insert new Table 6.8.1C as follows and renumber the remaining tables appropriately.

TABLE 6.8.1C Water Chilling Packages—Efficiency Requirements[a]

Equipment Type	Size Category	Units	Before 1/1/2010		As of 1/1/2010[c]				Test Procedure[b]
					Path A		Path B		
			Full Load	IPLV	Full Load	IPLV	Full Load	IPLV	
Air-Cooled Chillers	<528 kW	COP	≥2.802	≥3.050	≥2.802	≥3.664	NA[d]	NA[d]	ARI 550/590
	≥528 kW	COP			≥2.802	≥3.737	NA[d]	NA[d]	
Air-Cooled without Condenser, Electrically Operated	All Capacities	COP	≥3.100	≥3.450	Air-cooled chillers without condensers must be rated with matching condensers and comply with the air-cooled chiller efficiency requirements.				
Water Cooled, Electrically Operated, Reciprocating	All Capacities	COP	≥4.200	≥5.050	Reciprocating units must comply with water-cooled positive displacement efficiency requirements.				
Water Cooled, Electrically Operated, Positive Displacement	<264 kW	COP	≥4.450	≥5.200	≥4.509	≥5.582	≥4.396	≥5.861	
	≥264 kW and < 528 kW	COP	≥4.450	≥5.200	≥4.538	≥5.718	≥4.452	≥6.001	
	≥528 kW and < 1055 kW	COP	≥4.900	≥5.600	≥5.172	≥6.063	≥4.898	≥6.513	
	≥1055 kW	COP	≥5.500	≥6.150	≥5.672	≥6.513	≥5.504	≥7.177	
Water Cooled, Electrically Operated, Centrifugal	<528 kW	COP	≥5.000	≥5.250	≥5.547	≥5.901	≥5.504	≥7.815	
	≥528 kW and < 1055 kW	COP	≥5.550	≥5.900	≥5.547	≥5.901	≥5.504	≥7.815	
	≥1055 kW and < 2110 kW	COP	≥6.100	≥6.400	≥6.100	≥6.401	≥5.856	≥8.792	
	≥2110 kW	COP	≥6.100	≥6.400	≥6.170	≥6.525	≥5.961	≥8.792	
Air Cooled Absorption Single Effect	All Capacities	COP	≥0.600	NR[e]	≥0.600	NR[e]	NA[d]	NA[d]	ARI 560
Water-Cooled Absorption Single Effect	All Capacities	COP	≥0.700	NR[e]	≥0.700	NR[e]	NA[d]	NA[d]	
Absorption Double Effect Indirect-Fired	All Capacities	COP	≥1.000	≥1.050	≥1.000	≥1.050	NA[d]	NA[d]	
Absorption Double Effect Direct-Fired	All Capacities	COP	≥1.000	≥1.000	≥1.000	≥1.000	NA[d]	NA[d]	

a. The chiller equipment requirements do not apply for chillers used in low-temperature applications where the design leaving fluid temperature is < 4.4°C.
b. Section 12 contains a complete specification of the referenced test procedure, including the referenced year version of the test procedure.
c. Compliance with this standard can be obtained by meeting the minimum requirements of Path A or Path B. However, both the full load and IPLV must be met to fulfill the requirements of Path A or Path B.
d. *NA* means that this requirement is not applicable and cannot be used for compliance.
e. *NR* means that there are no minimum requirements for this category.

(This foreword is not part of this standard. It is merely informative and does not contain requirements necessary for conformance to the standard. It has not been processed according to the ANSI requirements for a standard and may contain material that has not been subject to public review or a consensus process. Unresolved objectors on informative material are not offered the right to appeal at ASHRAE or ANSI.)

FOREWORD

Variable-air-volume fan control is currently required in the standard for multiple-zone systems. This proposal extends these requirements for large single-zone units. Important aspects of this proposal include the following:

- *It applies to both unitary (packaged) equipment and chilled water air-handling units.*
- *It only applies to units with a cooling capacity greater than or equal to 110,000 Btu/h.*
- *The proposal can be met using either two-speed motors or variable-speed drives on the supply fan(s).*
- *The minimum speed requirement is set at 67% fan speed.*
- *It does not take effect until 1/1/2012.*

This proposal has achieved industry consensus through discussions with AHRI's Large Unitary Engineering (ULE) Group. Three of the criteria were critical to achieving that consensus:

- *The lower threshold of 10 tons for unitary equipment,*
- *The 2/3 minimum threshold for fan speed, and*
- *The delay in implementation to 2012.*

The significance of the two-thirds minimum speed threshold is to prevent coil frosting on DX coils (particularly for those units that are face split). The reasoning behind the delay in implementation to 2012 is to allow the AC unit manufacturers time to redesign and test their AC units. All of the manufacturers are currently redesigning their lines to meet the 2010 phase-out of certain refrigerants (R-22). Some have already completed this work for certain product lines. The volume of units being tested for refrigerant change outs is straining the available certified testing resources.

Although this requirement does not take effect until 2012, it is believed that manufacturers will begin introducing variable-volume signal-one units in advance of that date. Utility rebate programs, LEED certification, and other incentives should encourage wider demand for these units and will help this requirement to see real savings in advance of the 2012 date.

It should be noted that a second proposal is forthcoming to address the budget systems in the Energy Cost Budget Method (see Table 11.3.2A) to make the budget systems 5, 6, 7, 9, and 11 consistent with the requirements of this proposal.

Addendum n to Standard 90.1-2007

Include new item b in Section 6.3.2 as follows. Renumber subsequent section items as appropriate (I-P units).

6.3.2 Criteria. HVAC system must meet ALL of the following criteria:

a. The system serves a single HVAC zone.
b. The equipment must meet the variable flow requirements of Section 6.4.3.10
(c)~~(b)~~. Cooling (if any) shall be provided by a unitary packaged …

Add new Section 6.4.3.10 as follows (I-P units):

6.4.3.10 Single Zone Variable-Air-Volume Controls. HVAC systems shall have variable airflow controls as follows:

a. Effective January 1, 2010, air-handling and fan-coil units with chilled-water cooling coils and supply fans with motors greater than or equal to 5 hp shall have their supply fans controlled by two-speed motors or variable-speed drives. At cooling demands less than or equal to 50%, the supply fan controls shall be able to reduce the airflow to no greater than the larger of the following:

   1. One half of the full fan speed, or
   2. The volume of outdoor air required to meet the ventilation requirements of Standard 62.1.

b. Effective January 1, 2012, all air-conditioning equipment and air-handling units with direct expansion cooling and a cooling capacity at ARI conditions greater than or equal to 110,000 Btu/h that serve single zones shall have their supply fans controlled by two-speed motors or variable-speed drives. At cooling demands less than or equal to 50%, the supply fan controls shall be able to reduce the airflow to no greater than the larger of the following:

   1. Two-thirds of the full fan speed, or
   2. The volume of outdoor air required to meet the ventilation requirements of Standard 62.1.

Include new item b in Section 6.3.2 as follows. Renumber subsequent section items as appropriate (SI units).

6.3.2 Criteria. HVAC system must meet ALL of the following criteria:

a. The system serves a single HVAC zone.
b. The equipment must meet the variable flow requirements of Section 6.4.3.10
(c)~~(b)~~. Cooling (if any) shall be provided by a unitary packaged …

Add new Section 6.4.3.10 as follows (SI units):

6.4.3.10 Single Zone Variable-Air-Volume Controls. HVAC systems shall have variable airflow controls as follows:

a. Effective January 1, 2010, air-handling and fan-coil units with chilled-water cooling coils and supply fans with motors greater than or equal to 4 kW shall have their supply fans controlled by two-speed motors or variable-speed drives. At cooling demands less than or equal to 50%, the supply fan controls shall be able to reduce the airflow to no greater than the larger of the following:

   1. One half of the full fan speed, or

2. The volume of outdoor air required to meet the ventilation requirements of Standard 62.1.

b. Effective January 1, 2012, all air-conditioning equipment and air-handling units with direct expansion cooling and a cooling capacity at ARI conditions greater than or equal to 32.3 kW that serve single zones shall have their supply fans controlled by two-speed motors or variable-speed drives. At cooling demands less than or equal to 50%, the supply fan controls shall be able to reduce the airflow to no greater than the larger of the following:

1. Two-thirds of the full fan speed, or
2. The volume of outdoor air required to meet the ventilation requirements of Standard 62.1.

(This foreword is not part of this standard. It is merely informative and does not contain requirements necessary for conformance to the standard. It has not been processed according to the ANSI requirements for a standard and may contain material that has not been subject to public review or a consensus process. Unresolved objectors on informative material are not offered the right to appeal at ASHRAE or ANSI.)

FOREWORD

This addendum is the second phase of correcting the fan power limitation deficiencies of Standard 90.1-2004. The first phase was corrected by Addendum ac to the 2004 standard, which has been approved and is included in Standard 90.1-2007. That addendum addressed all fan systems with exception of those systems serving fume hoods. The reason for excluding fume hood systems was to allow Addendum ac to proceed, correcting a majority of the problems, and be included in the 2007 edition of Standard 90.1. This allowed time to assemble a lab working group that could properly address the needs of laboratory exhaust systems. This working group consisted of three individuals from Labs 21, three design engineers, and one person from the ECB subcommittee. This addendum provides the necessary pressure credits for laboratory exhaust systems that allow prescriptive compliance of these systems.

Addendum p to Standard 90.1-2007

Revise the standard as follows (I-P units):

Modify the exceptions to Section 6.5.3.1.1 as follows:

Exceptions to 6.5.3.1.1:

a. Hospital, vivarium and laboratory systems that utilize flow control devices on exhaust and/or return to maintain space pressure relationships necessary for occupant health and safety or environmental control may use variable-volume fan power limitation.

b. Individual exhaust fans with motor nameplate horsepower of 1 hp or less.

~~c. Fans exhausting air from fume hoods. (*Note:* If this exception is taken, no related exhaust side credits shall be taken from Table 6.5.3.1.1B and the Fume Exhaust Exception Deduction must be taken from Table 6.5.3.1.1B).~~

Modify Table 6.5.3.1.1B as follows:

TABLE 6.5.3.1.1B Fan Power Limitation Pressure Drop Adjustment

Device	Adjustment
Credits	
Fully ducted return and/or exhaust air systems	0.5 in. w.c. (2.15 in. w.c. for laboratory and vivarium systems)
Return and/or exhaust airflow control devices	0.5 in. w.c.
Exhaust filters, scrubbers, or other exhaust treatment	The pressure drop of device calculated at fan system design condition
Particulate Filtration Credit: MERV 9 through 12	0.5 in. w.c.
Particulate Filtration Credit: MERV 13 through 15	0.9 in. w.c.
Particulate Filtration Credit: MERV 16 and greater and electronically enhanced filters	Pressure drop calculated at 2× clean filter pressure drop at fan system design condition
Carbon and other gas-phase air cleaners	Clean filter pressure drop at fan system design condition
Heat recovery device, biosafety cabinet	Pressure drop of device at fan system design condition
Evaporative humidifier/cooler in series with another cooling coil	Pressure drop of device at fan system design condition
Sound Attenuation Section	0.15 in. w.c.
Exhaust system serving fume hoods	0.35 in. w.c.
Laboratory and vivarium exhaust systems in high-rise buildings	0.25 in. w.c./100 ft of vertical duct exceeding 75 ft
~~**Deductions**~~	
~~Fume Hood Exhaust Exception (required if 6.5.3.1.1 Exception [c] is taken)~~	~~–1.0 in. w.c.~~

Revise the standard as follows (SI units):

Modify the footnotes to Table 6.5.3.1.1A as follows:

TABLE 6.5.3.1.1A Fan Power Limitation*

	Limit	Constant Volume	Variable Volume
Option 1: Fan System Motor Nameplate kW	Allowable Nameplate Motor kW	$kW \le L/S_S \cdot 0.0017$	$kW \le L/S_S \cdot 0.0024$
Option 2: Fan System Input kW	Allowable Fan System Input kW	~~I~~$kW_i \le L/S_S \cdot 0.0015 + A$	~~I~~$kW_i \le L/S_S \cdot 0.0021 + A$

*where

L/S_S = the maximum design supply airflow rate to conditioned spaces served by the system in liters per second
kW = the maximum combined motor nameplate kW
~~I~~kW_i = the maximum combined fan ~~required~~ input kW
A = sum of ($PD \times L/S_D$/ 650000~~4131~~)

where

PD = each applicable pressure drop adjustment from Table 6.5.3.1.1B in Pa~~in. w.c.~~
L/S_D = the design airflow through each applicable device from Table 6.5.3.1.1B in liters per second

Modify the Exceptions to 6.5.3.1.1 as follows:

Exceptions to 6.5.3.1.1:

a. Hospital, vivarium and laboratory systems ~~systems~~ that utilize flow control devices on exhaust and/or return to maintain space pressure relationships necessary for occupant health and safety or environmental control may use variable-volume fan power limitation.
b. Individual exhaust fans with motor nameplate kW of 0.75 kW or less.
~~c. Fans exhausting air from fume hoods. (*Note:* If this exception is taken, no related exhaust side credits shall be taken from Table 6.5.3.1.1B and the Fume Exhaust Exception Deduction must be taken from Table 6.5.3.1.1B).~~

Modify Table 6.5.3.1.1B as follows:

TABLE 6.5.3.1.1B Fan Power Limitation Pressure Drop Adjustment

Device	Adjustment
Credits	
Fully ducted return and/or exhaust air systems	125 Pa (535 Pa for laboratory and vivarium systems)
Return and/or exhaust airflow control devices	125 Pa
Exhaust filters, scrubbers, or other exhaust treatment	The pressure drop of device calculated at fan system design condition
Particulate Filtration Credit: MERV 9 through 12	125 ~~in w.c.~~Pa
Particulate Filtration Credit: MERV 13 through 15	225 ~~in~~Pa
Particulate Filtration Credit: MERV 16 and greater and electronically enhanced filters	Pressure drop calculated at 2× clean filter pressure drop at fan system design condition
Carbon and other gas-phase air cleaners	Clean filter pressure drop at fan system design condition
Heat recovery device, biosafety cabinet	Pressure drop of device at fan system design condition
Evaporative humidifier/cooler in series with another cooling coil	Pressure drop of device at fan system design condition
Sound Attenuation Section	38 Pa
Exhaust system serving fume hoods	85 Pa
Laboratory and vivarium exhaust systems in high-rise buildings	60 Pa/30 meters of vertical duct exceeding 25 meters
~~**Deductions**~~	
~~Fume Hood Exhaust Exception (required if 6.5.3.1.1 Exception [c] is taken)~~	~~250 Pa~~

Modify Section 6.5.3.1.2 as follows:

6.5.3.1.2 Motor Nameplate kW. For each fan, the selected fan motor shall be no larger than the first available motor size greater than the input kW. The fan input kW must be indicated on the design documents to allow for compliance verification by the code official.

Exceptions to 6.5.3.1.2:

a.~~c.~~ For fans less than 4.5 kW, where the first available motor larger than the input kW has a nameplate rating within 50% of the input kW, the next larger nameplate motor size may be selected.
b.~~d.~~ For fans 4.5 kW and larger, where the first available motor larger than the input kW has a nameplate rating within 30% of the input kW, the next larger nameplate motor size may be selected.

(This foreword is not part of this standard. It is merely informative and does not contain requirements necessary for conformance to the standard. It has not been processed according to the ANSI requirements for a standard and may contain material that has not been subject to public review or a consensus process. Unresolved objectors on informative material are not offered the right to appeal at ASHRAE or ANSI.)

FOREWORD

This addendum modifies the vestibule requirements for Climate Zone 4.

Note: In this addendum, changes to the current standard are indicated in the text by underlining (for additions) and ~~strikethrough~~ (for deletions) unless the instructions specifically mention some other means of indicating the changes.

Addendum q to Standard 90.1-2007

Revise the standard as follows (I-P and SI units):

5.4.3.4 Vestibules. Building entrances that separate *conditioned space* from the exterior shall be protected with an enclosed vestibule, with all *doors* opening into and out of the vestibule equipped with self-closing devices. Vestibules shall be designed so that in passing through the vestibule it is not necessary for the interior and exterior *doors* to open at the same time. Interior and exterior *doors* shall have a minimum distance between them of not less than 7 ft (2.1 m) when in the closed position. The exterior envelope of conditioned vestibules shall comply with the requirements for a conditioned space. The interior and exterior envelope of unconditioned vestibules shall comply with the requirements for a semiheated space.

Exceptions:

a. *Building entrances* with revolving *doors*.

b. *Doors* not intended to be used as a *building entrance*.

c. *Doors* opening directly from a *dwelling unit*.

d. *Building entrances* in buildings located in climate zone 1 or 2.

e. *Building entrances* in buildings located in climate zone 3 ~~or 4~~ that are less than four stories above grade and less than 10,000 ft^2 (930 m^2) in area.

f. *Building entrances* in buildings located in climate zone 4, 5, 6, 7, or 8 that are less than 1000 ft^2 (90 m^2) in area.

g. *Doors* that open directly from a *space* that is less than 3000 ft^2 (280 m^2) in area and is separate from the *building entrance*.

(This foreword is not part of this standard. It is merely informative and does not contain requirements necessary for conformance to the standard. It has not been processed according to the ANSI requirements for a standard and may contain material that has not been subject to public review or a consensus process. Unresolved objectors on informative material are not offered the right to appeal at ASHRAE or ANSI.)

FOREWORD

In Summer 2005, ASHRAE approved addendum g to ASHRAE/IESNA Standard 90.1-2004, which increased the minimum energy efficiency standards of commercial air-cooled air conditioners and heat pumps greater than 65,000 Btu/h. EER and COP (at 47°F) were amended, with new levels taking effect on January 1, 2010. However, IPLV and COP at 17°F were left unchanged.

This addendum updates the COP at 17°F efficiency levels for commercial heat pumps and introduces a new part-load energy efficiency descriptor for all commercial unitary products above 65,000 Btu/h of cooling capacity. The new descriptor, Integrated Energy Efficiency Ratio (IEER), *is a replacement for IPLV. The IEER is a significant improvement over IPLV as it allows for uniform rating of all products including single- and multi-stage units. It is based on a weighted average of performance at 100%, 75%, 50%, and 25% of capacity. The new part-load metric is expected to more accurately rate the part-load performance of commercial unitary equipment.*

The IEER and COP at 17°F levels in Tables 6.8.1A and 6.8.1B were derived based on the expected performance of commercial unitary products meeting the new full-load EER and COP at 47°F requirements that will take effect on January 1, 2010. In addition, IEER values are now for product classes with cooling capacities between 65,000 and 240,000 Btu/h, which previously had no IPLV minimums.

Note: In this addendum, changes to the current standard are indicated in the text by <u>underlining</u> (for additions) and ~~strikethrough~~ (for deletions) unless the instructions specifically mention some other means of indicating the changes.

Addendum s to 90.1-2007

Revise the Standard as follows (I-P units).

Revise Tables 6.8.1A and 6.8.1B as follows:

TABLE 6.8.1A Electrically Operated Unitary Air Conditioners and Condensing Units—Minimum Efficiency Requirements

Equipment Type	Size Category	Heating Section Type	Sub-Category or Rating Condition	Minimum Efficiency[a]	Test Procedure[b]
Air Conditioners, Air Cooled	≥65,000 Btu/h and <135,000 Btu/h	Electric Resistance (or None)	Split System and Single Package	10.3 EER (before 1/1/2010) 11.2 EER (as of 1/1/2010) 11.4 IEER (as of 1/1/2010)	ARI 340/360
		All other	Split System and Single Package	10.1 EER (before 1/1/2010) 11.0 EER (as of 1/1/2010) 11.2 IEER (as of 1/1/2010)	
	≥135,000 Btu/h and <240,000 Btu/h	Electric Resistance (or None)	Split System and Single Package	9.7 EER (before 1/1/2010) 11.0 EER (as of 1/1/2010) 11.2 IEER (as of 1/1/2010)	
		All other	Split System and Single Package	9.5 EER (before 1/1/2010) 10.8 EER (as of 1/1/2010) 11.0 IEER (as of 1/1/2010)	
	≥240,000 Btu/h and <760,000 Btu/h	Electric Resistance (or None)	Split System and Single Package	9.5 EER (before 1/1/2010) 10.0 EER (as of 1/1/2010) 9.7 IPLV (before 1/1/2010) 10.1 IEER (as of 1/1/2010)	
		All other	Split System and Single Package	9.3 EER (before 1/1/2010) 9.8 EER (as of 1/1/2010) 9.5 IPLV (before 1/1/2010) 9.9 IEER (as of 1/1/2010)	
	≥760,000 Btu/h	Electric Resistance (or None)	Split System and Single Package	9.2 EER (before 1/1/2010) 9.7 EER (as of 1/1/2010) 9.4 IPLV (before 1/1/2010) 9.8 IEER (as of 1/1/2010)	
		All other	Split System and Single Package	9.0 EER (before 1/1/2010) 9.5 EER (as of 1/1/2010) 9.2 IPLV (before 1/1/2010) 9.6 IEER (as of 1/1/2010)	
Air Conditioners, Water and Evaporatively Cooled	<65,000 Btu/h	All	Split System and Single Package	12.1 EER 12.3 IEER (as of 1/1/2010)	ARI 210/240
	≥65,000 Btu/h and <135,000 Btu/h	Electric Resistance (or None)	Split System and Single Package	11.5 EER 11.7 IEER (as of 1/1/2010)	ARI 340/360
		All other	Split System and Single Package	11.3 EER 11.5 IEER (as of 1/1/2010)	
	≥135,000 Btu/h and <240,000 Btu/h	Electric Resistance (or None)	Split System and Single Package	11.0 EER 11.2 IEER (as of 1/1/2010)	
		All other	Split System and Single Package	10.8 EER 11.0 IEER (as of 1/1/2010)	
	≥240,000 Btu/h	Electric Resistance (or None)	Split System and Single Package	11.0 EER 10.3 IPLV (before 1/1/2010) 11.1 IEER (as of 1/1/2010)	
		All other	Split System and Single Package	10.8 EER 10.1 IPLV (before 1/1/2010) 10.9 IEER (as of 1/1/2010)	

The remainder of the table is left unchanged.

TABLE 6.8.1B Electrically Operated Unitary and Applied Heat Pumps—Minimum Efficiency Requirements

Equipment Type	Size Category	Heating Section Type	Sub-Category or Rating Condition	Minimum Efficiency[a]	Test Procedure[b]
Air Cooled (Cooling Mode)	≥65,000 Btu/h and <135,000 Btu/h	Electric Resistance (or None)	Split System and Single Package	10.1 EER (before 1/1/2010) 11.0 EER (as of 1/1/2010) 11.2 IEER (as of 1/1/2010)	ARI 340/360
		All other	Split System and Single Package	9.9 EER (before 1/1/2010) 10.8 EER (as of 1/1/2010) 11.0 IEER (as of 1/1/2010)	
	≥135,000 Btu/h and <240,000 Btu/h	Electric Resistance (or None)	Split System and Single Package	9.3 EER (before 1/1/2010) 10.6 EER (as of 1/1/2010) 10.7 IEER (as of 1/1/2010)	
		All other	Split System and Single Package	9.1 EER (before 1/1/2010) 10.4 EER (as of 1/1/2010) 10.5 IEER (as of 1/1/2010)	
	≥240,000 Btu/h	Electric Resistance (or None)	Split System and Single Package	9.0 EER (before 1/1/2010) 9.5 EER (as of 1/1/2010) 9.2 IPLV (before 1/1/2010) 9.6 IEER (as of 1/1/2010)	
		All other	Split System and Single Package	8.8 EER (before 1/1/2010) 9.3 EER (as of 1/1/2010) 9.0 IPLV (before 1/1/2010) 9.4 IEER (as of 1/1/2010)	
Air Cooled (Heating Mode)	≥65,000 Btu/h and <135,000 Btu/h (Cooling Capacity)	–	47°F db/43°F wb Outdoor Air	3.2 COP (before 1/1/2010) 3.3 COP (as of 1/1/2010)	ARI 340/360
			17°F db/15°F wb Outdoor Air	2.2 COP (before 1/1/2010) 2.25 COP (as of 1/1/2010)	
	≥135,000 Btu/h (Cooling Capacity)	–	47°F db/43°F wb Outdoor Air	3.1 COP (before 1/1/2010) 3.2 COP (as of 1/1/2010)	
			17°F db/15°F wb Outdoor Air	2.0 COP (before 1/1/2010) 2.05 COP (as of 1/1/2010)	

The remainder of the table is left unchanged.

Add the following in Section 3.2, just above IPLV:

integrated energy efficiency ratio (IEER): a single-number figure of merit expressing cooling part-load EER efficiency for commercial unitary air-conditioning and heat pump equipment on the basis of weighted operation at various load capacities for the equipment.

Modify the normative reference in Section 12 (under Air Conditioning and Refrigeration Institute) as follows:

Reference	Title
ARI 340/360-20047	Performance Rating of Commercial and Industrial Unitary Air-Conditioning and Heat Pump Equipment

Revise the Standard as follows (SI units).

Revise Tables 6.8.1A and 6.8.1B as follows:

TABLE 6.8.1A Electrically Operated Unitary Air Conditioners and Condensing Units—Minimum Efficiency Requirements

Equipment Type	Size Category	Heating Section Type	Sub-Category or Rating Condition	Minimum Efficiency[a]	Test Procedure[b]
Air Conditioners, Air Cooled	≥19 kW and <40 kW	Electric Resistance (or None)	Split System and Single Package	3.02 COP (before 1/1/2010) 3.28 COP (as of 1/ 1/2010) 3.34 ICOP (as of 1/1/2010)	ARI 340/360
		All other	Split System and Single Package	2.96 COP (before 1/1/2010) 3.22COP (as of 1/1/2010) 3.28 ICOP (as of 1/1/2010)	
	≥40 kW and <70 kW	Electric Resistance (or None)	Split System and Single Package	2.84 COP (before 1/1/2010) 3.22 COP (as of 1/1/2010) 3.28 ICOP (as of 1/1/2010)	
		All other	Split System and Single Package	2.78 COP (before 1/1/2010) 3.16 COP (as of 1/1/2010) 3.22 ICOP (as of 1/1/2010)	
	≥70 kW and <223 kW	Electric Resistance (or None)	Split System and Single Package	2.78 COP (before 1/1/2010) 2.93 COP (as of 1/1/2010) 2.84 IPLV (before 1/1/2010) 2.96 ICOP (as of 1/1/2010)	
		All other	Split System and Single Package	2.72 COP (before 1/1/2010) 2.87 COP (as of 1/1/2010) 2.78 IPLV (before 1/1/2010) 2.90 ICOP (as of 1/1/2010)	
	≥223 kW	Electric Resistance (or None)	Split System and Single Package	2.70 COP (before 1/1/2010) 2.84 COP (as of 1/1/2010) 2.75 IPLV (before 1/1/2010) 2.87 ICOP (as of 1/1/2010)	
		All other	Split System and Single Package	2.64 COP (before 1/1/2010) 2.78 COP (as of 1/1/2010) 2.69 IPLV (before 1/1/2010) 2.81 ICOP (as of 1/1/2010)	
Air Conditioners, Water and Evaporatively Cooled	<19 kW	All	Split System and Single Package	3.54 COP 3.60 ICOP (as of 1/1/2010)	ARI 210/240
	≥19 kW and <40 kW	Electric Resistance (or None)	Split System and Single Package	3.37 COP 3.43 ICOP (as of 1/1/2010)	ARI 340/360
		All other	Split System and Single Package	3.31 COP 3.37 ICOP (as of 1/1/2010)	
	≥40 kW and <70 kW	Electric Resistance (or None)	Split System and Single Package	3.22 COP 3.28 ICOP (as of 1/1/2010)	
		All other	Split System and Single Package	3.16 COP 3.22 ICOP (as of 1/1/2010)	
	≥70 kW	Electric Resistance (or None)	Split System and Single Package	3.22 COP 3.02 IPLV (before 1/1/2010) 3.25 ICOP (as of 1/1/2010)	
		All other	Split System and Single Package	3.16 COP 2.96 IPLV (before 1/1/2010) 3.19 ICOP (as of 1/1/2010)	

The remainder of the table is left unchanged.

TABLE 6.8.1B Electrically Operated Unitary and Applied Heat Pumps—Minimum Efficiency Requirements

Equipment Type	Size Category	Heating Section Type	Sub-Category or Rating Condition	Minimum Efficiency[a]	Test Procedure[b]
Air Cooled (Cooling Mode)	≥19 kW and <40 kW	Electric Resistance (or None)	Split System and Single Package	2.96 COP_C (before 1/1/2010) 3.22 COP_C (as of 1/1/2010) 3.28 ICOP (as of 1/1/2010)	ARI 340/360
		All other	Split System and Single Package	2.90 COP_C (before 1/1/2010) 3.16 COP_C (as of 1/1/2010) 3.22 ICOP (as of 1/1/2010)	
	≥40 kW and <70 kW	Electric Resistance (or None)	Split System and Single Package	2.72 COP_C (before 1/1/2010) 3.10 COP_C (as of 1/1/2010) 3.13 ICOP (as of 1/1/2010)	
		All other	Split System and Single Package	2.66 COP_C (before 1/1/2010) 3.04 COP_C (as of 1/1/2010) 3.08 ICOP (as of 1/1/2010)	
	≥70 kW	Electric Resistance (or None)	Split System and Single Package	2.64 COP_C (before 1/1/2010) 2.78 COP_C (as of 1/1/2010) 2.70 IPLV (before 1/1/2010) 2.81 ICOP (as of 1/1/2010)	
		All other	Split System and Single Package	2.58 COP_C (before 1/1/2010) 2.72 COP_C (as of 1/1/2010) 2.64 IPLV (before 1/1/2010) 2.75 ICOP (as of 1/1/2010)	
Air Cooled (Heating Mode)	≥19 kW and <70 kW (Cooling Capacity)	—	8.3°C db/6.1°C wb Outdoor Air	3.2 COP_H (before 1/1/2010) 3.3 COP_H (as of 1/1/2010)	ARI 340/360
			–8.3°C db/-9.4°C wb Outdoor Air	2.2 COP_H (before 1/1/2010) 2.25 COP_H (as of 1/1/2010)	
	≥70 kW (Cooling Capacity)	—	8.3°C db/6.1°C wb Outdoor Air	3.1 COP_H (before 1/1/2010) 3.2 COP_H (as of 1/1/2010)	
			–8.3°C db/-9.4°C wb Outdoor Air	2.0 COP_H (before 1/1/2010) 2.05 COP_H (as of 1/1/2010)	

The remainder of the table is left unchanged.

Add the following in Section 3.2, just above IPLV:

integrated coefficient of performance (ICOP): a single-number figure of merit expressing cooling part-load COP efficiency for commercial unitary air-conditioning and heat pump equipment on the basis of weighted operation at various load capacities for the equipment (analogous to IEER, but for SI or other consistent units).

Modify the normative reference in Section 12 (under Air Conditioning and Refrigeration Institute) as follows:

Reference	Title
ARI 340/360-20047	Performance Rating of Commercial and Industrial Unitary Air-Conditioning and Heat Pump Equipment

(This foreword is not part of this standard. It is merely informative and does not contain requirements necessary for conformance to the standard. It has not been processed according to the ANSI requirements for a standard and may contain material that has not been subject to public review or a consensus process. Unresolved objectors on informative material are not offered the right to appeal at ASHRAE or ANSI.)

FOREWORD

ASHRAE/IESNA Standard 90.1 established a product class for "replacement" packaged terminal equipment to distinguish products intended to replace existing equipment in existing constructions with nonstandard external wall openings from products intended for existing and new construction with standard wall openings (16 in. high × 42 in. wide). However, the term "replacement" has been misinterpreted to mean any packaged terminal equipment intended as a replacement unit regardless of the exterior wall openings it must fit in. Conversely, the term "new construction" has been interpreted as meaning a product intended for new constructions only, while in fact it applies equally to existing and new buildings with standard wall openings.

*This addendum removes the terms "replacement" and "new construction" from the product classes listed in Table 6.8.1D and replaces them with the terms "nonstandard size" and "standard size," respectively, to clarify that one product class is intended for applications with nonstandard size exterior wall openings while the other is intended for applications with standard size exterior wall openings. The addendum also amends Section 6.4.1.5.2 and footnote b to Table 6.8.1D to clarify that nonstandard size packaged terminal equipment have sleeves with an external wall opening less than 16 in. high **or** less than 42 in. wide to reflect existing applications where the wall opening is not necessarily less than 16 in. high **and** less than 42 in. wide. However, to avoid a potential abuse of the definition, nonstandard size packaged terminal equipment are required to have a cross-sectional area of the sleeves less than 670 in.² (less than 16 × 42 in.).*

Note: In this addendum, changes to the current standard are indicated in the text by underlining (for additions) and ~~strikethrough~~ (for deletions) unless the instructions specifically mention some other means of indicating the changes.

Addendum t to 90.1-2007

Revise the standard as follows (I-P units).

Revise Section 6.4.1.5.2 as follows:

6.4.1.5.2 Packaged Terminal Air Conditioners. Nonstandard size p~~P~~ackaged terminal air conditioners and heat pumps with existing sleeves ~~sizes~~ having an external wall opening of less than 16 in. high ~~and~~ or less than 42 in. wide and having a cross-sectional area less than 670 in.² shall be factory labeled as follows: *Manufactured for ~~replacement~~ nonstandard size applications only: not to be installed in new construction projects.*

Revise Table 6.8.1D as follows:

TABLE 6.8.1D Electrically Operated Packaged Terminal Air Conditioners, Packaged Terminal Heat Pumps, Single-Package Vertical Air Conditioners, Single-Package Vertical Heat Pumps, Room Air Conditioners, and Room Air Conditioner Heat Pumps—Minimum Efficiency Requirements

Equipment Type	Size Category (Input)	Subcategory or Rating Condition	Minimum Efficiency	Test Procedure[a]
PTAC (Cooling Mode) ~~New Construction~~ Standard Size	All Capacities	95.0°F db Outdoor air	12.5 – (0.213 × Cap/1000)[c] EER	
PTAC (Cooling Mode) ~~Replacements~~ Nonstandard Size[b]	All Capacities	95.0°F db Outdoor air	10.9 – (0.213 × Cap/1000)[c] EER	
PTHP (Cooling Mode) ~~New Construction~~ Standard Size	All Capacities	95.0°F db Outdoor air	12.3 – (0.213 × Cap/1000)[c] EER	
PTHP (Cooling Mode) ~~Replacements~~ Nonstandard Size[b]	All Capacities	95.0°F db Outdoor air	10.8 – (0.213 × Cap/1000)[c]EER	ARI 310/380
PTHP (Heating Mode) ~~New Construction~~ Standard Size	All Capacities		3.2 – (0.026 × Cap/1000)[c] COP	
PTHP (Heating Mode) ~~Replacements~~ Nonstandard Size[b]	All Capacities		2.9 – (0.026 × Cap/1000)[c] COP	

The remainder of the table is left unchanged.

Revise footnote b of Table 6.8.1D as follows:

b ~~Replacement~~ Nonstandard size units must be factory labeled as follows: "MANUFACTURED FOR ~~REPLACEMENT~~ NONSTANDARD SIZE APPLICATIONS ONLY; NOT TO BE INSTALLED IN NEW CONSTRUCTION PROJECTS." ~~Replacement~~ Nonstandard size efficiencies apply only to units being installed in ~~with~~ existing sleeves having an external wall opening of less than 16 in. high ~~and~~ or less than 42 in. wide and having a cross-sectional area less than 670 $in.^2$.

Revise the standard as follows (SI units).

Revise Section 6.4.1.5.2 as follows:

6.4.1.5.2 Packaged Terminal Air Conditioners. Nonstandard size p~~P~~ackaged terminal air conditioners and heat pumps with existing sleeves ~~sizes~~ having an external wall opening of less than 0.4 m high ~~and~~ or less than 1.0 m wide and having a cross-sectional area less than 0.4 m^2 shall be factory labeled as follows: *Manufactured for ~~replacement~~ nonstandard size applications only: not to be installed in new construction projects.*

Revise Table 6.8.1D as follows:

TABLE 6.8.1D Electrically Operated Packaged Terminal Air Conditioners, Packaged Terminal Heat Pumps, Single-Package Vertical Air Conditioners, Single -Package Vertical Heat Pumps, Room Air Conditioners, and Room Air Conditioner Heat Pumps–Minimum Efficiency Requirements

Equipment Type	Size Category (Input)	Subcategory or Rating Condition	Minimum Efficiency	Test Procedure[a]
PTAC (Cooling Mode) ~~New Construction~~ Standard Size	All Capacities	35.0°C db Outdoor air	3.66 – (0.213 × Cap/1000)[c] COP_C	ARI 310/380
PTAC (Cooling Mode) ~~Replacements~~ Nonstandard Size[b]	All Capacities	35.0°C db Outdoor air	3.19 – (0.213 × Cap/1000)[c] COP_C	
PTHP (Cooling Mode) ~~New Construction~~ Standard Size	All Capacities	35.0°C db Outdoor air	3.60 – (0.213 × Cap/1000)[c] COP_C	
PTHP (Cooling Mode) ~~Replacements~~ Nonstandard Size[b]	All Capacities	35.0°C db Outdoor air	3.16 – (0.213 × Cap/1000)[c] COP_C	
PTHP (Heating Mode) ~~New Construction~~ Standard Size	All Capacities		3.2 – (0.026 × Cap/1000)[c] COP_H	
PTHP (Heating Mode) ~~Replacements~~ Nonstandard Size[b]	All Capacities		2.9 – (0.026 × Cap/1000)[c] COP_H	

The remainder of the table is left unchanged.

Revise footnote b of Table 6.8.1D as follows:

b ~~Replacement~~ Nonstandard size units must be factory labeled as follows: "MANUFACTURED FOR ~~REPLACEMENT~~ NONSTANDARD SIZE APPLICATIONS ONLY; NOT TO BE INSTALLED IN NEW CONSTRUCTION PROJECTS." ~~Replacement~~ Nonstandard size efficiencies apply only to units being installed in ~~with~~ existing sleeves having an external wall opening of less than 0.4 m high ~~and~~ or less than 1.0 m wide and having a cross-sectional area less than 0.4 m^2.

(This foreword is not part of this standard. It is merely informative and does not contain requirements necessary for conformance to the standard. It has not been processed according to the ANSI requirements for a standard and may contain material that has not been subject to public review or a consensus process. Unresolved objectors on informative material are not offered the right to appeal at ASHRAE or ANSI.)

FOREWORD

Axial fan open-circuit cooling towers use approximately 50% of the energy consumed by centrifugal fan open-circuit cooling towers. Substantial energy can be saved by requiring centrifugal fan units over 1,100 US gpm at the rating conditions to meet the energy efficiency requirements for axial fan units found in Table 6.8.1G. These requirements are 38.0 gpm/hp for axial versus 20.0 gpm/hp for centrifugal, rated at 95°F entering, 85°F leaving, and 75°F entering wet-bulb temperature. This would encourage the current market trend towards lower energy axial fan designs. Exceptions are allowed for sound control and ducted installations (which might be used to reduce the potential for freezing in cold climates). Like-for-like replacements on existing buildings that would require extensive rework of the site (such as to the supporting steel) are permitted under Section 6.1.1.3, Exception b.

Note: In this addendum, changes to the current standard are indicated in the text by underlining (for additions) and ~~strikethrough~~ (for deletions) unless the instructions specifically mention some other means of indicating the changes.

Addendum u to 90.1-2007

Revise the standard as follows (I-P units).

Add a new section, Section 6.5.5.3, as follows:

6.5.5.3 Limitation on Centrifugal Fan Open-Circuit Cooling Towers. Centrifugal fan open-circuit cooling towers with a combined rated capacity of 1,100 gpm or greater at 95°F condenser water return, 85°F condenser water supply, and 75°F *outdoor air* wet-bulb temperature shall meet the energy efficiency requirement for axial fan open-circuit cooling towers listed in Table 6.8.1G.

Exception: Open-circuit cooling towers that are ducted (inlet or discharge) or require external sound attenuation.

Revise the standard as follows (SI units).

Add a new section, Section 6.5.5.3, as follows:

6.5.5.3 Limitation on Centrifugal Fan Open-Circuit Cooling Towers. Centrifugal fan open-circuit cooling towers with a combined rated capacity of 69.4 L/s or greater at 35°C condenser water return, 29°C condenser water supply, and 24°C *outdoor air* wet-bulb temperature shall meet the energy efficiency requirement for axial fan open-circuit cooling towers listed in Table 6.8.1G.

Exception: Open-circuit cooling towers that are ducted (inlet or discharge) or require external sound attenuation.

(This foreword is not part of this standard. It is merely informative and does not contain requirements necessary for conformance to the standard. It has not been processed according to the ANSI requirements for a standard and may contain material that has not been subject to public review or a consensus process. Unresolved objectors on informative material are not offered the right to appeal at ASHRAE or ANSI.)

FOREWORD

This addendum contains two changes. The first change to the footnote of Table G3.1.1A is to make it clear that Exception a to Section G3.1.1 also applies here. The second change is to the exception to G3.1.2.10 on Exhaust Air Energy Recovery for multifamily buildings because they are unlikely to have a centralized exhaust air system needed to effectively recover heat.

Note: In this addendum, changes to the current standard are indicated in the text by underlining (for additions) and ~~strikethrough~~ (for deletions) unless the instructions specifically mention some other means of indicating the changes.

Addendum w to 90.1-2007

Revise the standard as follows (I-P and SI units).

Modify the Notes to Table G3.1.1A as follows:

TABLE G3.1.1A *Baseline* HVAC System Types

Building Type	Fossil Fuel, Fossil/Electric Hybrid, and Purchased Heat	Electric and Other
Residential	System 1 – PTAC	System 2 – PTHP
Nonresidential & 3 Floors or Less & <25,000 ft^2	System 3 – PSZ-AC	System 4 – PSZ-HP
Nonresidential & 4 or 5 Floors & <25,000 ft^2 or 5 Floors or Less & 25,000 ft2to 150,000 ft^2	System 5 – Packaged VAV w/ Reheat	System 6 – Packaged VAV w/PFP Boxes
Nonresidential & More than 5 Floors or >150,000 ft^2	System 7 – VAV w/Reheat	System 8 – VAV w/PFP Boxes

Notes:
Residential building types include dormitory, hotel, motel, and multifamily. Residential space types include guest rooms, living quarters, private living space, and sleeping quarters. Other building and space types are considered nonresidential.
Where no heating system is to be provided or no heating energy source is specified, use the "Electric and Other" heating source classification.
Where attributes make a building eligible for more than one *baseline* system type, use the predominant condition to determine the system type for the entire building except as noted in Exception a to Section G3.1.1.
For laboratory spaces with a minimum of 5000 cfm of exhaust, use systems type 5 or 7 and reduce the exhaust and makeup air volume to 50% of design values during unoccupied periods. For all electric buildings the heating shall be Electric Resistance.

Add Exception i to the Exceptions to Section G3.1.2.10 as follows:

G3.1.2.10 Exhaust Air Energy Recovery

. . .

Exceptions: If any of these exceptions apply, exhaust air energy recovery shall not be included in the *baseline building design.*

. . .

i. Systems serving dwelling units in multifamily buildings.

(This foreword is not part of this standard. It is merely informative and does not contain requirements necessary for conformance to the standard. It has not been processed according to the ANSI requirements for a standard and may contain material that has not been subject to public review or a consensus process. Unresolved objectors on informative material are not offered the right to appeal at ASHRAE or ANSI.)

FOREWORD

A product class for heat pump pool heaters was first established in 2002 and was included in the 2004 version of ASHRAE 90.1. At that time, the minimum coefficient of performance (COP) was based on the test methods and rating conditions contained in ASHRAE Standard 146-1998. The rating conditions in Standard 146 used to rate heat pump pool heaters relied on an outdoor temperature of 80°F and an entering water temperature of 80°F.

Since then, the Air-Conditioning, Heating and Refrigeration Institute (AHRI) published ARI standard 1160 "Performance Rating of Heat Pump Pool Heaters," which establishes testing and rating requirements for heat-pump pool heaters. The standard makes reference to ASHRAE 146 for the test methods and provides standard rating conditions at high (80°F) and low (50°F) outdoor temperatures (the entering water temperature being at 80°F). In addition, AHRI has launched a third-party certification program to independently verify the performance ratings (heating capacity and coefficient of performance) of heat pump pool heaters claimed by manufacturers based on ARI 1160.

This proposal establishes ARI 1160 as the test procedure for heat-pump pool heaters and requires that the minimum coefficient of performance (COP) of 4 be met at the low outdoor temperature of 50°F (instead of the high outdoor temperature of 80°F currently required). These proposed changes significantly increase the stringency of ASHRAE Standard 90.1, as heat-pump pool heaters will now be required to deliver a COP of 4 at a higher temperature lift. Finally, it should be mentioned that the proposed requirements have been in place for over a year in the state of Florida, which has the largest heat pump pool heater market in the country (http://www.dca.state.fl.us/fbc/thecode/supp_051006icc_corrected0806_eff.pdf).

Addendum y to Standard 90.1-2007

Revise Table 7.8 as follows: (I-P and SI units).

TABLE 7.8 Performance Requirements for Water Heating Equipment

Equipment Type	Size Category (Input)	Subcategory or Rating Condition	Minimum Efficiency	Test Procedure[a]
Heat pump pool heaters	All	50.0°F [10.0°C] db 44.2°F [6.78°C] wb Outdoor air 80.0°F [26.7°C] Entering Water	4.0 COP	~~ASHRAE 146~~ ARI 1160

Remainder of table unchanged.

Add reference in Chapter 12 and modify as follows: (I-P and SI units).

Reference	Title
~~Air-Conditioning and Refrigeration Institute,~~ **Air Conditioning, Heating and Refrigeration Institute** **4100 North Fairfax Drive, Suite 200, Arlington, VA 22203**	
ARI 1160-2008	Performance Rating of Heat Pump Pool Heaters

(This foreword is not part of this standard. It is merely informative and does not contain requirements necessary for conformance to the standard. It has not been processed according to the ANSI requirements for a standard and may contain material that has not been subject to public review or a consensus process. Unresolved objectors on informative material are not offered the right to appeal at ASHRAE or ANSI.)

FOREWORD

Liquid-to-liquid heat exchangers are critical system components used in many buildings covered by ASHRAE Standard 90.1. Applications include, but are not limited to, free cooling with cooling towers, pressure interceptor, water-source heat pump loops, and heat recovery. The proper functioning of these heat exchangers helps to ensure that the energy efficiency of other certified equipment, such as chillers and cooling towers, is fully achieved.

A relatively new certification program for ARI Standard 400 is now being widely adopted by this industry. This certification program provides a sound engineering basis for rating the performance of liquid-to-liquid heat exchangers. Inclusion of certification requirements for this equipment will benefit both manufacturers and consumers, allow product comparisons, and provide incentives to manufacturers to improve heat exchanger efficiency in order to gain market share. This program also complements the recently adopted certification requirements for closed-circuit cooling towers (Addendum l to Standard 90.1-2007).

As these devices function to efficiently transfer heat between two fluids, no efficiency requirements are listed. Additionally, the cost for the ARI 400 certification program is similar to other ARI Certification Programs, involving thermal tests and the ARI program cost.

Lastly, the original Section 6.4.1.4f, addressing Table 6.81G must be deleted based on Addendum ak to Standard 90.1-2004, as requirements for CTI certification were added back in to Table 6.8.1G with that addendum, negating the original paragraph.

ARI Standard 400 can be downloaded from the ARI Web site at http://ari.org/NR/rdonlyres/C7CA14D8-B4DD-495C-B8F4-F5A68D8C63E3/0/4002001.pdf.

Addendum ak on cooling tower certification can be downloaded from the ASHRAE Web site at http://www.ashrae.org/doclib/20060815_200661121930_347.pdf.

Addendum ad to Standard 90.1-2007

Revise Section 6.8 as shown.

Add the following table to Section 6.8:

Revise Section 6.4.1.4 as shown.

TABLE 6.8.1K Heat Transfer Equipment

Equipment Type	Subcategory	Minimum Efficiency*	Test Procedure†
Liquid-to-liquid heat exchangers	Plate type	NR	ARI 400

* NR = No Requirement

† Section 12 contains a complete specification of the referenced test procedure, including the referenced year version of the test procedure.

6.4.1.4 Verification of Equipment Efficiencies. Equipment *efficiency* information supplied by *manufacturers* shall be verified as follows:

a. Equipment covered under EPACT shall comply with U.S. Department of Energy certification requirements.

b. If a certification program exists for a covered product, and it includes provisions for verification and challenge of equipment *efficiency* ratings, then the product shall be listed in the certification program, or

c. if a certification program exists for a covered product, and it includes provisions for verification and challenge of equipment *efficiency* ratings, but the product is not listed in the existing certification program, the ratings shall be verified by an independent laboratory test report, or

d. if no certification program exists for a covered product, the equipment *efficiency* ratings shall be supported by data furnished by the *manufacturer*, or

e. where components such as indoor or outdoor coils from different *manufacturers* are used, the system designer shall specify component efficiencies whose combined *efficiency* meets the minimum equipment *efficiency* requirements in Section 6.4.1.

~~f. Products covered in Table 6.8.1G shall have efficiency ratings supported by data furnished by the manufacturer.~~

f. Requirements for plate type liquid to liquid heat exchangers are listed in Table 6.8.1K.

Add the following reference to Section 12.

12. NORMATIVE REFERENCES

Reference	Title
~~**Air-Conditioning and Refrigeration Institute,**~~ **Air Conditioning, Heating and Refrigeration Institute** **4100 North Fairfax Drive, Suite 200, Arlington, VA 22203**	
ARI 400-2001 with Addendum 2	Liquid to Liquid Heat Exchangers

(This foreword is not part of this standard. It is merely informative and does not contain requirements necessary for conformance to the standard. It has not been processed according to the ANSI requirements for a standard and may contain material that has not been subject to public review or a consensus process. Unresolved objectors on informative material are not offered the right to appeal at ASHRAE or ANSI.)

FOREWORD

This change recognizes the practical design application of excluding bathroom lighting from "master" switch control in hotel/motel guest rooms and adds a requirement to eliminate wasted light in guest room bathrooms. Recent research shows that approximately 80% of the wasted guest room bathroom lighting can be saved with a 60-minute-limit control device. The 60-minute limit also provides ample time for any potential safety or convenience concerns related to bathrooms, such as the lights turning off too early while the bathroom is still occupied. The 5 W allowance for night lights recognizes the practical current design application of guest room bathroom night light use but at a reasonable low level.

Note: In this addendum, changes to the current standard are indicated in the text by underlining (for additions) and ~~strikethrough~~ (for deletions) unless the instructions specifically mention some other means of indicating the changes.

Addendum aw to Standard 90.1-2007

Revise as follows for I-P and SI versions.

9.4.1.4 Additional Control.

g. ~~*Hotel and Motel*~~ *Guest Room Lighting*~~—hotel and motel guest rooms and guest suites shall have a master *control device* at the main room entry that *controls* all *permanently installed luminaires* and switched receptacles~~ Guest rooms in hotels, motels, boarding houses, and similar buildings shall have one or more *control device(s)* at the entry door that collectively *control* all *permanently installed luminaires* and switched receptacles, except those in the bathroom(s). Suites shall have *control(s)* meeting these requirements at the entry to each room or at the primary entry to the suite. Bathrooms shall have a *control device* installed to automatically turn off the bathroom lighting, except for night lighting not exceeding 5 W, within 60 minutes of the occupant leaving the space.

(This appendix is not part of this standard. It is merely informative and does not contain requirements necessary for conformance to the standard. It has not been processed according to the ANSI requirements for a standard and may contain material that has not been subject to public review or a consensus process. Unresolved objectors on informative material are not offered the right to appeal at ASHRAE or ANSI.)

APPENDIX
18-MONTH SUPPLEMENT
ADDENDA TO ANSI/ASHRAE/IESNA STANDARD 90.1-2007

This supplement includes Addenda a, b, c, g, h, i, j, k, l, m, n, p, q, s, t, u, w, y, ad, and aw to ANSI/ASHRAE/IESNA Standard 90.1-2007. The following table lists each addendum and describes the way in which the standard is affected by the change. It also lists the ASHRAE, IESNA, and ANSI approval dates for each addendum.

Addendum	Section(s) Affected	Description of Changes*	ASHRAE Standards Committee Approval	ASHRAE BOD Approval	IESNA BOD Approval	ANSI Approval
a	6.8.1G	This addendum seeks to clarify that the current cooling tower requirements in the Standard apply to open-circuit cooling towers only.	6/23/2007	6/27/2007	6/12/2007	7/25/2007
b	6.5.2.3	This addendum updates the references for outdoor ventilation rates.	6/23/2007	6/27/2007	6/12/2007	7/25/2007
c	6.5.2.3	This addendum adds vivariums to the list of spaces that require specific humidity levels to satisfy process needs.	6/23/2007	6/27/2007	6/12/2007	7/25/2007
g	Section 5, Normative Appendix A2.3	This addendum updates the building envelope criteria for metal buildings.	6/21/2008	6/25/2008	6/30/2008	6/26/2008
h	6.5.2.1	This addendum adds a new exception that is geared toward zones with direct digital controls (DDC).	6/21/2008	6/25/2008	6/30/2008	6/26/2008
i	9.4.5	This addendum applies a four-zone lighting power density approach to exterior lighting requirements.	6/21/2008	6/25/2008	6/30/2008	6/26/2008
j	Section 12, Informative Appendix E	This addendum updates references in the Standard.	1/19/2008	1/23/2008	1/28/2008	1/26/2008
k	Table 6.8.1E, Table 7.8	This addendum specifies specific sections of reference standards in Tables 6.8.1E and 7.8.	1/19/2008	1/23/2008	1/28/2008	7/24/2008
l	Table 6.8.1G, Section 12	This Addendum adds minimum efficiency and certification requirements for both axial and centrifugal fan closed-circuit cooling towers (also known as *fluid coolers*) to Table 6.8.1G. In addition, a reference to ATC-105S, the Cooling Technology Institute test standard for closed-circuit cooling towers, has been added to Section 12, Normative References.	1/19/2008	1/23/2008	1/28/2008	7/24/2008
m	Section 6.4.1.2, Table 6.8.1C	This addendum establishes effective January 1, 2010, an additional path of compliance for water-cooled chillers and also combines all water-cooled positive displacement chillers into one category and adds a new size category for centrifugal chillers at or above 600 tons	10/12/2008	10/24/2008	10/10/2008	10/27/2008

Addendum	Section(s) Affected	Description of Changes*	ASHRAE Standards Committee Approval	ASHRAE BOD Approval	IESNA BOD Approval	ANSI Approval
n	6.4.3.10	This addendum extends variable air volume fan control requirements to large single-zone units.	6/21/2008	6/25/2008	6/30/2008	6/26/2008
p	6.5.3.1.1	This addendum addresses fan power limitations to all fan systems with exception to those serving fume hoods.	6/21/2008	6/25/2008	6/30/2008	6/26/2008
q	5.4.3.4	This addendum modifies the vestibule requirements for climate zone 4.	1/19/2008	1/23/2008	1/28/2008	7/24/2008
s	Table 6.8.1A, Table 6.8.1B	This addendum updates the COP at 17°F efficiency levels for commercial heat pumps and introduces a new part load energy efficiency descriptor for all commercial unitary products above 65,000 Btu/h of cooling capacity.	10/12/2008	10/24/2008	10/10/2008	10/27/2008
t	6.4.1.5.2, Table 6.8.1D	This addendum removes the terms "replacement" and "new construction" from the product classes listed in Table 6.8.1D and replaces them with the terms "non-standard size" and "standard size," respectively, to clarify that one product class is intended for applications with non-standard size exterior wall openings while the other is intended for applications with standard size exterior wall openings. The addendum also amends Section 6.4.1.5.2 and footnote b to Table 6.8.1D to clarify that non-standard size packaged terminal equipment have sleeves with an external wall opening less than 16 in. high <u>or</u> less than 42 in. wide to reflect existing applications where the wall opening is not necessarily less than 16 in. high <u>and</u> less than 42 in. wide.	10/12/2008	10/24/2008	10/10/2008	10/27/2008
u	6.5.5.3	This addendum adds requirements for axial fan open-circuit cooling towers.	10/12/2008	10/24/2008	10/10/2008	10/27/2008
w	Table G3.1.1A, Section G3.1.2.10	This addendum modifies requirements on exhaust air energy recovery for multifamily buildings in Appendix G.	10/12/2008	10/24/2008	10/10/2008	10/27/2008
y	Table 7.8, Section 12	This addendum establishes ARI 1160 as the test procedure for heat pump pool heaters and that the minimum COP be met at the low outdoor temperature of 50°F.	6/21/2008	6/25/2008	6/30/2008	6/26/2008
ad	Table 6.8.1K, Section 6.4.1.4, Section 12	This addendum adds requirements for liquid to liquid heat exchangers and adds a reference to AARI 400-2008.	6/21/2008	6/25/2008	6/30/2008	7/24/2008
aw	9.4.1.4	This change recognizes the practical design application of excluding bathroom lighting from "master" switch control in hotel/motel guest rooms and adds a requirement to eliminate wasted light in guest room bathrooms.	1/19/2008	1/23/2008	1/28/2008	6/26/2008

* These descriptions may not be complete and are provided for information only.

NOTE

When addenda, interpretations, or errata to this standard have been approved, they can be downloaded free of charge from the ASHRAE Web site at http://www.ashrae.org.

POLICY STATEMENT DEFINING ASHRAE'S CONCERN FOR THE ENVIRONMENTAL IMPACT OF ITS ACTIVITIES

ASHRAE is concerned with the impact of its members' activities on both the indoor and outdoor environment. ASHRAE's members will strive to minimize any possible deleterious effect on the indoor and outdoor environment of the systems and components in their responsibility while maximizing the beneficial effects these systems provide, consistent with accepted standards and the practical state of the art.

ASHRAE's short-range goal is to ensure that the systems and components within its scope do not impact the indoor and outdoor environment to a greater extent than specified by the standards and guidelines as established by itself and other responsible bodies.

As an ongoing goal, ASHRAE will, through its Standards Committee and extensive technical committee structure, continue to generate up-to-date standards and guidelines where appropriate and adopt, recommend, and promote those new and revised standards developed by other responsible organizations.

Through its *Handbook*, appropriate chapters will contain up-to-date standards and design considerations as the material is systematically revised.

ASHRAE will take the lead with respect to dissemination of environmental information of its primary interest and will seek out and disseminate information from other responsible organizations that is pertinent, as guides to updating standards and guidelines.

The effects of the design and selection of equipment and systems will be considered within the scope of the system's intended use and expected misuse. The disposal of hazardous materials, if any, will also be considered.

ASHRAE's primary concern for environmental impact will be at the site where equipment within ASHRAE's scope operates. However, energy source selection and the possible environmental impact due to the energy source and energy transportation will be considered where possible. Recommendations concerning energy source selection should be made by its members.

Most Widely Accepted and Trusted
ICC ES
ICC EVALUATION SERVICE
The leader in evaluating building products for code compliance & sustainable attributes
• Close to a century of experience in building product evaluation
• Most trusted and most widely accepted evaluation body
• In-house staff of licensed engineers
• Leader in innovative product evaluations
• Direct links to reports within code documents for easy access
Three signature programs:
1. ICC-ES Evaluation Report Program
2. Plumbing, Mechanical and Fuel Gas (PMG) Listing Program
3. Sustainable Attributes Verification and Evaluation (SAVE) Program
ICC-ES is a subsidiary of the International Code Council (ICC), the developer of the International Codes. Together, ICC-ES and ICC incorporate proven techniques and practices into codes and related products and services that foster safe and sustainable design and construction.
For more information, please contact us at es@icc-es.org or call 1.800.423.6587 (x42237).
www.icc-es.org
10-03908

ICC INTERNATIONAL CODE COUNCIL®

People Helping People Build a Safer World™

IECC TOOLS

ICC Code Resources help you learn, interpret and apply the IECC effectively

IN THE FIELD

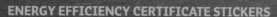

FITS IN YOUR POCKET!

ENERGY INSPECTOR'S GUIDE: BASED ON THE 2009 INTERNATIONAL ENERGY CONSERVATION CODE® AND ASHRAE/IESNA 90.1-2007

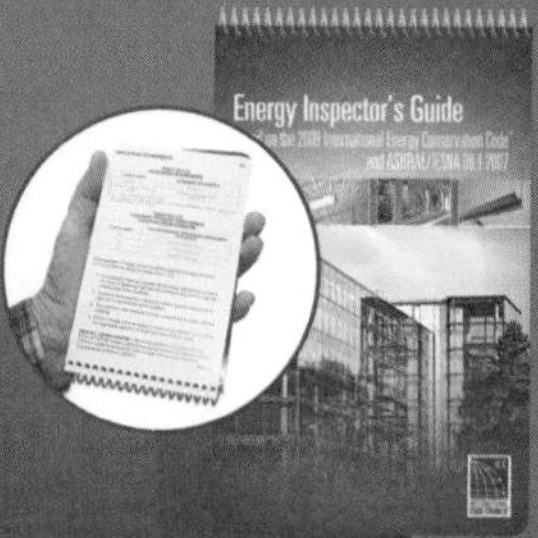

Your ideal resource for effective, accurate, consistent, and complete commercial and residential energy provisions. This handy pocket guide is organized in a manner consistent with the inspection sequence and process for easy use on site. Increase inspection effectiveness by focusing on the most common issues relevant to energy conservation. (76 pages)

SOFT COVER #7808S09

PDF DOWNLOAD #8886P09

ENERGY EFFICIENCY CERTIFICATE STICKERS

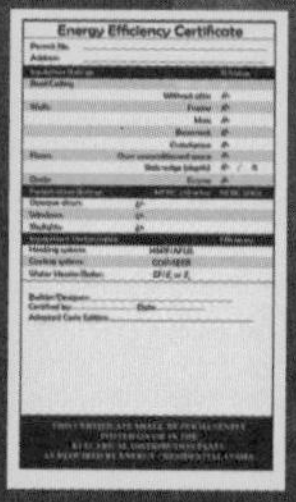

The energy provisions in IRC® Section N1101.8 and IECC Section 401.3 require a type of certificate be installed. This easy-to-use sticker clearly lists the general insulation, window performance, and equipment efficiency details. Sold in packets of 25.

#0726S

CODE SOURCE: ENERGY CONSERVATION CODE

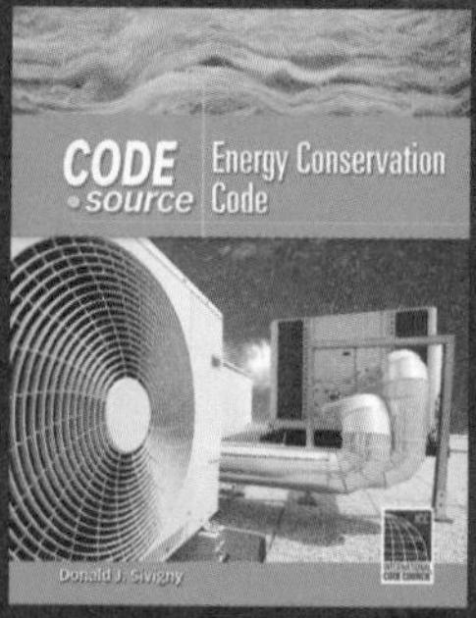

This value-packed resource will serve as a helpful in-the-field reference guide and as a critical component of the code enforcement and inspection process. Designed to assist field inspectors and plans examiners in the completion and performance of their duties, it will instill a solid knowledge of the practical application of the 2009 IECC and the standards set forth by the American Recovery and Reinvestment Act (ARRA).

Features:

- Lists the most common code items warranting examination for compliance.
- Organizes them in a manner that is both efficient and effective.
- Comprehensive coverage prepares building industry professionals and students for safe, accurate, and code-compliant work.
- The "quick tab" format allows easy access to critical information.
- Durable laminated pages withstand a variety of field conditions. (75 pages)

#4866S09

LEARN MORE ABOUT IECC TOOLS TODAY! 1-800-786-4452 | www.iccsafe.org/store